Lecture Notes in Mathematics

Edited by A. Dold and B. Eckm

1132

Operator Algebras and their Connections with Topology and Ergodic Theory

Proceedings of the OATE Conference held in Buşteni, Romania, Aug. 29 – Sept. 9, 1983

Edited by H. Araki, C. C. Moore,
Ş. Strătilă and D. Voiculescu
(with the assistance of Gr. Arsene)

Springer-Verlag
Berlin Heidelberg New York Tokyo

Editors

Huzihiro Araki
Research Institute for Mathematical Sciences
Kyoto University, Kyoto, Japan

Calvin C. Moore
Mathematical Sciences Research Institute
1000 Centennial Drive, Berkeley CA 94720, USA

Şerban-Valentin Strătilă
Dan-Virgil Voiculescu
INCREST, Department of Mathematics
Bd. Pacii 220, 79622 Bucharest, Romania

Mathematics Subject Classification (1980): primary: 46L05, 46L10
secondary: 28Dxx, 55N15

ISBN 3-540-15643-7 Springer-Verlag Berlin Heidelberg New York Tokyo
ISBN 0-387-15643-7 Springer-Verlag New York Heidelberg Berlin Tokyo

Printing and binding: Beltz Offsetdruck, Hemsbach / Bergstr.
2146/3140-543210

ERRATUM

Lecture Notes in Mathematics, vol. 1132
Operator Algebras and their Connections
with Topology and Ergodic Theory
Proceedings, Buşteni, Romania 1983

Edited by H. Araki, C.C. Moore, Ş. Strătilă and D. Voiculescu

BRATTELI, O.; DIGERNES, T.; ELLIOTT, G.A.:
Locality and differential operators on C*-algebras, II pp.46-83

Due to a mistake in pagination, pages 75-83 of this volume should be read in the following sequence:

pp. 75 - 82 - 76 - 77 - 78 - 79 - 80 - 81 - 83.

The Department of Mathematics of the National Institute for Scientific and Technical Creation organized a Conference on "Operator Algebras, Connections with Topology and Ergodic Theory" held in Buşteni, Romania, August 29 - September 8, 1983. The research contracts between the Mathematics Department of INCREST and the National Council for Science and Technology constituted the generous framework which made possible the organization of this conference. The organizing committee consisted of Zoia Ceauşescu, head of the department, Grigore Arsene, Radu Gologan, Mihai Pimsner, Sorin Popa, Şerban Strătilă and Dan Voiculescu.

The main topic of the conference was the recent progress in operator algebras, arising from the use of topological and ergodic theory ideas. The present volume contains the invited adresses and papers contributed by participants accepted on the basis of referees' reports. The volume has been edited by Huzihiro Araki, Calvin C.Moore, Şerban Strătilă and Dan Voiculescu, benefitting from the assistence of Grigore Arsene.

CONTENTS

ERGODIC PROPERTIES OF SOME C*-DYNAMICAL SYSTEMS

Huzihiro ARAKI
Research Institute for Mathematical Sciences
Kyoto University, Kyoto 606, JAPAN

Abstract: Return to equilibrium in the one-dimensional XY-model is discussed with emphasis on mathematical aspect. A related mathematical structure is discussed also in connection with the two-dimensional Ising model, especially using $\mathbb{Z}_2$-index for two projections.

§1. Asymptotic abelian property and its variant

We consider a C*-dynamical system consisting of a separable C*-algebra $\mathfrak{A}$ and a continuous one-parameter group α_t ($t \in \mathbb{R}$) of automorphisms of $\mathfrak{A}$. In general discussion of ergodic properties of such a C*-dynamical system, the following asymptotic abelian property is often assumed: for arbitrary a and b in $\mathfrak{A}$,

$$\lim_{t\to\infty} [\alpha_t(a), b] = 0 \tag{1.1}$$

where $[A,B] = AB-BA$ denotes the commutator.

If α_t is taken to be a realistic time translation of a quantum system, it is not easy to find out whether (1.1) holds or not in a specific model. We investigate this question for the one-dimensional XY-model in quantum statistical mechanics. The exact situation differs between the following two cases:

(I) The chain (i.e. one-dimensional lattice) extends to infinity in one direction but terminates at a finite point in the other direction. The chain is then identified with $\mathbb{N}$ (the natural numbers).

(II) The chain extends to infinity in both directions. The chain is then identified with $\mathbb{Z}$ (the integers).

In the XY-model, there is a parameter γ indicating the asymmetry between x and y spin interactions. For case (I) with $\gamma = 0$ (no asymmetry), we have the following result ([4] Lemma 1):

<u>Theorem 1.</u> <u>For case (I) with</u> $\gamma = 0$,

(1.2) $\lim_{t\to\infty}\|[a,\alpha_t(b)]\| = 0$ *if* $\Theta(b) = b$,

(1.3) $\lim_{t\to\infty}\|\Theta(a)\alpha_t(b)-\alpha_t(b)a\| = 0$ *if* $\Theta(b) = -b$,

where Θ *is an automorphism of* $\mathfrak{A}$ *satisfying* $\Theta\alpha_t = \alpha_t\Theta$ *for all* t *and* $\Theta^2 = \mathrm{id}$.

We call $a\in\mathfrak{A}$ Θ-even if $\Theta(a)=a$ and Θ-odd if $\Theta(a)=-a$. Proposition 1.1 says that if either a or b is even, then the asymptotic abelian property (1.1) holds, whilst if both a and b are odd, then the asymptotically anticommutating property

(1.4) $\lim_{t\to\infty}\| a\alpha_t(b)+\alpha_t(b)a\| = 0$

holds (and hence (1.1) does not hold for non-zero odd a and b). Thus we call the property in Proposition 1.1 the twisted asymptotic abelian property.

In this example, we find by chance an appropriate operator Θ explicitly and a corresponding twised form of the asymptotic abelian property. However we do not know how Θ is to be found for a given α_t in a general situation, even if it exists.

§2. Return to equilibrium

There is the following consequence (essentially [1] Proposition 4) of the asymptotic abelian property (1.1) for any representation π of $\mathfrak{A}$ for which the von Neumann algebra $\pi(\mathfrak{A})''$ is a factor (i.e. has a trivial center $\mathbb{C}\mathbf{1}$):

(2.1) $\text{w-}\lim_{t\to\infty}\{\pi(\alpha_t(A))-\omega_\Phi(\pi(\alpha_t(a)))\mathbf{1}\} = 0$

where Φ is any vector with the unit length in the representation space, $\omega_\Phi(A) = (A\Phi,\Phi)$ and w-lim is the limit in the weak operator topology.

If φ is an equilibrium state of the C*-dynamical system (satisfying the (α_t,β)-KMS condition), then it has an integral decomposition into pure phases (i.e. extremal (α_t,β)-KMS states), which induces an integral decomposition of the associated cyclic representation into factor representations. For the cyclic representation π_φ associated with a pure phase φ, one may take $\omega_\Phi = \varphi$, which is automatically α_t-

invariant. Thus one has the return to equilibrium [9]

(2.2) $$\underset{t\to\infty}{\text{w-lim}}\ \pi_{\varphi}(\alpha_t(a)) = \varphi(a)\mathbb{1}$$

or equivalently

(2.3) $$\lim_{t\to\infty} \psi(\alpha_t(A)) = \varphi(a)$$

for any vector state $\psi(a) = (\pi_{\varphi}(a)\Psi,\Psi)$ given by any unit vector Ψ in the cyclic representation space associated with the extremal equilibrium state φ. Such a state ψ is interpreted as an outcome of a local perturbation on φ and (2.3) means that the locally perturbed state ψ returns to equilibrium state φ by dynamical evolution in distant future $(t \to +\infty)$ as well as past.

With a modified mathematical argument, we obtain the following twisted version of such a result ([4] Lemma 2):

Theorem 2. *For any factor representation π of $\mathfrak{A}$ on a Hilbert space $\mathcal{H}$ containing a cyclic unit vector Φ such that the vector state $\varphi(a) = \omega_{\Phi}(\pi(a))$ $(= (\pi(a)\Phi,\Phi))$ is Θ-invariant (i.e. $\varphi(\Theta(a)) = \varphi(a)$ for any $a \in \mathfrak{A}$), then (1.2) and (1.3) imply*

(2.4) $$\underset{t\to\infty}{\text{w-lim}}\ \{\pi(\alpha_t(a))-\varphi(\alpha_t(a))\mathbb{1}\} = 0,$$

in which φ may be replaced by any vector state ω_{Ψ}.

In the XY-model (or in more general one-dimensional system, [3] and [8]), the equilibrium state φ_β for any given β is unique, which implies its Θ-invariance due to $\Theta\alpha_t = \alpha_t\Theta$. Thus the assumption of Theorem 2 is automatically satisfied for $\varphi = \varphi_\beta$ and we have the following conclusion:

Corollary 3. *The return to equilibrium (2.3) holds for the case (I) with* $\gamma=0$.

§3. Obstruction to the return to equilibrium

In case (I) with $\gamma \neq 0$, the fixed point algebra $\mathfrak{A}^\alpha$ under the dymanical evolution α_t is non trivial, i.e. a two-dimensional abelian algebra generated by $\mathbb{1}$ and a Θ-odd selfadjoint unitary element B_γ of $\mathfrak{A}$. The element B_γ gives an obstruction to the preceding type of results for the case (I) with $\gamma \neq 0$. Namely, for $b = B_\gamma$ (which satisfies $\alpha_t(b) = b$ and $\Theta(b) = -b$), (1.3) does not hold unless $\Theta(a)b = ba$ (there exist a plenty of a's for which $\Theta(a)b \neq ba$) and the return to equilibrium given by (2.3) does not hold for $a = B_\gamma$ and $\varphi = \varphi_\beta$ (for which $\alpha_t(a) = a$ and $\varphi(a) = 0$ due to $\varphi(a) = \varphi(\Theta(a)) = -\varphi(a)$) unless $\psi(B_\gamma) = 0$ (there exist a plenty of a vector state $\psi = \omega_\Psi$ for which $\psi(B_\gamma) \neq 0$).

On the other hand, this is the only obstruction, i.e. if we consider the twisted commutant

$$(3.1) \qquad (\mathfrak{A}^\alpha)^{tc} = \{a \in \mathfrak{A} : \Theta(a)B_\gamma = B_\gamma a\}$$

then

$$(3.2) \qquad \mathfrak{A} = (\mathfrak{A}^\alpha)^{tc} + B_\gamma (\mathfrak{A}^\alpha)^{tc}$$

and the following result holds ([4] Lemma 9):

Theorem 4. <u>The twisted asymptotic abelian property ((1.2) and (1.3)) holds if one of</u> a <u>and</u> b <u>is in</u> $(\mathfrak{A}^\alpha)^{tc}$ <u>in the case (I) with</u> $\gamma \neq 0$.

An arbitrary element $a \in \mathfrak{A}$ can be written as $a_1 + B_\gamma a_2$ with a_1 and a_2 in $(\mathfrak{A}^\alpha)^{tc}$. Then ([4] (6.21))

$$(3.3) \qquad \lim_{t\to\infty} \psi(\alpha_t(a)) = \varphi_\beta(a_1) + \varphi_\beta(a_2)\psi(B_\gamma)$$

for any vector state ψ in the cyclic representation associated with φ_β in the case (I) with $\gamma \neq 0$, whilst

$$(3.4) \qquad \varphi_\beta(a) = \varphi_\beta(a_1)$$

due to the KMS condition ([4] the last 3 lines of §6). As a consequence we have the following result:

Corollary 5. <u>In the case (I) with</u> $\gamma \neq 0$, <u>the return to equilibrium (2.3) holds for a vector state</u> ψ <u>given by a unit vector</u> Ψ <u>in the cyclic representation associated with the equilibrium state</u> $\varphi = \varphi_\beta$, <u>if either</u> $a \in (\mathfrak{A}^\alpha)^{tc}$ <u>or</u> $\Theta(a)=a$ <u>or</u> $\psi(B_\gamma)=0$.

§4. Return to equilibrium despite failure in twisted asymptotic abelian property

For case (II), we do not have an obstruction to return to equilibrium like B_γ in the case (I) and we have the following result for all values of γ ([5] Theorem 1):

Theorem 6. In case (II), the return to equilibrium (2.3) holds for all $a \in \mathfrak{A}$ and for any vector state $\psi = \omega_\Psi$ given by a unit vector Ψ in the cyclic representation associated with the equilibrium state $\varphi = \varphi_\beta$.

On the other hand, the asymptotic abelian property holds only partially ([5] Theorem 13):

Theorem 7. In the case (II), the following relations hold for $a, b \in \mathfrak{A}$:

(4.1) $$\lim_{t\to\infty} \| [a, \alpha_t(b)] \| = 0 \quad \text{if} \quad \Theta(a) = a,\ \Theta(b) = b,$$

(4.2) $$\lim_{t\to\infty} \| a\alpha_t(b) - \Xi(\alpha_t(b))a \| = 0 \quad \text{if} \quad \Theta(a) = -a,\ \Theta(b) = b,$$

(4.3) $$\lim_{t\to\infty} \| a\alpha_t(b) - \alpha_t(b)\Xi(a) \| = 0 \quad \text{if} \quad \Theta(a) = a,\ \Theta(b) = -b.$$

Here Ξ is an automorphism of the Θ-even part $\mathfrak{A}_+$ of $\mathfrak{A}$ such that $\Xi\alpha_t = \alpha_t\Xi$ for all t, $\Xi^2 = \text{id.}$ and $\Xi\Theta = \Theta\Xi$.

On the other hand it can be proved that there exists no extension of Ξ to an automorphism of $\mathfrak{A}$ and hence no straightforward extension of the twisted asymptotic abelian property above to the missing case of Θ-odd a and b ([5] Remark 15).

We note that the return to equilibrium given by Theorem 6 is the same as

(4.4) $$\underset{t\to\pm\infty}{\text{w-lim}}\ \pi_\beta(\alpha_t(a)) = \varphi_\beta(a)\mathbb{1}$$

for all $a \in \mathfrak{A}$, where π_β is the cyclic representation of $\mathfrak{A}$ associated with the equilibrium state φ_β, and hence implies the weak asymptotic abelian property

(4.5) $$\text{w-lim}[x, \pi_\beta(\alpha_t(a))] = 0$$

for any $a \in \mathfrak{A}$ and any operator x in the representation space of π_β.

§5. Concrete description of the dynamical system $(\mathfrak{A},\alpha_t)$

The C*-algebra $\mathfrak{A}$ under consideration can be described as follows. For each lattice point j (either in $\mathbb{N}$ or $\mathbb{Z}$ according as we deal with case (I) or (II)), there corresponds a full matrix algebra $\mathfrak{A}_j$ of 2×2 matrices linearly spanned by the identity $\mathbb{1}$ and

$$(5.1) \qquad \sigma_x^{(j)} = \begin{pmatrix} 0 & 1 \\ 1 & 0 \end{pmatrix}, \quad \sigma_y^{(j)} = \begin{pmatrix} 0 & -i \\ i & 0 \end{pmatrix}, \quad \sigma_z^{(j)} = \begin{pmatrix} 1 & 0 \\ 0 & -1 \end{pmatrix}.$$

It is to be identified with a subalgebra of $\mathfrak{A}$ such that $\mathfrak{A}_j$ and $\mathfrak{A}_k$ for $j\neq k$ commute elementwise and $\mathfrak{A}_j$ for all j together generate algebraically a dense subalgebra $\mathfrak{A}^{loc}$ of $\mathfrak{A}$.

The dynamical evolution α_t can be uniquely given in terms of its generator δ (i.e. $\alpha_t = e^{\delta}$), for which $\mathfrak{A}^{loc}$ is a core and

$$(5.2) \qquad \delta(a) = i \sum_j [h_j,a], \qquad a\in\mathfrak{A}^{loc}$$

$$(5.3) \qquad h_j = -J\{(1+\gamma)\sigma_x^{(j)}\sigma_x^{(j+1)}+(1-\gamma)\sigma_y^{(j)}\sigma_y^{(j+1)}\},$$

J is a positive constant, $-1 < \gamma < 1$ and the summation (5.2) is a finite sum because $[h_j,a]$ vanishes for all but a finite number of j if $a\in\mathfrak{A}^{loc}$.

The automorphism Θ is uniquely determined by the following requirement for all j:

$$(5.4) \qquad \Theta(\sigma_x^{(j)}) = -\sigma_x^{(j)}, \qquad \Theta(\sigma_y^{(j)}) = -\sigma_y^{(j)},$$

$$(5.5) \qquad \Theta(\sigma_z^{(j)}) = \sigma_z^{(j)}.$$

(It represents the 180° rotation of σ-spins at all sites around the z-axis.)

The element B_γ is given by

$$(5.6) \qquad B_\gamma = \sum_{j=1}^{\infty} (-\alpha)^{j-1}\sigma_z^{(1)}\cdots\sigma_z^{(2j-2)} \times \begin{cases} \sigma_x^{(2j-1)} & \text{if } 0<\gamma<1 \\ \sigma_y^{(2j-1)} & \text{if } -1<\gamma<0 \end{cases}$$

and $\alpha = (1-|\gamma|)/(1+|\gamma|)$.

Let Θ_- be the automorphism of $\mathfrak{A}$ in the case (II) uniquely determined by the requirement that

$$(5.7)\qquad \Theta_-(\sigma_x^{(j)}) = \begin{cases} \sigma_x^{(j)} \\ -\sigma_x^{(j)} \end{cases}, \quad \Theta_-(\sigma_y^{(j)}) = \begin{cases} \sigma_y^{(j)} & \text{if } j \geq 1, \\ -\sigma_y^{(j)} & \text{if } j \leq 0, \end{cases}$$

$$(5.8)\qquad \Theta_-(\sigma_z^{(j)}) = \sigma_z^{(j)} \quad \text{for all } j.$$

Then the following limit exists and defines Ξ for Θ-even a:

$$(5.9)\qquad \Xi(a) = \lim_{t\to\infty} \alpha_t \Theta_- \alpha_{-t}(a).$$

§6. Use of crossed product

For computational convenience, we introduce another C*-algebra $\mathfrak{A}^{CAR}$ described as follows: For each j, there is an element c_j such that

$$(6.1)\qquad c_j c_k + c_k c_j = 0,$$

$$(6.2)\qquad c_j c_k^* + c_k^* c_j = \begin{cases} 0 & \text{if } j \neq k \\ 1 & \text{if } j = k \end{cases}$$

(canonical anticommutation relations abbreviated as CAR) and c_j and c_j^* for all j together algebraically generate $\mathfrak{A}^{CAR}$.

In case (I), we can identify $\mathfrak{A}^{CAR}$ with $\mathfrak{A}$ through the following elementwise identification ($\mathfrak{A}^{CAR}$ is isomorphic to $\mathfrak{A}$ also in the case (II) but we are interested here in the concrete correspondence for the sake of a simple form for δ):

$$(6.3)\qquad c_j = \Big(\prod_{k=1}^{j-1} \sigma_z^{(k)}\Big)(\sigma_x^{(j)} - i\sigma_y^{(j)})/2,$$

$$(6.4)\qquad \sigma_z^{(j)} = 2c_j^* c_j - 1,$$

$$(6.5)\qquad \sigma_x^{(j)} = \Big[\prod_{k=1}^{j-1} (2c_k^* c_k - 1)\Big](c_j + c_j^*),$$

(6.6) $$\sigma_y^{(j)} = [\prod_{k=1}^{j-1}(2c_k^*c_k - 1)]i(c_j - c_j^*).$$

Then α_t is a quasifree automorphism of $\mathcal{A}^{CAR}$ and φ_β is a quasifree state of $\mathcal{A}^{CAR}$. The results in sections 2 and 3 are obtained by the spectral analysis of a selfadjoint operator on a Hilbert space (of the so-called test functions) representing the generator δ.

In case (II), we construct the crossed product of $\mathcal{A}$ by Θ_- action of the group $\mathbb{Z}_2$ (note that Θ_-^2=id. and Θ_- is an outer automorphism) and call it $\mathcal{B}$. Then $\mathcal{A}$ is a C*-subalgebra of $\mathcal{B}$ and

(6.7) $$\mathcal{B} = \mathcal{A} \oplus T\mathcal{A}$$

where T is an element of $\mathcal{B}$ satisfying $T^2 = 1$, $T^* = T$, and

(6.8) $$Ta = \Theta_-(a)T$$

for any $a \in \mathcal{A}$.

We then identify $\mathcal{A}^{CAR}$ with a C*-subalgebra of $\mathcal{B}$ through the following elementwise identification:

(6.9) $$c_j = TS^{(j)}(\sigma_x^{(j)} - i\sigma_y^{(j)})/2,$$

(6.10) $$S^{(j)} = \begin{cases} \sigma_z^{(1)}\cdots\sigma_z^{(j-1)} & \text{if } j > 1 \\ \mathbf{1} & \text{if } j = 1 \\ \sigma_z^{(0)}\cdots\sigma_z^{(j)} & \text{if } j < 1. \end{cases}$$

In comparison with the case (I), T plays the role of $\prod_{j=0}^{-\infty}\sigma_j^{(z)}$. Since this infinite product does not make sense in $\mathcal{A}$, we have introduced it by hand from outside.

Let $\mathcal{A}_+$ and $\mathcal{A}_-$ be Θ-even and Θ-odd parts of $\mathcal{A}$. Then

(6.11) $$\mathcal{A}^{CAR} = \mathcal{A}_+ + T\mathcal{A}_-.$$

We can extend Θ uniquely to an automorphism of $\mathcal{B}$ satisfying $\Theta(T) = T$. Then (6.11) gives the split of $\mathcal{A}^{CAR}$ into Θ-even and Θ-odd parts:

(6.12) $$\mathcal{A}_+^{CAR} = \mathcal{A}_+, \quad \mathcal{A}_-^{CAR} = T\mathcal{A}_-.$$

The converse equation is

(6.13) $$\mathfrak{A} = \mathfrak{A}_+^{CAR} + T\mathfrak{A}_-^{CAR}.$$

Since h_j belongs to $\mathfrak{A}_+ = \mathfrak{A}_+^{CAR}$, $\delta|\mathfrak{A}^{loc}$ can be extended through the equation (5.2) to

(6.14) $$\mathcal{B}^{loc} = \mathfrak{A}^{loc} + T\mathfrak{A}^{loc}$$

and α_t can be extended to $\mathcal{B}$ with its generator coinciding with the closure of the above mentioned extension of $\delta|\mathfrak{A}^{loc}$. The C*-subalgebra $\mathfrak{A}^{CAR}$ is α_t-invariant as a set and $\alpha_t|\mathfrak{A}^{CAR}$ is a continuous one-parameter group of quasifree automorphisms with the unique quasifree equilibrium state φ_β^{CAR} on $\mathfrak{A}^{CAR}$.

Let $\hat{\Theta}_-$ be the dual action of $\mathbb{Z}_2$ on $\mathcal{B}$, i.e. an automorphism of $\mathcal{B}$ uniquely determined by

(6.15) $$\hat{\Theta}_-(T) = -T, \qquad \hat{\Theta}_-(a) = a \qquad (a \in \mathfrak{A}).$$

Then $\mathfrak{A}^{CAR}$ is $\Theta\hat{\Theta}_-$ invariant as a set and the unique equilibrium state of $\mathfrak{A}$ is the restriction (to $\mathfrak{A}$) of the unique $\Theta\hat{\Theta}_-$ invariant extension of φ_β^{CAR} (to $\mathcal{B}$) given by

(6.16) $$\varphi_\beta(a_+ + a_-) = \varphi_\beta^{CAR}(a_+)$$

where $\Theta(a_\pm) = \pm a_\pm$ and $a_+ \in \mathfrak{A}_+ = \mathfrak{A}_+^{CAR}$.

The results in §4 is then again obtained by a spectral analysis together with a study of $\alpha_t(T)$, which is somewhat complicated.

§7. Application to two-dimensional Ising model

The equilibrium state of the two-dimensional Ising model is given as the limit $L \to \infty$, $M \to \infty$ of the expectation

(7.1) $$\langle F\rangle_{LM} = Z_{LM}^{-1}\sum_{\xi} F(\xi)e^{-\beta H^{LM}(\xi)},$$

(7.2) $$Z_{LM} = \sum_{\xi} e^{-\beta H^{LM}(\xi)},$$

(7.3) $$H^{LM}(\xi) = -J_1\sum_{i=-L}^{L-1}\sum_{j=-M}^{M}\xi_{ij}\xi_{i+1,j} - J_2\sum_{i=-L}^{L}\sum_{j=-M}^{M-1}\xi_{ij}\xi_{i,j+1},$$

where the spin variable ξ_{ij} at each lattice site $(i,j) \in \mathbb{Z}^2$ takes values ± 1, the sum is over all such values of $\xi = \{\xi_{ij}\}$, $J_1 \geq 0$, $J_2 \geq 0$, $\beta \geq 0$ and the observable F is a function of ξ_{ij} $(|i| \leq \ell, |j| \leq m$ for some $\ell \leq L$ and $m \leq M)$.

By the so-called transfer matrix method, we have

$$\psi_\beta(F_\beta) = \lim_{M\to\infty} \lim_{L\to\infty} \langle F \rangle_{LM} \tag{7.4}$$

where F_β is a certain element of $\mathfrak{A}$ determined by F, β and J's and ψ_β is a state of $\mathfrak{A}$. The computation of ψ_β proceeds by a technique similar to the preceding section and ψ_β turns out to be the Θ-invariant extension (to $\mathfrak{A}$) of the restriction (to $\mathfrak{A}_+ = \mathfrak{A}_+^{CAR}$) of a Θ-invariant pure state φ_E of $\mathfrak{A}^{CAR}$. In fact, φ_E is the Fock type state of $\mathfrak{A}^{CAR}$ determined by the so-called basis projection E on the double sized test function space ([2]).

A result about phase transition is as follows ([6] Theorem 1).

Theorem 8. The cyclic representation of $\mathfrak{A}$ associated with ψ_β is pure if $0 \leq \beta \leq \beta_c$, whilst it is of type I with 2-dimensional center (for the weak closure) if $\beta > \beta_c$ where β_c is the critical inverse temperature for the two-dimensional Ising model.

A method for judging whether ψ_β is primary or not is given by the following ([6] Theorem 2):

Theorem 9. ψ_β is not pure if and only if

(1) φ_E and $\varphi_E \circ \theta_-$ are equivalent and

(2) $\varphi_+ = \varphi|\mathfrak{A}_+$ and $\varphi_+ \circ \theta_-$ are not equivalent.

If ψ_β is not pure, it is a mixture of two non-equivalent pure states.

It turns out that the condition (1) is satisfied for $\beta \neq \beta_c$ and it is not satisfied for $\beta = \beta_c$. To find out about the condition (2), we use the following criterion ([6] Theorem 4):

Theorem 10. The restrictions of Fock type states φ_{E_1} and φ_{E_2} of $\mathfrak{A}^{CAR}$ to $\mathfrak{A}_+^{CAR}$ are equivalent if and only if

(i) $E_1 - E_2$ is in the Hilbert-Schmidt class

(ii) the $\mathbb{Z}_2$-index $\sigma(E_1, E_2) = 1$.

Here the first condition (i) is a necessary and sufficient condition for the condition (1) of Theorem 9 to hold. Under the condition (i), the $\mathbb{Z}_2$-index is defined as

$$\sigma(E_1,E_2) = (-1)^{\dim(E_1 \wedge (1-E_2))} . \tag{7.5}$$

This condition is easy to check by deformation due to the following result ([6] Theorem 3)

Theorem 11. $\sigma(E_1,E_2)$ is continuous in E_1 and E_2 with respect to the norm topology of E's as long as the condition (i) is satisfied.

E depends continuously on β, J_1 and J_2 as long as $\beta \neq \beta_c$. By choosing special values of parameters, the relevant $\mathbb{Z}_2$-index can be easily computed.

The validity of Theorem 11 depends only on the following properties of the basis projections E_1 and E_2: There is an involutive anti-unitary operator Γ (i.e. $\Gamma^2 = 1$, $(\Gamma f, \Gamma g) = (g,f)$) and $E = E_i$ $(i=1,2)$ is an orthogonal projection (i.e. $E^* = E = E^2$) satisfying $\Gamma E \Gamma = \mathbf{1} - E$.

References

[1] H. Araki, On the algebra of all local observables, Progr. Theoret. Phys. 32 (1964), 844-854.
[2] H. Araki, On quasifree states of CAR and Bogoliubov automorphisms, Publ. RIMS Kyoto Univ. 6 (1970), 384-442.
[3] H. Araki, On uniqueness of KMS states of one-dimensional quantum lattice systems, Commun. Math. Phys. 44 (1975), 1-7.
[4] H. Araki and E. Barouch, On the dynamical and ergodic properties of the XY model, J. Stat. Phys. 31 (1983), 327-345.
[5] H. Araki, On the XY-model on two-sided infinite chain, RIMS preprint 435. To appear in Publ. RIMS Kyoto Univ. 20 (1984), No.2.
[6] H. Araki and D. E. Evans, On a C*-algebra approach to phase transition in the two-dimensional Ising model, to appear in Commun. Math. Phys.
[7] O. Bratteli and D. W. Robinson, Operator algebras and quantum statistical mechanics II, Springer, 1981.
[8] A. Kishimoto, Dissipations and derivations, Commun. Math. Phys. 47 (1976), 167-170.
[9] D. W. Robinson, Return to equilibrium, Commun. Math. Phys. 31 (1973), 171-189.

FACTOR STATES ON C*-ALGEBRAS

R.J. Archbold and C.J.K. Batty

Unless stated otherwise, proofs of results mentioned below may be found in [2] and [3].

1. Prime C*-algebras, antiliminal C*-algebras.

Our starting point is the following result of Glimm, Tomiyama and Takesaki.

THEOREM 1 [6,10]

The set P(A) of pure states of a C*-algebra A is weak* dense in the state space S(A) if and only if A is prime and either antiliminal or one-dimensional.

Some of the methods used to prove this result (which is valid even if A is non-unital) were adapted in [2] to prove:

THEOREM 2

The set F(A) of factorial states of a C*-algebra A is weak* dense in S(A) if and only if A is prime.

This supplements an earlier result [1] (see also [5]) that a C*-algebra is prime if and only if its self-adjoint part is an antilattice.

Since

$$\overline{P(A)} \supseteq S(A) \iff \overline{F(A)} \supseteq S(A) \text{ and } \overline{P(A)} \supseteq F(A),$$

comparison of Theorems 1 and 2 suggests that the condition $\overline{P(A)} \supseteq F(A)$ might be related to antiliminality. In fact:

THEOREM 3

$\overline{P(A)} \supseteq F(A)$ if and only if either A is abelian or there exists an abelian ideal I such that A/I is antiliminal.

In comparing Theorems 1, 2 and 3, it should be noted that a prime C*-algebra, of dimension greater than one, cannot have a nonzero abelian ideal. Theorem 3 is in fact the case $k = 1$ of the following result, in which $F_k(A)$ $(1 \le k < \infty)$ denotes the set of factorial states ϕ for which the GNS representation π_ϕ gives rise to a commutant $\pi_\phi(A)'$ which is a factor of type I_n for some $n \le k$.

THEOREM 4

For any C*-algebra A, the following conditions are equivalent.

(i) $F_k(A)$ is weak* dense in $F(A)$

(ii) $F_k(A)$ is weak* dense in $F_{k+1}(A)$

(iii) Either A is k-subhomogeneous, or there is a k-subhomogeneous ideal I such that A/I is antiliminal.

2. Factorial states of type I.

Let $F_\infty(A) = \{\phi \in F(A) | \pi_\phi(A)' \text{ is type I}\}$ and let $F_f(A) = \{\phi \in F(A) | \pi_\phi(A)' \text{ is type I and finite}\}$ (so that $F_f(A) = \cup\{F_k(A) | 1 \le k < \infty\}$). Consideration of GNS representations reveals that $F_f(A)$ (respectively $F_\infty(A)$) is the set of all states which are convex (respectively σ-convex) combinations of equivalent pure states. In the original proof of Theorem 2 it was shown that if A is prime then $\overline{F_\infty(A)} \supseteq S(A)$. By decomposing a type I commutant $\pi_\phi(A)'$, it follows easily that if A is prime then $\overline{F_f(A)} \supseteq S(A)$. However, a more direct proof of this may be obtained by using the Krein-Milman theorem together with the following result which may be of independent interest.

PROPOSITION 5

If A is prime then any convex combination $\sum_{i=1}^{n} \lambda_i \phi_i$ of pure states of A may be approximated by states (necessarily in $F_f(A)$) of the form $\sum_{i=1}^{n} \lambda_i \psi_i$ where $\psi_1, \ldots, \psi_n$ are equivalent pure states of A.

If $\phi \in F(A)$ then $J = \ker \pi_\phi$ is a prime ideal of A and so $\overline{F_f(A/J)} \supseteq S(A/J)$. Since $\phi(J) = 0$, $\phi \in \overline{F_f(A)}$. Thus we have:

THEOREM 6

For any C*-algebra A, $\overline{F_f(A)} \supseteq F(A)$. Hence $\overline{F_f(A)} = \overline{F_\infty(A)} = \overline{F(A)}$.

3. Extension of states.

Let A be a C*-subalgebra of a C*-algebra B. A longstanding problem has been whether factorial states of A can be extended to factorial states of B [8; p.242]. However, it follows from a result of Sakai [4] (see also [11]) that $F_\infty(A) \subseteq F(B)|A$ (indeed, it can also be shown by various methods that $F_k(A) \subseteq F_k(B)|A$ $(1 \leq k \leq \infty)$). Thus Theorem 6, together with a simple compactness argument in B*, yields:

COROLLARY 7

If $A \subseteq B$ then $\overline{F(A)} \subseteq \overline{F(B)}|A$. In particular, any factorial state of A can be extended to a state in $\overline{F(B)}$.

This result is utilised in the next section.

4. The factorial state space $\overline{F(A)}$.

Suppose that $\phi \in S(A)$ and $\ker \pi_\phi$ contains a prime ideal J of A. Since ϕ factors through A/J and $\overline{F(A/J)} \supseteq S(A/J)$ (Theorem 2), we see that $\phi \in \overline{F(A)}$. If A is a von Neumann algebra then the converse holds:

THEOREM 8

Let A be a von Neumann algebra and let $\phi \in S(A)$. The following conditions are equivalent.

(i) $\phi \in \overline{F(A)}$

(ii) $\ker \pi_\phi$ contains a "Glimm ideal" of A [6; p.232]

(iii) $\ker \pi_\phi$ contains a prime ideal of A

(iv) $\ker \pi_\phi$ contains a primitive ideal of A.

For a general C*-algebra A, (i) does not imply (iii). However, if A is a central C*-algebra then (i), (iii) and (iv) are equivalent (and prime ideals are maximal).

Theorem 8 is related to the problem of describing $\overline{F(A)}$ for a general C*-algebra A by the following result:

THEOREM 9

Let A be a unital C*-algebra acting non-degenerately on a Hilbert space. Let $\bar{A}$ denote the von Neumann algebra generated by A. Then the factorial state space of $\bar{A}$ restricts to that of A i.e. $\overline{F(\bar{A})|A} = \overline{F(A)}$.

That $\overline{F(\bar{A})|A} \subseteq \overline{F(A)}$ is proved by using Glimm's analogous result for pure state spaces [6; Theorem 5], Theorem 6 above, and tensor products. Problems arising from the non-nuclearity of $\bar{A}$ are overcome in [2] by the use of $F_\infty(\bar{A})$. The technicalities of this are reduced in [3] by the use of $F_f(\bar{A})$ instead of $F_\infty(\bar{A})$. That $\overline{F(A)} \subseteq \overline{F(\bar{A})|A}$ is an immediate consequence of Corollary 7.

Finally we consider the possibility that $\overline{F(A)}$ consists entirely of (multiples of) factorial states.

THEOREM 10

For any C*-algebra A, the following conditions are equivalent.

(i) $\overline{F(A)} \subseteq \{\lambda\phi \mid 0 \le \lambda \le 1,\ \phi \in F(A)\}$

(ii) $\overline{P(A)} \cap S(A) \subseteq F(A)$

(iii) A is liminal and the spectrum $\hat{A}$ is Hausdorff.

This extends a result of Shultz [9; Proposition 9]. The equivalence of (i) and (iii) is analogous to Glimm's result concerning pure states [7; Theorem 6]. Indeed, the proof of Theorem 9 uses some of Glimm's arguments. The following result is analogous to [10; Theorem 1].

COROLLARY 11

F(A) is compact if and only if A is unital, liminal, and has Hausdorff spectrum.

References

1. R.J. Archbold, Prime C*-algebras and antilattices, Proc. London Math. Soc. (3), 24 (1972), 669-680.

2. R.J. Archbold, On factorial states of operator algebras, to appear in J. Functional Analysis.

3. C.J.K. Batty and R.J. Archbold, On factorial states of operator algebras II, in preparation.

4. J.W. Bunce, Stone-Weierstrass theorems for separable C*-algebras, Proc. Symp. Pure Math. 38 (1982), Part I, 401-408.

5. C-H. Chu, Prime faces in C*-algebras, J. London Math. Soc. (2), 7 (1973), 175-180.

6. J. Glimm, A Stone-Weierstrass theorem for C*-algebras, Ann. Math. 72 (1960), 216-244.

7. J. Glimm, Type I C*-algebras, Ann. Math. 73 (1961), 572-612.

8. S. Sakai, "C*-algebras and W*-algebras", Springer Verlag, Berlin, Heidelberg, New York, 1971.

9. F.W. Shultz, Pure states as a dual object for C*-algebras, Commun. Math. Phys. 82 (1982), 497-509.

10. J. Tomiyama and M. Takesaki, Applications of fibre bundles to the certain class of C*-algebras, Tohôku Math. J. (2), 13 (1961), 498-523.

11. S-K. Tsui, Factor state extension on nuclear C*-algebras, Yokohama Math. J. 29 (1981), 157-160.

Department of Mathematics
University of Aberdeen
The Edward Wright Building
Dunbar Street
Aberdeen AB9 2TY
Scotland

Department of Mathematics
University of Edinburgh
King's Buildings
Mayfield Road
Edinburgh EH9 3JZ
Scotland

CONTINUOUS NESTS AND THE ABSORPTION PRINCIPLE

William Arveson*

1. INTRODUCTION

In [2], a general absorption principle is established which provides a unification of theorems of Dan Voiculescu and Niels Toft Andersen (to be described presently). Andersen's theorem was subsequently generalized to a rather broad class of commutative subspace lattices. Since a substantial amount of work is required to set up this generalization, it is not made very clear in [2] that one can proceed in a simple way from the absorption principle to Andersen's theorem. The purpose of this note is to show how this can be done. We will discuss the absorption principle (without proof) and we will indicate (with proof) how one goes about deducing Andersen's theorem from it.

Throughout this paper, all Hilbert spaces will be separable, and the generic symbol K will denote the C*-algebra of compact operators on the appropriate Hilbert space.

Voiculescu's theorem [6] asserts that if $A \subseteq \mathcal{L}(\mathcal{H})$ is a *separable* C*-algebra of operators which contains the identity and σ is a non-degenerate representation of A which annihilates $A \cap K$, then

$$id \oplus \sigma \underset{a}{\sim} id \quad ,$$

where id denotes the identity representation of A. Here, $\underset{a}{\sim}$ is Voiculescu's notion of approximate equivalence: for two representations π_1, π_2 of A on spaces $\mathcal{H}_1, \mathcal{H}_2$, $\pi_1 \underset{a}{\sim} \pi_2$ means that there is a sequence W_n of unitary operators from $\mathcal{H}_1$ to $\mathcal{H}_2$ such that for each $A \in A$

*This research was supported by National Science Foundation grant MCS83-02061.

(i) $$W_n\pi_1(A)W_n^* - \pi_2(A) \in K ,$$

and

(ii) $$\lim_{n\to\infty} \|W_n\pi_1(A)W_n^* - \pi_2(A)\| = 0 .$$

Using this theorem, one can easily deduce

Corollary. *Let* $A_i \subseteq \mathcal{L}(\mathcal{H})$ *be two separable* C^**-algebras of operators which contain 1. Assume that*

i) A_1 *and* A_2 *are *-isomorphic, and*

ii) $A_i \cap K = \{0\}$, $i = 1,2$.

Then $A_1 + K$ *and* $A_2 + K$ *are unitarily equivalent.*

The corollary has a classical predecessor, due to Weyl and von Neumann. *Let* A_1, A_2 *be self-adjoint operators such that*

i) $sp(A_1) = sp(A_2)$,

ii) *neither* A_1 *nor* A_2 *has any isolated eigenvalue of finite multiplicity.*

Then A_1 *is unitarily equivalent to a compact perturbation* A_2+K *of* A_2; *moreover,* K *can be chosen so that its norm is arbitrarily small.* Actually, K can be chosen to be a small Hilbert-Schmidt operator, but that is not relevant to our purpose here (the essential step can be found on p.525 of [5]).

We want to point out that the corollary fails if one drops the separability hypothesis. Indeed, if A_1 is a nonatomic maximal abelian von Neumann algebra in $\mathcal{L}(\mathcal{H})$ and A_2 is the abelian von Neumann algebra on $\mathcal{H} \oplus \mathcal{H}$ defined by

$$A_2 = \{A \oplus A : A \in A_1\} ,$$

then A_1 and A_2 are *-isomorphic, $A_i \cap K = \{0\}$ for $i = 1,2$, but $A_1 + K$ and $A_2 + K$ are not unitarily equivalent. The argument can be found in the introduction of [2].

Let us recall Andersen's theorem ([1], 3.5.5) about continuous nests. By a continuous nest we mean here a projection-valued function $t \in [0,1] \mapsto P_t \in \mathcal{L}(\mathcal{H})$ satisfying

i) $P_0 = 0, \quad P_1 = 1$

ii) $s < t \Rightarrow P_s < P_t$

iii) $t \mapsto \langle P_t\xi, \eta \rangle$ *is continuous for every* ξ, η *in* $\mathcal{H}$.

Andersen's theorem asserts that if $\{P_t\}$ and $\{Q_t\}$ are two continuous nests, then there is a sequence W_n of unitary operators such that

i) $W_n P_t W_n^* - Q_t$ *is compact for all* t.

(1.1) ii) $\sup_{0 \leq t \leq 1} \|W_n P_t W_n^* - Q_t\|$ *tends to zero as* n *tends to* ∞.

iii) $t \mapsto W_n P_t W_n^* - Q_t$ *is a norm-continuous operator-valued function, for each* $n \geq 1$.

Notice that the assertions of (1.1) resemble the definitions of approximate equivalence of representations to some extent, but there are some essential differences. First, the commutative C*-algebras generated by $\{P_t\}$ and $\{Q_t\}$ are invariably inseparable. Second, the condition (1.1)ii) asserts that, for large n, the infinite set of norms

$$\{\|W_n P_t W_n^* - Q_t\| : \quad 0 \leq t \leq 1\}$$

are simultaneously small: the assertion of "approximate equivalence" would require only a finite number of these norms to be small. Finally, there is no counterpart whatsoever to the third property (1.1)iii) in the definition of approximate equivalence.

In the next section we will introduce a context appropriate for obtaining conclusions of this nature, and in section 3 we will derive Andersen's theorem from the main result of section 2.

2. THE ABSORPTION PRINCIPLE

By a *-semigroup we mean a second countable locally compact Hansdorff space X, on which there is defined a jointly continuous associative multiplication

$$(x,y) \in X \times X \mapsto x \cdot y \in X \quad ,$$

and a continuous involution $x \mapsto x^*$, i.e., a mapping of X satisfying $x^{**} = x$, $(xy)^* = y^* x^*$. For convenience, we also assume X has a unit e: $e \cdot x = x \cdot e = x$, $x \in X$.

To recover the context of Voiculescu's theorem, one chooses X to be a countable norm-dense subgroup of the unitary group of a separable C^*-algebra A, endowed with its discrete topology. The multiplication and involution in X are the obvious operations inherited from A. To recover the context of Andersen's theorem, take X to be the closed unit interval with its natural topology, and the operations

$$x \cdot y = \min(x,y) \quad ,$$

$$x^* = x \quad ,$$

$x,y \in [0,1]$. Other applications are described in [2].

A *representation* of a *-semigroup X is a mapping $x \mapsto U(x)$ from X to the operators on some Hilbert space $\mathcal{H}$, which is

(2.1)
i) *strongly continuous*

ii) *a homomorphism of unital *-semigroups*

iii) *bounded:* $\sup_{x \in X} \|U(x)\| < \infty$.

It is easy to see that, in fact, we must have

$$\|U(x)\| \leq 1$$

for all $x \in X$. We will also write $\mathcal{H}_U$ for the Hilbert space on which a given representation U acts.

Definition 2.2 (Norm equivalence). *If* U,V *are representations of* X, *we will write* $U \sim V$ *if for every compact subset* K *of* X *and* $\varepsilon > 0$, *there is a unitary operator* W *from* $\mathcal{H}_U$ *to* $\mathcal{H}_V$ *such that*

$$\sup_{x \in K} \|WU(x)W^* - V(x)\| \leq \varepsilon \quad .$$

This is clearly an equivalence relation in the collection of all representations of X. This relation has a simple definition and is easy to work with. But what we are really interested in is the following much stronger relation.

Definition 2.3 (Approximate equivalence). *For two representations* U,V *of* X, $U \underset{a}{\sim} V$ *means that for every compact subset* K *of* X *and* $\varepsilon > 0$, *there is a unitary operator* W *from* $\mathcal{H}_U$ *to* $\mathcal{H}_V$ *satisfying*

i) $\sup_{x \in K} \|WU(x)W^* - V(x)\| \leq \varepsilon$, *and*

ii) $x \to WU(x)W^* - V(x)$ *is a norm-continuous function from* X *to the compact operators.*

Let U,V be two representations of X. We require some criteria for determining when V is "absorbed" by U in the following sense,

$$U \oplus V \sim U \quad .$$

These criteria should involve the action of U and V on their respective spaces, and should involve properties that can be checked in specific examples. We will see that such criteria exist, but that they involve not only U and V but a sequence of representations associated with U and V.

This sequence is defined as follows. Let X be a *-semigroup. For each positive integer n, let G_n be a finite subgroup of the unitary group of the C^*-algebra M_n of all $n \times n$ matrices, such that

$$M_n = \text{span } G_n \quad .$$

For instance, one may take G_n to be the group of all $n \times n$ matrices having exactly one nonzero entry, consisting of ±1, in each row and each column. G_n is considered to be fixed throughout the remainder of the discussion.

G_n is a *-semigroup in its discrete topology. So for each $n > 1$ we may form the Cartesian product of *-semigroups $G_n \times X$. Finally, if U is a representation of X on $\mathcal{H}$ then we can form a sequence of representations $U_n: G_n \times X \to \mathcal{L}(\mathbb{C}^n \otimes \mathcal{H})$ by

$$U_n(u,x) = u \otimes U(x) ,$$

$u \in G_n$, $x \in X$. The process whereby one considers the sequence of representations $U_1, U_2, \ldots$ along with U is somewhat analogous to the process of considering, along with a completely positive linear map of C^*-algebras

$$\phi: A \to \mathcal{L}(\mathcal{H}) ,$$

its associated sequence of completely positive maps

$$\mathrm{id} \otimes \phi: \; M_n \otimes A \to \mathcal{L}(\mathbb{C}^n \otimes \mathcal{H}) ,$$

$n = 1,2,\ldots$.

Finally, we will say that a representation V is *subordinate* to a representation U if, for every normal state ρ of $\mathcal{L}(\mathcal{H}_V)$, there is a sequence ξ_n of unit vectors in $\mathcal{H}_u$ such that

(2.4)

i) $\xi_n \to 0$ *weakly in* $\mathcal{H}_u$, *and*

ii) $\rho(V(x)) = \lim_{n\to\infty} \langle U(x)\xi_n, \xi_n\rangle$ *uniformly on compact subsets of* X.

Roughly speaking, (2.4) says that normal states of V can be approximated by vector states of U, where the approximating vectors are "near infinity".

We can now state the main result.

Theorem 2.5. *Let* U,V *be representations of* X. *The following are equivalent:*

i) $U \oplus V \sim U$.

ii) $U \oplus V \underset{a}{\sim} U$.

iii) V_n *is subordinate to* U_n, *for every* $n = 1,2,\ldots$.

3. CONTINUOUS NESTS

We now prove the following theorem from ([1], 3.5.5).

Theorem 3.1. *Let* $\{P_t : 0 \le t \le 1\}$, $\{Q_t : 0 \le t \le 1\}$ *be continuous nests. Then there is a unitary operator* W *such that the properties* 1.1 *are satisfied.*

We require the following variation of the continuity theorem of probability theory.

Lemma. *Let* μ_n *be a sequence of positive finite measures on* [0,1] *which converges to a nonatomic measure* μ *in the weak* topology of* C[0,1]. *Then*

$$\sup_{0 \le t \le 1} |\mu_n([0,t]) - \mu([0,t])|$$

tends to zero as n *tends to* ∞.

Proof of Lemma. We may clearly assume that $\mu_n([0,1]) = \mu([0,1]) = 1$ for every n. Let

$$F_n(t) = \mu_n([0,t])$$

$$F(t) = \mu([0,t]) \quad , \qquad 0 \le t \le 1 \quad .$$

By ([4], theorem 1, p.249), the sequence F_n converges *pointwise* to F. We have to show that this convergence is actually uniform.

For that, fix $\varepsilon > 0$. Since μ is nonatomic, F is continuous and therefore uniformly continuous. So we may find points

$$0 = t_0 < t_1 < \dots < t_N = 1$$

in $[0,1]$ such that $|F(t_j) - F(t_{j-1})| \le \varepsilon$ for all $1 \le j \le N$. Choose n large enough so that

$$\max_{0 \le j \le N} |F_k(t_j) - F(t_j)| \le \varepsilon \quad ,$$

for all $k \ge n$. Then for every such k and every $s \in [0,1]$, say $t_{j-1} \le s \le t_j$, we can write

$$F_k(s) \le F_k(t_j) \le F(t_j) + \varepsilon$$

$$\le F(t_{j-1}) + 2\varepsilon \le F(s) + 2\varepsilon \quad .$$

Hence, $F_k(s) \le F(s) + 2\varepsilon$. Similarly, $F_k(s) \ge F(s) - 2\varepsilon$, and so $\|F_k - F\|_\infty \le \varepsilon$. □

To prove theorem 3.1, we consider the *-semigroup structure

$$X = [0,1] \qquad (\textit{usual topology})$$

$$x \cdot y = \min(x,y)$$

$$x^* = x \quad .$$

For $t \in X = [0,1]$, put

$$U(t) = P_t , \qquad V(t) = Q_t .$$

U and V are representations of X, and we have to show that $U \underset{a}{\sim} V$. It is enough to prove $U \oplus V \underset{a}{\sim} U$ and $V \oplus U \underset{a}{\sim} V$.

We will prove that $U \oplus V \underset{a}{\sim} U$; the rest will follow by symmetry. By theorem 2.5, we need only prove that V_n is subordinate to U_n for every $n = 1,2,\ldots$.

For that, fix $n \geq 1$ and let ρ be a normal state of $\mathcal{L}(\mathbb{C}^n \otimes \mathcal{H}_V)$. We have to find a sequence of unit vectors $\xi_1, \xi_2, \ldots$ in $\mathbb{C}^n \otimes \mathcal{H}_u$ such that $\xi_p \to 0$ weakly, as $p \to \infty$, and

(3.2) $$\sup_{0 \leq t \leq 1} |\rho(u \otimes Q_t) - \langle u \otimes P_t \xi_p, \xi_p \rangle| \longrightarrow 0$$

as $p \to \infty$, for every u in G_n. We will actually prove (3.2) for every u in M_n and, for that, it is enough to prove it for $u \geq 0$.

So fix $u \geq 0$ in M_n, and consider the representations π, σ of $C[0,1]$ defined by

$$\pi(f) = \int_0^1 f(t)\, dP_t ,$$

$$\sigma(f) = \int_0^1 f(t)\, dQ_t .$$

Now the range of π contains no nonzero compact operators because the spectral measure defined by $\{P_t : 0 \leq t \leq 1\}$ is nonatomic. It follows that the C^*-algebra of operators

$$\mathbb{A} = id_n \otimes \pi(M_n \otimes C[0,1]) ,$$

being essentially the $n \times n$ operator matrices over $\pi(C[0,1])$, contains no compact operators either.

Note, too, that π and σ are both faithful representations of $C[0,1]$, because of the conditions $s < t \Rightarrow P_s < P_t$ and $Q_s < Q_t$. It follows that

$$\|id_n \otimes \pi(z)\| = \|id_n \otimes \sigma(z)\| ,$$

for every z in $M_n \otimes C[0,1]$.

It follows that the linear functional $\lambda: \mathbb{A} \to \mathbb{C}$ defined by

$$\lambda(id_n \otimes \pi(a \otimes f)) = \rho(id_n \otimes \sigma(a \otimes f)) ,$$

is a well-defined state of $\mathbb{A}$. Since $\mathbb{A} \cap K = \{0\}$, we may extend λ to a state $\tilde{\lambda}$ of the perturbed algebra $\mathbb{A} + K$ in the obvious way,

$$\tilde{\lambda}(A + K) = \lambda(A) , \qquad A \in \mathbb{A} , \quad K \in K$$

and we now have a state of $\mathbb{A} + K$ which annihilates all compact operators.

Now Glimm's lemma ([3], 11.2.1) implies that $\tilde{\lambda}$ is a weak* limit on $\mathbb{A} + K$ of vector states. Since $\mathbb{A} + K$ is separable, we obtain a *sequence* of unit vectors $\xi_p \in \mathbb{C}^n \otimes \mathcal{H}_u$ such that

$$(3.3) \qquad \lambda(A) = \tilde{\lambda}(A + K) = \lim_{p \to \infty} \langle (A + K)\xi_p, \xi_p \rangle ,$$

for every $A \in \mathbb{A}$, $K \in K$. Take $A = 0$ to obtain $\langle K\xi_p, \xi_p \rangle \to 0$ for every compact K, and hence

$$(3.4) \qquad \xi_p \to 0 \text{ is the weak topology of } \mathbb{C}^n \otimes \mathcal{H}_u .$$

Taking $K = 0$ and $A = u \otimes \pi(f)$ for $f \in C[0,1]$, we obtain from (3.3):

(3.5) $$\rho(u \otimes \sigma(f)) = \lim_{p \to \infty} \langle u \otimes \pi(f)\, \xi_p, \xi_p \rangle \quad .$$

Now we can define measures $\mu, \mu_1, \mu_2, \ldots$ on $[0,1]$ by

$$\int_0^1 f(t)\, d\mu(t) = \rho(u \otimes \sigma(f)) \quad ,$$

$$\int_0^1 f(t)\, d\mu_p(t) = \langle u \otimes \pi(f)\, \xi_p, \xi_p \rangle$$

for all $f \in C[0,1]$, by the Riesz-Markov theorem. All of these are finite positive measures, and $\mu_p \to \mu$ is the weak* topology of $C[0,1]$ because of (3.5). Finally, since μ is a nonatomic measure (because the spectral measure defined by $\{Q_t\}$ is nonatomic), we may conclude from the lemma that

(3.6) $$\sup_{0 \le t \le 1} |\mu([0,t]) - \mu_p([0,t])|$$

tends to zero as $p \to \infty$. Since we clearly have

$$\mu([0,t]) = \rho(u \otimes Q_t)$$

$$\mu_p([0,t]) = \langle u \otimes P_t\, \xi_p, \xi_p \rangle$$

by definition of μ and μ_p, we have established (3.2). □

References

[1] Andersen,N.T., Compact perturbations of reflexive algebras, J. Funct. Anal. **38** (1980), 366-400.

[2] Arveson,W., Perturbation theory for groups and lattices, J. Funct. Anal., to appear.

[3] Dixmier,J. "les C*-algèbres et leurs représentations," Gauthier-Villars, Paris (1964).

[4] Feller,W., "An introduction to probability theory and its applications, II," 2nd edition, Wiley, New York (1970).

[5] Kato,T., "Perturbation theory for linear operators," 2nd edition, Springer-Verlag, Grund. der Math. Wiss. **132**, Berlin (1976).

[6] Voiculescu,D., A non-commutative Weyl-von Neumann theorem, Rev. Rom. Pure et Appl. **21** (1976), 97-113.

Department of Mathematics
University of California
Berkeley, California 94720

BANACH BIMODULE ASSOCIATED TO AN ACTION OF A DISCRETE GROUP ON A COMPACT SPACE

J. BION-NADAL
ECOLE NORMALE SUPERIEURE
1, rue Maurice Arnoux
92120 Montrouge, France

ABSTRACT

We associate to an action of a countable discrete group Γ *on a compact space* X *a normal dual Banach* N *bimodule* Z, *with separable predual, where* N *is the group Von Neumann algebra of* Γ *; and a canonical class of derivations from* M *to* $Z_{\circ}$ *which are inner if and only if there is on* X *a* Γ *invariant probability measure.*

Let Γ be a countable discrete group. Let λ be the left regular representation of Γ in $L^2(\Gamma)$ and $N = \lambda(\Gamma)'' \subset L(l^2(\Gamma))$ the group Von Neumann algebra. Let Tr be the normalized faithful normal trace on N. Note that if every element $\gamma \neq e$ of Γ has an infinite conjugacy class, N is a type II_1 factor.

Let X be a compact space on which Γ acts. The action of Γ on $C(X)$ is then given by $\alpha_g(f)(x) = (g.f)(x) = f(g^{-1}.x)$. Let $(\Pi_{\circ},U,H)$ be a faithful covariant representation of $(C(X),\Gamma,\alpha)$ where H is a separable Hilbert space. We let $C(X)'' = \Pi_{\circ}(C(X))''$ in $L(H)$ and we still note α the extension of α in a homomorphism $\alpha : \Gamma \to \mathrm{Aut}(C(X)'')$ continuous when $\mathrm{Aut}(C(X)'')$ is gifted with the topology of pointwise norm convergence in the predual $C(X)''_*$ of $C(X)''$. Then we consider the crossed product $C(X)'' \rtimes_{\alpha} \Gamma \subset L(H \otimes l^2(\Gamma))$.

<u>Remark</u> : each element of $C(X)'' \rtimes_\alpha \Gamma$ can be written as a sum

$\sum_{\gamma \in \Gamma} x_\gamma \lambda(\gamma)$ (strong convergence) where $x_\gamma \in C(X)''$.([6] 7.11.)

<u>1. Definition</u> : We define Y as the norm closure in $L(H \otimes l^2(\Gamma))$ of the N bimodule generated by the elements of the form $\sum_{\text{finite}} f_\gamma \lambda(\gamma)$; $f_\gamma \in C(X)$.

<u>Remark</u> : $Y \subset C(X)'' \rtimes_\alpha \Gamma$

Let $E = \{\sum_{\text{finite}} f_\gamma \lambda(\gamma)\}$ and $\mathcal{E} = \{\sum_{\text{finite}} x_i T_i y_i \; ; \; x_i, y_i \in N \; ; \; T_i \in E\}$

<u>2. lemma</u> : Y is a Banach N bimodule containing N and $C(X)$, in which $\mathcal{E}$ is dense. Fürthermore $Y \subset W$ the Woronowicz subspace of $L(H \otimes l^2(\Gamma))$ defined by $W = \{T \in L(H \otimes l^2(\Gamma)) \; / \; (1 \otimes \varphi)(T) \in C(X) \; \forall \varphi \in L(l^2)_*\}$.

proof : It is clear that $\mathcal{E}$ is the N bimodule generated by the $\sum_{\text{finite}} f_\gamma \lambda(\gamma)$; the inequality $||xTy|| \leq ||x|| \, ||T|| \, ||y|| \; \forall x,y \in N, \forall T \in Y$ is trivial. The inclusion $E \subset W$ is trivial ; then from the equality $(1 \otimes \varphi) \; (xTy) = (1 \otimes \underbrace{y\varphi x}_{\in l^2_*}) \; (T) \; \forall x,y \in N$ and $T \in E$, we get the inclusion $\mathcal{E} \subset W$, thus $Y \subset W$.

q.e.d.

□ <u>Definition of the Banach N bimodule associated to the action of Γ on X.</u>

Let M be a Von Neumann algebra with a finite trace Tr ; we consider on M the norm $|| \; ||_2$ defined by $||x||_2 = \mathrm{Tr}(x^* x)^{1/2}$.

Let X be a M bimodule ; let then $Z(X)$ defined by :

$Z(X) = \{ \varphi$ linear form on X for which there exists $K > 0$ such that

$$|\varphi(xTy)| \leq K \, ||x||_2 ||T|| \, ||y||_2 \; \forall x,y \in M \; \forall T \in X \}$$

On $Z(X)$ we consider the norm $\|\ \|'$ given by : $\|\varphi\|'$ is the smallest $K \geq 0$ such that the inequality $|\varphi(xTy)| \leq K \|x\|_2 \|T\| \|y\|_2$ is true for each $x,y \in M$ and each $T \in X$.

We use the notion of normal dual Banach M bimodule as it is defined in [5]or[1]

<u>3. lemma</u> : $(Z(X), \|\ \|')$ is a Banach M-bimodule. The unit ball of $(Z(X), \|\ \|')$ is $\sigma(Z(X),X)$ compact and $Z(X)$ is a normal dual Banach M bimodule with predual $Z(X)_*$ in which X is dense.

proof : - From the inequality $\|xy\|_2 \leq \|x\| \|y\|_2$ for each $x,y \in M$, we deduce that $Z(X)$ is a bimodule on M and that the inequality $\|x\varphi y\|' \leq \|x\| \|\varphi\|' \|y\|$ is true for every $x,y \in M$ and $\varphi \in Z(X)$

- Let $(\varphi_n)_{n \in N}$ be a Cauchy sequence on $Z(X)$ with respect to $\|\ \|'$; $(\varphi_n)_{n \in N}$ is then a Cauchy sequence with respect to $\|\ \|$, so there exists a linear form φ on X such that $\|\varphi_n - \varphi\| \to o$ as $n \to \infty$.

From $|(\varphi_n - \varphi_m)(xTy)| \leq \varepsilon \|x\|_2 \|T\| \|y\|_2$ $\forall\, T \in X, \forall x,y \in M$ it follows that $\varphi \in Z(X)$ and that $\|\varphi_n - \varphi\|' \to 0$.

- In a similar manner, we deduce the compactness of the unit ball of $(Z(X), \|\ \|')$ with respect to the $\sigma(Z(X),X)$ topology from the compactness of the unit ball of $(X^*, \|\ \|)$ (where $\|\varphi\| = \sup_{\substack{T \in X \\ \|T\| \leq 1}} |\varphi(T)|$) with respect to the $\sigma(Z(X),X)$ topology.

- From what precedes it is then easy to verify that $Z(X)$ is the dual of the Banach space $\bar{X}$ which is the completion of X with respect to the norm

$\| \ \|'$ defined by $\|T\|' = \sup\limits_{\substack{\varphi \in Z(X) \\ \|\varphi\|' \leq 1}} |\varphi(T)|$

- The σ strong continuity of the linear forms : $x \in M \mapsto \varphi(Tx)$, $x \in M \mapsto \varphi(xT)$ results from the inequality $|\varphi(xTy)| \leq \|\varphi\|' \|T\| \, \|x\|_2 \, \|y\|_2$.

- The applications $\varphi \in Z(X) \mapsto x\varphi \in Z(X)$ and $\varphi \in Z(X) \mapsto \varphi x \in Z(X)$ are obviously $\sigma(Z(X),X)$ continuous.

q.e.d.

__4. proposition__ : Let Y be as in definition 1. Let $Z = Z(Y)$ defined as above. Then $(Z, \| \ \|')$ is a normal dual Banach N bimodule with separable predual $(\bar{Y}, \| \ \|')$, where $\bar{Y}$ is the closure of Y with respect to the norm $\| \ \|'$.

proof: The first part is already shown in the previous lemma. It only remains to show that $(Z_*, \| \ \|')$ is separable.

From the lemma 3 and the inequality $\|x\|' \leq \|x\| \ \forall x \in Y$, we obtain the density of $\mathcal{E}$ in Z_* with respect to the norm $\| \ \|'$. The density of E in $\mathcal{E}$ with respect to the norm $\| \ \|'$ results from the inequality $\|xTy\|' \leq \|x\|_2 \, \|T\|' \, \|y\|_2$. $\forall x,y \in N \ \forall T \in Y$, and from the density in N of the finite sums with respect to $\| \ \|_2$.

q e.d

□ __Convex cone in Z__

We shall define it by its intersection Z_1^+ with the unit sphere of Z.

__5. definition.__ $Z_1^+ = \{\varphi$ positive linear form on Y such that $\varphi_{|N} = Tr_N\}$.

<u>6. proposition.</u> Z_1^+ is a compact convex set of the unit sphere of $(Z, \|\ \|')$ with respect to the weak topology.

proof: The only non trivial thing we have to show is that Z_1^+ is contained in the unit sphere of Z.

Y is a selfadjoint subspace of $L(H \otimes l^2(\Gamma))$ containing 1, so that using a variant of the Hahn Banach theorem as in [2], every positive linear form φ on Y extends to a positive linear form $\bar{\varphi}$ on $L(H \otimes l^2(\Gamma))$. $\bar{\varphi}$ is positive and $\bar{\varphi}(1) = 1$. We then know that $\bar{\varphi}$ is continuous and that $\|\bar{\varphi}\| = 1$. From the Cauchy Schwartz inequality we obtain $|\varphi(xTy)| \leq \|x\|_2 \|T\| \|y\|_2$ $\forall x, y \in N \cdot \forall T \in Y$.

So $\varphi \in Z$ and $\|\varphi\|' = 1$ (because $\|\varphi\|' \leq 1$ and $\varphi(1) = 1$).

q.e.d.

<u>remark</u> : Z_1^+ is invariant under the unitary group of N (i.e $\forall \varphi \in Z_1^+$ $\forall u \in U(N)$ $u\varphi u^* \in Z_1^+$).

We denote $M(X)$ the set of bounded measures on X and $M_1^+(X)$ the set of probability measures on X.

<u>7.definition.</u> Let $\varphi_\circ$ be the normal state on $L(l^2(\Gamma))$ associated to ε_e ($\varepsilon_e(t) = 0$ if $t \neq e; \varepsilon_e(e) = 1$) $\varphi_\circ(x) = (x\varepsilon_e | \varepsilon_e)$. For each $T \in Y$, $(1 \otimes \varphi_\circ)(T) \in C(X)$. We define then for each $\mu \in M(X)$ the linear form φ_μ on Y by $\varphi_\mu(T) = \mu((1 \otimes \varphi_\circ)(T))$

8. proposition. The application $\mu \in \mathcal{M}(X) \mapsto \varphi_\mu$ is an isometry from $\mathcal{M}(X)$ to Z Moreover if $\mu \in \mathcal{M}_1^+(X)$, $\varphi_\mu \in Z_1^+$.

proof: $\varphi_\mu(\lambda(e)f) = \varphi_\mu(f) = \mu(f)$; so $||\mu|| \le ||\varphi_\mu|| \le ||\varphi_\mu||'$.

conservely : $\varphi_\mu(xTy) = \mu((1 \otimes \varphi_\circ)(xTy) = \mu(1 \otimes y\varphi_\circ x)(T)$.

We want to show that $||\varphi_\mu||' \le ||\mu||$

Now $||y\varphi_\circ x|| \le ||y||_2\ ||x||_2$ and $|\varphi_\mu(xTy)| \le ||\mu|| \times ||(1 \otimes y\varphi_\circ x)(T)||$

SO, to get the result, it suffices to prove the inequality $||(1\otimes\psi)(T)|| \le ||\psi||\ ||T||$ for each $\psi \in L(l^2(T))_*$ and each $T \in Y$. By [3]I we can suppose that ψ is of the form $\psi(x) = (x\ \xi|\eta)$, then $((1\otimes\psi)(T)\ \xi'|\eta') = (T(\xi'\otimes\xi)|\eta'\otimes\eta)$.

So $||(1\otimes\psi)(T)|| \le ||T||\ ||\xi||\ ||\eta||$ .i.e. $||(1\otimes\psi)(T)|| \le ||\psi||\ ||T||$

- Let $\mu \in \mathcal{M}_1^+(X)$ $\varphi_\mu \ge 0$ and for each $x \in N$ $\varphi_\mu(x) = \mu(1)(x\varepsilon_e|\varepsilon_e)$
$= Tr(x)$

q.e.d.

□ Characterization of the existence of a Γ invariant measure on X.

9.definition: $Z_\circ = \{\varphi \in Z \ /\ \varphi_{|N} = 0\}$

remark: It is obvious that $Z_\circ$ is a sub N bimodule of Z. Furthermore $Z_\circ$ is $\sigma(Z,Z_*)$ closed by construction. So finally $Z_\circ$ is a normal dual Banach N bimodule.

10.definition: To each measure $\mu \in \mathcal{M}_1^+(X)$, we associate a derivation δ_μ from N to $Z_\circ$ defined by $\delta_\mu(x) = x\varphi_\mu - \varphi_\mu x$.

remark: Let μ,ν be two probability measures on X ; $\varphi_\mu - \varphi_\nu \in Z_\circ$, so the derivation δ_μ is inner if and only if the derivation δ_ν is inner.

11.proposition: The derivation δ_μ is a coboundary, i.e. is inner, if and only if there is a Γ invariant probability measure on X.

proof :- Let $\mu \in M_1^+(X)$. Suppose that δ_μ is a coboundary.

Let $\varphi \in Z_\circ$ such that $x\varphi_\mu - \varphi_\mu x = x\varphi - \varphi x$ for each $x \in N$. Let $\psi = \varphi_\mu - \varphi$.

$\psi \in Z$ $\psi_{|_N} = Tr_N$. The restriction of ψ to $C(X)$ gives a measure ν on X such that for each $g \in \Gamma$ and $f \in C(X)$, $(g.\nu)(f) = \lambda(g)\psi\lambda(g^{-1})f = \psi(f) = \nu(f)$, $\nu(1) = 1$, and $||\nu|| < \infty$ Replacing ν by $|\nu|$, we obtain a positive Γ invariant measure $|\nu|$ on X such that $|\nu|(1) \geq 1$. Then $\frac{|\nu|}{|\nu|(1)} \in M_1^+(X)$ is Γ invariant.

- Let μ be a Γ invariant measure on X, $\mu \in M_1^+(X)$.

Let $f \in C(X)$ $\lambda(g)\varphi_\mu\lambda(g^{-1})(f) = (g.\mu)(f) = \mu(f) = \varphi_\mu(f)$;

and $\lambda(g)\varphi_\mu\lambda(g^{-1})(\lambda(k)f) = 0 = \varphi_\mu(\lambda(k)f)$ if $k \neq e$;

so $\lambda(g)\varphi_\mu = \varphi_\mu\lambda(g)$ $\forall g \in \Gamma$.

N is the Von Neumann algebra generated by $\lambda(\Gamma)$ and the application $x \in N \mapsto x\varphi_\mu \in Z$ is σ weakly continuous. Then by linearity and density, we obtain the equality $x\varphi_\mu = \varphi_\mu x$ for each $x \in N$. So $\delta_\mu = 0$; and δ_ν is a coboundary for each $\nu \in M_1^+(X)$.

q.e.d.

Recall the following characterization of amenability which results easily from [4] theorem 3.3.5. :

Let G be a locally compact group ; G is amenable if and only if for each action of G on a compact space X , there is on X a G invariant measure.

As an immediate consequence of the proposition 11, we then obtain :

12.corollary: Γ is amenable if and only if for each action of Γ on a compact space X , and for each $\mu \in M_1^+(X)$, δ_μ is a coboundary.

remark: As in [1] a Von Neumann algebra M is called amenable when all derivations of M in normal dual Banach M bimodules are coboundaries. It is then obvious from corollary 12 that the amenability of $N = \lambda(\Gamma)''$ implies the amenability of the group Γ.

ACKNOWLEDGEMENTS.

It's a pleasure to thank Alain Connes who suggested me this work.

References

[1] A. CONNES On the cohomology of operator algebras, Journal of functional analysis. Vol.28, N°2.

[2] J. DIXMIER Les moyennes invariantes dans les semi-groupes et leurs applications, Acta Sci.Math.12.

[3] J. DIXMIER Les algèbres d'opérateurs dans l'espace hilbertien, Gauthier-Villars, Paris.

[4] F-P. GREENLEAF Invariant means on topological groups, Math. studies N°16, van Nostand-Reinhold, New York

[5] B. JOHNSON, R-V KADISON and J. RINGROSE, Cohomology of operator algebras, III, Bull.soc.Math. France 100.

[6] G-K PEDERSON C^* algebras and their automorphism groups, Academic Press.

NONCOMMUTATIVE SHAPE THEORY

Bruce Blackadar

Most of the "noncommutative topology" developed so far has been "noncommutative algebraic topology," the process of extending to noncommutative C*-algebras the standard functors of topology, such as the K-groups, regarded as functors from the category of commutative C*-algebras (equivalent to the category of pointed compact Hausdorff spaces), or some suitable subcategory, to abelian groups. In contrast, the theory described in this article can be regarded as an aspect of "noncommutative general topology."

Shape theory has played an important role in topology in recent years. Roughly speaking, the goal of shape theory is to separate out the global properties of a topological space X which can be measured by the homology or cohomology groups of X from the possibly pathological local structure of X. The idea is to write X as a projective limit $X = \varprojlim X_n$ of "nice" spaces, and then consider only those topological properties of X which can be determined from the homotopy type of the X_n and the connecting maps. A general reference for shape theory is [2].

Shape theory for C*-algebras was first introduced by Effros and Kaminker [3]. The idea here is to write a general C*-algebra A as an inductive limit $A = \varinjlim A_n$ of "nice" C*-algebras and classify the algebras up to homotopy equivalence of the associated inductive systems.

There are two drawbacks to the theory developed in [3]. First, it is not a direct noncommutative analog of topological shape theory; secondly, and more importantly, it seems to be applicable to only a very restricted class of C*-algebras.

A shape theory for general (separable) C*-algebras, which exactly restricts to topological shape theory in the commutative case, was developed in [1]; this theory overcomes both drawbacks of [3]. It is hoped that this shape theory will play a role in noncommutative topology similar to that played by ordinary shape theory in the commutative case.

In this article, we give a survey of topological shape theory, the noncommutative theory of [1], and some applications, examples, and open problems.

In section 1, we briefly review the basic facts about absolute neighborhood retracts and topological shape theory. Section 2 contains the definitions of projective and semiprojective C*-algebras and some results analogous to the commutative ones. Then in Section 3 shape equivalence of C*-algebras is defined and the fundamental theorem given. Section 4 outlines some relationships with K-theory and Kasparov theory.

Throughout, we will assume that all C*-algebras are <u>separable</u>. Most of the results do carry through, with obvious modifications, to the nonseparable case. The term "ideal" will mean "closed two-sided ideal."

This work was done while I was on sabbatical at the Mathematics Institute, Universität Tübingen, West Germany, supported by a Forschungsstipendium from the Alexander von Humboldt-Stiftung.

1. TOPOLOGICAL SHAPE THEORY

For simplicity, we will work in the category of compact metrizable spaces, which will just be called "spaces"; the theory for pointed compact metrizable spaces is similar. See [2] and [5] for a complete exposition.

<u>Definition 1.1.</u> A space X is an absolute retract or AR [resp. an absolute neighborhood retract or ANR] if, whenever X is embedded as a (closed) subspace of a space Y, there is a retraction of Y onto X [resp. there is a neighborhood U of X in Y and a retraction of U onto X].

AR's and ANR's are characterized by the Tietze Extension Property:

<u>Theorem 1.2.</u> X is an AR [resp. ANR] if and only if, for any space Y and closed subspace Z, every continuous map $\phi: Z \longrightarrow X$ extends to Y [resp. given ϕ, there is a neighborhood U of Z for which ϕ extends to U].

A simple compactness argument yields:

<u>Corollary 1.3.</u> X is an ANR if and only if, for any space Y and decreasing sequence (Z_n) of closed subspaces with $Z = \bigcap Z_n$, every continuous $\phi: Z \longrightarrow X$ extends to Z_n for sufficiently large n.

An AR is homeomorphic to a retract of the Hilbert cube I^∞, and is contractible; an ANR is homeomorphic to a (compact) retract of an open set in I^∞.

The converses are also true. Any product of intervals is an AR. Every polyhedron or compact manifold is an ANR; every ANR is locally an AR.

The three basic "finiteness" or "rigidity" properties of maps into ANR's which will be relevant for our purposes are:

Theorem 1.4. Let X be an ANR and $Y = \varprojlim Y_n$. Let $\phi: Y \longrightarrow X$ be continuous. Then for sufficiently large n, there is a map $\psi_n: Y_n \longrightarrow X$ such that $\psi_n \circ \pi_n \simeq \phi$ (homotopic), where $\pi_n: Y \longrightarrow Y_n$ is the canonical map.

Theorem 1.5. Let X,Y be as in 1.4. If ϕ_n, ψ_n are maps from Y_n to X such that $\phi_n \circ \pi_n \simeq \psi_n \circ \pi_n$ as maps from Y to X, then for sufficiently large $m>n$ $\phi_n \circ \pi_{m,n} \simeq \psi_n \circ \pi_{m,n}$, where $\pi_{m,n}: Y_m \longrightarrow Y_n$ is the canonical map.

Theorem 1.6. Let X be an ANR. Then there is an open cover $\mathcal{U} = \{U_1, \ldots, U_k\}$ of X such that, whenever Y is any space and $\phi, \psi: Y \longrightarrow X$ are $\mathcal{U}$-close (i.e. for any $y \in Y$ there is an i for which $\phi(y)$ and $\psi(y)$ are both in U_i), then $\phi \simeq \psi$.

Definition 1.7. A shape system for a space X is an inverse system (X_n) with $X = \varprojlim X_n$ and each X_n an ANR.

Two inverse systems (X_n) and (Y_n) are (homotopy) equivalent if there is an increasing sequence $k_1 < n_1 < k_2 < n_2 < \ldots$ and maps $\alpha_i: Y_{n_i} \longrightarrow X_{k_i}$ and $\beta_i: X_{k_{i+1}} \longrightarrow Y_{n_i}$ for which $\alpha_i \circ \beta_i \simeq \sigma_{k_{i+1}, k_i}$ and $\beta_i \circ \alpha_{i+1} \simeq \psi_{n_{i+1}, n_i}$ for all i, where $\sigma_{m,n}: X_m \quad X_n$ and $\pi_{m,n}: Y_m \to Y_n$ are the canonical maps. That is, the triangles in the following diagram commute up to homotopy.

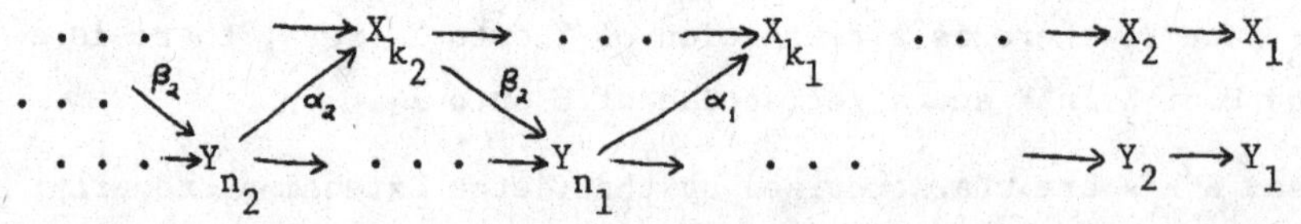

Theorem 1.8. Every space X has a shape system (in fact, $X = \varprojlim X_n$ with X_n a polyhedron). Any two shape systems for X are equivalent.

Definition 1.9. Two spaces X and Y have the same shape or are shape equivalent, written Sh(X) = Sh(Y), if they have equivalent shape systems.

If X and Y are homotopy equivalent, then they are shape equivalent. But shape equivalent spaces need not be homotopy equivalent.

Example 1.10. The "Warsaw Circle" WS^1 is the union of the closure of the graph of $y = \sin \frac{1}{x}$ for $0 < x \leq \frac{2}{\pi}$ and a semicircle connecting (0,1) and

$(\frac{2}{\pi},1)$. WS^1 is a projective limit of circles (at the n'th stage, project the part of WS^1 to the left of the line $x = \frac{2}{(4n+3)\pi}$ horizontally onto the line), with connecting maps homotopic to the identity. Thus WS^1 is shape equivalent to the circle S^1. But any map of S^1 to WS^1 is homotopic to a constant, so S^1 and WS^1 are not homotopy equivalent.

Proposition 1.11. If H* is a (possibly extraordinary) cohomology theory [7], then H* is a shape invariant, i.e. if Sh(X) = Sh(Y), then $H^*(X) \cong H^*(Y)$.

This follows immediately from the homotopy invariance and continuity axioms.

2. PROJECTIVE AND SEMIPROJECTIVE C*-ALGEBRAS

We imitate 1.2 and 1.3 by "turning the arrows around" in the standard way. It is convenient to generalize the definition slightly.

Definition 2.1. Let A and B be C*-algebras, $\phi: A \longrightarrow B$ a homomorphism. Then ϕ is projective if for every C*-algebra C and ideal J of C, and homomorphism $\alpha: B \longrightarrow C/J$, there is a homomorphism $\psi: A \longrightarrow C$ with $\pi \circ \psi = \alpha \circ \phi$ (where $\pi: C \longrightarrow C/J$ is the quotient map). ϕ is semiprojective if for every C and ideals $J_1 \subseteq J_2 \subseteq \ldots$ with $J = [\bigcup J_n]^-$, and $\alpha: B \longrightarrow C/J$, there is a homomorphism $\psi: A \rightarrow C/J_n$ for sufficiently large n with $\pi \circ \psi = \alpha \circ \phi$, where π is regarded as map from C/J_n to C/J. A is [semi-]projective if the identity map on A is semi-]projective.

Although this definition is analogous to the topological one, it is not the same: the square I^2 is an AR, but $C(I^2)$ is not projective, in fact not even semiprojective. This is because commuting self-adjoint elements in a quotient need not lift to commuting self-adjoints. Our definition of semiprojectivity also does not agree with the one in [3]; a C*-algebra which is semiprojective in our sense is semiprojective in the sense of [3], but the converse is false.

Using a long series of fairly simple arguments, one can show that the following "standard" C*-algebras are semiprojective: M_n, $C(S^1)$, the Toeplitz algebra $\mathcal{T}$, the Cuntz-Krieger algebras $\mathcal{O}_A$, and Brown's G_n^{nc}. None of these (except M_1) is projective. The class of unital semiprojective C*-algebras is also closed under strong Morita equivalence and finite direct sums.

The fact that there exist simple semiprojective C*-algebras which are not projective shows that, unlike the commutative case, semiprojective C*-algebras are not "locally projective." We can consequently only obtain the following weak analog of 1.6:

Proposition 2.2. Let $\phi:A\longrightarrow B$ be semiprojective and $\beta_n,\beta:B\longrightarrow D$ with $\beta_n\longrightarrow\beta$ pointwise. Then for sufficiently large n $\beta_n\circ\phi\simeq\beta\circ\phi$.

Proof(sketch): Let $C = C([0,1],D)$, $J_n = \{f\in C \mid f(1/k) = 0 \ \forall k,$ $f\equiv 0$ on $[0,1/n]\}$, $J = \{f \in C \mid f(1/k) = 0 \ \forall k\}$. C/J is isomorphic to the C*-algebra of convergent sequences from D. / /

There are exact analogs of 1.4 and 1.5:

Theorem 2.3. Let $\phi:A\longrightarrow B$ be semiprojective, and $D = \varinjlim D_n$. If $\beta:B\longrightarrow D$, then for sufficiently large n there is a homomorphism $\alpha_n:A\longrightarrow D_n$ with $\beta\circ\phi\simeq\alpha_n\circ\gamma_n$, where $\gamma_n:D_n\longrightarrow D$ is the canonical map.

Theorem 2.4. Let A,B,D,ϕ be as in 2.3. If $\beta_0,\beta_1:B\longrightarrow D_n$ for some n with $\gamma_n\circ\beta_0\simeq\gamma_n\circ\beta_1$, then for sufficiently large m > n $\gamma_{n,m}\circ\beta_0\circ\phi\simeq\gamma_{n,m}\circ\beta_1\circ\phi$, where $\gamma_{n,m}:D_n\longrightarrow D_m$ is the canonical map.

In 2.3 and 2.4 it is important to note that the connecting maps from D_n to D_{n+1} in the inductive limit are not required to be injective.

3. SHAPE THEORY FOR C*-ALGEBRAS

Definition 3.1. A shape system for a C*-algebra A is an inductive system (A_n) with $A = \varinjlim A_n$, where the connecting maps $\gamma_{n,n+1}:A_n\longrightarrow A_{n+1}$ are semiprojective. (A_n) is a strong shape system if each A_n is semiprojective, and is a faithful shape system if each connecting map is injective.

Theorem 3.2. Every C*-algebra has a shape system.

A shape system for A can be constructed as follows. Write A as the "universal C*-algebra" on a countable set of generators $\{x_1,x_2,\ldots\}$ and relations of the form $(\|p_k(x_1,\ldots,x_n,x_1^*,\ldots x_n^*)\|\leq\eta_k)$, where p_k is a polynomial in 2n noncommuting variables, with complex coefficients. Then A_n is the universal C*-algebra with generators $\{x_1,\ldots,x_n\}$ and relations $\{(\|p_k\|\leq\eta_k+1/n) \mid 1\leq k\leq n\}$. The fact that the constants in the relations strictly decrease makes the obvious map from A_n to A_{n+1} semiprojective.

It is not clear that every C*-algebra has a strong shape system.

Not every C*-algebra has a faithful shape system, making necessary the consideration of inductive systems with noninjective connecting maps.

Equivalence is defined as in the topological situation:

Definition 3.3. Let (A_n) and (B_n) be inductive systems of C*-algebras. Then $(A_n) \sim (B_n)$ if there are $k_1 < n_1 < k_2 < n_2 < \ldots$ and maps $\alpha_i : A_{k_i} \longrightarrow B_{n_i}$ and $\beta_i : B_{n_i} \longrightarrow A_{k_{i+1}}$ for which $\beta_i \circ \alpha_i \simeq \gamma_{k_i,k_{i+1}}$ and $\alpha_{i+1} \circ \beta_i \simeq \theta_{n_i,n_{i+1}}$ for each i, where $\gamma_{n,m} : A_n \longrightarrow A_m$ and $\theta_{n,m} : B_n \longrightarrow B_m$ are the connecting maps.

The following is the fundamental theorem of shape theory.

Theorem 3.4. Let (A_n) and (B_n) be shape systems for A and B respectively. If there are any inductive systems (C_n) and (D_n) with $A = \varinjlim C_n$, $B = \varinjlim D_n$, and $(C_n) \sim (D_n)$, then $(A_n) \sim (B_n)$.

The proof is a fairly intricate inductive construction using repeated applications of 2.3 and 2.4.

Thus any two shape systems for A are equivalent.

Definition 3.5. Sh(A) = Sh(B) if A and B have equivalent shape systems.

By 3.4, Sh(A) = Sh(B) if and only if A and B have equivalent inductive systems of any kind. It follows that homotopy equivalence implies shape equivalence. Also, $Sh(C_o(X)) = Sh(C_o(Y))$ if and only if Sh(X) = Sh(Y), so this shape theory is an exact generalization of topological shape theory even though the semiprojectives are different.

Another consequence of 3.4 is that shape equivalence commutes with maximal and minimal tensor products: if Sh(A) = Sh(C) and Sh(B) = Sh(D), then $Sh(A \otimes_{max} B) = Sh(C \otimes_{max} D)$ and $Sh(A \otimes_{min} B) = Sh(C \otimes_{min} D)$.

Two AF algebras are shape equivalent if and only if they are isomorphic, generalizing the topological fact that shape-equivalent 0-dimensional spaces are homeomorphic. The story is completely different for the C*-algebras of the form $\varinjlim (C(S^1) \otimes F_n)$, F_n finite-dimensional, considered in [4]. It is hopeless to try to classify such algebras up to isomorphism or even homotopy equivalence; but they can be elegantly classified up to shape equivalence.

4. RELATIONS WITH K-THEORY

It is easily seen that K-theory is a shape invariant:

<u>Proposition 4.1.</u> Let A and B be shape-equivalent C*-algebras. Then $K_o(A) \cong K_o(B)$ as scaled preordered groups, and $K_1(A) \cong K_1(B)$.

In connection with the ordering, it is interesting to note that if Sh(A) = Sh(B) and A is stably finite, then so is B.

Shape theory may have some applications in computing K-theory. The insensitivity of semiprojective maps to small perturbations may facilitate computations along the lines of the computations for free products and other universal constructions. Specifically, if $\phi: A \longrightarrow B$ is semijective, there should be a reasonable way of determining the group $\phi_*(K_o(A)) \subseteq K_o(B)$.

Stable shape equivalence of A and B should imply Kasparov equivalence, i.e. the existence of an invertible element in KK(A,B). This follows easily from the results of [6] if A and B have sufficiently nice equivalent inductive systems, but there should be a direct proof valid in general.

Note that Kasparov equivalence is much weaker than shape equivalence in general. If A and B are AF algebras, then it follows from [6] that $KK(A,B) \simeq Hom(K_o(A), K_o(B))$, and the intersection product corresponds to composition of functions, so A and B are Kasparov equivalent if and only if $K_o(A) \simeq K_o(B)$ as groups, ignoring the order structure completely. There should be some way of building an order structure into the Kasparov groups so that "ordered Kasparov equivalence" is something more closely related to shape equivalence.

REFERENCES

1. B. Blackadar, Shape theory for C*-algebras, to appear.

2. K. Borsuk, <u>Theory of Shape</u>, Polska Akademia Nauk Monografie Matematyczne v. 59, Warsaw 1975.

3. E. Effros and J. Kaminker, Homotopy continuity and shape theory for C*-algebras, to appear.

4. E. Effros and J. Kaminker, to appear.

5. S. Hu, Theory of Retracts, Wayne State University Press, Detroit 1965.

6. J. Rosenberg and C. Schochet, The classification of extensions of C*-algebras, Bull. Amer. Math. Soc. (2) 4 (1981), 105-110.

7. J. Taylor, Banach algebras and topology, Algebras in Analysis, ed. J. H. Williamson, Academic Press, 1975.

Bruce Blackadar
Department of Mathematics
University of Nevada, Reno
Reno, Nevada 89557
USA

LOCALITY AND DIFFERENTIAL OPERATORS ON C*-ALGEBRAS, II

Ola Bratteli,
Trond Digernes,
Institute of Mathematics,
University of Trondheim,
N-7034 Trondheim-NTH, Norway,

and

George A. Elliott,
Mathematics Institute,
Universitetsparken 5,
DK-2100 Copenhagen Ø, Denmark

ABSTRACT

A characterization in terms of locality is given of certain differential operators on a C*-algebra. These differential operators are polynomials in the generator of a one-parameter group of automorphisms of the C*-algebra, with coefficients in the centre of the algebra. (More precisely, the coefficients may be central multipliers of a certain two-sided ideal of the algebra.)

1. INTRODUCTION

It is well known that differential operators are local, in the sense that if a function is zero on an open set, then also all its derivatives are zero there. Peetre has shown that differential operators on $\mathbb{R}^n$ are characterized by this property of locality: if K is a linear operator on the space $C_0^\infty(\mathbb{R}^d)$ of C^∞ functions on $\mathbb{R}^d$ with compact support, such that the support of $K(f)$ is always contained in the support of f, then for each relatively compact open subset Ω of $\mathbb{R}^d$, the restriction of K to $C_0^\infty(\Omega)$ is a differential operator of finite order with C^∞ coefficients ([Pee]; see also [AN], [Nar]).

More recently, the generator problem for derivations of C*-algebras has led to the study of locality conditions for linear operators. In [Bat 1], Batty considered two derivations δ_0, δ on an abelian C*-algebra $A = C_0(X)$ satisfying the following condition:

$$\text{support } (\delta(f)) \subseteq \text{support } (\delta_0(f)), \; f \in \mathcal{D} = D(\delta_0) \cap D(\delta).$$

Another condition, somewhat stronger than this (also considered by Batty), is the following:

$$\delta_0(f)(x) = 0 \Rightarrow \delta(f)(x) = 0, \; x \in X \; , \; f \in \mathcal{D} \; .$$

In [BDR], this latter condition was generalized to the setting of noncommutative C*-algebras as follows:

$$\omega(\delta_0(A)^*\delta_0(A)) = 0 \Rightarrow \omega(\delta(A)^*\delta(A)) = 0, \; A \in \mathcal{D} \; , \; \omega \in E_A \; ,$$

where E_A denotes the state space of the C*-algebra A. In this case δ is said to be strongly δ_0-local.

Both [Bat 1] and [BDR] considered the case that δ_0 is a generator, and typically δ_0-locality of δ forces δ to have the form

$$\delta(A) = L\delta_0(A) \quad , \quad A \in \mathcal{D} \; ,$$

for a (generally unbounded) operator L in the centre of the multiplier algebra of the minimal dense ideal in the closed ideal in A generated by the range of δ_0. More precisely, if A is abelian and Batty's weak condition holds, this relation is valid on the open subset of the spectrum of A consisting of points ω such that $\delta_0(f)(\omega) \neq 0$ for some $f \in D(\delta_0)$ [Bat 1]. For general A, the relation is valid if $\mathcal{D}$ is a dense *-subalgebra of δ_0 with $\delta_0(\mathcal{D}) \subseteq \mathcal{D}$, $D(\delta) = \mathcal{D}$, and δ is completely strongly δ_0-local [BDR] (for explanation of complete strong locality, see below; if A is abelian this concept reduces to strong locality).

Locality conditions arise in the study of the generator problem for derivations. For a compact group action τ a sufficient condition for a derivation δ commuting with τ to be a generator is that $\delta(A^\tau) = 0$, where A^τ denotes the fixed point algebra of τ ([BJ], [BGJ]). In general the condition $\delta(A^\tau) = 0$ is weaker than the above locality conditions, and is not sufficient for δ to be a generator when τ represents a noncompact group such as $\mathbb{R}$ (see [BJ] Example 6.5, and [BDR] Example 2.3). One reason for this is that the condition $\delta(A^\tau) = 0$ in the compact case forces δ to be tangential to the orbits of the τ-action. In the noncompact case, the orbits are not necessarily closed and the fixed point algebra A^τ does not give enough information about the orbits (it may even be zero).

Recently, another locality condition, specializing to the stronger of Batty's conditions in the commutative case, occurred in [BEE], where dissipations were studied. If H, K are linear operators on a C*-algebra A, K is said to be strictly H-local if the following implication holds:

$$\omega(H(A)^*B^*BH(A)) = 0 \Rightarrow \omega(K(A)^*B^*BK(A)) = 0,$$
$$A \in \mathcal{D} = D(H) \cap D(K),\ B \in A,\ \omega \in E_A .$$

In the present paper we introduce a "purified" version of a linear locality condition briefly touched upon in [BDR]. With H and K as above, we say that K is purely H-local if the following implication holds:

$$\omega(H(A)) = 0 \Rightarrow \omega(K(A)) = 0,\ A \in \mathcal{D},\ \omega \in P_A ,$$

where P_A denotes the space of pure states of A.

In the definition of strong or strict locality, E_A could be replaced by P_A with no change in the property. In contrast with this, the preceding implication becomes much stronger if stated for E_A instead of P_A. It was shown in [BDR] that if $D(H) = D(K)$ and if $\omega(H(A)) = 0$ implies $\omega(K(A)) = 0$ for any $\omega \in E_A$, then $K = \lambda H$ for some scalar λ.

In Section 2 we describe the relation between pure and strict locality. All of the above notions of locality can be stated also for $K \otimes 1$ relative to $H \otimes 1$ on $A \otimes M_r$; we shall refer to these conditions as r-pure locality, r-strict locality, etc. If K is r-purely H-local for all $r = 2,3,\ldots$ we shall say that K is completely purely H-local and similarly for strict locality, etc. The main result of Section 2 is that strict locality is equivalent to complete strict locality, and also to 2-pure locality (and hence also to complete pure locality). We give conditions on H, satisfied if H is a derivation, or if A is commutative, which are sufficient for 2-pure locality (and hence strict locality) to be equivalent to pure locality.

It turns out that by using the concept of pure locality it is possible to extend one of the results of [BEE]: the characterization, in terms of strict locality, of second order differential operators on a C*-algebra which are of the form $L\delta + M\delta^2$, where δ is the generator of a one-parameter group, and L and M are central multipliers of the ideal generated by the range of δ. In Section 3, we extend this to differential operators (of this type) of arbitrary finite order. While the characterization is still in terms of strict locality (with respect to a finite number of powers of δ), pure locality plays an important rôle in the proof.

If B is a C*-algebra, $\mathrm{Ped}(B)$ denotes the minimal dense ideal in B, and $M(\mathrm{Ped}(B))$ denotes the *-algebra of (possibly unbounded) multipliers of $\mathrm{Ped}(B)$. (See [Ped].)

For elementary material on derivations, generators, etc. see Chapter 3 of [BR]. For algebraic variations on the theme of locality see [Bat 2].

2. LOCALITY WITH RESPECT TO A LINEAR OPERATOR

2.1 THEOREM. Let A be a C*-algebra, let $\mathcal{D}$ be a vector space, let π be an irreducible representation of A, and let H and K be linear maps from $\mathcal{D}$ into $\pi(A)$.

The following three conditions are equivalent.

1. K is 2-purely H-local.
2. K is strictly H-local.
3. There exists $\lambda \in \mathbb{C}$ such that $K = \lambda H$.

Suppose that either $\dim H_\pi = 1$, or $\dim H(\mathcal{D}) \neq 1$. Then the preceding three conditions are also equivalent to the following one.

4. K is purely H-local.

Proof. $3 \Rightarrow 1$ and $3 \Rightarrow 2$ are immediate. $1 \Rightarrow 3$ follows from Lemma 2.4 below. $2 \Rightarrow 3$ is essentially a special case of 3.1 of [BEE], but let us give here the simple argument needed in this case.

By strict locality (used just for π-normal pure states),

$$\pi(B)H(A)\xi = 0 \Rightarrow \pi(B)K(A)\xi = 0$$

whenever $B \in A$, $A \in \mathcal{D}$, and $\xi \in H_\pi$. Hence by Kadison transitivity, there exists $\lambda(A, \xi) \in \mathbb{C}$ such that

$$K(A)\xi = \lambda(A, \xi)H(A)\xi \ , \ A \in \mathcal{D} \ , \ \xi \in H_\pi \ .$$

Hence by Lemma 2.3, there exists $\lambda \in \mathbb{C}$ such that

$$K = \lambda H \ .$$

Finally, $3 \Rightarrow 4$ is immediate; let us show that $4 \Rightarrow 3$ holds in the presence of the additional hypothesis. If $\dim H_\pi = 1$, then 3 holds by Lemma 2.2. In any case, by Lemma 2.2, for any pure state ω of $\pi(A)$ there exists $\lambda(\omega) \in \mathbb{C}$ such that

$$\omega K = \lambda(\omega)\omega H \ ,$$

and clearly $\lambda(\omega)$ is unique if $\omega H \neq 0$. If $\dim H(\mathcal{D}) \neq 1$, then we shall show as follows that $\lambda(\omega)$ is independent of ω when $\omega = \omega_\xi$ is a vector state of $\pi(A)$ and $\omega H \neq 0$. We may assume that $\dim H(\mathcal{D}) \geq 2$

Let ω_1 be any vector state of $\pi(A)$ such that $\omega_1 H \neq 0$. Let ω_2 be any vector state of $\pi(A)$ such that $\omega_1 H$ and $\omega_2 H$ are independent. Such a vector state ω_2 exists, since the weak*-closed subspace generated by the vector states of $\pi(A)$ is the whole dual of $\pi(A)$, and $\dim H(\mathcal{D}) \geq 2$. Denote by P the supremum in $B(H_\pi)$ of the support projections of the vector states ω_1 and ω_2 of $B(H_\pi)$; P is two-dimensional. Considering the maps

$$PHP: A \mapsto PH(A)P, \ PKP: A \mapsto PK(A)P$$

as maps from $\mathcal{D}$ into $B(PH_\pi)$, by Lemma 2.5 (with $A = B(PH_\pi)$ and π the identity) we obtain $\lambda(P) \in \mathbb{C}$ such that

$$PKP = \lambda(P)PHP .$$

Hence

$$\lambda(\omega_1) = \lambda(P) = \lambda(\omega_2) .$$

Let now ω be any vector state of $\pi(A)$ such that $\omega H \neq 0$. Then at least one of the pairs $\{\omega_1 H, \omega H\}$, $\{\omega_2 H, \omega H\}$ are independent. In either case, by what precedes,

$$\lambda(\omega) = \lambda(\omega_1) = \lambda(\omega_2) .$$

This shows that $\lambda(\omega)$ is the same for any vector state ω of $\pi(A)$ with $\omega H \neq 0$, say $\lambda(\omega) = \lambda$. It follows that $\omega K = \lambda\omega H$ for every vector state ω of $\pi(A)$, and hence that $K = \lambda H$. This ends the proof of $4 \Rightarrow 3$.

<u>2.2 LEMMA</u>. Let $H, K: \mathcal{D} \to \pi(A)$ be as in 2.1. If K is purely H-local then for every pure state ω of $\pi(A)$ there exists $\lambda(\omega) \in \mathbb{C}$ such that

$$\omega K = \lambda(\omega)\omega H .$$

Furthermore, $\lambda(\omega)$ is unique if $\omega H \neq 0$, and $\omega \mapsto \lambda(\omega)$ is weak*-continuous on $\{\omega \in P_{\pi(A)} | \omega H \neq 0\}$.

<u>Proof</u>. ωH and ωK are linear functionals on $\mathcal{D}$ such that the kernel of the first is contained in the kernel of the second; hence the existence of $\lambda(\omega)$.

Suppose that $\omega H(x) \neq 0$ for some $x \in D$. Then $\lambda(\omega)$ is determined uniquely by the condition $\omega K(x) = \lambda(\omega)\omega H(x)$. Furthermore, if ω' is close to ω on $H(x)$ then also $\lambda(\omega')$ is determined uniquely, and if ω' is also close to ω on $K(x)$, $\lambda(\omega')$ must be close to $\lambda(\omega)$.

<u>2.3 LEMMA</u>. Let $H, K: \mathcal{D} \to \pi(A)$ be as in 2.1. Suppose that for each $A \in \mathcal{D}$ and $\xi \in H_\pi$ there exists $\lambda(A, \xi) \in \mathbb{C}$ such that

$$K(A)\xi = \lambda(A, \xi)H(A)\xi .$$

Then there exists $\lambda \in \mathbb{C}$ such that

$$K = \lambda H .$$

Proof. The proof is similar to the proof of Proposition 1.3 of [BDR]. First fix $A \in \mathcal{D}$. We have $K(A)\mathcal{H}_\pi \subseteq H(A)\mathcal{H}_\pi$, so if $\dim H(A)\mathcal{H}_\pi \leq 1$, then $H(A)$ and $K(A)$ may be considered as linear functionals with kernel $H(A) \subseteq$ kernel $K(A)$, and so $\lambda(A, \xi)$ may be chosen to be independent of ξ -cf. Lemma 2.2. If there exist ξ_1 and ξ_2 in $\mathcal{H}_\pi$ with $H(A)\xi_1$ and $H(A)\xi_2$ independent, then from

$$\lambda(A,\xi_1 + \xi_2)H(A)(\xi_1 + \xi_2) = K(A)(\xi_1 + \xi_2)$$

$$= \lambda(A, \xi_1)H(A)\xi_1 + \lambda(A, \xi_2)H(A)\xi_2$$

we deduce that

$$\lambda(A, \xi_1) = \lambda(A, \xi_1 + \xi_2) = \lambda(A, \xi_2) .$$

Let ξ be any vector in $\mathcal{H}_\pi$ with $H(A)\xi \neq 0$. Then at least one of the pairs $\{H(A)\xi_1, H(A)\xi\}$, $\{H(A)\xi_2, H(A)\xi\}$ are independent. In either case,

$$\lambda(A, \xi) = \lambda(A, \xi_1) = \lambda(A, \xi_2) .$$

It follows that, even if $H(A)\xi = 0$, we may choose $\lambda(A, \xi)$ to be independent of ξ, say $\lambda(A, \xi) = \lambda(A)$. Now

$$K(A) = \lambda(A)H(A) , A \in \mathcal{D} ,$$

and a similar argument shows that we may choose $\lambda(A)$ to be independent of A, say $\lambda(A) = \lambda$. Now $K = \lambda H$.

2.4 LEMMA. Let $H, K: \mathcal{D} \to \pi(A)$ be as in 2.1. If K is 2-purely H-local then there exists $\lambda \in \mathbb{C}$ such that

$$K = \lambda H .$$

Proof. By hypothesis, $K \otimes 1: \mathcal{D} \otimes M_2 \to \pi(A) \otimes M_2$ is purely local with respect to $H \otimes 1: \mathcal{D} \otimes M_2 \to \pi(A) \otimes M_2$. By Lemma 2.2, for every pair $(\xi_1, \xi_2) \in \mathcal{H}_\pi \oplus \mathcal{H}_\pi$ there exists $\lambda(\xi_1, \xi_2) \in \mathbb{C}$ such that

$$\omega_{(\xi_1,\ \xi_2)}(K \otimes 1) = \lambda(\xi_1,\ \xi_2)\omega_{(\xi_1,\ \xi_2)}(H \otimes 1) \ ,$$

where $\omega_{(\xi_1,\ \xi_2)}$ denotes the vector state determined by $(\xi_1,\ \xi_2)$. Evaluating both sides at $\begin{pmatrix} 0 & 0 \\ A & 0 \end{pmatrix}$ where $A \in \mathcal{D}$, we obtain

$$(K(A)\xi_1|\xi_2) = \lambda(\xi_1,\ \xi_2)(H(A)\xi_1|\xi_2) \ .$$

Fix $\xi \in H_\pi$. With $\xi_1 = \xi$ and ξ_2 arbitrary, the preceding equation implies that the kernel of the functional on H_π determined by (the inner product with) $H(A)\xi$ is contained in the kernel of that determined by $K(A)\xi$, so for some $\lambda(A,\ \xi) \in \mathbb{C}$,

$$K(A)\xi = \lambda(A,\ \xi)H(A)\xi \ .$$

Hence by Lemma 2.3, we may choose $\lambda(A,\ \xi)$ to be independent of A and ξ , say $\lambda(A,\ \xi) = \lambda$. Then $K = \lambda H$.

2.5 LEMMA. Let $H, K\colon \mathcal{D} \to \pi(A)$ be as in 2.1, and assume that $\dim H_\pi = 2$. Suppose that $\dim H(\mathcal{D}) \neq 1$. If K is purely H-local then there exists $\lambda \in \mathbb{C}$ such that

$$K = \lambda H \ .$$

Proof. We shall use the observation of C. Davis ([Dav], page 246, lines 12 and 13) that the set of pure states of the C*-algebra $M_2 = M_2(\mathbb{C})$ of 2×2 complex matrices is a two-sphere S in the four-dimensional real linear space E of selfadjoint complex linear functionals on M_2 , with its natural Euclidean norm (in which the norm of a functional is the Hilbert-Schmidt norm of the corresponding density matrix). Since the linear span of the two-sphere S is the whole four-dimensional space E , the three-flat which S spans affinely, which we shall denote by F, does not contain 0 . (The corresponding density matrices are in fact those with trace 1.) Thus, affine independence of a subset of F is the same as linear independence of this subset in E . We note that any three distinct elements of S are affinely independent, and hence linearly independent in E .

We may suppose that $H \neq 0$. Let us show that the subset $N = \{\omega \in S | \omega H \neq 0\}$ of $S = P_{\pi(A)}$ is dense in S, in a strong sense. The set of $\varphi \in E$ such that $\varphi H = 0$ is a real linear subspace of E , so it intersects S in either a set with at most two points, or a circle, or all of S , but in this last case H would have to be 0 , which we are supposing not to be the case. Thus, the subset N of S is the relative complement either of at most two points, or of a circle. (In particular, N is dense in S.)

By Lemma 2.2, for each $\omega \in N$ there is a unique scalar $\lambda(\omega) \in \mathbb{C}$ such that

$$\omega K = \lambda(\omega)\omega H .$$

We shall show that $\lambda(\omega)$ is independent of $\omega \in N$.

Let ω_1, ω_2, and ω_3 be three distinct elements of N; the affine span of $\{\omega_1, \omega_2, \omega_3\}$ in $F \subseteq E$ is then a two-flat in F. Consider the intersection of this two-flat with S; this intersection is a circle since ω_1, ω_2 , and ω_3 all belong to S and are distinct, and S is a two-sphere. This circle does not entirely lie in the complement of N, as $\omega_i \in N$. Since $S\backslash N$ is either at most two points, or a circle, in either case at most two points of the affine span of $\{\omega_1, \omega_2, \omega_3\}$ lie in $S\backslash N$. Thus, the intersection of the affine span of $\{\omega_1, \omega_2, \omega_3\}$ with N is a circle with possibly one or two points missing.

Let ω be an element of N such that ω belongs to the affine span of $\{\omega_1, \omega_2, \omega_3\}$, that is,

$$\omega = \mu_1\omega_1 + \mu_2\omega_2 + \mu_3\omega_3$$

with $\mu_i \in \mathbb{R}$ and $\mu_1 + \mu_2 + \mu_3 = 1$. Write $\lambda(\omega_i) = \lambda_i$, $i = 1, 2, 3$. We have

$$\sum\lambda(\omega)\mu_i\omega_i H = \lambda(\omega)\omega H = \omega K = \sum\mu_i\omega_i K = \sum\lambda_i\mu_i\omega_i H ,$$

that is,

$$\sum(\lambda(\omega) - \lambda_i)\mu_i\omega_i H = 0 .$$

Suppose first that $\dim H(\mathcal{D}) \geq 3$. ($H(\mathcal{D})$, as a complex linear subspace of M_2 , has dimension at most four, so we are considering the case that its dimension is either three or four.) Then ω_1, ω_2 , and ω_3 may be chosen such that their restrictions $\omega_i|H(\mathcal{D})$ to $H(\mathcal{D})$ are linearly independent as complex linear functionals on $H(\mathcal{D})$, i.e.

such that the complex linear functionals $\omega_i H$ on $\mathcal{D}$ are independent. (The complex linear span of N is equal to the whole dual space of $\pi(A) = B(H_\pi)$.) In fact, the set of such choices for the triple $(\omega_1, \omega_2, \omega_3)$ is dense in $N \times N \times N$. To see this, first let ω_1 be equal to an arbitrary element of N. The set of $\omega' \in S$ such that the complex linear functional $\omega' H$ is a complex scalar multiple of the functional $\omega_1 H$ is either at most two points or a circle. (The set of $\varphi \in E$ such that φH is a (complex) multiple of $\omega_1 H$ is a real linear subspace of E, not equal to all of E since $\dim H(\mathcal{D}) > 1$, and therefore not containing S; the intersection of this set with S is therefore either at most two points or a circle.) Therefore ω_2 may be chosen close to a second arbitrary element of N and such that $\omega_1 H$, $\omega_2 H$ are independent functionals on $\mathcal{D}$. Now in the same way we see that the set of $\omega' \in S$ such that $\omega' H$ is a complex linear combination of $\omega_1 H$ and $\omega_2 H$ is either at most two points or a circle. (The set of $\varphi \in E$ such that φH is in the complex linear span of $\omega_1 H$ and $\omega_2 H$ is a real linear subspace of E, not equal to all of E since, in the case we are considering, $\dim H(\mathcal{D}) > 2$, and therefore not containing S.) Therefore ω_3 may be chosen close to a third arbitrary element of N (although we shall not use this) and such that $\omega_1 H$, $\omega_2 H$, and $\omega_3 H$ are independent complex linear functionals on $\mathcal{D}$. In this case we of course obtain that, when $\omega = \mu_1\omega_1 + \mu_2\omega_2 + \mu_3\omega_3$ is as in the preceding paragraph,

$$(\lambda(\omega) - \lambda_1)\mu_1 = (\lambda(\omega) - \lambda_2)\mu_2 = (\lambda(\omega) - \lambda_3)\mu_3 .$$

Choosing ω such that none of μ_1, μ_2, μ_3 is equal to 0, equivalently (as ω_1, ω_2, and ω_3 are distinct), such that ω is different from each of ω_1, ω_2, and ω_3, we conclude that

$$\lambda(\omega) - \lambda_1 = \lambda(\omega) - \lambda_2 = \lambda(\omega) - \lambda_3 = 0 ,$$

and hence that $\lambda_1 = \lambda_2 = \lambda_3$. This shows that, with ω_1 arbitrary in N, $\lambda(\omega_1) = \lambda(\omega_2)$ where ω_2 is close to a second arbitrary element of N, say ω'. Hence by continuity (Lemma 2.2), $\lambda(\omega_1) = \lambda(\omega')$. This shows that the map $N \ni \omega' \mapsto \lambda(\omega')$ is constant; in other words, there exists a fixed $\lambda \in \mathbb{C}$ such that

$$\omega K = \lambda \omega H$$

whenever $\omega \in N$. Since N is dense in S this holds for any $\omega \in S$.

Hence $K = \lambda H$.

Now let us consider the case that $\dim H(\mathcal{D}) = 2$. Let us consider the different cases for $S \setminus N$. We have shown that $S \setminus N$ is either a circle or at most two points. Let us note now that when $\dim H(\mathcal{D}) > 1$, $S \setminus N$ cannot be a circle; indeed, three distinct points of $S \setminus N$ are three independent functionals on M_2, which, if they vanish on $H(\mathcal{D})$, force $\dim H(\mathcal{D}) \leq 1$.

Suppose, first, that $S \setminus N$ is not empty ; thus, $S \setminus N$ consists of either one or two points. It follows that $H(\mathcal{D})$ contains a nonzero selfadjoint element which is either unique up to a scalar multiple or may be chosen to be invertible. To see this, reason as follows. Fix $\omega_\xi \in S \setminus N$ and change notation so that $\xi = \begin{pmatrix} 1 \\ 0 \end{pmatrix}$, so that $H(\mathcal{D})$ is contained in the space of matrices

$$\left\{ \begin{pmatrix} 0 & a \\ b & c \end{pmatrix} \middle| a, b, c \in \mathbb{C} \right\} .$$

$H(\mathcal{D})$ is then determined by a single nontrivial linear relation

$$\alpha a + \beta b + \gamma c = 0 ,$$

where $\alpha, \beta, \gamma \in \mathbb{C}$. We may suppose that γ is equal to 0 or to 1.

Consider the case that $S \setminus N$ consists of two points. In this case the second point gives a relation on $H(\mathcal{D})$, which by uniqueness must be the relation above. Thus, the triple (α, β, γ) is equal to $(\bar{\delta}, \delta, 1)$ for some $\delta \in \mathbb{C}$. Then $H(\mathcal{D})$ contains the invertible selfadjoint elements $\begin{pmatrix} 0 & 1 \\ 1 & -(\delta + \bar{\delta}) \end{pmatrix}$, $\begin{pmatrix} 0 & i \\ -i & i(\delta - \bar{\delta}) \end{pmatrix}$.

Consider now the case that $S \setminus N$ consists of one point. If $\gamma = 1$, then $H(\mathcal{D})$ contains the invertible selfadjoint element $\begin{pmatrix} 0 & a \\ \bar{a} & -(\alpha a + \beta \bar{a}) \end{pmatrix}$ where $a \neq 0$ is chosen such that $(\beta - \bar{\alpha})\bar{a}$ is real (so $\alpha a + \beta \bar{a} = \alpha a + \overline{\alpha a} + (\beta - \bar{\alpha})\bar{a}$ is real). (In this case, as $\beta - \bar{\alpha} \neq 0$, the solution is unique up to a real multiple.) If $\gamma = 0$ and $|\alpha| = |\beta|$, then there are two selfadjoint solutions, $\begin{pmatrix} 0 & 0 \\ 0 & 1 \end{pmatrix}$ and the invertible matrix $\begin{pmatrix} 0 & a \\ \bar{a} & 0 \end{pmatrix}$ where $a \neq 0$ is chosen so that $\bar{a} = -\frac{\alpha}{\beta} a$. If $\gamma = 0$ and $|\alpha| \neq |\beta|$, then $\begin{pmatrix} 0 & 0 \\ 0 & 1 \end{pmatrix}$ is the unique selfadjoint solution, up to a real multiple.

Let us now prove that $K = \lambda H$ in the case that $H(\mathcal{D})$ contains an invertible selfadjoint element $H(x)$. In this case, the set of $\omega \in S$

uch that $\omega H(x) = 0$ is a circle, not equal to a single point. (The umerical range of $H(x)$ is an interval in $\mathbb{R}$, containing 0 since \N $\neq \emptyset$, but not having 0 as an endpoint since 0 is not an eigenalue; thus, the affine function on F determined by $H(x)$ has both trictly positive and strictly negative values on S and is therefore qual to 0 on a circle in S with nonzero radius.) Then by pure ocality, $\omega K(x) = 0$ for every point ω on this circle. It follows hat $K(x)$ is a scalar multiple of $H(x)$, say $K(x) = \lambda H(x)$. (Use irst that the numerical range of $K(x)$, the set of all $\omega K(x)$ with $\in S$, is an affine image of S zero on a circle, so is a line segent through 0 in $\mathbb{C}$, and therefore a scalar multiple of a subset f $\mathbb{R}$. Deduce that a nonzero scalar multiple of $K(x)$ is real-valued n S, and, since it is equal to 0 on the same circle as $H(x)$, is herefore a real multiple of $H(x)$ (first as an affine real-valued unction on F, and hence as an element of M_2).) Now for any $\omega \in N$, ith $x \in \mathcal{D}$ as above,

$$\lambda\omega H(x) = \omega(\lambda H(x)) = \omega(K(x)) = \lambda(\omega)\omega H(x) ,$$

hence $\lambda(\omega) = \lambda$ for all $\omega \in N$ with $\omega H(x) \neq 0$, and hence by coninuity (note that the complement of a circle in N is dense in N), $(\omega) = \lambda$ for all $\omega \in N$. Thus,

$$\omega K = \lambda\omega H$$

or all $\omega \in N$, and hence by continuity for all $\omega \in S$. This shows hat $K = \lambda H$ (in the case that $H(\mathcal{D})$ contains an invertible selfadjoint lement).

If $H(\mathcal{D})$ does not contain an invertible selfadjoint element, and f S\N is nonempty, then, as shown above, there is up to a real scalar ultiple a unique nonzero selfadjoint element of $H(\mathcal{D})$, say $H(x)$. ince $H(x)$ is not zero and not invertible, and $\dim H_\pi = 2$, $\omega H(x) = 0$ or a unique $\omega \in S$. Furthermore, for any $H(y) \in H(\mathcal{D})$ not a scalar ultiple of $H(x)$, there are two distinct points of S which are ero on $H(y)$; in the notation introduced above, we must have $\alpha| \neq |\beta|$ and $\gamma = 0$, so $H(y)$ is a scalar multiple of $\begin{pmatrix} 0 & \beta \\ -\alpha & c \end{pmatrix}$ here $c \in \mathbb{C}$ and, as $|\alpha| \neq |\beta|$, the equation

$$\bar{\delta}\beta - \delta\alpha + c = 0$$

as a unique solution for δ (the determinant of the 2×2 matrix of real coefficients of the associated system of two linear equations in the real and imaginary parts of δ is $|\alpha|^2 - |\beta|^2 \neq 0$).

On the other hand, if $K(\mathcal{D})$ does not contain an invertible selfadjoint element, then either $\dim K(\mathcal{D}) \leq 1$ (in which case, see below), or the same conclusion applies to $K(\mathcal{D})$; that is, $K(\mathcal{D})$ contains a unique nonzero selfadjoint element (up to a real multiple), and this is the unique element (up to a scalar multiple) which is zero on exactly one point of S . By pure locality, which says that, if $y \in \mathcal{D}$ and $\omega \in S$, then $\omega H(y) = 0$ implies $\omega K(y) = 0$, it follows that, if as above $H(x)$ denotes the element of $H(\mathcal{D})$ with a single zero on S , then $K(x)$ must be the element of $K(\mathcal{D})$ with a single zero on S . (If this element is $K(y)$, then $H(y)$ has only a single zero on S and must therefore be a nonzero scalar multiple of $H(x)$, so $K(y)$ is a nonzero multiple of $K(x)$.) In particular, $K(x)$ is a scalar multiple of a selfadjoint element, so its numerical range is a line segment, in $\mathbb{C}$, containing 0. The conclusion that we can usefully draw from this is the following: by consideration of the equation $\omega K(x) = \lambda(\omega)\omega H(x)$, $\omega \in N$, remembering that both $\{\omega H(x) | \omega \in S\}$ and $\{\omega K(x) | \omega \in S\}$ are line segments in $\mathbb{C}$ containing 0 , corresponding to the same single point ω , it follows that the set $\{\lambda(\omega) | \omega \in N\}$ is a line segment in $\mathbb{C}$ (contained in a line through 0) . The final conclusion, that we shall use later, is that there must exist two distinct points ω_1 and ω_2 in N such that $\lambda(\omega_1) = \lambda(\omega_2)$. (No continuous map of an open subset of a two-sphere into a line segment can be injective, by Brouwer's theorem on invariance of domain.)

We must still consider the case $\dim K(\mathcal{D}) \leq 1$. If $\dim K(\mathcal{D}) = 1$, then $K(\mathcal{D}) = \mathbb{C}K(y)$ where y is a fixed element of $\mathcal{D}$ such that $K(y) \neq 0$. Let $y' \in \mathcal{D}$ be such that $H(y')$ is not a scalar multiple of a selfadjoint. Recall that we are assuming that neither $H(\mathcal{D})$ nor $K(\mathcal{D})$ contains an invertible selfadjoint element, and that $S \setminus N$ is nonempty. Then, as shown above, $H(y') = \begin{pmatrix} 0 & \beta \\ -\alpha & c \end{pmatrix}$, where $|\alpha| \neq |\beta|$, and $\omega H(y') = 0$ where ω is the pure state determined by the vector $\begin{pmatrix} \delta \\ 1 \end{pmatrix}$ with $\bar{\delta}\beta - \delta\alpha + c = 0$. It follows by locality that $\omega K(y') = 0$, and, if $K(y') \neq 0$ (so that $\mathbb{C}K(y') = \mathbb{C}K(y)$) , that $\omega K(y) = 0$. Thus, if $H(y')$ belongs to the complement of the union of two one-dimensional subspaces of $H(\mathcal{D})$, then $\omega K(y) = 0$. Since, as inspection of the dependence of δ on c shows, the map $c \mapsto \delta$ is injective, it follows that $\omega K(y) = 0$ for infinitely many $\omega \in S$. Thus, $K(y)$ is zero on a nontrivial circle in S , and so is a scalar multiple of an invertible selfadjoint element; this is contrary to assumption.

It remains to consider the case that $K(\mathcal{D})$ does contain a nonzero element zero on a nontrivial circle in S. If $K(y)$ is such an element of $K(\mathcal{D})$, then from $\omega K(y) = \lambda(\omega)\omega H(y)$, and the assumption that $\omega H(y)$ is zero for at most two points $\omega \in S$, we deduce that $\lambda(\omega) = 0$ for all points $\omega \in N$ lying on a nontrivial circle (not contained in $S \backslash N$, as this has at most two points). Hence, for any $x \in \mathcal{D}$, $\omega K(x) = 0$ for all points ω on this circle. Hence (cf. above) $K(x)$ is a scalar multiple of $K(y)$, so $\dim K(\mathcal{D}) = 1$. On the other hand, as the map $c \mapsto \delta(c)$ considered above from $\mathbb{C}$ into $\mathbb{C}$ is injective, and continuous, by Brouwer's theorem on invariance of domain the image of this map restricted to the dense open subset of points $c \in \mathbb{C}$ such that the linear map $H(x) \mapsto K(x)$ is not zero on $\begin{pmatrix} 0 & \beta \\ -\alpha & c \end{pmatrix}$ is not contained in a one-dimensional submanifold of $\mathbb{C}$. Therefore there exist $x \in \mathcal{D}$ such that $K(x) \neq 0$, and $\omega \in S$ such that $\omega H(x) = 0$ and ω does not lie on the above circle, i.e. $\omega K(x) \neq 0$. This contradicts the hypothesis of pure locality; therefore this case does not arise.

Next, let us note that the same conclusion holds if $S \backslash N$ is empty: in this case $\lambda : N \to \mathbb{C}$ is a continuous map on a two-sphere and (again by algebraic topology) therefore cannot be injective.

In summary, we have shown that in all cases there exist distinct points ω and ω' in N with $\lambda(\omega) = \lambda(\omega')$. Set $\omega = \omega_1$, and choose $\omega_2 \in N$ such that the functionals $\omega_1 H$ and $\omega_2 H$ on $\mathcal{D}$ are independent (this uses only that $\dim H(\mathcal{D}) > 1$, and that N is dense in S). Choose ω_2 such that, moreover, $\omega_2 \neq \omega'$. Choose any point $\omega_3 \in N$ such that ω_3 is on the circle in S through ω_1, ω_2, and ω' (recall that $S \backslash N$ is either a circle or at most two points). With now ω denoting an arbitrary affine combination $\mu_1\omega_1 + \mu_2\omega_2 + \mu_3\omega_3$ lying in N, consider the identity

$$\sum(\lambda(\omega) - \lambda_i)\mu_i\omega_i H = 0$$

derived above, where $\lambda_i = \lambda(\omega_i)$. Since $\dim H(\mathcal{D}) = 2$, the complex linear span of the functionals $\{\omega_1 H, \omega_2 H, \omega_3 H\}$ has dimension two, and we deduce that the triple

$$((\lambda(\omega) - \lambda_1)\mu_1 , (\lambda(\omega) - \lambda_2)\mu_2 , (\lambda(\omega) - \lambda_3)\mu_3)$$

in $\mathbb{C}^3$ lies in a fixed one-dimensional subspace of $\mathbb{C}^3$, independently of $\omega = \sum \mu_i \omega_i \in N$. (This triple is orthogonal in $\mathbb{C}^3$ to the subspace

$$\{(\overline{\omega_1 H(x)}\ ,\ \overline{\omega_2 H(x)}\ ,\ \overline{\omega_3 H(x)})\,|\,x \in \mathcal{D}\} \subseteq \mathbb{C}^3$$

which has dimension two since, as $\omega_1 H$ and $\omega_2 H$ are independent, we may choose x_1 and x_2 in $\mathcal{D}$ such that the matrix $(\omega_i H(x_j))$ has rank two.) It follows by taking $\omega = \omega'$, since $\omega' = \mu_1'\omega_1 + \mu_2'\omega_2 + \mu_3'\omega_3$ with all of μ_1', μ_2', μ_3' nonzero, and since $\lambda(\omega') = \lambda(\omega_1) = \lambda_1$, that either

$$\lambda(\omega') - \lambda_2 = \lambda(\omega') - \lambda_3 = 0 ,$$

i.e. $\lambda_1 = \lambda_2 = \lambda_3$, or

$$\lambda(\omega) - \lambda_1 = 0$$

whenever $\omega = \sum \mu_i \omega_i \in N$ and $\mu_1 \neq 0$, and hence by continuity whenever $\omega = \sum \mu_i \omega_i \in N$, and so in any case $\lambda_1 = \lambda_2 = \lambda_3$. But ω_2 is an arbitrary element of N such that the functionals $\omega_1 H$ and $\omega_2 H$ are independent, and so ω_2 is any element of N except possibly for points on a circle in S containing ω_1. Thus, as $\lambda(\omega_2) = \lambda(\omega_1)$, we obtain by continuity that $\lambda(N) = \lambda(\omega_1)$, and hence, denoting $\lambda(\omega_1)$ by just λ, as above we have $K = \lambda H$.

<u>2.6 COROLLARY</u>. Let A be a C*-algebra, let $\mathcal{D}$ be a vector space, and let H and K be linear maps from $\mathcal{D}$ into A. Denote by $\mathcal{B}$ the closed two-sided ideal of A generated by $H(\mathcal{D})$.

The following three conditions are equivalent.

1. K is 2-purely H-local.
2. K is strictly H-local.
3. $K(\mathcal{D}) \subseteq \mathcal{B}$, and there exists $L \in$ centre $(M(\mathrm{Ped}(\mathcal{B})))$ such that

$$K = LH$$

in the sense that $K(A) = LH(A)$ in $M(\mathrm{Ped}(\mathcal{B}))$, $A \in \mathcal{D}$.

Suppose that for every irreducible representation π of A, either $\dim H_\pi = 1$ or $\dim \pi H(\mathcal{D}) \neq 1$. Then the preceding three conditions are also equivalent to the following one.

4. K is purely H-local.

<u>Proof</u>. The implications $1 \Leftrightarrow 2$ and, under the additional hypothesis

. ⇔ 4 are immediate from Theorem 2.1.

Condition 3 implies in particular that for every irreducible rep-
esentation π of A there exists $\lambda(\pi) \in \mathbb{C}$ such that $\pi K = \lambda(\pi)\pi H$;
y Theorem 2.1, this property is equivalent to Condition 1 (and to Con-
ition 2).

Suppose, finally, that for every irreducible representation π of
there exists $\lambda(\pi) \in \mathbb{C}$ such that $\pi K = \lambda(\pi)\pi H$. In particular,
H = 0 implies $\pi K = 0$, and it follows that

$$K(\mathcal{D}) \subseteq \bigcap\{\ker\pi \mid \pi(\mathcal{B}) = 0\} = \mathcal{B} .$$

et us identify $\hat{\mathcal{B}}$ with $\{\pi \in \hat{A} \mid \pi(\mathcal{B}) \neq 0\}$. Note that $\lambda(\pi)$ is unique
f $\pi \in \hat{\mathcal{B}}$ (that is, if $\pi H \neq 0$). To construct an element L of centre
M(Ped(B))) such that $\pi(L) = \lambda(\pi)$, $\pi \in \mathcal{B}$, it is sufficient, as
hown in the proof of 4.1 of [BEE], to show that $\lambda(\pi)$ depends only
n $\ker\pi$ for $\pi \in \hat{\mathcal{B}}$, and that the function $\pi \mapsto \lambda(\pi)$ is continuous
it then induces a continuous function on Prim $\mathcal{B}$, and the Dauns-
ofmann theorem can be applied locally to show that this function multi-
lies Ped(B) into itself - see 4.1 of [BEE]). First, the statement
$\pi K = \lambda(\pi)\pi H$ is clearly identical to the statement $(K - \lambda(\pi)H)(\mathcal{D}) \subseteq \ker\pi$,
which, together with uniqueness, shows that $\lambda(\pi)$ depends only on $\ker\pi$,
or each $\pi \in \hat{\mathcal{B}}$. To show that $\pi \mapsto \lambda(\pi)$ is continuous on $\hat{\mathcal{B}}$, let
(π_i) be a net in $\hat{\mathcal{B}}$ converging to $\pi \in \hat{\mathcal{B}}$, choose $A \in \mathcal{D}$ such that
$\pi H(A) \neq 0$, and choose a pure state ω of $\mathcal{B}$ associated with π
such that $\omega H(A) \neq 0$. By 3.4.11 (or 3.4.2) of [Dix 2] there exists a
net (ω_i) of pure states of $\mathcal{B}$ converging to ω such that $\pi_{\omega_i} = \pi_i$.
Then $\omega_i K(A) \to \omega K(A)$, so $\lambda(\pi_i)\omega_i H(A) \to \lambda(\pi)\omega H(A)$, and also
$\omega_i H(A) \to \omega H(A)$. Since $\omega H(A) \neq 0$, it follows that $\lambda(\pi_i) \to \lambda(\pi)$.

2.7 REMARKS. The additional hypothesis on H in 2.1, or in 2.6,
s satisfied if $\mathcal{D}$ is a dense subalgebra of A and H is a derivation.
For otherwise, we have an irreducible representation π of A with
$\dim H_\pi > 1$, and a derivation $H:\mathcal{D} \to \pi(A)$ such that for some
$0 \neq C \in B(H_\pi)$ and some nonzero linear functional φ on $\mathcal{D}$,

$$H(A) = \varphi(A)C , A \in \mathcal{D} .$$

Hence by the derivation law,

$$\varphi(AB)C = H(AB) = H(A)\pi(B) + \pi(A)H(B) = \varphi(A)C\pi(B) + \varphi(B)\pi(A)C .$$

Fix $B \in \mathcal{D}$ such that $\varphi(B) \neq 0$. This equation implies that

$$\pi(\mathcal{D})C \subsetneq \mathrm{span}\{\mathbb{C}, C\pi(B)\},$$

and hence since $\mathcal{D}$ is dense in A,

$$B(H_\pi)C \subsetneq \mathrm{span}\{C, C\pi(B)\} .$$

If $\pi(B) = 0$ for some B with $\varphi(B) \neq 0$, or if $C\pi(B_1)$ and $C\pi(B_2)$ are independent for some B_1, B_2 with $\varphi(B_i) \neq 0$, then we conclude that $\dim B(H_\pi)C = 1$, which contradicts $\dim H_\pi > 1$. If neither of these cases holds, that is, if $\dim C\pi(\mathcal{D}) = 1$, then by density of $\mathcal{D}$ in A, $\dim CB(H_\pi) = 1$, i.e. rank $C = 1$, and again by the above inclusion $\dim B(H_\pi)C = 1$, which contradicts $\dim H_\pi > 1$.

For any space $\mathcal{D}$ and map H the additional hypothesis in 2.1 or in 2.6 is satisfied by $H \otimes 1$ on $\mathcal{D} \otimes M_2$. This yields a different, but longer proof of $1 \Rightarrow 3$.

If, in 2.6, it is assumed that $B = A$, or more generally, that $K(\mathcal{D}) \subsetneq B$, then the additional hypothesis on H may be replaced by the weaker one that $\dim H_\pi = 1$ or $\dim \pi H(\mathcal{D}) \neq 1$ holds for all π in a dense subset X of $\hat{A}$. For then, by 2.1, for all $\pi \in X$, $\pi K = \lambda(\pi)\pi H$ for a unique $\lambda(\pi) \in \mathbb{C}$ such that $\lambda(\pi) = 0$ if $\pi H = 0$. If (π_i) is a net in X converging to $\pi \in \hat{A}$, then clearly $\pi K = 0$ in the case that ultimately $\lambda(\pi_i) = 0$, or, more generally, if $\pi H = 0$, i.e. $\pi(B) = 0$. If this is not the case, then, with $A \in \mathcal{D}$ and ω a pure state of $\pi(B)$ such that $\omega H(A) \neq 0$, and (see proof of 2.6) with (ω_i) a net of pure states of B such that $\pi_{\omega_i} = \pi_i$ and $\omega_i \to \omega$, we have $\lambda(\pi_i)\omega_i H(A) = \omega_i K(A) \to \omega K(A)$, and $\omega_i H(A) \to \omega H(A)$, so $\lambda(\pi_i)$ converges, say to $\lambda(\pi)$, and then $\omega K(A) = \lambda(\pi)\omega H(A)$. Since this relation holds whenever $A \in \mathcal{D}$ and ω is a pure state of $\pi(A)$ such that $\omega H(A) \neq 0$, we deduce that $\pi K = \lambda(\pi)\pi H$. (If ω is a pure state of $\pi(A)$ with $\omega H(A) = 0$, then with $\omega_i \to \omega$, $\pi_{\omega_i} = \pi_i$ as above, we have $\lambda(\pi_i) \to \lambda(\pi)$ and $\omega_i H(A) \to 0$, so $\omega_i K(A) = \lambda(\pi_i)\omega_i H(A) \to 0$, $\omega K(A) = 0$. This shows that in any case $\omega K(A) = \lambda(\pi)\omega H(A)$, i.e. $\pi K(A) - \lambda(\pi)\pi H(A)$ has numerical range $\{0\}$, i.e. $\pi K(A) - \lambda(\pi)\pi H(A) = 0$.)

By 2.1, it is sufficient in 2.6 to assume that $\dim \pi H(\mathcal{D}) \neq 1$ for just a single element π which is dense in $\hat{A}$ (i.e., such that er $\pi = 0$) , if such an element of $\hat{A}$ exists (i.e., if A is primitive).

Finally, pure locality does not imply 2-pure locality if the additional hypothesis in 2.1 is dropped. In M_2 , if $\mathcal{D}$ is the space of multiples of the unit, if H is the identity map $\mathcal{D} \to \mathcal{D} \subseteq M_2$, then any linear map $K:\mathcal{D} \to M_2$ is purely H-local.

3. LOCALITY WITH RESPECT TO A DERIVATION

3.1 THEOREM. Let A be a C*-algebra, let $t \mapsto \tau_t = e^{t\delta}$ be a strongly continuous one-parameter group of *-automorphisms of A , let $n \in \{0, 1, 2,\ldots\}$, let $\mathcal{D}$ be a linear subspace of $D(\delta^n)$, let π be an irreducible representation of A , and let K be a linear map from $\mathcal{D}$ into $\pi(A)$. Denote the restriction $\pi\delta^i|\mathcal{D}$ by H_i, $i = 0,\ldots,n$. Assume that $\mathcal{D}$ is a core for δ^n .

The following two conditions are equivalent.

1. K is strictly $(H_0,\ldots,H_n)$-local.
2. There exist $\lambda_0,\ldots,\lambda_n \in \mathbb{C}$ such that

$$K = \sum_{i=0}^{n} \lambda_i H_i .$$

Suppose that all the maps $\pi\delta^0$, $\pi\delta^1$, $\pi\delta^2,\ldots$ are linearly independent. (This is automatic except in the case that all representations $\pi\tau_t$ are unitarily equivalent.) Then the preceding two conditions are also equivalent to the following one.

3. K is purely $(H_0,\ldots,H_n)$-local.

Proof. We first prove $2 \Rightarrow 1 \Rightarrow 3$, then $3 \Rightarrow 2$ in the presence of the additional hypothesis, and finally $1 \Rightarrow 2$.

$2 \Rightarrow 1$ is immediate. (So is $2 \Rightarrow 3$, but we shall not use this.)

Let us prove $1 \Rightarrow 3$ (without the additional hypothesis on π and τ). Let ω be a pure state of $\pi(A)$ and let $A \in \mathcal{D}$, and suppose that $H_i(A) = 0$, $i = 0,\ldots,n$. To show that $\omega K(A) = 0$ we may pass to the irreducible representation determined by ω (which factors through π) and suppose that ω is a vector state of $\pi(A)$, i.e. $\omega = \omega_\xi$ with $\xi \in H_\pi$. By strict locality,

$$\pi(B)H_i(A)\xi = 0 \text{ , } i = 0,\ldots,n \Rightarrow \pi(B)K(A)\xi = 0$$

whenever $B \in A$. By Kadison transitivity, it follows that there exist $\lambda_i(A, \xi) \in \mathbb{C}$, $i = 0,\ldots,n$ such that

$$K(A)\xi = \sum_{i=0}^n \lambda_i(A, \xi)H_i(A)\xi \quad .$$

Taking inner products with ξ yields

$$\omega K(A) = \sum_{i=0}^n \lambda_i(A, \xi)\omega H_i(A) = 0 \text{ .}$$

Next, let us prove $3 \Rightarrow 2$ in the presence of the additional hypothesis.

Case 1: $\dim H_\pi = 1$.

Then K and $H_0,\ldots,H_n$ are essentially linear functionals, and since by hypothesis the kernel of K contains the intersection of the kernels of $H_0,\ldots,H_n$, we have immediately $K = \sum_{i=0}^n \lambda_i H_i$ with $\lambda_i \in \mathbb{C}$.

Case 2: $\dim H_\pi > 1$.

Case 2.1: Not all the representations $\pi\tau_t$, $t \in \mathbb{R}$, are unitarily equivalent.

Let ω_1 and ω_2 be distinct vector states of $\pi(A)$, $\omega_1 = \omega_{\xi_1}$ and $\omega_2 = \omega_{\xi_2}$ with $\xi_1, \xi_2 \in H_\pi$. By pure locality, there exist scalars

$$\lambda_0^1,\ldots,\lambda_n^1, \lambda_0^2,\ldots, \lambda_n^2 \in \mathbb{C}$$

such that

$$\omega_j K = \sum_{i=0}^n \lambda_i^j \omega_j H_i \text{ , } j = 1, 2 \text{ .}$$

By Lemma 3.2, the scalars λ_i^j are unique. We shall show that $\lambda_i^1 = \lambda_i^2$, $i = 0,\ldots,n$. It follows easily that, with $\lambda_i = \lambda_i^1$, $K = \sum_{i=0}^n \lambda_i H_i$. (For each $A \in D$, $K(A) = \sum_{i=0}^n \lambda_i H_i(A)$ has numerical range $\{0\}$.)

To show that $\lambda_i^1 = \lambda_i^2$, choose a unit vector ξ_3 in the linear span of ξ_1 and ξ_2 in H_π, but not proportional to either ξ_1 or ξ_2. Then $\omega_3 = \omega_{\xi_3}$ is distinct from both ω_1 and ω_2. As in the proof of Lemma 2.5, we shall use that the set of vector states ω_ξ with ξ in the linear span of ξ_1 and ξ_2 is a two-sphere ([Dav]). In particular, the set of vector states in the affine span of $\{\omega_1, \omega_2, \omega_3\}$ is a circle. Choose a vector state $\omega = \sum \mu_j \omega_j$ in the affine span of $\{\omega_1, \omega_2, \omega_3\}$ and distinct from ω_1, ω_2, and ω_3; then μ_1, μ_2, and μ_3 are all nonzero. As for ω_1 and ω_2, we have scalars

$$\lambda_0^3, \ldots, \lambda_n^3, \lambda_0(\omega), \ldots, \lambda_n(\omega) \in \mathbb{C}$$

such that

$$\omega_3 K = \textstyle\sum_{i=0}^n \lambda_i^3 \omega_3 H_i \ , \ \omega K = \sum_{i=0}^n \lambda_i(\omega)\omega H_i \ .$$

We now have, on the one hand,

$$\omega K = \textstyle\sum_i \lambda_i(\omega)\omega H_i = \sum_{i,j} \lambda_i(\omega)\mu_j \omega_j H_i \ ,$$

and on the other hand,

$$\omega K = \textstyle\sum \mu_j \omega_j K = \sum_{i,j} \mu_j \lambda_i^j \omega_j H_i \ ;$$

hence,

$$\textstyle\sum_{i=0}^n (\sum_{j=1}^3 (\lambda_i(\omega) - \lambda_i^j)\mu_j \omega_j) H_i = 0 \ .$$

By Lemma 3.2 and Lemma 3.4, together with the hypothesis that $\mathcal{D}$ is a core for δ^n, it follows that

$$\textstyle\sum_{j=1}^3 (\lambda_i(\omega) - \lambda_i^j)\mu_j \omega_j = 0 \ , \ i = 0, \ldots, n \ .$$

Since ω_1, ω_2, and ω_3 are linearly independent, and all μ_j are different from 0, we have

$$\lambda_i(\omega) - \lambda_i^j = 0, \ i = 0, \ldots, n; \ j = 1, 2, 3 \ .$$

In particular, we have $\lambda_i^1 = \lambda_i^2$, $i = 0, \ldots, n$, as desired.

Case 2.2: All the representations $\pi\tau_t$, $t \in \mathbb{R}$, are unitarily equivalent.

We shall proceed in this case by reducing the problem to the case that A is separable. Then we shall show that, if A is separable, the hypothesis that each representation $\pi\tau_t$ is unitarily equivalent to π implies that in fact $\pi\tau$ is weakly inner, i.e. is determined by a strongly continuous unitary group $t \mapsto U_t = e^{itH}$ in $B(H_\pi)$, in the sense that $\pi\tau_t = (\mathrm{Ad}U_t)\pi$, $t \in \mathbb{R}$.

Case 2.2.1: A is not separable.

Let us reduce the problem to the case that A is separable. It follows from the additional hypothesis, as $\mathcal{D}$ is a core for δ^n, and therefore also by Lemma 3.4 a joint core for $\delta,\ldots,\delta^n$, that $H_0,\ldots,H_n$ are linearly independent. To show that $K = \sum\lambda_i H_i$, it is sufficient to show that this holds on every subspace of $\mathcal{D}$ of countable (linear) dimension on which $H_0,\ldots,H_n$ are still independent, for then the λ's for each such subspace are unique, and therefore independent of the subspace. One must of course note that there exists such a subspace. To see this, suppose that $H_0,\ldots,H_n$ are dependent on every subspace of $\mathcal{D}$ of countable dimension, and for each such subspace of $\mathcal{D}$ consider the nonzero space of linear relations among the restrictions of $H_0,\ldots,H_n$ to this subspace of $\mathcal{D}$; this space of relations is a nonzero subspace of $\mathbb{C}^n$. In this way one obtains a decreasing net of nonzero subspaces of $\mathbb{C}^n$, indexed by the upward directed set of subspaces of $\mathcal{D}$ of countable dimension. Since $\mathbb{C}^n$ is of finite dimension, this decreasing net must be eventually constant; this yields a nonzero space of linear relations among $H_0,\ldots,H_n$ on $\mathcal{D}$, which was shown not to exist (under the additional hypothesis).

To reduce $3 \Rightarrow 2$ to the separable case it is therefore sufficient to prove the following statement. For each subspace $\mathcal{D}_1$ of $\mathcal{D}$ of countable dimension, and each projection P_1 in $B(H_\pi)$ such that $P_1 H_\pi$ is separable, there exists a subspace $\mathcal{D}_0$ of $\mathcal{D}$ of countable dimension containing $\mathcal{D}_1$, with the following properties, where A_0 denotes the sub-C*-algebra of A generated by $\mathcal{D}_0$, and P denotes the smallest projection in $\pi(A_0)'$ containing P_1:

(i) A_0 is invariant under τ (i.e. $\tau_t(A_0) = A_0$);
(ii) $\mathcal{D}_0$ is a core for $\delta^n|A_0$;
(iii) $K(\mathcal{D}_0) \subseteq \pi(A_0)$;
(iv) $P\pi$ is an irreducible representation of A_0 on PH_π;
(v) $P\pi\delta^0|\mathcal{D}_0$, $P\pi\delta^1|\mathcal{D}_0$, $P\pi\delta^2|\mathcal{D}_0$, ... are independent.

Then A_0, $\tau|A_0$, $\mathcal{D}_0$, $P\pi$, and $PK|\mathcal{D}_0$ verify the hypotheses of the heorem, including the additional hypothesis, and if K is purely $H_0,\dots,H_n$)-local, then $PK|\mathcal{D}_0$ is purely $(P\pi\delta^0|\mathcal{D}_0,\dots,P\pi\delta^n|\mathcal{D}_0)$-local; urthermore, A_0 is separable.)

Let $\mathcal{D}_1$ be a subspace of $\mathcal{D}$ of countable dimension, let P_1 be a rojection in $B(H_\pi)$ such that P_1H_π is separable, and let us construct subspace $\mathcal{D}_0$ of $\mathcal{D}$ to the above specifications. Choose a separable ub-C*-algebra A_1 of A containing $\mathcal{D}_1$, such that $\pi(A_1)$ contains $(\mathcal{D}_1)$ and contains $P_1B(H_\pi)P_1$ in its weak closure. Choose A_1 such hat, moreover, if P_2 denotes the smallest projection in $\pi(A_1)'$ con- aining P_1, then the maps $P_2\pi\delta^0|A_1$, $P_2\pi\delta^1|A_1,\dots$ are independent. Do this first for the first k of these maps, for each $k = 1, 2,\dots,$ nd then put all these separable subalgebras into a single one; to see hat such a subalgebra must exist for each k , just modify the argument iven above: consider the decreasing net of subspaces of $\mathbb{C}^k$ indexed y the upward directed set of separable sub-C*-algebras, associating o each such subalgebra the space of linear relations among the first of the above maps with this subalgebra in place of A_1 .) Clearly ny larger separable sub-C*-algebra will do as well. Therefore we may ssume that A_1 is τ-invariant. Then $\delta^n|A_1$ has a core $\mathcal{D}_2'$ of count- ble dimension, for example, the subspace of $D(\delta^n|A_1)$ generated by $f_p(t)\tau_t(A_q)dt$, $p, q = 1, 2,\dots,$ where (A_q) is a dense sequence in $(\delta^n|A_1)$, and (f_p) is a sequence of functions in $C^n(\mathbb{R})$ with integral and arbitrarily small supports contained in $[-1, 1]$. (We shall how below that this subspace is a core for $\delta^n|\mathcal{D}_1$.) Choose a subspace $_2$ of $\mathcal{D}$ of countable dimension, containing $\mathcal{D}_1$ and such that the losure of the graph of $\delta^n|\mathcal{D}_2$ contains the graph of $\delta^n|\mathcal{D}_2'$, and herefore also the graph of $\delta^n|A_1$. With P_2 denoting the smallest rojection in $\pi(A_1)'$ containing P_1 , choose a separable τ-invariant ub-C*-algebra A_2 of A containing A_1 , such that $\pi(A_2)$ contains $(\mathcal{D}_2)$ and contains $P_2B(H_\pi)P_2$ in its weak closure. Continue in this ay to obtain increasing sequences

$$\mathcal{D}_1 \subseteq \mathcal{D}_2 \subseteq \dots \subseteq \mathcal{D} \ , \ A_1 \subseteq A_2 \subseteq \dots \subseteq A$$

uch that $\mathcal{D}_k$ is of countable dimension, A_k is separable and τ-in- ariant, the closure of the graph of $\delta^n|\mathcal{D}_{k+1}$ contains the graph of $^n|A_k$, and $\pi(A_{k+1})$ contains $\pi(\mathcal{D}_k)$ and contains $P_{k+1}B(H_\pi)P_{k+1}$ in ts weak closure, where P_{k+1} is the smallest projection in $\pi(A_k)'$ ontaining P_k . Denote $\cup_{k\geq 1}\mathcal{D}_k$ by $\mathcal{D}_0$, denote the closure of $\cup_{k\geq 1}A_k$

by A_0 , and denote $\vee_{k\geq 1}P_k$ by P . Then P is the smallest projection in $\pi(A_0)'$ containing P_1, and $\mathcal{D}_0$ is dense in A_0 (since the closure of $\mathcal{D}_{k+1}$ contains A_k). Furthermore, this choice of $\mathcal{D}_0$ verifies properties (i) to (v) specified above. The only one which requires comment is property (ii) . By construction, the closure of the graph of $\delta^n|\mathcal{D}$ contains the graph of $\delta^n|A_k$ for each $k = 1, 2, \ldots$, so it is enough to prove that $\cup_{k\geq 1}D(\delta^n|A_k)$ is a core for $\delta^n|A_0$. To prove this it is enough to show that if (A_q) and (f_p) are as above with A_1 replaced by A_0, then the linear span of elements $\int f_p(t)\tau_t(A_q)dt$, $p, q = 1, 2, \ldots$, is a core for $\delta^n|A_0$. (This is of course what was asserted above, for A_1; we shall now prove it.)

Let A be an arbitrary element of $D(\delta^n|A_0)$. Then for each $i = 0,\ldots,n$, $\delta^i\int f_p(t)\tau_t(A)dt$ is equal to $(-1)^i\int f_p^{(i)}(t)\tau_t(A)dt$ and converges to $\delta^i(A)$ in the weak Banach space topology of A_0 . Hence with f a suitable convex combination of f_p's , $\delta^i\int f(t)\tau_t(A)dt$ is close in norm to $\delta^i(A)$, $i = 0,\ldots,n$. Choose q so that A_q is close in norm to A , close enough that $(-1)^i\int f^{(i)}(t)\tau_t(A - A_q)dt$ is small in norm. This says that $\delta^i\int f(t)\tau_t(A_q)dt$ is close in norm to $\delta^i\int f(t)\tau_t(A)dt$; it is therefore also close to $\delta^i A$.

Case 2.2.2: A is separable.

Assume now that A is separable, and that $\pi\delta^0, \pi\delta^1, \pi\delta^2,\ldots$ are independent. As shown in Case 2.1 , $3 \Rightarrow 2$ holds if the representations $\pi\tau_t$ are not all unitarily equivalent, so to prove $3 \Rightarrow 2$ we may assume that they are unitarily equivalent. (This is just to remark that our reduction to the separable case might not preserve this property.) Let us show that there exists a strongly continuous unitary group $t \mapsto U_t$ in $B(H_\pi)$ such that $\pi\tau_t = (\mathrm{Ad}U_t)_\pi$, $t \in \mathbb{R}$. The only consequence of separability of A that we shall use is that H_π is separable.

That $\pi\tau_t$ is unitarily equivalent to π , for fixed $t \in \mathbb{R}$, just says that for some unitary $V \in B(H_\pi)$, $\pi\tau_t = (\mathrm{Ad}V)\pi$. It follows that, for each $t \in \mathbb{R}$, there exists a unique automorphism $\overline{\tau}_t$ of $B(H_\pi)$ such that $\pi\tau_t = \overline{\tau}_t\pi$. Then the map $t \mapsto \overline{\tau}_t$ is a group homomorphism, and since H_π is separable and the map $t \mapsto \overline{\tau}_t\pi(A) = \pi\tau_t(A)$ is norm continuous for each $A \in A$, it follows that for each ξ , $\eta \in H_\pi$ and each $T \in B(H_\pi)$, the map $t \mapsto (\overline{\tau}_t(T)\xi|\eta)$ is Borel. (Approximate T strongly by a sequence $(\pi(A_k))$ in $\pi(A)$, and note that for each k, $t \mapsto (\overline{\tau}_t(A_k)\xi|\eta)$ is continuous.) Fix a projection $E \in B(H_\pi)$ of rank ne. By [Dix 1] Lemme 3, there is a Borel subset B of the unitary roup of $B(H_\pi)$ (which is a Polish group in the weak operator topology) onsisting of exactly one representative of each left coset of the losed subgroup $\{U \text{ unitary} \mid UEU^* = E\}$. In other words, for every rojection $F \in B(H_\pi)$, there is a unique $U \in B$ such that $UEU^* = F$. urthermore, the map $U \mapsto UEU^*$ is a Borel bijection from B, a stanard Borel space, onto the space of projections of rank one in $B(H_\pi)$, ith the strong operator topology. Since this latter topology has a ountable basis (H_π being separable), which in particular separates oints and generates the Borel structure, it follows from Theorem 3.2 f [Mac] (see also Appendix B21 of [Dix 2]) that the map $U \mapsto UEU^*$ is Borel isomorphism of B onto the space of rank one projections. ince the map $t \mapsto \bar{\tau}_t(E)$ is Borel, the map

$$t \mapsto V_t \quad \text{where} \quad V_t \in B \quad \text{is such that} \quad V_t E V_t^* = \bar{\tau}_t(E)$$

s Borel. Now, the Banach space $B(H_\pi)E$ is identifiable with H_π, y the map $T \mapsto T\xi$ where ξ is a fixed unit vector in EH_π, so for ach $t \in \mathbb{R}$ the isometry $(\mathrm{Ad}V_t^*)\bar{\tau}_t$ of $B(H_\pi)E$ determines a unitary W_t on H_π, and it is easy to see that $\bar{\tau}_t = \mathrm{Ad}V_tW_t$, and that the ap $t \mapsto W_t$ and also the map $t \mapsto Y_t = V_tW_t$ are Borel, in the strong perator topology. The argument of page 197, lines 3 to 28 of [Kad] s now applicable and shows that there exists a strongly continuous nitary group $t \mapsto U_t$ such that $\bar{\tau}_t = \mathrm{Ad}U_t$. Hence $\pi\tau_t = \bar{\tau}_t\pi = (\mathrm{Ad}U_t)\pi$ s desired.

We must now prove $3 \Rightarrow 2$, under the additional hypothesis, in the ase that $\pi\tau_t = (\mathrm{Ad}U_t)\pi$ where $t \mapsto U_t = e^{itH}$ is a strongly continuous nitary group in $B(H_\pi)$. The additional hypothesis is now just that he spectrum of the selfadjoint operator H is not finite. It turns ur that there are still many pairs of vector states ω_1, ω_2 of $\pi(A)$ or which we can argue as above, but now we must do this without using emma 3.2. A sufficient replacement for Lemma 3.2, for this purpose, s Lemma 3.3.

Denote by X the set of vectors $\xi \in H_\pi$ for which the functionals $\omega_\xi H_0, \ldots, \omega_\xi H_n$ are independent. Then by pure locality, for each $\xi \in X$, here exist unique scalars $\lambda_0(\xi), \ldots, \lambda_n(\xi)$ such that

$$\omega_\xi K = \textstyle\sum_{i=0}^n \lambda_i(\xi)\,\omega_\xi H_i \;.$$

Furthermore, an easy consequence of linear independence is that X is open and the functions $\xi \mapsto \lambda_i(\xi)$ are continuous on X. (If $\xi \in X$, then $\omega_\xi H_i$ are independent, and hence independent on some finite-dimensional subspace $\mathcal{D}_0$ of $\mathcal{D}$, and if η is close to ξ, then $\omega_\eta H_i$ is close to $\omega_\xi H_i$ in the finite-dimensional vector space dual of $\mathcal{D}_0$, and so $\omega_\eta H_0,\ldots,\omega_\eta H_n$ are independent on $\mathcal{D}_0$, and as also $\omega_\eta K$ is close to $\omega_\xi K$ in this finite-dimensional space, it follows that the coefficients $\lambda_i(\eta)$ of $\omega_\eta K$ with respect to the basis $\omega_\eta H_0,\ldots,\omega_\eta H_n$ of this finite-dimensional space are close to the coefficients $\lambda_i(\xi)$ of $\omega_\xi K$ with respect to the nearby basis $\omega_\xi H_0,\ldots,\omega_\xi H_n$.)

If $\mathcal{H}_\pi$ is spanned by the eigenspaces of the selfadjoint operator H, then the application of Lemma 3.3 is quite straightforward. By Lemma 3.3, if ξ is a vector with nonzero components in at least $n + 1$ eigenspaces of H, the functionals $\omega_\xi H_0,\ldots,\omega_\xi H_n$ are independent, i.e. $\xi \in X$. In particular, this shows that X is dense. Furthermore, if ξ_1 and ξ_1' are such vectors, with only finitely many nonzero components, then we can find another such vector ξ_2 with zero component in each eigenspace of H in which the component of either ξ_1 or ξ_1' is nonzero. It follows by Lemma 3.3 that, if $\varphi_0,\ldots,\varphi_n$ are normal functionals on $\pi(A)$ with support contained in the projection onto the subspace spanned by ξ_1 and ξ_2, then

$$\textstyle\sum_{i=0}^n \varphi_i H_i = 0 \Rightarrow \varphi_i = 0,\ i = 0,\ldots,n\ .$$

Hence as above, $\lambda_i(\xi_1) = \lambda_i(\xi_2)$. Similarly, $\lambda_i(\xi_1') = \lambda_i(\xi_2)$. In particular, $\lambda_i(\xi_1) = \lambda_i(\xi_1')$. Since ξ_1 and ξ_1' belong to X, and also are arbitrary elements of the dense subset of $\mathcal{H}_\pi$ consisting of vectors with nonzero components in at least $n + 1$ of the eigenspaces of H, and in at most finitely many, it follows by continuity that $\lambda_i(\xi)$ is independent of $\xi \in X$, say $\lambda_i(\xi) = \lambda_i$. We have, then,

$$\omega_\xi K = \textstyle\sum_{i=0}^n \lambda_i \omega_\xi H_i\quad ,$$

first for $\xi \in X$, and hence by continuity and density of X, for all $\xi \in \mathcal{H}_\pi$. It follows that $K = \sum \lambda_i H_i$, as desired.

Let us now pass to the general case, in which the eigenspaces of H do not necessarily span $\mathcal{H}_\pi$. Let us first show that, as before, X is dense in $\mathcal{H}_\pi$. Let $\xi \in \mathcal{H}_\pi$ be given, and let $\varepsilon > 0$. For any $\delta > 0$, by the Weyl-von Neumann theorem we can decompose H as $H' + H''$ where H'' is a bounded selfadjoint operator on $\mathcal{H}_\pi$ with $\|H''\| < \delta$, and H' is a selfadjoint operator with the same spectrum as H but such that the eigenspaces of H' do span $\mathcal{H}_\pi$. Assume, as we may, that the eigenspaces of H do not span $\mathcal{H}_\pi$, and fix $n + 1$ distinct points $\beta_0,\ldots,\beta_n$ in the spectrum of H which are not eigenvalues, and in particular not isolated. These are then nonisolated points of the spectrum of H'. We can then find points $\beta_0',\ldots,\beta_n'$ in the spectrum of H' which are near to $\beta_0,\ldots,\beta_n$ and are such that the components of ξ in the $n + 1$ eigenspaces corresponding to $\beta_0',\ldots,\beta_n'$ have norm at most $\varepsilon(n + 1)^{-\frac{1}{2}}$. We can then change each of these components by at most $\varepsilon(n + 1)^{-\frac{1}{2}}$ to make its norm equal to $\varepsilon(n + 1)^{-\frac{1}{2}}$. The resulting vector η is within distance ε of ξ. By Lemma 3.3, the functionals $\omega_\eta H_0',\ldots,\omega_\eta H_n'$ are independent, where $H_k' = (\mathrm{adi}H')^k|\mathcal{D}$. At this point, we note that we may suppose that H' commutes with the spectral projection of H corresponding to some fixed bounded open interval containing $\beta_1,\ldots,\beta_n$, and that the spectrum of H' cut down by this projection is the same as the spectrum of H cut down by this projection. Furthermore, we may suppose that ξ and η belong to the range of this spectral projection. Then $\omega_\eta H_0',\ldots,\omega_\eta H_n'$ are weak operator continuous. We note next that since $\beta_0,\ldots,\beta_n$ are fixed, by Lemma 3.3 it follows that the degree of independence of the functionals $\omega_\eta H_0',\ldots,\omega_\eta H_n'$, measured by the norm of their unique extensions to normal functionals on $B(\mathcal{H}_\pi)$, which is just the trace class norm of the corresponding density matrices, while it does depend on ε, does not depend on δ. Therefore, if δ is sufficiently small, the functionals $\omega_\eta H_0,\ldots,\omega_\eta H_n$, which by the choice of H' and η must be close to $\omega_\eta H_0',\ldots,\omega_\eta H_n'$, must be independent. In other words, if δ is small, $\eta \in X$. This shows that X is dense in $\mathcal{H}_\pi$.

Let us now show that $\lambda_i(\xi)$ is independent of $\xi \in X$. As above, we use the Weyl-von Neumann decomposition $H = H' + H''$ and imitate the argument in the case considered earlier, that the eigenspaces of H span $\mathcal{H}_\pi$. Thus, given vectors ξ_1 and ξ_1' in X, and $\varepsilon > 0$, $\delta > 0$, we approximate these vectors by η_1 and η_1' as above, with respect to a decomposition as above with $\|H''\| < \delta$. Let us choose η_1 and η_1' so that in addition they have nonzero components in only finitely many of the spectral spaces of H'. Then we can find a vector η_2, inside the spectral projection corresponding to the bounded open interval introduced above, with nonzero components in at least $n + 1$ spectral subspaces of H', and in at most finitely many, and such that the component of η_2 is zero in any spectral subspace of H' where the component of η_1 or of η_1' is nonzero. Furthermore, if δ is sufficiently small, and the $n + 1$ eigenvalues for η_2 are (approximately) fixed, we can conclude that $\eta_2 \in X$. By Lemma 3.3, if $\varphi_0,\ldots,\varphi_n$ are normal functionals on $\pi(A)$ with support contained in the projection onto the subspace spanned by η_1 and η_2, then

$$\sum_{i=0}^{n}\varphi_i H_i' = 0 \Rightarrow \varphi_0 = \ldots = \varphi_n = 0 .$$

Furthermore, the degree of independence of the functionals $\varphi_0 H_0',\ldots,\varphi_n H_n'$ for fixed nonzero functionals $\varphi_0,\ldots,\varphi_n$ is independent of δ. Therefore, if δ is sufficiently small, the functionals $\varphi_0 H_0,\ldots,\varphi_n H_n$, which by the choice of H' and of η_1 and η_2 must be close to $\varphi_0 H_0',\ldots,\varphi_n H_n'$, must be independent. As in the case that Lemma 3.2 was applied to earlier, take

$$\varphi_i = \sum_{j=1}^{3}(\lambda_i(\omega) - \lambda_i^j)\mu_j\omega_j ,$$

where now $\omega_1 = \omega_{\eta_1}$, $\omega_2 = \omega_{\eta_2}$, $\omega_3 = \omega_\eta$ with $\eta = (\eta_1 + \eta_2)/\sqrt{2}$, so that also $\eta \in X$ if δ is small, and $\omega = \sum_{j=1}^{3}\mu_j\omega_j$ where μ_1, μ_2, μ_3 are in $\mathbb{R}$, have sum 1, are nonzero, and, furthermore, such that ω is sufficiently close to ω_1 that the unit vector determining ω belongs to X (recall that X is open, and $\eta_1 \in X$). As before, $\lambda_i(\omega)$ is determined by $\omega K = \sum_{i=0}^{n}\lambda_i(\omega)\omega H_i$, and $\lambda_i^j = \lambda_i(\omega_j)$, $j = 1, 2, 3$. As before, $\sum\varphi_i H_i = 0$, and we conclude that $\varphi_i = 0$, and hence $\lambda_i^j = \lambda_i(\omega)$; in particular, $\lambda_i^1 = \lambda_i^2$. In other words, $\lambda_i(\eta_1) = \lambda_i(\eta_2)$. Similarly, $\lambda_i(\eta_1') = \lambda_i(\eta_2)$. This shows that $\lambda_i(\eta_1) = \lambda_i(\eta_1')$. By continuity, as η_1 is close to ξ_1 and η_1' is close to ξ_1', $\lambda_i(\xi_1) = \lambda_i(\xi_2)$. This shows that $\lambda_i(\xi)$ is independent of $\xi \in X$, say $\lambda_i(\xi) = \lambda_i$. Thus,

$$\omega_\xi K = \sum_{i=0}^{n}\lambda_i\omega_\xi H_i$$

holds for $\xi \in X$, and hence as before for all $\xi \in H_\pi$; hence $K = \sum\lambda_i H_i$, as desired.

Finally, let us prove $1 \Rightarrow 2$. We have $1 \Rightarrow 3$, and, in the presence of the additional hypothesis, we have $3 \Rightarrow 2$. Therefore, to prove $1 \Rightarrow 2$ it remains only to consider the case that the additional hypothesis is not satisfied. It follows from Lemma 3.2 that, if the representations $\pi\tau_t$ are not all unitarily equivalent, then the maps $\pi\delta^0$, $\pi\delta^1$, $\pi\delta^2$,... are independent. Since, in the case that we are considering, this does not hold, the representations $\pi\tau_t$ must be unitarily equivalent, and in particular have the same kernel. Thus, we may pass to $\pi(A)$ and replace π by the inclusion mapping, in which case the egation of the hypothesis is just that the maps $\delta^0, \delta^1, \delta^2, \ldots$ must e dependent, i.e., δ satisfies a polynomial equation. This implies hat δ is bounded, and hence $\delta = (\mathrm{adi}H)|\mathcal{A}$ for some selfadjoint $\in B(\mathcal{H}_\pi)$. To see that δ is bounded, note that the restriction of to the invariant subspace $\cap_{k\geq 1} D(\delta^k)$ is a linear map satisfying a olynomial equation, and therefore this space is a finite sum of eigenpaces of δ . Since any finite sum of eigenspaces of δ in $B(\mathcal{H}_\pi)$ s closed in the weak operator topology, and the subspace $\cap_{k\geq 1} D(\delta^k)$ is ense in $B(\mathcal{H}_\pi)$ in this topology (it contains $\int f(t)\tau_t(A)dt$ for any $\in C^\infty(\mathbb{R})$ of compact support), it follows that $B(\mathcal{H}_\pi)$ is a finite sum f eigenspaces of δ and hence δ is bounded and in fact $\delta = (\mathrm{adi}H)|\mathcal{A}$ or some $H = H^* \in B(\mathcal{H}_\pi)$ with finite spectrum.

If $\beta_1,\ldots,\beta_m$ denote the eigenvalues of iH , and $P_1,\ldots,P_m$ the orresponding spectral projections of H , then the eigenspace of δ n $B(\mathcal{H}_\pi)$ corresponding to the eigenvalue $\beta_j - \beta_k$ of δ is

$$\sum\{P_{j'}B(\mathcal{H}_\pi)P_{k'}|\beta_{j'} - \beta_{k'} = \beta_j - \beta_k\} \quad .$$

he corresponding eigenspace of δ on $\mathcal{A}$ is of course the intersection f this subspace with $\mathcal{A}$. As shown in the proof of $1 \Rightarrow 3$, for every $\in \mathcal{H}_\pi$ and $A \in \mathcal{D}$, $K(A)\xi$ belongs to the linear span of $\xi, \delta(A)\xi,\ldots,\delta^n(A)\xi$. In particular, if $\xi \in P_k$ we conclude that $(A)\xi$ is in the span of $P_jA\xi$, $j = 1,\ldots,m$. Thus, for every A and , $P_jK(A)P_k\xi$ is proportional to $P_jAP_k\xi$. By Lemma 2.3, with $(A) = P_jAP_k$ and $K(A) = P_jK(A)P_k$, $A \in \mathcal{D}$, there exists $\lambda_{jk} \in \mathbb{C}$ uch that

$$P_jK(A)P_k = \lambda_{jk}P_jAP_k \ , \quad \text{all} \quad A \in \mathcal{D} \ .$$

hus, $K(A) = \sum\lambda_{jk}P_jAP_k$, $A \in \mathcal{D}$. Using that K is purely $(\delta^0,\ldots,\delta^n)$-ocal (we have $1 \Rightarrow 3$) , let us show that K is a linear combination f $\delta^0,\ldots,\delta^n$. For each $\xi \in \mathcal{H}_\pi$, $\omega_\xi K$ is a linear combination f $\omega_\xi\delta^0,\ldots,\omega_\xi\delta^n$ on $\mathcal{D}$. Hence we may extend K by weak operator ontinuity so that this holds on $B(\mathcal{H}_\pi)$. K multiplies the functional $_k\omega_\xi P_j$ by λ_{jk} . If p is a polynomial then $p(\delta)$ multiplies the unctional $P_k\omega_\xi P_j$ by the eigenvalue $p(\beta_j - \beta_k)$. Choose $\xi \in \mathcal{H}_\pi$ o that $P_k\omega_\xi P_j \neq 0$ for all k and j (this means that $P_k\xi \neq 0$ for ll k). There exists by locality a polynomial p of degree at most such that $\omega_\xi K = \omega_\xi p(\delta)$. It follows that $\lambda_{jk} = p(\beta_j - \beta_k)$, all and k ; this implies that $K = p(\delta)$, i.e. $K = \sum_{i=0}^n \lambda_i\delta^i$, as desired.

3.2 LEMMA. Let A be a C*-algebra, let $t \mapsto \tau_t = e^{t\delta}$ be a strongl continuous one-parameter group of *-automorphisms of A, and let π be an irreducible representation of A. Suppose that not all the representations $\pi\tau_t$, $t \in \mathbb{R}$, are unitarily equivalent. If $\varphi_0,\dots,\varphi_n$ are π-normal linear functionals on A, then

$$\sum_{i=0}^{n}\varphi_i\delta^i = 0 \Rightarrow \varphi_0 = \dots = \varphi_n = 0 .$$

Proof. Since δ preserves adjoints, it is enough to prove the implication for adjoint-preserving functionals, Let $\varphi_0,\dots,\varphi_n$ be adjoint-preserving π-normal functionals on A, of norm at most one, such that $\sum_{i=0}^{n}\varphi_i\delta^i = 0$. Let A be a selfadjoint element of $D(\delta^n)$, of norm at most one, and set

$$\varphi_i e^{t\delta}(A) = f_i(t),\ t \in \mathbb{R} ;\ i = 0,\dots,n .$$

We have $\varphi_i\delta^i e^{t\delta}(A) = f_i^{(i)}(t)$, $i = 0,\dots,n$, and therefore

$$\sum_{i=0}^{n} f_i^{(i)} = 0 .$$

Integrating successively over the n intervals $[0, s_1],\dots,[0,s_{n-1}]$, $[0, s]$ yields

$$\int_0^s ds_{n-1}\dots\int_0^{s_1} f_0(t)dt + \dots + \int_0^s f_{n-1}(t)dt + f_n(s) = p(s)$$

where p is a polynomial of degree at most $n - 1$, with real coefficients. Since $f_0,\dots,f_{n-1}$ have absolute value at most one, it follows that

$$f_n(s) = p(s) + O(s) ,$$

where the estimate on the error is given by replacing $f_0,\dots,f_{n-1}$ in the integrals above by 1, and is therefore independent of the choice of A. Of course, the polynomial p depends on the choice of A. Choose $\varepsilon > 0$ sufficiently small that

$$|f_n(s) - p(s)| \le \tfrac{1}{3}\|\varphi_n\| ,\ s \in [0, \varepsilon] .$$

Note that the set

$$G = \{t \in \mathbb{R} \mid \pi e^{t\delta} \text{ is unitarily equivalent to } \pi\}$$

s a subgroup of $\mathbb{R}$. By hypothesis, the subgroup G is dense in $\mathbb{R}$, nd hence each coset $s + G$ is dense in $\mathbb{R}$. Furthermore, each nonzero uotient group of $\mathbb{R}$ is divisible and in particular infinite. In par- icular there are infinitely many distinct cosets $s + G$, and as each f these is dense we can choose $s_0 < s_1 < \ldots < s_n$ in the interval $0, \varepsilon]$ such that the cosets $s_0 + G, \ldots, s_n + G$ are distinct. In other ords, $s_0, \ldots, s_n \in [0, \varepsilon]$ and the representations $\pi\tau_{s_0}, \ldots, \pi\tau_{s_n}$ are airwise unitarily inequivalent. Then the image of A in the represen- ation $\pi\tau_{s_0} \oplus \ldots \oplus \pi\tau_{s_n}$ is ultraweakly dense in $B(H_\pi) \oplus \ldots \oplus B(H_\pi)$, nd furthermore this holds also for the selfadjoint parts of the unit alls of these algebras. It follows that there exists a selfadjoint lement A of $D(\delta^n)$ of norm one such that, with $i \in \{0, 1, \ldots, n\}$,

$$|\varphi_n\tau_{s_i}(A) - \|\varphi_n\|| \le \tfrac{1}{3}\|\varphi_n\| \quad \text{if } i \text{ is even,}$$

$$|\varphi_n\tau_{s_i}(A) + \|\varphi_n\|| \le \tfrac{1}{3}\|\varphi_n\| \quad \text{if } i \text{ is odd.}$$

ecalling that $\varphi_n\tau_s(A) = f_n(s)$, we deduce from this and the above nequality $|f_n(s) - p(s)| \le \frac{1}{3}\|\varphi_n\|$, $s \in [0, \varepsilon]$, that, with $\in \{0, 1, \ldots, n\}$,

$$p(s_i) \ge \tfrac{1}{3}\|\varphi_n\| \quad \text{if } i \text{ is even,}$$

$$p(s_i) \le -\tfrac{1}{3}\|\varphi_n\| \quad \text{if } i \text{ is odd.}$$

ince the polynomial p is real valued on $\mathbb{R}$, it follows by continuity hat p has at least n zeroes on the interval $[0, \varepsilon]$ (at least one etween each consecutive pair in the sequence $s_0 < s_1 < \ldots < s_n$). Since has degree at most $n - 1$, it follows that $p = 0$. Since $(s_0) \ge \frac{1}{3}\|\varphi_n\|$, $\varphi_n = 0$. The lemma follows by induction.

3.3 LEMMA. Let $n \in \{1, 2, \ldots\}$ and let H be a selfadjoint op- rator on the Hilbert space $\mathbb{C}^{2n+2}$ which is diagonal with respect to he canonical basis and has $2n + 2$ distinct eigenvalues. Denote the perators $\mathrm{adi}H, (\mathrm{adi}H)^2, \ldots, (\mathrm{adi}H)^n$ on the C*-algebra $B(\mathbb{C}^{2n+2})$ by $_1, \ldots, H_n$, and denote the identity operator on this algebra by H_0 . et ξ be a vector in $\mathbb{C}^{2n+2}$ with first $n + 1$ components nonzero nd last $n + 1$ components zero. Let η be a vector in $\mathbb{C}^{2n+2}$ with irst $n + 1$ components zero and last $n + 1$ components nonzero. If

Since H_i is $(\mathrm{adi}H)^i$, to get the corresponding entries of the density matrix of $\varphi_i H_i$, one multiplies the entries in the first columns of $a_i A$, $b_i B$, and $c_i C$ by the ith powers of these numbers. In particular the entries of the first column of $a_i A$ are multiplied by the numbers

$$0^i, (\beta_1 - \beta_2)^i, \ldots, (\beta_1 - \beta_{n+1})^i .$$

The point is that the $(n + 1) \times (n + 1)$ matrix whose columns are obtained by multiplying the entries of the first column of A by the $n + 1$ ith powers above is invertible; its determinant is the Vandermonde determinant of the $n + 1$ distinct numbers $0, \beta_1 - \beta_2, \ldots, \beta_1 - \beta_{n+1}$ multiplied by the product of the $n + 1$ nonzero entries of the first column of A. From $\sum \varphi_i H_i = 0$ we therefore deduce that $a_0 = \ldots = a_n = 0$. Similarly one obtains that the b's and the c's are zero. In other words, $\varphi_0 = \ldots = \varphi_n = 0$, as desired.

3.4 LEMMA. Let $\mathcal{A}$ be a C*-algebra, let $t \mapsto \tau_t = e^{t\delta}$ be a strongly continuous one-parameter group of *-automorphisms of $\mathcal{A}$, let $n \in \{2, 3, \ldots\}$, and let $\mathcal{D}$ be a linear subspace of $D(\delta^n)$. If $\mathcal{D}$ is a core for δ^n then $\mathcal{D}$ is a joint core for $\delta^1, \ldots, \delta^n$.

Proof. By induction, it is sufficient to show that if A and $\delta^n(A)$ are small in norm then $\delta^{n-1}(A)$ is small in norm. Since δ preserves adjoints, it is sufficient to consider the case that A is selfadjoint. It is sufficient to produce a number $\gamma = \gamma(n) > 0$ depending only on n such that, if $\|\delta^{n-1}(A)\| > 1$ and $\|\delta^n(A)\| < \gamma$, then $\|A\| > 1$. It is sufficient to produce $\gamma > 0$ such that if $f \in C^n(\mathbb{R})$ is real-valued, and $\|f^{(n-1)}\|_\infty > 1$ and $\|f^{(n)}\|_\infty < \gamma$, then $\|f\|_\infty > 1$. (Take $f(t) = \varphi\tau_t(A)$ where φ is an adjoint-preserving linear functional on $\mathcal{A}$ of norm one such that $\varphi\delta^{n-1}(A) > 1$; then $\|f^{(n-1)}\|_\infty > 1$ and $\|f^{(n)}\|_\infty \le \|\delta^n(A)\|$, $\|f\|_\infty \le \|A\|$.) If f is a real-valued function and $\|f^{(n-1)}\|_\infty > 1$, and if $\|f^{(n)}\|_\infty$ is sufficiently small (independently of f), then either $f^{(n-1)}$ is strictly greater than 1 over an arbitrarily long interval, or $f^{(n-1)}$ is strictly less than -1 over an arbitrarily long interval. If $n = 1$, so that $f^{(n-1)} = f$, we are finished. If $n > 1$, then we deduce that $f^{(n-2)}$ has absolute value strictly bigger than 1 at some point of this interval, if it is long enough (again independently of f), and in fact at all points of an arbitrarily long subinterval. Hence by induction, $\|f\|_\infty > 1$.

3.5 COROLLARY. Let $\mathcal{A}$ be a C*-algebra, let $t \mapsto \tau_t = e^{t\delta}$ be a trongly continuous one-parameter group of *-automorphisms of A, let $\in \{0, 1, 2, \ldots\}$, let $\mathcal{D}$ be a linear subspace of $D(\delta^n)$, and let be a linear map from $\mathcal{D}$ into A. Denote the restriction $\delta^i|\mathcal{D}$ y H_i, $i = 0,\ldots,n$. Suppose that for each irreducible representation of A such that $\pi\delta \neq 0$, the maps $\pi H_0,\ldots,\pi H_n$ are linearly in-ependent. Denote by $\mathcal{B}$ the closed two-sided ideal of A generated by $(\mathcal{D})$, and denote by $\dot{K}$ the map $K: \mathcal{D} \to A$ composed with the canonical ap $A \to A/\mathcal{B}$. Assume that $\mathcal{D}$ is a core for δ^n.

The following two conditions are equivalent.

1. K is strictly $(H_0,\ldots,H_n)$-local.
2. There exists $\dot{L}_0 \in$ centre $(M(\mathrm{Ped}(A/\mathcal{B})))$ such that

$$\dot{K} = \dot{L}_0 ,$$

n the sense that for each $A \in \mathcal{D}$, $\dot{K}(A)$ is equal to $(A + \mathcal{B})\dot{L}_0$ as an lement of $M(\mathrm{Ped}(A/\mathcal{B}))$, and there exist $L_0,\ldots,L_n \in \mathrm{centre}(M(\mathrm{Ped}(\mathcal{B})))$ uch that, with φ denoting the canonical map from A into $M(\mathrm{Ped}(\mathcal{B}))$ this is an injective map only if $\hat{\mathcal{B}}$ is dense in $\hat{A}$),

$$\varphi K = \sum_{i=0}^{n} L_i H_i ,$$

n the sense that for each $A \in \mathcal{D}$, $\varphi K(A)$ is equal to $\sum_{i=0}^{n} L_i H_i(A)$ in $(\mathrm{Ped}(\mathcal{B}))$.

Suppose that, for each irreducible representation π of A, all he maps $\pi\delta^0, \pi\delta^1, \pi\delta^2,\ldots$ are linearly independent. Then the pre-eding two conditions are also equivalent to the following one.

3. K is purely $(H_0,\ldots,H_n)$-local.

Proof. Ad $2 \Rightarrow 1$. If $\pi \in A\backslash\mathcal{B}$, so that $\pi(\mathcal{B}) = 0$, then by 2, $K = \pi(\dot{L}_0)\pi H_0$. If $\pi \in \mathcal{B}$ then by 2, $\pi K = \sum_{i=0}^{n} \pi(L_i)\pi H_i$.

Ad $1 \Rightarrow 3$ and $3 \Rightarrow 1$ (with hypothesis). This follows from 3.1.

Ad $1 \Rightarrow 2$. Given 1, it is sufficient to construct $L_0,\ldots,L_n$ uch that $\varphi K = \sum_{i=0}^{n} L_i H_i$; the existence of $\dot{L}_0$ such that $\dot{K} = \dot{L}_0$ is special case of this with $n = 0$ (note that $K(\mathcal{B} \cap \mathcal{D}) \subseteq \mathcal{B}$, and ass to the quotient $A/\mathcal{B}$).

By $1 \Rightarrow 2$ of 3.1, if 1 holds then for each $\pi \in A$ there exist

$\lambda_0(\pi),\ldots,\lambda_n(\pi) \in \mathbb{C}$ such that

$$\pi K = \sum_{i=0}^{n} \lambda_i(\pi)\pi H_i .$$

By the hypothesis of the theorem, such scalars $\lambda_i(\pi)$ are unique. In particular, it follows that, for each i, $\lambda_i(\pi)$ depends only on $\ker \pi$, i.e. on the image of π in $\mathrm{Prim}\,\mathcal{B}$. Let us show that, for each i, the function $\pi \mapsto \lambda_i(\pi)$ is continuous on $\hat{\mathcal{B}}$.

Fix $\pi \in \hat{\mathcal{B}}$, and let us show that there exist a pure state ω of $\pi(\mathcal{B})$ and elements $A_0,\ldots,A_n \in \mathcal{D}$ such that the $(n+1) \times (n+1)$ matrix $(\omega H_i(A_j))$ is invertible. It is of course sufficient to construct a pure state ω of $\pi(\mathcal{B})$ such that, when ω is also considered as a pure state of $\pi(A)$, the $n+1$ functionals $\omega H_0,\ldots,\omega H_n$ on $\mathcal{D}$ are independent. If not all unitary representations $\pi\tau_t$ are unitarily equivalent, then by Lemma 3.2, together with Lemma 3.4 and the hypothesis that $\mathcal{D}$ is a core for δ^n, the functionals $\omega H_0,\ldots,\omega H_n$ are independent for any vector state ω of $\pi(A)$, which is then of course also a pure state of $\pi(\mathcal{B})$ since in this case $\pi(\mathcal{B}) \neq 0$. In the general case, let us show that we may suppose that A is separable, provided that we produce a pure state ω of $\pi(A)$ as above which is a vector state. As shown in the proof of 3.1, as $\pi H_0,\ldots,\pi H_n$ are independent on $\mathcal{D}$ there is a subspace $\mathcal{D}_1$ of $\mathcal{D}$ of countable linear dimension on which $\pi H_0,\ldots,\pi H_n$ are still independent. Hence by the proof of 3.1 there exist a subspace $\mathcal{D}_0$ of $\mathcal{D}$ of countable dimension containing $\mathcal{D}_1$, and a nonzero projection $P \in \pi(A_0)'$ where A_0 is the sub-C*-algebra of A generated by $\mathcal{D}_0$, such that properties (i) to (v) listed in the proof of 3.1 are verified. In particular, if ξ is a unit vector in the image of P such that the functionals $\omega_\xi H_0|\mathcal{D}_0,\ldots,\omega_\xi H_n|\mathcal{D}_0$ are independent, then ω_ξ is a pure state of $\pi(\mathcal{B})$ and $\omega_\xi H_0,\ldots,\omega_\xi H_n$ are independent.

Thus, to produce such a state ω_ξ, we may suppose that A is separable. Since the case that not all representations $\pi\tau_t$ are unitarily equivalent has already been dealt with, we may suppose that all these representations are equivalent. As shown in the proof of 3.1, this in conjunction with separability of A implies that there exists a strongly continuous unitary group $t \mapsto U_t = e^{itH}$ in $B(\mathcal{H}_\pi)$ such that $\pi\tau_t = (\mathrm{Ad}U_t)\pi$, $t \in \mathbb{R}$. Furthermore, a slight modification of that part of the proof of 3.1 dealing with this case shows that there exists a vector pure state ω_ξ of $\pi(A)$ (i.e. of $\pi(\mathcal{B})$) such that $\omega_\xi H_0,\ldots,\omega_\xi H_n$

re independent. The only modification needed is to handle the case hat the spectrum of H has fewer then $n+1$ elements. If there are ewer, then as $\pi H_0,\ldots,\pi H_n$ are independent we know that at least the pectrum of $\mathrm{ad}H$ has $n+1$ elements (or more). Then the $n+1$ owers $(\mathrm{ad}H)^0,\ldots,(\mathrm{ad}H)^n$ applied to a density matrix of rank one decribed as follows (in analogy with Lemma 3.3) yield $n+1$ independent atrices, and so one has a vector state ω_ξ with $\omega_\xi H_0,\ldots,\omega_\xi H_n$ inependent. To define a suitable density matrix, choose one unit vector rom each of the (finitely many) eigenspaces of H, and choose the ensity matrix which is supported in the span of these vectors, and has ll of its entries equal to $\frac{1}{k}$ where k is the number of eigenspaces f H. As in the proof of Lemma 3.3 a Vandermonde matrix argument hows that $(\mathrm{ad}H)^0,\ldots,(\mathrm{ad}H)^n$ acting on this matrix give independent atrices.

This shows that there exist a pure state ω of $\pi(\mathcal{B})$ and elements $_0,\ldots,A_n \in \mathcal{D}$ such that the matrix $(\omega H_i(A_j))$ is invertible. Then if $'$ is a pure state of A close to ω, also the matrix $(\omega' H_i(A_j))$ s invertible, and moreover, the inverse is close to the inverse of $\omega H_i(A_j))$. Since the inverse of $(\omega H_i(A_j))$ multiplies the vector $\omega K(A_j)) \in \mathbb{C}^{n+1}$ into the vector $(\lambda_i(\omega)) \in \mathbb{C}^{n+1}$ (this just says that $K(A_j) = \sum_{i=0}^n \lambda_i(\omega)\omega H_i(A_j))$, and similarly with ω' in place of ω, t follows that $(\lambda_i(\omega'))$ is close to $(\lambda_i(\omega))$. It follows by 3.4.11 f [Dix 2] that if $\pi' \in \hat{\mathcal{B}}$ is close to π then $(\lambda_i(\pi'))$ is close to $\lambda_i(\pi))$.

This shows that the functions $\pi \mapsto \lambda_i(\pi)$ on $\hat{\mathcal{B}}$ are continuous, and etermine continuous functions on the quotient space Prim $\mathcal{B}$. Hence y an application of the Dauns-Hofmann theorem locally, as in Theorem .1 of [BEE], there exist unique central multipliers $L_0,\ldots,L_n$ of ed$(\mathcal{B})$ such that $\pi(L_i) = \lambda_i(\pi)$, $\pi \in \hat{\mathcal{B}}$, $i = 0,\ldots,n$. Since then, or each $\pi \in \hat{\mathcal{B}}$,

$$\pi K = \sum_{i=0}^n \pi(L_i)\pi H_i ,$$

e have

$$\varphi K = \sum L_i H_i ,$$

s desired.

3.6 REMARKS. The results 3.1 and 3.5 above should be compared with the results 3.1 and 4.1 of [BEE]. In 3.1 and 3.5 above, the set of maps $\{H_0,\dots,H_n\}$ could be replaced by any subset, as inspection of the proofs reveals. If we take $n = 2$, and consider the subset $\{H_1, H_2\}$ of $\{H_0, H_1, H_2\}$, then from 3.1 and 3.5 above we recover the statements of 3.1 and 4.1 of [BEE]. (See, however, Remark 4.2 of [BEE].) To see that 3.5 above yields 4.1 of [BEE], note that by 3.3 of [BEE], if $\pi\delta$ and $\pi\delta^2$ are linearly dependent, then $\pi\delta = 0$. This shows that the hypothesis of 3.5 that the subset of $\{\pi H_0,\dots,\pi H_n\}$ being considered is independent, unless $\pi\delta = 0$, is automatic in 4.1 of [BEE] where this subset is $\{\pi H_1, \pi H_2\}$.

This hypothesis of 3.5 above is in fact automatic (i.e. vacuous) for any subset of $\{\pi H_0,\dots,\pi H_n\}$ when $n = 2$ (use Lemma 3.2 above, and the fact that if $\ker \pi$ is τ-invariant then either $\pi\delta = 0$ or the derivation of $\pi(A)$ induced by δ has at least three points in its spectrum). If $n > 2$, however, this hypothesis is not vacuous. Furthermore, it is not dispensable, as the following example shows. Take for A the C*-algebra of continuous functions from the interval $[-1, 1]$ into M_3 which at the point 0 have zero entries in the last row and last column, and take for δ the derivation $\mathrm{ad}iH$ where H is the multiplier given by the constant function equal to the matrix with diagonal $(0, 1, 2)$ at each point of $[-1, 1]$; then $\pi_0\delta = \pi_0\delta^3$, where π_0 denotes evaluation at the point $0 \in [-1, 1]$. Take for K the map from A into A such that for each $A \in A$, $\pi K(A) = \pi\delta(A)$ if π is evaluation at a point of the left hand interval $[-1, 0]$, and $\pi K(A) = \pi\delta^3(A)$ if π is evaluation at a point of the right hand interval $[0, 1]$. K has property 1 of 3.5 above, but not property 2.

While 3.1 and 4.1 of [BEE] are concerned only with strict (H_1, H_2)-locality, the concept of pure locality is essential for the proof of 3.1 and 3.5 above. And if we omitted that part of the statement pertaining to pure locality, i.e. Property 3, the proof of $1 \Leftrightarrow 2$ in 3.1 or 3.5 would be no different - it would still proceed by way of Property 3. In other words, to generalize 3.1 and 4.1 of [BEE] to involve more powers of the derivation, as we have done, seems to require introducing the idea of pure locality.

On the other hand, as noted in Remark 4.2 of [BEE], the methods of [BEE] lead also to a result, analogous to 3.1 or 4.1 of [BEE], involving a single power of each of two different derivations, generating two ommuting one-parameter automorphism groups. We have not been able to xtend the methods of the present paper to deal with more than one arameter. These methods therefore only supplement the methods of [BEE]; hey do not supplant them.

Finally, let us point out that, while the additional hypothesis made n 3.1 and 3.5 above to ensure that pure locality implies strict locality ight conceivably be weakened, it may not just be dropped. In the C^*-lgebra M_2 with $\delta = \mathrm{ad}\begin{pmatrix}1 & 0\\ 0 & 0\end{pmatrix}$, the transpose operation is purely $\delta^0, \delta^1, \delta^2$)-local, but it is clearly not a polynomial in δ, and is herefore not strictly local.

4. INVARIANT LOCAL DISSIPATIONS ARE GENERATORS

This subject will be treated in a forthcoming paper.

ACKNOWLEDGEMENTS

We are indebted to the organizers of the INCREST International onference "Operator Algebras, Topology, and Ergodic Theory"; it was uring this conference that we developed the main method of this paper - he exploitation of pure locality using the geometry of the two-sphere.

We are indebted to Ryszard Nest for pointing out to us that if an rreducible representation of a separable C^*-algebra is a fixed point f the spectrum under a continuous one-parameter automorphism group, hen the representation is covariant. This is an important part of the roof of $1 \Rightarrow 2$ of 3.1.

Two of us (O.B. and G.A.E.) are indebted to Derek W. Robinson for nvitations to visit the Australian National University, where this aper was finished and typed.

for each $i = 0,\ldots,n$ φ_i is a linear functional on $B(\mathbb{C}^{2n+2})$ which is a linear combination of vector states ω_ζ where ζ is a linear combination of ξ and η, then

$$\sum_{i=0}^{n}\varphi_i H_i = 0 \Rightarrow \varphi_0 = \ldots = \varphi_n = 0 .$$

Proof. The density matrix for the functional ω_ζ where $\zeta = \lambda\xi + \mu\eta$ is

$$\begin{pmatrix} \lambda\bar{\lambda}A & \lambda\bar{\mu}B^* \\ \mu\bar{\lambda}B & \mu\bar{\mu}C \end{pmatrix}$$

where $A_{ij} = \xi_i\bar{\xi}_j$, $B_{ij} = \eta_i\bar{\xi}_j$, and $C_{ij} = \eta_i\bar{\eta}_j$. It follows that, if φ_i is adjoint-preserving, as we may assume, then the density matrix of φ_i is

$$\begin{pmatrix} a_iA & (b_iB)^* \\ b_iB & c_iC \end{pmatrix}$$

where a_i, b_i, and c_i are (arbitrary) real numbers. (It simplifies notation to restrict to this case.)

Denote the eigenvalues of iH, in order of the corresponding eigenvectors in the canonical basis, by $\beta_1,\ldots,\beta_{2n+2}$. Then the density matrix of $\varphi_1 H_1$, i.e. the negative of the commutator of H with the density matrix of φ_1, is obtained by multiplying the (i, j)th entry by $-(\beta_i - \beta_j)$. In particular, the entries of the first column of a_1A are multiplied by the $n + 1$ distinct numbers

$$0,\ \beta_1 - \beta_2,\ldots,\beta_1 - \beta_{n+1} ,$$

the entries of the first column of b_1B are multiplied by the $n + 1$ distinct numbers

$$\beta_1 - \beta_{n+2},\ldots,\beta_1 - \beta_{2n+2} ,$$

and the entries of the first column of c_1C are multiplied by the $n + 1$ distinct numbers

$$0,\ \beta_{n+2} - \beta_{n+3},\ldots,\beta_{n+2} - \beta_{2n+2} .$$

REFERENCES

AN] E. Albrecht and M. Neumann, "Local operators between spaces of ultradifferentiable functions and ultradistributions", *Manuscripta Math.* 38 (1982) 131-161.

Bat 1] C.J.K. Batty, "Derivations on compact spaces", *Proc. London Math. Soc. (3)* 42 (1981) 299-330.

Bat 2] C.J.K. Batty, "Local operators and derivations on C*-algebras", *preliminary version* (1983).

BDR] O. Bratteli, T. Digernes, and D.W. Robinson, "Relative locality of derivations", *preprint* (1983).

BEE] O. Bratteli, G.A. Elliott, and D.E. Evans, "Locality and differential operators on C*-algebras", *preprint* (1983).

BGJ] O. Bratteli, F. Goodman, and P.E.T. Jørgensen, " Unbounded derivations tangential to compact groups of automorphisms II ", *preprint* (1983).

BJ] O. Bratteli and P.E.T. Jørgensen, "Unbounded derivations tangential to compact groups of automorphisms", *J. Functional Analysis* 48 (1982) 107-133.

BR] O. Bratteli and D.W. Robinson, "*Operator algebras and quantum statistical mechanics, I and II* ", Springer-Verlag, Berlin-Heidelberg-New York, 1979 and 1981.

Dav] C. Davis, "The Toeplitz-Hausdorff theorem explained", *Canad. Math. Bull.* 14 (1971) 245-246.

Dix 1] J. Dixmier, "Dual et quasi-dual d'une algèbre de Banach involutive" *Trans. Amer. Math. Soc.* 104 (1962) 278-283.

Dix 2] J. Dixmier, "*Les C*-algèbres et leurs représentations*", Gauthier-Villars, Paris, 1964.

Kad] R.V. Kadison, "Transformations of states in operator theory and dynamics" *Topology 3, Suppl. 2* (1965) 177-198.

Mac] G.W. Mackey, "Borel structure in groups and their duals", *Trans. Amer. Math. Soc.* 85 (1957) 134-165.

Nar] R. Narasimhan, "*Analysis on Real and Complex Manifolds*", North Holland, Amsterdam, 1968.

Ped] G.K. Pedersen, "*C*-algebras and their Automorphism Groups*" , Academic Press, London-New York-San Francisco , 1979.

Pee] J. Peetre, "Rectification à l'article 'Une caractérisation abstraite des opérateurs différentiels'", *Math. Scand.* 8 (1960) 116-120.

STRONG ERGODICITY AND FULL II_1-FACTORS

Marie Choda
Department of Mathematics
Osaka Kyoiku University
Tennoji, Osaka 543/JAPAN

1. Introduction.

A type II_1-factor N is called full if the group $Int(N)$ of all inner automorphisms of N is closed in the group $Aut(N)$ of all automorphisms of N - this notion introduced by A. Connes[5]. One of the methods of constructing full II_1-factors is the so-called group measure space construction of Murray and von Neumann. The aim of this paper is to give examples of non-isomorphic full II_1-factors constructed by this method out of actions of the same group.

Full II_1-factors are subdivided into the following three (mutually disjoint) classes :

(1) Those of Haagerup type (see definition 1, below)

(2) Those with property T (see [7] for definition)

(3) The rest.

In section 4, we discuss the condition on the group G and its action on a von Neumann algebra A of Haagerup type so that the crossed product of A by G is of Haagerup type. In Theorem 1, this is shown to be the case if G has " property H " introduced earlier in [4] (see Definition 1 below) and if its action commutes with a canonical mapping p_s (see Definition 1 below) associated with any von Neumann algebra of Haagerup type.

In section 5, we apply this theorem to the group measure space construction (where A is taken to be $L^\infty(X,\mu)$ for the measure space (X, μ)) and construct examples of full II_1-factors of Haagerup type and non-Haagerup type out of similar actions of two different groups.

In scetion 6, we construct examples of full II_1-factors of the class (2) and (3) out of different actions (on the non atomic measure space) of the same group $SL(3,Z)$. Lemma 4 is the key lemma for the proof of this result and guarantees that $Int(N)$ is not open (and hence N is not in the class (2)) for the crossed product N under a certain general condition.

The author would like to express her hearty thanks to professor Araki for taking the pains reading her manuscript carefully.

2. Basic Notions.

i) Group measure space construction and crossed product.

Let (X,μ) be a non-atomic probability space, G a countably in-inite discrete group and α a μ-preserving ergodic free action of on (X,μ). Then we shall denote by $W^*(X,G,\alpha)$ the von Neumann alge-ra obtained by the group measure space construction from the triple (X,μ), G, α}

Let A be a finite von Neumann algebra with the separable pre-ual. Fix a faithful normal tracial state τ on A. Any homomor-hism of G to the group of μ-preserving automorphisms of A will e called an action of G on (A,τ). For an action α of G on $(A,$, we shall denote by $R(G,A,\alpha)$ the crossed product of A by G w. . t. α. Let π and v be the canonical imbedding of A into (G,A,α) and the canonical unitary representation of G in $R(G,A,\alpha)$. $R(G,A,\alpha)$ is generated by $\pi(A)$ and $v(G)$).

If $A = L^\infty(X,\mu)$, τ is the trace on A defined by the measure and α is the action of G on (A,τ) defined by the action α of on (X,μ), then $R(G,A,\alpha) = W^*(X,G,\alpha)$.

ii) Full II_1-factors

A II_1-factor N is called full ([5]) if the inner automorphism roup $Int(N)$ is closed in the group $Aut(N)$ of all automorphisms on . Here the relevant topology on $Aut(N)$ is as follows : α_ι con-erges to α if $\|\alpha_\iota(x) - \alpha(x)\|_2 \longrightarrow 0$ for all $x \in N$, where $\|x\|_2 =$ $(x^*x)^{1/2}$ for the trace τ on N.

Effros[8] has introduced the notion of inner amenable groups. All roups used in our examples do not have this property. For such groups, I_1-factor $N = W^*(X,G,\alpha)$ given by the group measure space construction s full if and only if the group action has the property called " strong-y ergodic " to be defined immediately below by results in [1] and [2]. ll group actions in our examples actually have a stronger property alled "s-strongly ergodic" and have the resulting.

iii) Ergodic actions.

An action α of a group G on a von Neumann algebra A is said o be ergodic if the fixed point algebra A^α is trivial. For a pair $A,\tau)$ (τ being a faithful normal tracial state as before), an action is said to be strongly ergodic if every sequence (a_n) in A (ounded in the operator norm) satisfying the asymptotic $\alpha(G)$-invariance .e. $\|\alpha_g(a_n) - a_n\|_2 \longrightarrow 0$ for all $g \in G$) tends to multiple of the dentity in the sense that $\|a_n - \tau(a_n)\|_2 \longrightarrow 0$.

Each automorphism α_g $(g \in G)$ as a linear mapping of A in $L^2(A,\tau)$ is naturally extended to a unitary operator on $L^2(A,\tau)$, which we shall denote by the same notation α_g. An action α of G on (A,τ) is said to be s-strong ergodic if every sequence (ξ_n) of unit vectors in $L^2(A,\tau)$ satisfying the asymptotic $\alpha(G)$-invariance (on $L^2(A,\tau)$) is asymptotically one-dimensional in the sense that $\|\xi_n - \langle\xi_n,1\rangle 1\|_2 \longrightarrow 0$.

Every s-strongly ergodic action is always strongly ergodic. If G has Kazhdan's property T, then every ergodic action of G on (A,τ) is s-strongly ergodic.

(iv) Haagerup type.

Definition 1. A pair (A,τ) is said to be of Haagerup type if there exists a net (p_s) of completely positive, $\|\cdot\|_2$-compact, normal linear maps on A such that $p_s(x)$ converges to x for all $x \in A$ in the strong topology.

Definition 2. A group G is said to have property H if there exists a net (ϕ_t) of positive definite functions on G with compact support such that $\phi_t(g)$ converges to 1 for all $g \in G$.

We note that if G is an ICC group then the group von Neumann algebra $R(G)$ is of Haagerup type if and only if G has property H ([4]).

3. Examples of strongly ergodic actions.

The following two actions β and γ of a group G on some measure space (X,μ) are examples of s-strongly ergodic actions, which are easy to deal with.

(3.1) Let $X_\beta = T^G$ be the space of functions on G with values in the torus T, ν be the Haar measure on T and $\mu = \nu^G$ be the infinite product of ν's indexed by G. We denote β to be the action of G on (X_β,μ) defined by $(\beta_g x)(h) = x(g^{-1}h)$ for $g,h \in G$ and $x \in X_\beta$.

If G is not amenable, then the action β is s-strongly ergodic by [11].

(3.2) We define the action γ of $SL(3,Z)$ on (T^3,ν^3) by

$$\gamma_g(x) = (s^{g_{11}}t^{g_{12}}u^{g_{13}},\ s^{g_{21}}t^{g_{22}}u^{g_{23}},\ s^{g_{31}}t^{g_{32}}u^{g_{33}})$$

for $g = (g_{ij}) \in SL(3,Z)$ and $(s,t,u) \in T^3$.

The action γ of $SL(3,Z)$ on (T^3,ν^3) is ergodic, so that it is s-strongly ergodic, because $SL(3,Z)$ has property T (see [3] for F_2).

4. Haagerup type and crossed product.

Theorem 3. Let α be an action of G on (A,τ). Assume that (A,τ) is of Haagerup type. If G has property H and if α_g commutes with p_s in Definition 1 for all $g \in G$, then the pair $(R(G,A,\alpha), \tau \cdot E)$ is of Haagerup type, where E is the conditional expectation of $R(G,A,\alpha)$ onto $\pi(A)$ such that $E(v(g)) = 0$ for all $g (\neq 1) \in G$.

Proof. Let (ϕ_t) be a net of positive definite functions on G with compact support such that $\phi_t(g) \longrightarrow 1$ for all $g \in G$. We may assume that $p_s(1) = 1$ and $\phi_t(1) = 1$ for all s and t. Then for each t, there exists a sequence $(a^t_i) \subset l^\infty(G)$ such that

$$\Sigma_i \, |a^t_i(g)|^2 = 1, \qquad \Sigma_i \, a^t_i(k)\overline{a^t_i(g^{-1}k)} = \phi_t(g)$$

for all $g, k \in G$ (cf. [9]). The crossed product $R(G,A,\alpha)$ is acting standardly on the Hilbert space $H = l^2(G, L^2(A,\tau))$. We define the bounded operator a^t_i on H by

$$(a^t_i\xi)(g) = a^t_i(g)\xi(g), \qquad \text{for } \xi \in H \text{ and } g \in G.$$

The crossed product $R(G,A,\alpha)$ is a von Neumann subalgebra of the tensor product $A \otimes B(l^2(G))$. Let $p_s \otimes id$ be the tensor product of the map p_s and the identity map id on $B(l^2(G))$. We define the map $q_{s,t}$ on $R(G,A,\alpha)$ by

$$q_{s,t}(x) = (p_s \otimes id)\, \Sigma_i \, a^t_i x \, a^t_i{}^* \qquad \text{for } x \in R(G,A,\alpha).$$

Then the linear map $q_{s,t}$ satisfies

$$q_{s,t}(x) = \Sigma_{g \in G} \, \pi(p_s(x_g))\phi_t(g)v(g),$$

where $x = \Sigma_{g \in G} \, \pi(x_g)v(g) \in R(G,A,\alpha) \quad (x_g \in A)$.

Hence each $q_{s,t}$ is a completely positive, normal linear map on $R(G,A,\alpha)$. The net $(q_{s,t})$ satisfies that $\| q_{s,t}(x) - x \|_2 \longrightarrow 0$, for all $x \in R(G,A,\alpha)$, because the nets (p_s) and (ϕ_t) satisfy

$$\| p_s(a) \|_2 \leqq \| a \| , \qquad \| p_s(a) - a \|_2 \longrightarrow 0 \qquad (a \in A)$$

and

$$|\phi_t(g)| \leqq 1, \qquad \phi_t(g) \longrightarrow 1 \qquad (g \in G).$$

Since each p_s is $\| \cdot \|_2$-compact and the support of each ϕ_t is compact, each $q_{s,t}$ is $\|\cdot\|_2$-compact.

Hence the net $(q_{s,t})$ satisfies the properties of (p_s) in the definition 1 of Haagerup type for the pair $(R(G,A,\alpha), \tau\cdot E)$.

5. Examples of Haagerup type and non-Haagerup type.

We now show that the two full II_1-factors $W^*(X_\beta, F_2, \beta)$ and $W^*(X_\beta, SL(3,Z), \beta)$ are not isomorphic, the former being of Haagerup type (i.e. the class (1)) and the latter being of non-Haagerup type (in the next section, we show that the latter belongs to the class (3)).

Proof. The group $SL(3,Z)$ and the free group F_2 are not inner amenable. Hence $W^*(X_\beta, SL(3,Z), \beta)$ and $W^*(X_\beta, F_2, \beta)$ are full II_1-factors by [2].

The non atomic abelian von Neumann algebra $A = L^\infty(T,\nu)$ is approximately infinite dimensional, so that the pair (A,ν) is of Haagerup type. Let $|g|$ be the length of $g \in F_2$ and

$$\phi_t(g) = e^{t|g|} \qquad \text{for } g \in F_2.$$

Then by [8] the net (ϕ_t) satisfies the properties in the definition 2 of property H.

We now prove the other premises of Theorem 3 for the present case. We can then conclude that $W^*(X_\beta, F_2, \beta)$ is of Haagerup type.

Since (A,ν) is of Haagerup type, we can take a net (p_s) of completely positive, $\| \ \|_2$-compact, normal linear maps on A such that $p_s(x) \to x$ for all $x \in A$. We then consider the infinite tensor product $\otimes_{g\in G} p_{s,g}$ of p_s's acting on $L^\infty(X_\beta)$ and denote it by the same notation p_s. Then the new net (p_s) satisfies the properties in the definition 2 of Haagerup type for $(L^\infty(X_\beta,\mu),\mu)$ and commutes with the group action : $p_s\beta_g = \beta_g p_s$ for all s and $g \in F_2$. Hence $W^*(X_\beta, F_2, \beta)$ is of Haagerup type.

Next we show that $W^*(X_\beta, SL(3,Z), \beta)$ is not Haagerup type. Assume the contrary. Then the factor $v(SL(3,Z))''$ is of Haagerup type, which is a contradiction to the result obtained in [4]. (Also see [7]).

6. Non-isomorphic factors arising from ergodic actions of $SL(3,Z)$.

We first prove the following general result.

Lemma 4. Let α be an ergodic action of an ICC group G on (A,τ). If there exists a net (θ_s) of non-trivial automorphisms on such that $\|\theta_s(a) - a\|_2 \longrightarrow 0$ for all $a \in A$ and each θ_s commutes ith α_g for all $g \in G$, then $\mathrm{Int}(R(G,A,\alpha))$ is not open.

<u>Proof</u>. We denote $R(G,A,\alpha)$ by N. Since G is an ICC group and is an ergodic action of G on (A,τ), we have that $v(G)' \cap N$ is he scalar multiples of the identity. Hence N is a finite factor. ince each θ_s commutes with α_g for all $g \in G$, each θ_s can be xtended to an automorphism σ_s of N such that $\sigma_s(\pi(a)) = \pi(\theta_s(a))$ nd $\sigma_s(v(g)) = v(g)$ for $a \in A$ and $g \in G$. If σ_s is an inner auto- orphism of N, then there exists a nonzero $y \in N$ such that $yx =$ $_s(x)y$ for all $x \in N$. Since y commutes with v_g for all $g \in G$, $= cl$ for some scalar c, which contradicts that θ_s is a non- rivial automorphism. Hence all σ_s are outer automorphisms of N. urther, the net (σ_s) converges to the identity by the assumption $\theta_s(a) - a\|_2 \longrightarrow 0$ for all $a \in A$ and the property $\sigma_s(v(g)) = v(g)$ or all s and $g \in G$. Hence $\mathrm{Int}(N)$ is not open.

We now claim that $W^*(X_\beta, SL(3,Z),\beta)$ <u>and</u> $W^*(T^3, SL(3,Z),\gamma)$ <u>are</u> <u>on-isomorphic</u> <u>full</u> II_1-<u>factors</u>. This is the first example of two trongly ergodic actions of a group on the nonatomic probability space hich give non-isomorphic factors.

<u>Proof</u>. The group $G = SL(3,Z)$ is not inner amenable and two ctions β and γ are strongly ergodic. Hence $W^*(X_\beta,G,\beta)$ and $*(T^3,G,\gamma)$ are full II_1-factors by [2]. Fix $\theta \in T$ with $(\theta/2\pi)$ rrational. Put $(\theta(x))(g) = x(g)\theta$ for all $x \in X_\beta$ and $g \in G$. hen $\beta_g \cdot \theta = \theta \cdot \beta_g$ for $g \in G$. Put $(\theta(f))(x) = f(\theta(x))$ for all f $L^\infty(X_\beta,\mu)$, then the map θ is an aperiodic automorphism of $A =$ $^\infty(X_\beta,\mu)$ and θ commutes with β_g for all $g \in G$. There exists a ubnet (θ_s) of the sequence (θ^n) which converges to the identity ecause of irrationality of $\theta/2\pi$. We can now use Lemma 4 to see that nt$(W^*(X_\beta,G,\beta))$ is not open.

On the other hand, using the Fourier transformation between $l^2(Z^3)$ nd $L^2(T^3)$, we can show that $W^*(T^3,SL(3,Z),\gamma)$ is isomorphic to the roup von Neumann algebra of the semi-direct product $SL(3,Z)\times_s Z^3$, here $SL(3,Z)$ acts on Z^3 in the natural manner. The group $SL(3,Z)$ $_sZ^3$ is an ICC group and has property T by [12]. Hence $\mathrm{Int}(R(SL(3,Z)$ $_sZ^3))$ is open by [8]. Therefore $W^*(X_\beta,SL(3,Z),\beta)$ is not isomorphic o $W^*(T^3,SL(3,Z),\gamma)$.

References.

[1] Choda, M.; Property T and fullness of the group measure space construction, Math. Japonica, 27(1982),535-539.

[2] Choda, M.; Inner amenability and fullness, Proc. Amer. Math. Soc., 86(1982), 663-666.

[3] Choda, M.; Effect of inner amenability on strong ergodicity, Math. Japonica, 28(1983), 109-115.

[4] Choda, M.; Group factors of the Haagerup type, Proc. Japan Acad., 59(1983), 174-177.

[5] Connes, A.; Almost periodic states and factors of type III_1, J. Funct. Anal., 16(1974), 415-445.

[6] Connes, A.; A factor of type II_1 with countable fundamental groups, J. Operator Theory, 4(1980), 151-153.

[7] Connes, A. and Jones, V.; Property T for von Neumann algebras, Preprint.

[8] Effros, E. G.; Property Γ and inner amenability, Proc. Amer. Math. Soc., 47(1975), 483-486.

[9] Haagerup, U.; An example of a non-nuclear C*-algebra which has the metric approximation property, Invent. Math., 50(1978/79), 279-293.

[10] Kazhdan, D. A.; Connection of the dual space of a group with the structure of its closed subgroups, Funct. Anal. Appl., 1(1967), 63-65.

[11] Schmidt, K.; Amenability, Kazhdan's property T, strong ergodicity and invariant means for ergodic group actions, Ergod. Th. and Dynam. Sys., 1(1981), 223-236.

[12] Wang, P. S.; On isolated points in the dual space of locally compact groups, Math. Ann., 218(1975), 19-34.

DIAMETERS OF STATE SPACES OF TYPE III FACTORS

by

Alain Connes

Uffe Haagerup

Erling Størmer

1. Introduction. Let M be a von Neumann algebra and $S_0(M)$ the norm closed set of its normal states. For each $\omega \in S_0(M)$ let $[\omega]$ be the norm closure of its orbit under the action of the inner $*$-automorphisms, $\mathrm{Int}(M)$, by $\omega \to u\omega u^* = \omega \circ \mathrm{Ad}u$. The orbit space $S_0(M)/\mathrm{Int}(M)$ is a metric space with metric

$$d([\omega],[\phi]) = \inf\{\|\omega'-\phi'\| : \omega' \in [\omega], \phi' \in [\phi]\}.$$

If M is not a factor the diameter of $S_0(M)/\mathrm{Int}(M)$ is clearly equal to 2. However, if M is a factor it may be different.

Powers proved in [8] that if M is a factor of type I_n, $n<\infty$, and $\phi = \mathrm{Tr}(h\cdot)$, $\psi = \mathrm{Tr}(k\cdot)$ are states then

$$d([\phi],[\psi]) = \sum_{i=1}^{n} |\lambda_i - \mu_i|,$$

where $\lambda_1 \geq \lambda_2 \geq \ldots \geq \lambda_n$ are the eigenvalues of h, and $\mu_1 \geq \mu_2 \geq \ldots \geq \mu_n$ are the eigenvalues of k. From this one easily gets that

$$\mathrm{diam}(S_0(M)/\mathrm{Int}(M)) = 2(1-\frac{1}{n}).$$

The value $2(1-\frac{1}{n})$ is attained when ϕ is the tracial state and ψ is a pure state.

The arguments of Powers can be extended to the case when M is a semifinite factor with faithful normal semifinite trace τ. If $\phi = \tau(h\cdot)$, $\psi = \tau(k\cdot)$ are two positive normal functionals given by two positive operators h and k in M, which have "joint diagonalization"

$$h = \sum_{i=1}^{n} \lambda_i p_i, \quad k = \sum_{i=1}^{n} \mu_i p_i,$$

where $p_1,\dots,p_n$ are orthogonal projections with sum 1 and $\lambda_1 > \lambda_2 > \dots > \lambda_n$, $\mu_1 > \mu_2 > \dots > \mu_n$, then

$$d([\phi],[\psi]) = \sum_{i=1}^{n} |\lambda_i - \mu_i| \tau(p_i) = \|\phi - \psi\| .$$

From this one derives easily that if ϕ,ψ are two states of the form

$$\phi(x) = \frac{1}{\tau(p)} \tau(px), \quad \psi(x) = \frac{1}{\tau(q)} \tau(qx),$$

where p and q are two nonzero finite projections in M, and $p < q$, then

$$d([\phi],[\psi]) = 2\left(1 - \frac{\tau(p)}{\tau(q)}\right).$$

Hence for a factor of types I_∞ or II we have

$$\mathrm{diam}(S_0(M)/\mathrm{Int}(M)) = 2.$$

The main result of the present paper is a formula for the diameter when M is of type III. The result will be a characterization of factors of type III_λ, $\lambda \in [0,1]$, purely in terms of the geometry of the state space and independent of Tomita-Takesaki theory.

<u>Theorem</u>. Let M be a σ-finite factor of type III_λ, $\lambda \in [0,1]$.

Then

$$\operatorname{diam}(S_0(M)/\operatorname{Int}(M)) = 2\,\frac{1-\lambda^{\frac{1}{2}}}{1+\lambda^{\frac{1}{2}}}.$$

In particular for a factor of type III_0 the diameter is 2 and for a factor of type III_1 it is 0. The last statement was previously proved by two of us in [6]. In the case when $0<\lambda<1$ it was shown by Bion-Nadal [2] that $2(1-\lambda^{\frac{1}{2}})$ is an upper bound for the diameter, a result which inspired the present work. Our proof will be divided into two parts, namely to show the inequalities $\operatorname{diam}(S_0(M)/\operatorname{Int}(M)) \gtrless 2\,\frac{1-\lambda^{\frac{1}{2}}}{1+\lambda^{\frac{1}{2}}}$ for $\lambda\in[0,1)$.

2. Proof of the inequality $\leq$.

The number $2\,\frac{1-\lambda^{\frac{1}{2}}}{1+\lambda^{\frac{1}{2}}}$ that gives the diameter appears as a consequence of the following function theoretic lemma.

Lemma 2.1. Let $0<a<b$ be real numbers, and let $K_{a,b}$ denote the convex set of nonnegative decreasing functions f on $[a,b]$ such that $\int_a^b f\,dt = 1$ and $af(a) = bf(b)$. Then we have

$$\sup_{f,g\in K_{a,b}} \int_a^b f\vee g\,dt = 2\,\frac{b^{\frac{1}{2}}}{a^{\frac{1}{2}}+b^{\frac{1}{2}}}$$

Proof. In order to show the lemma it suffices to consider step functions in $K_{a,b}$. If $\alpha\in[0,1]$ and $f_1,f_2,\in K_{a,b}$ then we have

$$(\alpha f_1+(1-\alpha)f_2)\vee f\leq\alpha(f_1\vee f)+(1-\alpha)(f_2\vee f).$$

Hence it suffices to prove the lemma for extremal step functions in $K_{a,b}$. Let

$$f = \sum_{i=1}^{n-1} c_i \chi_{[a_i, a_{i+1})} + c_n \chi_{[a_n, a_{n+1}]} \in K_{a,b},$$

where $a = a_1 < a_2 < \ldots < a_{n+1} = b$, $c_1 > c_2 > \ldots > c_n = \frac{a}{b} c_1$. If $n > 3$ we can find $\varepsilon > 0$ and $\eta > 0$ such that $(1-\varepsilon)c_1 > (1+\eta)c_2$, $(1-\eta)c_2 > c_3$, $c_{n-1} > (1+\varepsilon)c_n$ and such that the two functions

$$f_\pm = (1\pm\varepsilon)c_1\chi_{[a_1,a_2)} + (1\mp\eta)c_2\chi_{[a_2,a_3)} + \sum_{i=3}^{n-1} c_i\chi_{[a_i,a_{i+1})} + (1\pm\varepsilon)c_n\chi_{[a_n,a_{n+1}]}$$

belong to $K_{a,b}$. Since $f = \frac{1}{2}(f_+ + f_-)$, f is not extremal in $K_{a,b}$. Therefore it suffices to show the lemma for step functions of the form

$$f_s = \frac{b}{s(b-a)} \chi_{[a,s)} + \frac{a}{s(b-a)} \chi_{[s,b]},$$

where $s \in (a,b]$. If $a < r < s < b$ we find

$$\int_a^b f_r \vee f_s \, dt = \frac{1}{b-a} \left(2b - b\frac{r}{s} - a\frac{s}{r}\right).$$

Since the maximum of this function of $\frac{s}{r}$ is obtained for $\frac{s}{r} = (\frac{b}{a})^{\frac{1}{2}}$ the proof is complete.

Since for two functions f and g, $|f-g| = 2f\vee g - f - g$, we have:

<u>Corollary 2.2</u>. In the above notation, if $0<\lambda<1$ we have

$$\sup_{f,g \in K_{\lambda,1}} \int_\lambda^1 |f-g| \, dt = 2\,\frac{1-\lambda^{\frac{1}{2}}}{1+\lambda^{\frac{1}{2}}}.$$

<u>Lemma 2.3</u>. Let M be a σ-finite factor of type III_λ, $0<\lambda<1$, and let $T = -\frac{2\pi}{\log\lambda}$. Let ϕ_0 be a faithful normal state on M for

which $\sigma_T^{\phi_0}$ is the identity. Then for any faithful normal state ϕ on M there exists a positive operator h in the centralizer M_{ϕ_0} of ϕ_0 such that

(i) $\operatorname{Sp} h \subset [\lambda a, a]$ for some $a>0$,

(ii) There exists a unitary $u \in M$ such that $\phi(uxu^*) = \phi_0(hx)$, $x \in M$.

<u>Proof:</u> Put $v = (D\phi : D\phi_0)_T$, see [4]. Then for $x \in M$

$$\sigma_T^{\phi}(x) = v\sigma_T^{\phi_0}(x)v^* = vxv^*,$$

so in particular $\phi(vxv^*) = \phi(\sigma_T^{\phi}(x)) = \phi(x)$. Thus $v \in M_\phi$. By spectral theory and the Riesz representation theorem there is a unique probability measure μ on $\mathbb{T} = \{z \in \mathbb{C} : |z| = 1\}$ for which

$$\int_{\mathbb{T}} f(z)\,d\mu(z) = \phi(f(v))$$

for any Borel function f on $\mathbb{T}$. Let ν be the positive Borel measure on $\mathbb{R}$ obtained by "rewinding" μ, i.e. ν is determined by

$$\nu(B) = \mu(\exp(iB)), \quad B \subset [0, 2\pi), B \text{ Borel},$$

and

$$\nu(B+2\pi) = \nu(B), \quad B \subset \mathbb{R}, B \text{ Borel}.$$

Note that $\nu([s, s+2\pi)) = 1$ for all $s \in \mathbb{R}$. Put

$$g(s) = \int_{[s,s+2\pi)} \exp(-\tfrac{t}{T})\,d\nu(t), \quad s \in \mathbb{R}.$$

Since $\exp(-\frac{2\pi}{T}) = \lambda$ we have

$$\int_{[s,\infty)} \exp(-\tfrac{t}{T})\,d\nu(t) = \sum_{n=0}^{\infty} \int_{[s+n2\pi, s+(n+1)2\pi)} \exp(-\tfrac{t}{T})\,d\nu(t)$$

$$= \Big(\sum_{n=0}^{\infty} \lambda^n\Big) g(s) = \frac{1}{1-\lambda}\, g(s).$$

Hence we also have

(1) $$g(s) = (1-\lambda) \int_{[s,\infty)} \exp(-\frac{t}{T})d\nu(t).$$

This shows that g is a decreasing function on $\mathbb{R}$, continuous from left. Let $g(s+)$ (resp. $g(s-)$) denote the limits of $g(s')$ for $s' \to s$ from right (resp. left). Then

$$g(0+) = \int_{(0,2\pi]} \exp(-\frac{t}{T})d\nu(t) < 1,$$

and

$$g((-2\pi)-) = \int_{[-2\pi,0)} \exp(-\frac{t}{T})d\nu(t) > 1.$$

Hence we can choose $r \in [-2\pi, 0]$ such that

$$g(r+) \leq 1 \leq g(r-).$$

By (1) we have

$$g(r-)-g(r+) = (1-\lambda)\exp(-\frac{r}{T})\nu(\{r\})$$

$$= (1-\lambda)\exp(-\frac{r}{T})\mu(\{e^{ir}\}).$$

This shows that r is a point of continuity for g if and only if e^{ir} is not an eigenvalue for V.

Moreover

$$g(r-)-g(r+) = (1-\lambda)\exp(-\frac{r}{T})\phi(p),$$

where p is the projection on the eigenspace of the vectors ξ such that $v\xi = e^{ir}\xi$. There are two cases to be considered.

Case 1. Assume first that e^{ir} is not an eigenvalue for v. Let

$$\mathrm{Arg}_r : \mathbb{T}\setminus\{e^{ir}\} \to (r, r+2\pi)$$

be the branch of the argument functions that takes values in $(r, r+2\pi)$, and put

$$a = \mathrm{Arg}_r(v)$$

$$k = \exp(\tfrac{1}{T}a).$$

Since $v \in M_\phi$ so are a and k. Moreover, a and k are self-adjoint, and their spectra satisfy

$$\mathrm{Sp}a \subset [r, r+2\pi]$$

$$\mathrm{Sp}k \subset [\exp(\tfrac{r}{T}), \lambda^{-1}\exp(\tfrac{r}{T})].$$

Furthermore, since r is a continuity point for g,

$$\phi(k^{-1}) = \int_{\mathbf{T}} \exp(-\tfrac{1}{T}\,\mathrm{Arg}_r(z))d\mu(z)$$

$$= \int_r^{r+2\pi} \exp(-\tfrac{t}{T})d\nu(t)$$

$$= 1.$$

Put $\psi(x) = \phi(k^{-1}x)$, $x \in M$. Then ψ is a faithful normal state on M. Since $k^{iT} = \exp(ia) = v$, we get, see [4],

$$\sigma_T^\psi(x) = k^{-iT}\sigma_T^\phi(x)k^{iT} = v^*(vxv^*)v = x, \quad x \in M,$$

and

$$(D\psi : D\phi_0)_T = (D\psi : D\phi)_T (D\phi : D\phi_0)_T = k^{-iT}v = 1.$$

Since σ^ψ and σ^{ϕ_0} both have period T we can conclude as in the proof of [4, 4.3.2] that there exists a unitary $u \in M$ such that $\psi(uxu^*) = \phi_0(x)$ for $x \in M$. Hence, if $h = u^*ku$ we have

$$\phi(uxu^*) = \psi(kuxu^*) = \psi(uhxu^*) = \phi_0(hx).$$

Since $\mathrm{Sp}h = \mathrm{Sp}k \subset [\exp(\tfrac{r}{T}), \lambda^{-1}\exp(\tfrac{r}{T})]$, h and u satisfy the conditions in the lemma.

Case 2. Assume next that e^{ir} is an eigenvalue for v, and let

p be the projection on the corresponding eigenspace. Clearly $p \in M_\phi$. Since

$$g(r+) < 1 < g(r-)$$

we can choose $\alpha \in [0,1]$ such that

$$1 = (1-\alpha)g(r+) + \alpha g(r-).$$

Now $\sigma_T^\phi(x) = vxv^*$ for $x \in M$ and $pv = e^{ir}p$. Thus the restriction of σ_T^ϕ to the reduced algebra pMp is trivial. Since M is σ-finite of type III, $pMp \cong M$, so is also a factor of type III_λ. Thus, as in the proof of [4, 4.2.6] the centralizer of the restriction $\phi|pMp$ is a factor of type II_1. Therefore we can choose a projection $p' < p$, $p' \in M_\phi$, such that $\phi(p') = \alpha\phi(p)$. Define now self-adjoint operators a and k in M_ϕ by

$$a = \mathrm{Arg}_r(v(1-p)) + rp' + (r+2\pi)(p-p')$$

$$k = \exp(\tfrac{1}{T}a).$$

The operators are well defined since e^{ir} is not in the point spectrum of $v(1-p)$. Clearly $\mathrm{Sp}(a) \subset [r, r+2\pi]$; hence

$$\mathrm{Sp}(k) \subset [\exp(\tfrac{r}{T}), \lambda^{-1}\exp(\tfrac{r}{T})].$$

Moreover, $k^{iT} = e^{ia} = v(1-p) + e^{ir}p = v$. Computing we find the following formulas:

$$\phi(k^{-1}) = \int_{(r,r+2\pi)} \exp(-\tfrac{t}{T})d\nu(t) + \alpha\phi(p)\exp(-\tfrac{r}{T}) + (1-\alpha)\phi(p)\exp(-\tfrac{r+2\pi}{T}),$$

$$g(r+) = \int_{(r,r+2\pi]} \exp(-\tfrac{t}{T})d\nu(t) = \int_{(r,r+2\pi)} \exp(-\tfrac{t}{T})d\nu(t) + \phi(p)\exp(-\tfrac{r+2\pi}{T}),$$

$$g(r-) = \int_{[r,r+2\pi)} \exp(-\tfrac{t}{T})d\nu(t) = \int_{(r,r+2\pi)} \exp(-\tfrac{t}{T})d\nu(t) + \phi(p)\exp(-\tfrac{r}{T}).$$

Adding we obtain $\phi(k^{-1}) = (1-\alpha)g(r+)+\alpha g(r-) = 1$. The proof can now be completed as in Case 1.

<u>Proof of the inequality diam $(S_0(M)/\mathrm{Int}(M)) \leq 2\,\frac{1-\lambda^{1/2}}{1+\lambda^{1/2}}$.</u>

It suffices to show the inequality for faithful states. Let ϕ and ψ be faithful normal states on the factor M of type III_λ, $0<\lambda<1$. Let ϕ_0 be a faithful normal state such that $\sigma_T^{\phi_0}$ is the identity map. By Lemma 2.3 there are $\phi' \in [\phi]$, $\psi' \in [\psi]$ such that $\phi'(x) = \phi_0(hx)$, $\psi'(x) = \phi_0(kx)$, $x \in M$, where $h,k \in M_{\phi_0}$ and $\lambda a \leq h \leq a$, $\lambda b \leq k \leq b$ for some $a,b>0$.

If $\delta>0$ we can by spectral theory find an integer n and orthogonal families $\{p_1,\ldots,p_n\}, \{q_1,\ldots,q_n\}$ of projections in M_{ϕ_0} with $\phi_0(p_i) = \phi_0(q_i) = \frac{1}{n}$, $i = 1,\ldots,n$, and constants $\alpha_1 \geq \alpha_2 \geq \ldots \geq \alpha_n = \lambda\alpha_1$, $\beta_1 \geq \beta_2 \geq \ldots \geq \beta_n = \lambda\beta_1$ satisfying $\sum\alpha_i = \sum\beta_i = n$ such that

$$\|h-\sum_1^n\alpha_i p_i\|_1<\delta, \quad \|k-\sum_1^n\beta_i q_i\|_1<\delta,$$

where $\|x\|_1 = \phi_0(|x|)$ for $x \in M_{\phi_0}$. In order to show the desired estimate we may assume h and k are of this form, i.e. $h = \sum\alpha_i p_i$, $k = \sum\beta_i q_i$. Since M_{ϕ_0} is a factor of type II_1 there is a unitary $u \in M_{\phi_0}$ such that $uq_iu^* = p_i$ for all i, hence $uku^* = \sum_1^n\beta_i p_i$. Thus the state ψ'' defined by

$$\psi''(x) = \phi_0(uku^*x) = \phi_0(ku^*xu)$$

belongs to $[\psi]$.

Let f and g be functions on the interval $[\lambda,1]$ defined by $f = (1-\lambda)^{-1}\sum_{i=1}^n\alpha_i\chi_{I_i}$, $g = (1-\lambda)^{-1}\sum_{i=1}^n\beta_i\chi_{I_i}$, where

$$I_i = \begin{cases} [\lambda+(i-1)\frac{1-\lambda}{n}, \lambda+i\frac{1-\lambda}{n}) & \text{for } i = 1,\dots,n-1, \\ [\lambda+(n-1)\frac{1-\lambda}{n}, 1] & \text{for } i = n. \end{cases}$$

Then f and g are decreasing step functions with integrals 1 and satifying $f(1) = \lambda f(\lambda)$, $g(1) = \lambda g(\lambda)$, i.e. f,g belong to the set $K_{\lambda,1}$ of Lemma 2.1. Thus by Corollary 2.2 we have,

$$\|\phi'-\psi''\| = \|h-uku^*\|_1 = \sum_{i=1}^{n} |\alpha_i-\beta_i|\phi(p_i) = \int_\lambda^1 |f-g|dt \leq 2\frac{1-\lambda^{\frac{1}{2}}}{1+\lambda^{\frac{1}{2}}},$$

completing the proof. The case $\lambda = 0$ is trivial.

3. Proof of the inequality $\geq$.

The proof of the inequality

$$\operatorname{diam}(S_0(M)/\operatorname{Int}(M)) \geq 2\frac{1-\lambda^{\frac{1}{2}}}{1+\lambda^{\frac{1}{2}}}$$

for a factor of type III_λ is based on the following theorem.

Theorem 3.1. Let M be a von Neumann algebra, let ϕ,ψ be two faithful normal positive functionals on M, and let $0<a<b$ be real numbers. Suppose

(i) ϕ and ψ commute and $a\phi \leq \psi \leq b\phi$,

(ii) $Sp(\Delta_\phi) \cap (\frac{a}{b}, \frac{b}{a}) = \{1\}$,

where Δ_ϕ is the modular operator of ϕ.

Then $\|u\phi u^* - \psi\| \geq \|\phi-\psi\|$ for all unitary operators u in M.

The proof of the above theorem will be divided into three steps:

Step 1: M is finite,

Step 2: $T(M) = \{t : \sigma_t^\phi \in \mathrm{Int}(M)\}$ is dense in $\mathbb{R}$,

Step 3: The general case.

In order to prove Step 1 we assume M is finite and that ϕ, ψ, a, b satisfy the above conditions (i) and (ii). Since M has a faithful normal state it also has a faithful normal tracial state τ. There exist two positive operators h and k affiliated with M such that

$$\phi = \tau(h\cdot) \quad \text{and} \quad \psi = \tau(k\cdot).$$

By the usual identification of M_* and $L^1(M,\tau)$ the inequality stated in Theorem 3.1 is equivalent to

$$\|uhu^* - k\|_1 \geq \|h-k\|_1$$

for all unitary operators $u \in M$. To prove this we shall need

Lemma 3.2. Let M be a finite von Neumann algebra with a faithful normal tracial state τ and let $h, k \in M$ be two positive operators with bounded inverses such that

(i) h and k commute and $ah \leq k \leq bh$,

(ii) with $\phi = \tau(h\cdot)$, $\mathrm{Sp}(\Delta_\phi) \cap (\frac{a}{b}, \frac{b}{a}) = \{1\}$.

Then $\|uhu^* - k\|_1 \geq \|h-k\|_1$ for all unitary operators $u \in M$.

Proof. The modular automorphism group asociated with ϕ is, see [10],

$$\sigma_t^\phi(x) = h^{it} x h^{-it}, \quad x \in M.$$

Moreover M acts standardly on $L^2(M,\tau)$. Let $\mathrm{Sp}(\sigma^\phi)$ denote the Arveson spectrum of the one parameter group σ^ϕ. We shall consider $\mathrm{Sp}(\sigma^\phi)$ as a subset of the multiplicative group $\mathbb{R}_+$. Since

h is bounded and has bounded inverse, $0 \notin Sp(\Delta_\phi)$ and therefore

$$Sp(\sigma^\phi) = Sp(\Delta_\phi).$$

By [10] if J is the conjugation on $L^2(M,\tau)$ defined by σ^ϕ such that $JMJ = M'$, we have $\Delta_\phi = hJh^{-1}J$. We first assume M is a factor; then

$$Sp(\Delta_\phi) = Sp(h)\cdot Sp(h)^{-1}.$$

By condition (ii) we therefore get that if $\mu_1, \mu_2 \in Sp(h)$ and $\mu_1 > \mu_2$ then

$$\frac{\mu_2}{\mu_1} < \frac{a}{b}.$$

Since $Sp(h)$ is a compact subset of $(0,\infty)$ it follows that $Sp(h)$ is finite.

By (i) we have $k = mh$, where $m \in M$ commutes with h, and

$$a1 < m < b1.$$

By continuity it is enough to prove the inequality $\|uhu^* - k\|_1 > \|h-k\|_1$ in the case when the spectrum of m is a finite subset of the interval $[a,b]$. In this case k also has finite spectrum, and h and k have a "joint diagonalization"

$$h = \sum_{i=1}^{n} \lambda_i p_i, \quad k = \sum_{i=1}^{n} \mu_i p_i,$$

where $p_1,\ldots,p_n$ are nonzero orthogonal projections with sum 1. By permuting the indices $\{1,\ldots,n\}$ we may assume that

$$\lambda_1 > \lambda_2 > \ldots > \lambda_n.$$

Let $i_1 < i_2 < \ldots < i_q$ be the values of i for which $\lambda_i > \lambda_{i+1}$. By permuting the indices inside each of the $q+1$ sets on which the

λ_k's are constant we may also obtain that

$$\mu_1 > \dots > \mu_{i_1}, \mu_{i_1+1} > \dots > \mu_{i_2}, \dots, \mu_{i_q+1} > \dots > \mu_n.$$

However, since

$$\lambda_{i_k+1} < \frac{a}{b} \lambda_{i_k},$$

and since by (i)

$$a\mu_i < \lambda_i < b\mu_i,$$

we also have

$$\mu_{i_1} > \mu_{i_1+1}, \mu_{i_2} > \mu_{i_2+1}, \dots, \mu_{i_q+1} > \mu_{i_q}.$$

Hence by the extension of Powers' result mentioned in the introduction, we get

$$\| uhu^* - k \|_1 > \sum_{i=1}^{n} |\lambda_i - \mu_i| \tau(p_i) = \| h-k \|_1$$

for all unitary operators $u \in M$. This completes the proof in the case when M is a factor.

Let now M be general, and let $T: M \to Z$ be the center valued trace on M, where Z denotes the center of M. For every pure state ω on Z

$$\tau_\omega = \omega \circ T$$

is a (possibly nonnormal) tracial state on M. Put

$$I_\omega = \{ x \in M : \tau_\omega(x^* x) = 0 \}.$$

Then I_ω is a maximal ideal in M, and

$$M_\omega = M / I_\omega$$

is a finite factor, see [9, Ch. II]. The tracial state on M_ω will also be denoted by τ_ω. Let π_ω be the quotient map

$\pi_\omega : M \to M_\omega$, put

$$h_\omega = \pi_\omega(h), \quad k_\omega = \pi_\omega(k),$$

and put $\phi_\omega = \tau_\omega(h_\omega \cdot)$. By Arveson's definition of $Sp(\sigma^\phi)$, see [1], we have

$$\int_{-\infty}^{\infty} f(t)h^{it}xh^{-it}dt = 0 \quad \text{for every} \quad x \in M$$

if $f \in L^1(\mathbb{R})$ and $supp(\hat{f}) \cap Sp(\sigma^\phi) = \emptyset$, where the Fourier transform $\hat{f}$ of f is considered as a function on $(\mathbb{R}_+, \cdot)$. Since $t \to h^{it}$ is norm continuous it follows that under the same condition on f,

$$\int_{-\infty}^{\infty} f(t)h_\omega^{it}yh_\omega^{-it}dt = 0 \quad \text{for every} \quad y \in M_\omega .$$

Hence $Sp(\sigma^{\phi_\omega}) \subset Sp(\sigma^\phi)$. Therefore h_ω and k_ω satisfy the conditions of Lemma 3.2, so by the first part of the proof

$$\| vh_\omega v^* - k_\omega \|_1 \geq \| h_\omega - k_\omega \|_1$$

for every unitary $v \in M_\omega$. By the spectral theorem $Z \cong C(\hat{Z})$. Thus if ν is the probability measure on $\hat{Z}$ which corresponds to the restriction of τ to Z, we have for $x \in M$:

$$\tau(x) = \tau \circ T(x) = \int_{\hat{Z}} \tau_\omega \circ T(x) d\nu(\omega) = \int_{\hat{Z}} \tau_\omega(x) d\nu(\omega).$$

Hence for any unitary operator $u \in M$,

$$\| uhu^* - k \|_1 = \int_{\hat{Z}} \| \pi_\omega(u) h_\omega \pi_\omega(u)^* - k_\omega \|_1 d\nu(\omega)$$

$$\geq \int_{\hat{Z}} \| h_\omega - k_\omega \|_1 d\nu(\omega)$$

$$= \| h - k \|_1 .$$

This completes the proof of Lemma 3.2.

<u>Completion of Step 1</u>. To complete the proof of Theorem 3.1 in the case when M is finite we need to extend Lemma 3.2 to the case when h and k are (possibly unbounded) positive operators in $L^1(M,\tau)$ with trivial nullspaces.

Let p_n be the spectral projection of h corresponding to the interval $[\frac{1}{n},n]$, $n\in\mathbb{N}$. Then $h_n = p_n h$ and $k_n = p_n k$ satisfy the conditions of Lemma 3.2 with respect to the von Neumann algebra $p_n M p_n$. For every unitary $u\in M$ we can find a sequence of partial isometries $u_n \in M$ with support and range projections equal to p_n such that $u_n \to u$ in the strong-* topology (for instance write u in the form $u = \exp(ia)$ and put $u_n = p_n \exp(ip_n a p_n)$). Then

$$\|uhu^* - k\|_1 = \lim_{n\to\infty} \|u_n h_n u_n^* - k_n\|_1$$

$$\geq \lim_{n\to\infty} \|h_n - k_n\|_1 = \|h-k\|_1 .$$

This completes the proof of Step 1.

<u>Step 2</u>. For any faithful normal positive functional ϕ on a von Neumann algebra M we let $\|\cdot\|_\phi^\#$ be the norm

$$\|x\|_\phi^\# = \phi(\tfrac{1}{2}(x^* x + xx^*))^{\frac{1}{2}}.$$

Note that if ϕ is a state and u is unitary then $\|u\|_\phi^\# = 1$.

<u>Lemma 3.3</u>. Let M be a von Neumann algebra for which $T(M)$ is dense in $\mathbb{R}$. Let ϕ be a faithful normal state on M, and let u be a unitary operator in M. For every $\varepsilon>0$ there exist a faithful normal state ω on M and a unitary operator $v\in M$ such that

(a) ϕ and ω commute,

(b) $M_\phi \subset M_\omega$,

(c) $v\in M_\omega$ and $\|u-v\|_\phi^\# < \varepsilon$.

<u>Proof</u>. Let $\delta>0$. Since the function $t \to \sigma_t^\phi(u)$ is strong-* continuous there is $t_1>0$ such that

$$\|\sigma_t^\phi(u)-u\|_\phi^\# < \delta \quad \text{for} \quad |t|<t_1.$$

Since $T(M)$ is dense in $\mathbb{R}$ we can therefore choose $t_0>0$, $t_0 \in T(M)$ such that

$$\|\sigma_t^\phi(u)-u\|_\phi^\# < \delta \quad \text{for} \quad |t|<t_0.$$

Let $w \in M$ be a unitary operator such that

$$\sigma_{t_0}^\phi(x) = wxw^*, \quad x \in M.$$

By [4, 1.3.2] w belongs to the center of M_ϕ. Hence

$$\|uw-wu\|_\phi^\# = \|u-wuw^*\|_\phi^\# < \delta.$$

Let Arg be the branch of the argument function on $\mathbb{C}\sim\{0\}$ that takes values in the half-open interval $[0,2\pi)$. Then for $\theta \in \mathbb{R}$

$$\mathrm{Arg}_\theta(z) = \mathrm{Arg}(e^{-i\theta}z)+\theta$$

is the branch of the argument function that takes values in $[\theta, 2\pi+\theta)$. Put

$$a_\theta = \mathrm{Arg}_\theta(w), \quad \theta \in \mathbb{R}.$$

We shall show that θ can be chosen such that

$$\|ua_\theta - a_\theta u\|_\phi^\# < (2\pi\delta)^{\frac{1}{2}}.$$

Let H_ϕ denote the completion of M with respect to the norm $\|\ \|_\phi^\#$. Let

$$\langle x,y\rangle_\phi^\# = \tfrac{1}{2}\phi(y^*x+xy^*)$$

be the corresponding inner product on M. Define a unitary representation π of $\mathbf{Z}^2$ on H_ϕ by

$$\pi(n,m)x = w^n x w^m$$

(the representation is unitary since $w \in M_\phi$). By Bochner's theorem there exists a probability measure μ on $\mathbf{T}^2 = (\mathbf{Z}^2)^\wedge$ such that

$$\langle w^n u w^m, u\rangle_\phi^\# = \iint_{\mathbf{T}^2} \alpha^n \beta^m d\mu(\alpha,\beta).$$

Hence for any pair of bounded Borel functions f and g on $\mathbf{T}$

$$\langle f(w)ug(w), u\rangle_\phi^\# = \iint_{\mathbf{T}^2} f(\alpha)g(\beta)d\mu(\alpha,\beta).$$

From this equality we obtain that

(1) $$(\| f(w)u-uf(w)\|_\phi^\#)^2 = \iint_{\mathbf{T}^2} |f(\alpha)-g(\beta)|^2 d\mu(\alpha,\beta)$$

for every bounded Borel function f on $\mathbf{T}$ (compare with the proof of Proposition 1.1 in [5]). In particular

$$\iint_{\mathbf{T}^2} |\alpha-\beta|^2 d\mu(\alpha,\beta) = (\| wu-uw\|_\phi^\#)^2 < \delta^2.$$

Moreover,

(2) $$(\| a_\theta u-ua_\theta\|_\phi^\#)^2 = \iint_{\mathbf{T}^2} |\mathrm{Arg}(e^{-i\theta}\alpha)-\mathrm{Arg}(e^{-i\theta}\beta)|^2 d\mu(\alpha,\beta).$$

Therefore

$$\frac{1}{2\pi}\int_0^{2\pi} (\| a_\theta u-ua_\theta\|_\phi^\#)^2 d\theta = \iint_{\mathbf{T}^2} h(\alpha,\beta)d\mu(\alpha,\beta),$$

where

$$h(\alpha,\beta) = \frac{1}{2\pi}\int_0^{2\pi} |\mathrm{Arg}(e^{-i\theta}\alpha)-\mathrm{Arg}(e^{-i\theta}\beta)|^2 d\theta.$$

For $\alpha = 1$ and $\beta = e^{i\sigma}$, $0<\sigma<2\pi$, we have

$$\operatorname{Arg}(e^{-i\theta}\alpha) = 2\pi-\theta, \quad 0<\theta<2\pi,$$

$$\operatorname{Arg}(e^{-i\theta}\beta) = \begin{cases} \sigma-\theta, & 0<\theta<\sigma \\ \sigma-\theta+2\pi, & \sigma<\theta<2\pi \end{cases}.$$

Now the function

$$f(\sigma) = 4\pi \sin\frac{\sigma}{2} - \sigma(2\pi-\sigma)$$

is continuous on the interval $[0,2\pi]$ and $f(0) = f(2\pi) = 0$. Moreover, its derivative

$$f'(\sigma) = 2\pi(\cos\frac{\sigma}{2} - (1-\frac{\sigma}{\pi}))$$

is positive for $0<\sigma<\pi$ and negative for $\pi<\sigma<2\pi$, because $\cos\frac{\sigma}{2}$ is concave on $[0,\pi]$ and convex on $[\pi,2\pi]$. Hence

$$4\pi \sin\frac{\sigma}{2} - \sigma(2\pi-\sigma)>0 \quad \text{for} \quad 0<\sigma<2\pi.$$

We therefore find

$$\begin{aligned} h(1,e^{i\sigma}) &= \frac{1}{2\pi}\left(\int_0^\sigma (2\pi-\sigma)^2 d\theta + \int_\sigma^{2\pi} \sigma^2 d\theta\right) \\ &= \sigma(2\pi-\sigma) \\ &< 4\pi \sin\frac{\sigma}{2} \\ &= 2\pi|1-e^{i\sigma}|. \end{aligned}$$

Thus

$$h(1,\beta)<2\pi|1-\beta|, \quad \beta\in\mathbf{T}.$$

It is clear that $h(e^{it}\alpha,e^{it}\beta) = h(\alpha,\beta)$, $t\in\mathbf{R}$. Therefore

$$h(\alpha,\beta) = h(1,\frac{\beta}{\alpha})<2\pi|1-\frac{\beta}{\alpha}| = 2\pi|\alpha-\beta|, \quad \alpha,\beta\in\mathbf{T}.$$

Using that $\mu(1) = 1$ we therefore get

$$\frac{1}{2\pi}\int_0^{2\pi}(\|a_\theta u-ua_\theta\|_\phi^\#)^2\,d\theta \le 2\pi\iint_{\mathbf{T}^2}|\alpha-\beta|\,d\mu(\alpha,\beta)$$

$$\le 2\pi\Big(\iint_{\mathbf{T}^2}|\alpha-\beta|^2\,d\mu(\alpha,\beta)\Big)^{\frac{1}{2}}$$

$$= 2\pi\|wu-uw\|_\phi^\#$$

$$< 2\pi\delta .$$

Hence we can choose $\theta\in[0,2\pi)$ such that with $a = a_\theta$

$$(\|au-ua\|_\phi^\#)^2 < 2\pi\delta .$$

For $\sigma_1,\sigma_2\in\mathbf{R}$, $|e^{i\sigma_1}-e^{i\sigma_2}|\le|\sigma_1-\sigma_2|$. Using formulas (1), (2) and the fact that $a = \mathrm{Arg}_\theta(w)$ we therefore have

$$\|\exp(isa)u-u\exp(isa)\|_\phi^\# =$$

$$= \Big(\iint_{\mathbf{T}^2}|\exp(is\mathrm{Arg}_\theta(\alpha))-\exp(is\mathrm{Arg}_\theta(\beta))|^2\,d\mu(\alpha,\beta)\Big)^{\frac{1}{2}}$$

$$\le |s|\Big(\iint_{\mathbf{T}^2}|\mathrm{Arg}_\theta(\alpha)-\mathrm{Arg}_\theta(\beta)|^2\,d\mu(\alpha,\beta)\Big)^{\frac{1}{2}}$$

$$= |s|\,\|au-ua\|_\phi^\#,$$

for all $s\in\mathbf{R}$.

Put $h = \exp(\frac{1}{t_0}a)$ and

$$\omega(x) = \frac{1}{\phi(h^{-1})}\,\phi(h^{-1}x), \quad x\in M.$$

Since w belongs to the center of M_ϕ so does h. Therefore ω is a faithful normal state on M, ω commutes with ϕ, and

$$M_\phi \subset M_\omega .$$

Moreover, we have

$$\sigma_t^\omega(x) = h^{-it}\sigma_t^\phi(x)h^{it} = \sigma_t^\phi(h^{-it}xh^{it}), \quad x\in M.$$

Since $h^{it_0} = w$ we get in particular

$$\sigma^{\omega}_{t_0}(x) = x, \quad x \in M.$$

Therefore we can define a conditional expectation E_{ω} of M onto M_{ω} by

$$E_{\omega}(x) = \frac{1}{t_0}\int_0^{t_0} \sigma^{\omega}_t(x)dt, \quad x \in M.$$

Since $\sigma^{\omega}_t(u)-u = \sigma^{\phi}_t(h^{-it}uh^{it}-u)+\sigma^{\phi}_t(u)-u$, and since $h^{-it} = \exp(-i\frac{t}{t_0}a)$, we get for $0 \le t \le t_0$,

$$\begin{aligned}\|\sigma^{\omega}_t(u)-u\|^{\#}_{\phi} &\le \|h^{-it}uh^{it}-u\|^{\#}_{\phi}+\|\sigma^{\phi}_t(u)-u\|^{\#}_{\phi}\\ &= \|h^{-it}u-uh^{-it}\|^{\#}_{\phi}+\|\sigma^{\phi}_t(u)-u\|^{\#}_{\phi}\\ &\le \frac{t}{t_0}\|au-ua\|^{\#}_{\phi}+\delta\\ &\le (2\pi\delta)^{\frac{1}{2}}+\delta .\end{aligned}$$

Therefore we also have

$$\|E_{\omega}(u)-u\|^{\#}_{\phi} \le (2\pi\delta)^{\frac{1}{2}}+\delta .$$

Put $y = E_{\omega}(u)$ and $\delta' = (2\pi\delta)^{\frac{1}{2}}+\delta$. Since M_{ω} is a finite von Neumann algebra the partial isometry in the polar decomposition of y can be extended to a unitary operator $v \in M_{\omega}$. Clearly $y = v|y| = |y^*|v$. Using the inequality $(1-t)^2 \le 1-t^2$ for $t \in [0,1]$ we get

$$\phi((v-y)^*(v-y)) = \phi((1-|y|)^2) \le \phi(1-|y|^2),$$

and

$$\phi((v-y)(v-y)^*) = \phi((1-|y^*|)^2) \le \phi(1-|y^*|^2).$$

Hence

$$(\|v-y\|^{\#}_{\phi})^2 \le \tfrac{1}{2}\phi(2-y^*y-yy^*) = 1-(\|y\|^{\#}_{\phi})^2 .$$

On the other hand

$$\|y\|_\phi^\# \geq \|u\|_\phi^\# - \|u-y\|_\phi^\# \geq 1-\delta'.$$

Thus

$$(\|v-y\|_\phi^\#)^2 \leq 1-(1-\delta')^2 \leq 2\delta'.$$

Therefore

$$\|u-v\|_\phi^\# \leq \|u-y\|_\phi^\# + \|y-v\|_\phi^\# \leq \delta' + (2\delta')^{\frac{1}{2}}.$$

Since δ was arbitrary we have proved Lemma 3.3.

Completion of step 2. Assume that $T(M)$ is dense in $\mathbb{R}$. Let ϕ and ψ be commuting faithful normal positive functionals on M such that there are positive real numbers a and b with

$$a\phi \leq \psi \leq b\phi,$$

and such that

$$Sp(\Delta_\phi) \cap (\tfrac{a}{b}, \tfrac{b}{a}) = \{1\}.$$

We shall prove that

$$\|u\phi u^* - \psi\| \geq \|\phi - \psi\|$$

for every unitary operator $u \in M$. Clearly it is enough to prove the inequality for a strongly dense set of unitaries. Hence by Lemma 3.3 we may assume that there exists a faithful normal state ω on M, ϕ and ω commute, $M_\phi \subset M_\omega$, and such that $u \in M_\omega$. Let ϕ_1 and ψ_1 be the restrictions of ϕ and ψ to M_ω. Since $\omega \circ \sigma_t^\phi = \omega$, M_ω is a σ_t^ϕ-invariant subalgebra of M, and therefore $\sigma_t^{\phi_1}$ is simply the restriction of σ_t^ϕ to M_ω. In particular

$$Sp(\Delta_{\phi_1}) \subset Sp(\Delta_\phi),$$

hence

$$Sp(\Delta_{\phi_1}) \cap (\tfrac{a}{b}, \tfrac{b}{a}) = \{1\}.$$

We have $\psi = \phi(m\cdot)$ for some positive operator $m \in M_\phi$. Since $M_\phi \subset M_\omega$, $\psi_1 = \phi_1(m\cdot)$, so ϕ_1 and ψ_1 also commute. Clearly $a\phi_1 \leq \psi_1 \leq b\phi_1$, so by step 1

$$\|u\phi_1 u^* - \psi_1\| \geq \|\phi_1 - \psi_1\|.$$

Let $E_\omega : M \to M_\omega$ be the conditional expectation for which $\omega \circ E_\omega = \omega$. Since ϕ and ψ can be written in the form

$$\phi = \omega(h\cdot), \quad \psi = \omega(k\cdot),$$

where h and k are positive operators affiliated with M_ω, we have

$$\phi = \phi_1 \circ E_\omega, \quad \psi = \psi_1 \circ E_\omega.$$

Therefore

$$\|\phi - \psi\| = \|(\phi_1 - \psi_1) \circ E_\omega\| = \|\phi_1 - \psi_1\|,$$

which implies that

$$\|u\phi u^* - \psi\| \geq \|\phi - \psi\|.$$

This completes the proof of step 2.

<u>Step 3</u>. Let now M be an arbitrary von Neumann algebra and let ϕ and ψ be normal positive functionals on M which satisfy the condition of Theorem 3.1. We can assume that M acts on a Hilbert space H with a separating and cyclic vector ξ_0 such that $\phi(x) = (x\xi_0, \xi_0)$, $x \in M$. Let G be a countable dense subgroup of $\mathbb{R}$ and let

$$N = M \times_{\sigma^\phi} G$$

be the crossed product of M with the discrete group $\{\sigma_t^\phi : t \in G\}$ of automorphisms. N is the von Neumann algebra on $\ell^2(G,H)$ generated by $\pi(M)$ and $\lambda(G)$, where

$$(\pi(x)\xi)(t) = \sigma_{-t}^{\phi}(x)\xi(t), \quad x \in M, \quad \xi \in \ell^2(G,H),$$

$$(\lambda(s)\xi)(t) = \xi(t-s), \qquad s \in G, \quad \xi \in \ell^2(G,H).$$

For this and the following the reader may consult [7] and [3], see also [11]. Since G is discrete there is a faithful normal conditional expectation ε of N onto $\pi(M)$ such that

$$\varepsilon(\lambda(s)\pi(x)) = \begin{cases} \pi(x) & \text{if } s = 0 \\ 0 & \text{if } s \neq 0 \end{cases}.$$

Put $\tilde{\phi} = \phi \circ \pi^{-1} \circ \varepsilon$. Then $\tilde{\phi}$ is the "dual weight" of ϕ, so we have

$$\sigma_t^{\tilde{\phi}}(\pi(x)) = \pi(\sigma_t^{\phi}(x)), \quad x \in M,$$

$$\sigma_t^{\tilde{\phi}}(\lambda(s)) = \lambda(s), \qquad s \in G.$$

Moreover, the vector $\tilde{\xi}_0 \in \ell^2(G,H)$ given by

$$\tilde{\xi}_0(t) = \begin{cases} \xi_0 & \text{if } t = 0 \\ 0 & \text{if } t \neq 0' \end{cases}$$

is cyclic and separating for N,

$$\tilde{\phi}(y) = (y\xi_0, \xi_0), \quad y \in N,$$

and

$$(\Delta_{\tilde{\phi}}^{it}\xi)(t) = \Delta_{\phi}^{it}\xi(t), \quad \xi \in \ell^2(G,H),$$

where $\Delta_{\tilde{\phi}}^{it}$ is computed with respect to $\tilde{\xi}_0$.

From the above formulas it follows that

$$\sigma_t^{\tilde{\phi}}(y) = \lambda(t)y\lambda(t)^*, \quad t \in G, \quad y \in N.$$

Hence $G \subset T(N)$, whence $T(N)$ is dense in $\mathbb{R}$, and step 2 is applicable. Since $\Delta_{\tilde{\phi}}$ is just an amplification of Δ_{ϕ} it is clear that $sp(\Delta_{\tilde{\phi}}) = sp(\Delta_{\phi})$, so also

$$\operatorname{sp}(\Delta_{\tilde{\phi}}) \cap (\tfrac{a}{b},\tfrac{b}{a}) = \{1\}.$$

Put $\tilde{\psi} = \psi \circ \pi^{-1} \circ \varepsilon$. Then clearly $a\tilde{\phi} \leq \tilde{\psi} \leq b\tilde{\phi}$. Moreover one verifies easily that

$$\pi^{-1} \circ \varepsilon \circ \sigma_t^{\tilde{\phi}} = \sigma_t^{\phi} \circ \pi^{-1} \circ \varepsilon.$$

Indeed, it is easily checked that the formula holds on elements in N of the form $\lambda(s)\pi(x)$, $s \in G$, $x \in M$. Since $\psi \circ \sigma_t^{\phi} = \psi$ it follows that $\tilde{\psi} \circ \sigma_t^{\tilde{\phi}} = \tilde{\psi}$, i.e. $\tilde{\phi}$ and $\tilde{\psi}$ commute. Therefore $\tilde{\phi}$ and $\tilde{\psi}$ also satisfy the conditions of the theorem, whence by step 2 we have

$$\| v\tilde{\phi}v^* - \tilde{\psi} \| \geq \| \tilde{\phi} - \tilde{\psi} \|$$

for all unitaries $v \in N$.

Let $u \in M$ be a unitary operator. Then

$$\pi(u)\tilde{\phi}\pi(u)^* - \tilde{\psi} = (u\phi u^* - \psi) \circ \pi^{-1} \circ \varepsilon.$$

Thus

$$\| u\phi u^* - \psi \| \geq \| \pi(u)\tilde{\phi}\pi(u)^* - \tilde{\psi} \| \geq \| \tilde{\phi} - \tilde{\psi} \| = \| \phi - \psi \|.$$

This completes the proof of Theorem 3.1.

The proof of the main theorem follows from section 2 and the following result.

__Corollary 3.4__. Let M be a σ-finite factor of type III_λ, $0 < \lambda \leq 1$. Then

$$\operatorname{diam}(S_0(M)/\operatorname{Int}(M)) \geq 2\,\frac{1-\lambda^{\frac{1}{2}}}{1+\lambda^{\frac{1}{2}}}.$$

__Proof__. For $\lambda = 1$ there is nothing to prove.

Suppose $0 < \lambda < 1$. Then we can choose a faithful normal state

ϕ on M such that

$$Sp(\Delta_\phi) = \{\lambda^n : n \in \mathbb{Z}\} \cup \{0\}.$$

Thus $Sp(\Delta_\phi) \cap (\lambda, \lambda^{-1}) = 1$. Moreover, the centralizer M_ϕ of ϕ is a type II_1 factor [4, 4.2.6]. Hence we can choose a projection $p \in M_\phi$ such that

$$\phi(p) = \frac{1}{1+\lambda^{\frac{1}{2}}}.$$

Put $m = \lambda^{\frac{1}{2}} p + \lambda^{-\frac{1}{2}}(1-p) \in M_\phi$. Then $\lambda^{\frac{1}{2}} 1 \leq m \leq \lambda^{-\frac{1}{2}} 1$, and $\phi(m) = 1$. Thus

$$\psi(x) = \phi(mx), \quad x \in M$$

defines a normal state on M such that ϕ and ψ commute, and $\lambda^{\frac{1}{2}}\phi \leq \psi \leq \lambda^{-\frac{1}{2}}\phi$. By Theorem 3.1 it follows that

$$\| u\phi u^* - \psi \| \geq \| \phi - \psi \|$$

for every unitary operator u in M. Let ϕ_1 and ψ_1 be the restrictions of ϕ and ψ to M_ϕ. Since ϕ is a trace on M_ϕ we can identify $(M_\phi)_*$ with $L^1(M_\phi, \phi_1)$. Therefore

$$\| \phi - \psi \| \geq \| \phi_1 - \psi_1 \| = \phi_1(|1-m|) = 2\,\frac{1-\lambda^{\frac{1}{2}}}{1+\lambda^{\frac{1}{2}}},$$

proving the corollary when $0 < \lambda < 1$.

Finally if $\lambda = 0$ we can for every $\mu \in (0,1)$ choose a faithful normal state ϕ such that

$$Sp(\Delta_\phi) \cap (\mu, \mu^{-1}) = \{1\}.$$

As in [4, 3.2.7] one gets that the centralizer of ϕ is a type II_1 von Neumann algebra with diffuse center. Hence we can choose a projection $p \in M_\phi$ such that

$$\phi(p) = \frac{1}{1+\lambda^{\frac{1}{2}}}.$$

Arguing as above we get that

$$\operatorname{diam}(S_0(M)/\operatorname{Int}(M)) \geq 2\,\frac{1-\mu^{\frac{1}{2}}}{1+\mu^{\frac{1}{2}}},$$

so in the limit as $\mu \to 0$ we find that the diameter is (at least) 2. The proof is complete.

References

1. W. Arveson, On groups of automorphisms of operator algebras, J. Fnal. Anal., 15 (1974), 217-243.
2. J. Bion-Nadal, Espace des états normaux d'une facteur de type III_λ, $0<\lambda<1$, et d'un facteur de type III_0, Canadian J. Math. (to appear).
3. O. Bratteli and U. Haagerup, Unbounded derivations and invariant states, Comm. Math. Phys., 59 (1978), 79-95.
4. A. Connes, Une classification des facteurs de type III, Ann. Ec. Norm. Sup., 6 (1973), 133-252.
5. A. Connes, Classification of injective factors, Ann. Math., 104 (1976), 73-115.
6. A. Connes and E. Størmer, Homogeneity of the state space of factors of type III_1, J. Fnal. Anal., 28 (1978), 187-196.
7. U. Haagerup, On the dual weights for crossed products of von Neumann algebras, II, Math. Scand., 43 (1978), 119-140.
8. R. Powers, Representations of uniformly hyperfinite algebras and their associated von Neumann rings, Ann. Math., 86 (1967), 138-171.
9. S. Sakai, The theory of W^*-algebras, Lecture notes, Yale University Press, New Haven, 1962.
10. M. Takesaki, Tomita's theory of modular Hilbert algebras and its applications, Lecture Notes in Math., No. 128, 1970.
11. M. Takesaki, Duality for crossed products and the structure of von Neumann algebras of type III, Acta Math., 131 (1973), 249-310.

Alain Connes
Institut des Hautes Etudes Scientifiques
35, route de Chartres
91440-Bures-sur-Yvette, France

Uffe Haagerup
Matematisk Institut
Odense Universitet
DK-5230 Odense, Denmark

Erling Størmer
Matematisk institutt
Universitetet i Oslo
Blindern, Oslo 3, Norway

PROBLEMS IN QUANTUM FIELD THEORY AND IN OPERATOR ALGEBRAS (*)

S. Doplicher

Dipartimento di Matematica, Università di Roma "La Sapienza"

I - 00185 Roma / Italy

ABSTRACT

We review briefly the theory of superselection quantum numbers and exact inter-
al symmetries in the algebraic formulation of Quantum Field Theory; we discuss some
roblems, and the developments they lead to in the theory of Operator Algebras.

. Local Quantum Theory and Superselection Structure.

In Quantum Field Theory the basic first principles can be formulated in terms of observable quantities. Most characteristic is the principle of locality, asserting that we can assign to each bounded region in space-time [1)] a C*-algebra, generated by all observables that can be measured within that region.

The correspondence so obtained

1) $\mathcal{O} \to \mathfrak{A}(\mathcal{O})$

preserves inclusions and maps mutually spacelike double cones into commuting subalgebras of $\mathfrak{A} = \overline{\bigcup_{\mathcal{O}} \mathfrak{A}(\mathcal{O})}$.

Relativistic invariance is expressed by an action α of the inhomogeneous Lorentz group (the Poincaré group) $\mathcal{P}$ by automorphisms of $\mathfrak{A}$ s.t.

$$\alpha_L(\mathfrak{A}(\mathcal{O})) = \mathfrak{A}(L\mathcal{O}).$$

It is also assumed that there is a pure state ω_0 on $\mathfrak{A}$ describing the physical vacuum in the sense that ω_0 is Poincaré invariant, i.e. $\omega_0 \circ \alpha_L = \omega_0$, and, if (π_0, U_0) is the covariant representation of $\{\mathfrak{A}, \alpha\}$ determined by ω_0, U_0 is strongly continuous and its restriction to translation subgroup $\mathbb{R}^4$ has spectrum in the forward light cone $\overline{V}^+$ (i.e. the energy is positive and the vacuum is a ground state) [1].

The vector states of π_0, i.e. the states $\{\omega_{0B} = \omega_0(B^* \cdot B) / B \in \mathfrak{A}, \omega_0(B^*B) = 1\}$ form the so called vacuum superselection sector. The set of all superselection sectors is a subset $\hat{\mathfrak{A}}_S$ of the spectrum $\hat{\mathfrak{A}} = \mathrm{Irr}(\mathfrak{A})/\simeq$ of $\mathfrak{A}$ obtained as follows.

---

1) for convenience we consider only double cones, i.e. regions $\mathcal{O} = x_1 + \overline{V}^+ \wedge x_2 - \overline{V}^+$ with $x_2 - x_1$ in the future light cone $V^+ = \{x \in \mathbb{R}^4 / x_0 > (x_1^2 + x_2^2 + x_3^2)^{1/2}\}$.

(*) Research supported by Ministero della Pubblica Istruzione and CNR-gnafa.

One selects by general requirements a subset $S \subset S(\mathfrak{a})$ of elementary states (which in the end should describe the collisions of finitely many elementary particles); then $\hat{\mathfrak{a}}_S = S \cap P(\mathfrak{a})/\approx$.

A first natural restriction is to consider only states which are normal relative to π_0 when restricted to each subalgebra $\pi(\mathcal{O})$, $\mathcal{O}$ a double one. It is then equivalent to assume that $\pi_0(\mathfrak{a}(\mathcal{O}))$ is weakly closed for each double cone $\mathcal{O}$.

Dropping the symbol π_0, it is often assumed, moreover that

(2) $\mathfrak{a}(\mathcal{O}) = \mathfrak{a}(\mathcal{O}')'$

where $\mathfrak{a}(\mathcal{O}')$ is the C*-subalgebra of $\mathfrak{a}$ generated by all observables located in double cones spacelike separated from the given double cone $\mathcal{O}$. The duality assumption (2) may also be replaced [24] by the weaker requirement that $\mathfrak{a}(\mathcal{O}')'$ is a local net; in this form it can be proved to hold whenever there is an underlying Wightman field theory [2].

Next, we consider states which are essentially localized perturbations of the vacuum (i.e. $\|\omega-\omega_0|\mathfrak{a}(\mathcal{O}')\|$ is small for sufficiently large $\mathcal{O}$), and have continuous orbits under the action α of Poincaré transformations.

It has been shown that these requirements lead to the following rich structure for $\hat{\mathfrak{a}}_S$ [3].

Consider those endomorphisms ρ of $\mathfrak{a}$ which are localized (in the sense that $\rho|\mathfrak{a}(\mathcal{O}'_\rho)$ = identity, for some double cone $\mathcal{O}_\rho$), irreducible, and Poincaré covariant; the subset of $\hat{\mathfrak{a}}$ they determine is precisely $\hat{\mathfrak{a}}_S$.

Saturating this set of endomorphisms with products and subrepresentations of products one obtains a semigroup Δ_S of localized morphism with the following properties (here $I = \Delta_S \cap \mathrm{Int}(\mathfrak{a})$ is the group of all automorphisms of $\mathfrak{a}$ induced by a unitary in some $\mathfrak{a}(\mathcal{O})$.

(i) Δ_S/I is a commutative semigroup;

(ii) each $\xi \in \Delta_S/I$ is the direct sum of finitely many irreducible components in Δ_S/I;

(iii) there is a unique involution of Δ_S/I, $\xi \to \bar{\xi}$, which commutes with direct sum decompositions and s.t. $\xi \cdot \bar{\xi}$ contains the identity morphism as a subrepresentation;

(iv) there is a remarkable homomorphism of Δ_S/I into the multiplicative semigroup $\mathbb{Z}\setminus\{0\}$

$$\xi \in \Delta_S/I \to \lambda(\xi)^{-1} = \pm d(\xi)$$

such that $\lambda(\xi) = \lambda(\bar{\xi})$ and, if $\xi \in \Delta_S/I$, $\xi = \oplus_j \xi_j$, $\xi_j \in \Delta_S/I$, then $\mathrm{sign}\,\lambda(\xi_j) = \mathrm{sign}\,\lambda(\xi)$ and $d(\xi) = \sum_j d(\xi_j)$.

(v) $d(\xi) = 1$ iff the endomorphisms in the class ξ are automorphisms.

As a consequence of the spectrum condition in the vacuum sector it follows that each sector (each representation in Δ_S) obeys the spectrum condition.

In physical terms, $\xi \to \bar{\xi}$ is the symmetry between particle-antiparticle quantum numbers, the integer $d(\xi)$ is the order of the parastatistics in the sector ξ, and the sign of $\lambda(\xi)$ determines its Bose or Fermi character; $d(\xi) = 1$ iff ξ obeys the ordinary Bose or Fermi statistics. (cf. [3]).

To be more precise, Lorentz covariance is assumed here to simplify the exposition, but plays essentially no role. To prove the absence of sectors ξ with $\lambda(\xi) = 0$ (infinite statistics) requires to assume that $\bar{\xi}$ (which, if $\lambda(\xi) = 0$, a priori is a representation in a wider class[3,I]) is also covariant and fulfills the spectrum condition[3,II]. Alternatively, it suffices to assume that Δ_s/I is generated by those representations where massive particles isolated from the continuum appear[4,5].

It is a remarkable result [4] that the latter assumption [1], without any a priori hypothesis on the space behaviour of states, selects a family of representations of $\mathfrak{a}$, described by endomorphisms of a covering C*-algebra, which are localized in spacelike cones (that is, in infinitely extended regions rather than in double cones), and for which all the mentioned results apply.

This extension of the theory is physically relevant since it may cover massive non abelian gauge theories. Charges in this wider class which do not belong to the former one might be viewed as topological charges and their appearence might be related to the confinement problem in the gauge theories of nuclear forces [4].

It might be remarked that the set Δ_s/I has a structure analogous to that of the dual object of a compact group. Indeed, assume that a "field algebra" F is given, together with an action α of a compact group G (the gauge or internal symmetry group) by automorphisms of F and of each $F(\mathcal{O})$ [2] in such a way that, for each double cone $\mathcal{O}$, $\mathfrak{a}(\mathcal{O})$ is the set of fixed points in $F(\mathcal{O})$ under α_G i.e.

3) $$\pi(\mathfrak{a}(\mathcal{O})) = F(\mathcal{O})^{\alpha_G}$$

π being a representation of $\mathfrak{a}$ associated to F, which contains all the superselections sectors, cf.[6,I]). The vacuum state and the action of the (covering of the) Poincaré group are assumed to extend to F with the same properties as above; the vacuum is a pure gauge invariant state over F and the associated unitary repesentation U of G is continuous.

Then one can prove that Δ_s/I is isomorphic to the set of unitary equivalence classes of those representations of G which are obtained from the unitary irreducible

---

1. these authors take the different point of view that the covariance of the representation and the spectrum condition on the associated energy-momentum operator is the basic selection criterium. For attempts in this direction earlier to the developments mentioned here see [27] (cf., however,[6..I]).

2. it is assumed here too that $F(\mathcal{O}_1) \subset F(\mathcal{O}_2)$ if $\mathcal{O}_1 \subset \mathcal{O}_2$, but no commutativity at spacelike distances is required a priori, since generic elements of F are not observable quantities.

ones by tensor products and subrepresentations. The isomorphism carries products in Δ_s/I onto tensor products, the order onto the dimensionality, the involution (iii) onto complex conjugation, and intertwines direct sum decompositions.

Moreover, for each $\rho\varepsilon\Delta_s$ localized in O, with class $\hat{\rho}=\xi\varepsilon\Delta_s/I$, there is a $d(\xi)$-dimensional Hilbert space of isometries $H(\rho)\varepsilon I(\mathcal{O})$ such that

(4) $\psi\pi(A)=\pi(\rho(A))\psi;\ \psi\ H(\rho),\ A\varepsilon\mathfrak{a}.$

The family of all such $H(\rho)$'s, with ρ localized in $\mathcal{O}$, generates $F(\mathcal{O})$. The isomorphism between Δ_s/I and $\hat{G}$ assigns to $\xi\varepsilon\Delta_s/I$ the unitary equivalence class of the (unitary!) representation $\alpha|H(\rho)$ for any $\rho\varepsilon\Delta$ of class ξ. Actually there is a functorial correspondence between Δ_s and Rep G carrying the intertwining operators between ρ_1 and $\rho_2, \rho_i\varepsilon\Delta_s$ (these intertwining operator belong always to some local algebra in $\mathfrak{a}$ [3]) to the intertwining operators between $\alpha|H(\rho_1)$ and $\alpha|H(\rho_2)$ [7].

An outstanding problem is the following: is it always possible to construct from the net $\mathcal{O}\to\mathfrak{a}(\mathcal{O})$ and the associated semigroup Δ_s, a "field algebra" F and a compact "gauge" group G, with an action α of G on F, having all the properties above and normal commutation or anticommutation of spacelike fields? The latter property means that, if $\rho_1, \rho_2\ \varepsilon\Delta_s$ are localized in spacelike separated double cones,

$$\psi_1\psi_2=\pm\psi_2\psi_1$$

for each $\psi_i\varepsilon H(\rho_i)$, the (-) sign occuring if both $\lambda(\hat{\rho}_i)$ are negative, the (+) sign occuring otherwise.

It was shown in [6.II] that this problem has a unique solution whenever $d(\xi)=1$ for all $\xi\varepsilon\Delta_s/I$. In this case G is the dual of the discrete abelian group $\Delta_s/I=\hat{G}$; in the special case where $\hat{G}$ is finitely generated, F is the covariance algebra of $\mathfrak{a}$ with an action α of G by localized automorphisms. For general abelian $\hat{G}$, F can be viewed as a covariance algebra of $\mathfrak{a}$ by the action of an extension of $\hat{G}$ by a unitary group of $\mathfrak{a}$ [6.II].

A solution to the problem above exists also in general. The C*-algebra F is obtained by a new construction, which extends that of a covariance algebra to some actions by endomorphisms, and provides a concept of crossed product of a C*-algebra by a Cuntz algebra [8]. These results provide an example where duality theorems for compact groups play a constructive role.

We conclude this section mentioning some stability problems. The semigroup homomorphism

5) $\rho\varepsilon\Delta_s \to \hat{\rho}\varepsilon\Delta_s/I \to d(\hat{\rho})\varepsilon\mathbb{N}\setminus\{0\}$

btained by (iv) above modulo $\mathbb{Z}_2$, can be viewed as a non commutative version of he index of a Fredholm operator. Are there deformation invariance properties of the ap (5) - can we identify it with topological invariants?

We could also ask the same question for the map

6) $\rho\varepsilon\Delta_s \to \hat{\rho}\varepsilon\Delta_s/I \to \mathrm{sign}(\lambda(\hat{\rho}))\varepsilon\mathbb{Z}_2$,

hich distinguishes (para)Bose from (para)Fermi elements.

From the physical point of view, natural candidates for deformations leaving 6) fixed are to be searched within the class of perturbation of the dynamics. amely the local algebras and the representation α of the Poincaré group should ndergo a deformation induced by the perturbation $P_o \to P_o+\lambda V$ of the generator f U_o in the x_o -direction, λ a real parameter. Very little is known about how he theory would change , but for a characterization of the perturbed α' for ounded V's [25]. However, it is clear that the correspondence $\mathcal{O} \to \mathfrak{A}(\mathcal{O})$ is hanged drastically ; such deformations are likely to change the map (5).

Since the map (5) is determined only by the correspondence $\mathcal{O} \to \mathfrak{A}(\mathcal{O})$ one can sk for deformations like local homotopic equivalence of the maps $\mathcal{O}\to\mathfrak{A}(\mathcal{O})$. . These eformations would have a less transparent physical interpretation than the perturations of dynamics, but are likely to leave the map (5) invariant. One could also sk for KK theoretic generalization of local homotopic equivalence [26].

A geometric interpretation of the superselection structure as the first <u>local ohomology</u> of the theory has been given by J.E. Roberts [24], motivated by the search, ong proposed by R. Haag, of structural distinctions between theories with only hort range forces, to which our former discussion applies, and Quantum Electrodyamics or general Gauge Theories.

The first local cohomology might be thought of as the analogue in our context of he topological index of an elliptic pseudodifferential operator. Unfortunately othing is known about its homotopic invariance properties.

. <u>Local Charges</u>

We may think of the superselection structure described in Section 1, namely the rreducible elements in Δ_s/I , as the point spectrum of global charge operators. In agrangean theories, if e.g. Δ_s/I is a finitely generated torsion free group (i.e. here are no parastatistics and $\hat{G}$ is a direct product of finitely many copies f $\mathbb{Z}$; in physical terms, Δ_s/I is generated by finitely many additive independent harges) then such global observables are the space integral of local densities, hich are observable (Wightman) fields. Can we derive a similar structure from first principles (without postulating any Lagrangian, etc.)?

As a first step in that direction we discuss the problem of constructing observables which are the analogues of space integrals over finite volumes of regularized densities.

An additonal basic assumption, that we discuss in the next section, allows to solve the problem above and to find the much richer structure we now describe [9,10].

Suppose the superselection structure is embed in a field net $\mathcal{O} \to F(\mathcal{O})$ as described in Section 1.

Say that the gauge group action α has the local implementation property if, for all double cones $\mathcal{O}_1$, $\mathcal{O}_2$ with $\mathcal{O}_1 \subset \dot{\mathcal{O}}_2$ (the interior of $\mathcal{O}_2$), we can find unitaries $V_g \varepsilon F(\mathcal{O}_2)$, $g \varepsilon G$, such that

(i) $g \varepsilon G \to V_g \varepsilon F(\mathcal{O}_2)$ is a unitary continuous representation;

(ii) $\mathrm{Ad} V_g | F(\mathcal{O}_1) = \alpha_g | F(\mathcal{O}_1)$, $g \varepsilon G$;

(iii) $\alpha_h (V_g) = V_{hgh^{-1}}$; $g, h \varepsilon G$.

The global charge operators generate the center of $\{U_G\}''$, where $U_g F\Omega = \alpha_g(F)\Omega, F \varepsilon F; (\Omega;\Omega) = \omega_o$. By (ii) the desired local charge operators are the correponding generating elements of the center A of $\{V_G\}''$. Indeed, by (i) and (iii) $A \subset F(\mathcal{O}_2) \cap U_G' = \mathfrak{a}(\mathcal{O}_2)$, i.e. these local charges are local observables; moreover by (ii) $A \subset \mathfrak{a}(\mathcal{O}_1)' \cap \mathfrak{a}(\mathcal{O}_2)$.

The case where G is a connected Lie group is physically important. This will be the case whenever Δ_s/I has no finite stable subset and is finitely generated (under the operations of taking products and subrepresentations of products); for, these conditions on G characterize the closed connected subgroups of U(n) [11; 28.21, 30, 38] and closed subgroups of Lie groups are Lie groups [12; II, Thm 2.3].

In this case properties (i) (ii) (iii) have an infinitesimal version. Namely the generators of V provide a representation J of the Lie algebra L(G) by self adjoint operators affiliated to $F(\mathcal{O}_2)$ which is gauge covariant and induces on $F(\mathcal{O}_1)$ the appropriate derivations, the infinitesimal gauge transformations; these relations hold on the appropriate dense domains [10]. This structure is a rigorous version of the <u>current algebra hypothesis</u> of Particle Physics. Conversely if there is a local current algebra in the above sense, for each pair of double cones $\mathcal{O}_1, \mathcal{O}_2, \mathcal{O}_1 \subset \dot{\mathcal{O}}_2$, the local implementation property holds for the action $\tilde{\alpha}$ of the simply connected covering of G [10].

The Casimir operators associated to the local current algebra J provide the local charge operators.

In the next section we discuss the general assumptions which imply the local implementation property. We end this section discussing a special case.

In a theory with no parastatistic, Δ_s/I is a (discrete) commutative group (cf. Section 1) and its dual G is a commutative compact group. If the unitaries V_g, $g \varepsilon G$, fulfill conditions (i) (ii) (iii) above, they provide a continuous unitary epresentation of G with values in $\mathfrak{a}(\mathcal{O}_1)' \cap \mathfrak{a}(\mathcal{O}_2)$. The existence of the esired representation is characterized in this case entirely in terms of the net of lgebras of observables; we briefly sketch this point.

The automorphisms localized in $\mathcal{O}_1$ induce automorphisms of $\mathfrak{a}(\mathcal{O}_1)' \cap \mathfrak{a}(\mathcal{O}_2)$ nd this action is trivial for inner automorphisms. If $\Delta_s(\mathcal{O}_1)$ meets all classes of $_s/I$ (i.e. if charges carry no "minimal length"), we get an action τ of $=\Delta_s/I$ on $\mathfrak{a}(\mathcal{O}_1)' \cap \mathfrak{a}(\mathcal{O}_2)$; the desired representation $V_g \in \mathfrak{a}(\mathcal{O}_1)' \cap \mathfrak{a}(\mathcal{O}_2)$, $g \in G$, can be characterized by

$$\tau_\xi(V_g) = \langle \xi, g \rangle V_g; \qquad g \in G, \ \xi \in \hat{G}.$$

Therefore the local implementation property is in this case equivalent to the ssertion that for each pair of double cones $\mathcal{O}_1 \subset \mathcal{O}_2$, the action τ is dominant in he sense of Connes and Takesaki [28] (the relative commutant $\mathfrak{a}(\mathcal{O}_1)' \cap \mathfrak{a}(\mathcal{O}_2)$ is lways properly infinite [29,9]).

So far it has been possible to prove that τ is dominant only using the assump-ions of the next section [9], which allow also the derivation of the local implemen-ation property, hence of local charges and current algebra, also in presence of arastatistics [10]. An important application of the same technique yields also the xistence of a local implementation of the Poincaré group transformations for a uitable neighbourhood of the identity in the group, by a unitary representation of with values in a local algebra.

This provides the existence of the analogues of the space integrals of regula-ized energy-momentum density fields [13] (cf. next Section).

. Standard and Split Inclusions of von Neumann Algebras and their Applications.

Let $R_1 \subset R_2$ be von Neumann algebras acting on a Hilbert space H and $\Omega \in H$ unit vector; the triple $\Lambda = (R_1, R_2, \Omega)$ is called a standard W* inclusion if Ω s cyclic and separating for R_1, R_2 and $R_1' \cap R_2$; Λ is called split if there is ype I factor N such that

$$R_1 \subset N \subset R_2.$$

The theory of standard and split W* inclusions has been developed in [14], moti-ated by the problems described in the foregoing Section [10]. An important appli-ation to solve some classical problems in operator algebras has been given in [15]. ere we limit ourselves to mentioning some of the results and developments, and how hey apply to Quantum Field Theory.

The first result is that for each standard and split W* inclusion Λ we can hoose an intermediate type I factor N_Λ so that the correspondence $\Lambda \to N_\Lambda$ is unctorial. It follows that $\mathrm{Aut}(\Lambda)$ leaves N_Λ globally stable and is a compact etrizable subgroup of $\mathrm{Aut}(R_2)$, equipped with the topology of strong convergence n the predual; it is a discrete subgroup in the uniform topology.

The natural factor N_Λ can be constructed in several ways. Let $\omega(B)= =(\Omega,B\Omega)$, $B\in B(H)$. By the standard split assumption, there is a faithful normal state φ over $R_1 \vee R_2'$ s.t. $\varphi(ab')=\omega(a)\omega(b')$, $a\in R_1$, $b'\in R_2'$. Let η_Λ be the unique vector in the natural cone $P^{\#}_{R_1'\wedge R_2,\Omega}$ (see e.g. [16]) inducing φ; η_Λ is cyclic and separating for $R_1'\wedge R_2$ and the correspondence $\Lambda \to \eta_\Lambda$ is functorial. Let U_Λ be the unitary defined by

$$(1) \qquad U_\Lambda ab'\eta_\Lambda=(a\otimes b')\Omega\otimes\Omega; \qquad a\in R_1,\ b'\in R_2';$$

we define a normal endomorphism ψ_Λ of $B(H)$ by

$$(2) \qquad \psi_\Lambda(B)=U_\Lambda^{-1}(B\otimes I)U_\Lambda, \qquad B\in B(H),$$

then N_Λ is defined by

$$(3) \qquad N_\Lambda=\psi_\Lambda(B(H)).$$

Alternatively, N_Λ can be defined as $R_1 \vee \{P_{\eta_\Lambda}\}$ where P_{η_Λ} is the orthogonal projection onto $[R_2'\eta_\Lambda]$ (compare [17]).

Moreover, let J be the canonical involution associated to $R_1'\wedge R_2,\Omega$ by the Tomita-Takesaki theory. If R_1 and R_2 are factors, we have also

$$(4) \qquad N_\Lambda=R_1 \vee JR_1J=R_2 \wedge JR_2J.$$

By varying the standard vector Ω for a fixed split inclusion $R_1\subset R_2$, the factor N_Λ runs through all the intermediate type I factors N s.t. $R_1'\wedge N$ and $N'\wedge R_2$ are properly infinite. The case where two standard vectors Ω_1 and Ω_2 yield the same natural factor can be explicitely characterized.

One can see that N_Λ is not the only type one factor globally stable under $\mathrm{Aut}(\Lambda)$; let, however, $\mathrm{Aut}(\Lambda)$ denote the group of all automorphisms of R_2 leaving R_1 globally stable which are induced by a unitary which maps $P^{\#}_{R_1'\wedge R_2,\Omega}$ onto itself. Then, N_Λ is globally stable under $\mathrm{Aut}(\Lambda)$ also and, if R_1, R_2 are factors not both of type I, N_Λ is the only intermediate factor N with this property and such that $N'\wedge R_2$, $R_1'\wedge N$ are properly infinite. Under the above restriction, $\mathrm{Aut}(\Lambda)$ has trivial centralizer in $\mathrm{Aut}(R_1,R_2)$.

If Λ is standard but not split, the second equality in (3) might still hold; we say then that Λ is pseudonormal and define N_Λ by (3). One can see that N_Λ can be a von Neumann algebra of any type.

An outstanding problem is to characterize pseudonormalcy by purely algebraic roperties of the inclusion $R_1 \subset R_2$ ([14] ; see 18 for further developments).

If Λ is split and standard, so are the inclusions (R_1, N_Λ, Ω) and N, R_2, Ω); by iteration we get a monotone family of type I factors N_d, d a iadic rational, interpolating between R_1 and R_2 and canonically associated to Λ here $N_0 = N_\Lambda$). Defining

5) $N_{-\infty} = \wedge N_d \quad N_{+\infty} = \vee N_d$,

e get two AFD von Neumann algebras, also canonically associated to Λ, s.t.

6) $$R_1 \subset N_{-\infty} \subset \cdots \subset N_d \subset \cdots \subset N_{+\infty} \subset R_2.$$

The same process is possible if Λ is standard, non split, but, in an obvious sense, pseudonormal at each step. In this case $N_{\pm\infty}$ are not necessarily AFD unless ne of R_1, R_2 is. However, R. Longo has proved the remarkable result that if $R_1' \wedge R_2$ s a factor, $R_1' \wedge N_{-\infty} = N_{+\infty}' \wedge R_2 = \mathbb{C}I$.

This theorem entails the existence of an AFD factor with trivial relative commutant in any infinite factor with separable predual; this solves important problems in the theory of operator algebras ([15] ; see also [19] for independent related results).

With Λ a standard W* inclusion, which is pseudonormal at each step, when does the chain (6) extend to a continuous one parameter family, with $N_{-\infty} = R_1$, $N_{+\infty} = R_2$? This would provide a sort of deformation equivalence between R_1 and R_2. It is interesting to note that in some examples from free field theory this happens, and the interpolating von Neumann algebras N_λ, $\lambda \in \mathbb{R}$, have a natural interpretation (in these examples, R_1, R_2 and $N_\lambda, \lambda \in \mathbb{R}$, are isomorphic to the III_1 Krieger factor).

We conclude this survey coming back to the problems of Section 2. We say that a theory has the split property if, for each pair of double cones O_1, O_2 with $O_1 \subset \dot{O}_2$, the W* inclusion $(F(O_1), F(O_2), \Omega)$ is standard and split.

The "standard" part of this assumption is related to field commutation or anti-commutation at spacelike distances and to the classical Reeh-Schlieder theorem [10].

The existence of intermediate type I factors is the real restriction. It has been proved to hold for free field theories ([17]; see also [20]; other references in [14]).

For a general theory, D. Buchholz and E. Wichmann have shown that the split property follows from physically meaningful conditions on the spectral properties of localized states [21] . These conditions are related to the Haag-Swieca compactness criterion [22] and are fulfilled e.g. by the free massive field theories [21]. It would be important to know to what extent they are implied, in a general theory, by assumptions like asymptotic completeness and the occurrence of finitely many stable particles [1)] (or else: the sequence of particle masses should diverge sufficiently

1. incidentally, such assumptions would also imply that the gauge group is a Lie group ([6,I] and comments above).

fast).

The split property has the virtue of excluding physically undesirable models, as the infinite tensor products of copies of the same theory, or the generalized free fields (unless the Källen-Lehmann measure is purely atomic and the sequence of masses diverges sufficiently fast, accumulating at most at infinity) [14].

At the level of observables, the split property can be equivalently formulated as a principle of local preparation of states. Namely, $(\mathfrak{a}(\mathcal{O}_1), \mathfrak{a}(\mathcal{O}_2), \Omega)$ is split for each $\mathcal{O}_1$, $\mathcal{O}_2$ as above, iff the following holds. Given any normal state ω on $\mathfrak{a}(\mathcal{O}_1)$ there is a projection (i.e. yes-no measurement) E_ω in $\mathfrak{a}(\mathcal{O}_2)$ such that filtering any normal state φ of $\mathfrak{a}(\mathcal{O}_2)$ (s.t. $\varphi(E_\omega)\neq 0$) with the question E_ω yields a state φ_{E_ω} (where as known $\varphi_{E_\omega}(B)=\varphi(E_\omega)^{-1}\varphi(E_\omega BE_\omega)$ which coincides with ω when restricted to $\mathfrak{a}(\mathcal{O}_1)$ [13]. For the relations between the split properties for fields resp. for observables, see [9,14,13].

It would of great importance to see whether, conversely to some of the above comments, the split property has consequences for the validity of asymptotic completness.

Coming back to the local implementation property, we sketch how it follows from the split property [10]

With $\Lambda=(F(\mathcal{O}_1),F(\mathcal{O}_2),\Omega)$, $\mathcal{O}_1\subset\dot{\mathcal{O}}_2$ double cones, let U_g, $g\in G$ be the global implementation of the gauge group and ψ_Λ as in equation (2). We have by the canonicity of the construction of ψ_Λ

(7) $\quad \psi_\Lambda\circ\alpha=\alpha\circ\psi_\Lambda, \quad \alpha\in \mathrm{Aut}(\Lambda).$

Moreover, by definition

$$\psi_\Lambda(F)=F, \quad F\in F(\mathcal{O}_1);$$

(8)

$$\psi_\Lambda(B(H))=N_\Lambda\subset F(\mathcal{O}_2).$$

By equations (7,8) one can see that the desired representation V fulfilling the conditions (i), (ii), (iii) of Section 2 is given by

$$V_g=\psi_\Lambda(U_g), \quad g\in G.$$

Similarly, if the theory is Poincaré invariant and $\tilde{U}_L$, $L\in\tilde{P}$ is the continuous unitary representation of the covering of the Poincaré group which implements the transformations of the fields, one gets a local implementation of $\tilde{P}$ as follows. Let $\mathcal{O}_1$, $\mathcal{O}_2$ be double cones as above and $\mathcal{O}$ a double cone, $N(e)$ a neighbourhood of e in $\tilde{P}$, such that $L\mathcal{O}\subset\mathcal{O}_1$, $L\in N(e)$. Define

(9) $\quad \tilde{V}_L=\psi_\Lambda(\tilde{U}_L), \quad L\in\tilde{P}.$

ince $\tilde{U}_L$ commutes with the gauge transformations U_g (cf. eg. [14]), by (7)
$_g(V_L)=\tilde{V}_L$; $L\in\tilde{P}$, $g\in G$; hence

$$\tilde{V}_L \in \mathfrak{a}(O_2)$$

nd clearly, $\tilde{V}$ is a strongly continuous representation of $\tilde{P}$ s.t.

$$\mathrm{Ad}\ \tilde{V}_L | F(O) = \alpha_L | F(O), \quad L \in N(c).$$

The existence of the functorial choice of the intermediate type I factors allows o reformulate local quantum theories with the split property, in terms of "quasi-ocal nets" of type I factors. The new description is equivalent to the usual one 10].

In this connection one should recall that the local von Neumann algebras are xpected to be always III_1 factors - at least, this is the case for double cones in ree field theory and, for special regions like a wedge, in a general theory (see 23] for a survey of such results).

cknowledgements. This report was written while visiting the II. Institut für heoretische Physik, University of Hamburg. It is a pleasure to thank Professors). Buchholz and R. Haag for the kind hospitality.

eferences

1 R. Haag, D. Kastler: "An Algebraic Approach to Quantum Field Theory", Journ. Math. Phys. 5, 848 (1964)

2 J.J. Bisognano, E. Wichmann: "On the Duality Condition for a Hermitian Scalar Field", Journ. Math. Phys. 16, 907 (1975)

3 S. Doplicher, R. Haag, J.E. Roberts: "Local Observables and Particle Statistics" I, Commun. Math. Phys. 23, 199 (1971), II, Commun. Math. Phys. 35, 49 (1974)

4 D. Buchholz, K. Fredenhagen: "Locality and the Structure of Particle States", Commun. Math. Phys. 84, 1 (1982)

5 K. Fredenhagen: "On the Existence of Antiparticles", Commun. Math. Phys. 79, 141 (1981)

6 S. Doplicher, R. Haag, J.E. Roberts: "Field, Observables and Gauge Transformations", I, Commun. Math. Phys. 13, 1 (1969); II, Commun. Math. Phys. 15, 173 (1969)

7 S. Doplicher, J.E. Roberts: "Field, Statistics and Non Abelian Gauge Groups", Commun. Math. Phys. 28, 331 (1972)

8 S. Doplicher, J.E. Roberts: in preparation

9 S. Doplicher: "Local Aspects of Superselection Rules", Commun. Math. Phys. 85, 73 (1982)

10 S. Doplicher, R. Longo: "Local Aspects of Superselection Rules", II, Commun. Math. Phys. 88, 399 (1983)

11 E. Hewitt, K.A. Ross: "Abstract Harmonic Analysis", Vol. II, Springer-Verlag, Berlin - Heidelberg - New York, 1970

12 S. Helgason: "Differential Geometry and Symmetric Spaces", Academic Press, New York and London, 1962

13 D. Buchholz, S. Doplicher, R. Longo: in preparation

14 S. Doplicher, R. Longo: "Standard and Split Inclusions of von Neumann Algebras", Inventiones Math., to appear

15 R. Longo: "Solution of the Factorial Stone-Weierstrass conjecture. An Application of the Theory of Standard Split W*-inclusions", Inventiones Math., to appear

16 H. Araki: "Recent Developments in the Theory of Operator Algebras", Symposia Mathematicea XX, Istituto Nazionale di Alta Mathematica; Academic Press, New York 1976

17 D. Buchholz: "Product States for Local Algebras", Commun. Math. Phys. 36, 287 (1974)

18 R. Longo: "Remarks on Pseudonormaley", these Proceedings.

19 S. Popa: "Constructing MASA's in Factors", Inventiones Math., to appear

20 C. D'Antoni, R. Longo: "Interpolation by Type I Factors and the Flip Automorphisms", Journ. Functional Analysis 51, 361 (1983)

21 D. Buchholz, E. Wichmann: "Causal Independence and Energy-Level Density of Localized States in Quantum Field Theory", preprint

22 R. Haag, A. Swieca: "When does a Theory describe Particles?", Commun. Math. Phys. 1, 308 (1965)

23 R. Longo: "Algebraic and Modular Structure of von Neumann Algebras of Physics", Proceedings A.M.S. Summer Institute on Operator Algebras, Kingston 1981

24 J.E. Roberts: "Local Cohomology and Superselection Structure", Commun. Math. Phys. 51, 107 (1976);
"Net Cohomology and its Applications to Field Theory". In: "Quantum Fields, Algebras, Processes", L. Streit ed.; Springer, Wien, New York (1980)

25 D. Buchholz, J.E. Roberts: "Bounded Perturbations of Dynamics", Commun. Math. Phys. 49, 161 (1976)

26 J. Cuntz: "Generalized Homomorphisms between C*-Algebras and KK-Theory", preprint

27 H.J. Borchers: "Local Rings and the connection between Spin and Statistics", Commun. Math. Phys. 1, 281 (1965)

28 A. Connes, M. Takesaki: "The Flow of Weights on Factors of Type III", Tohoku Math. J. 29, 473 (1977)

29 H.J. Borchers: "A Remark on a Theorem of Misra", Commun. Math. Phys. 4, 315 (1967)

QUASI-PRODUCT STATES ON C*-ALGEBRAS

David E. Evans

Mathematics Institute, University of Warwick,
Coventry CV4 7AL, England

BSTRACT

We introduce and study a class of Markov measures, which we call quasi-product easures, on compact totally disconnected path spaces, and consider the induced states, alled quasi-product states on the associated unital AF algebras and the infinite *-algebras $\mathcal{O}_A$ associated with a topological Markov chain A. For product spaces, and HF algebras these are precisely product measures and product states respectively. n particular, we give sufficient conditions which ensure that the gauge group is eakly outer in certain quasi-product weights on the stablised C*-algebra of $\mathcal{O}_A$.

1. INTRODUCTION

Here we introduce a notion of quasi-product state on AF algebras, which generalises that of product states on UHF algebras. They induce a class of states on the *-algebras $\mathcal{O}_A$ by projecting onto the gauge invariant AF subalgebra. We consider the uestion of weak implementability of gauge automorphisms in the stablised algebras of $_A$ in associated weights, as was done in [1] for the algebras of the full shift [4,6]. his necessitates a discussion on factoriality and equivalence of quasi-product states t the AF-level which is done in §§3 and 4. Most of these results were announced withut proof in [8].

2. PRELIMINARIES

A vector $u = (u_i)$ in $\mathbb{R}^q$ will be regarded as a column vector, so that u^{tr} denotes he corresponding row vector. If $v = (v_i)$, $u = (u_i) \in \mathbb{R}^q$, $\langle v,u\rangle$ will denote $v^{tr}u = \Sigma v_i u_i$. If $B = [b_{ij}]$ is a matrix (not necessarily square) we say B is positive and rite $B \geq 0$ (respectively B is strictly positive and write $B > 0$) if $B(i,j) \geq 0$ for ll i,j (respectively $B(i,j) > 0$ for all i,j). A square matrix $A = [a_{ij}]$ with $A \geq 0$ s said to be aperiodic if there exists k such that $A^k > 0$. If A is an aperiodic atrix, θ the spectral radius, then by the Perron Frobenius theory [12] the right and eft eigenvectors u,v associated with (A,θ) (i.e. $Au = \theta u$, $A^{tr}v = \theta v$) are unique up to calar multiplication. Moreover, u,v can be taken strictly positive in which case hey are called right and left Perron eigenvectors respectively, and if they are ormalised so that $\langle u,v\rangle = 1$, then

$$A^k/\theta^k \to uv^{tr} \quad \text{as } k \to \infty . \tag{2.1}$$

§3. QUASI-PRODUCT MEASURES

Let $\{q_j\}_{j=1}^{\infty}$ be a sequence of strictly positive integers, and $\Sigma_j = \{1,\ldots,q_j\}$ denote the jth state space, with $|\Sigma_j| = q_j$. If $1 \le m \le n \le \infty$, let

$$\Sigma_{[m,n]} = \prod_{j=m}^{n} \Sigma_j, \quad \Sigma_{[m]} = \Sigma_{[1,m]}.$$

Let $A = \{A_j\}_{j=1}^{\infty}$ be a sequence of matrices, where A_j is a $q_j \times q_{j+1}$ matrix with zero-one entries. For $1 \le m \le n \le \infty$, let

$$\Omega_{[m,n]} = \Omega^A_{[m,n]} = \{x = (x_i)_{i=m}^{n} \in \Sigma_{[m,n]} : A_j(x_j,x_{j+1}) = 1,\ j = m,\ldots,n-1\}$$

$$\Omega_m = \Omega_{[1,m]}$$

and $\qquad \Omega = \Omega^A = \Omega^A_{\infty}$

all endowed with the relative product topology. If $m' \le m$, there are canonical chopping off maps from $\Sigma_{[m]}$ to $\Sigma_{[m']}$ and Ω_m to $\Omega_{m'}$, all denoted by ω. Then $\Omega = \varprojlim \Omega_m$, a direct limit over finite indices.

Let $\Gamma = \Gamma_A$ denote the group of uniformly finite dimensional homeomorphisms which damage only finitely many coordinates. Thus if $\ell \ge 1$, let Γ_ℓ denote the homeomorphisms ϕ of Ω such that

$$\phi(x)_j = x_j \text{ if } j > \ell, \quad x = (x_j) \in \Omega.$$

Then $\qquad \Gamma = \bigcup_\ell \Gamma_\ell.$

If $q_j = q$, a constant integer, and $A_j = A$, a constant matrix, we write $\Sigma = \Sigma_j$, $\Omega_A = \Omega_A$, $\Gamma_A = \Gamma_A$, and let $\sigma = \sigma_A$ denote the shift on Ω:

$$\sigma(x)_j = x_{j+1}, \quad x = (x_j) \in \Omega.$$

For each $j \ge 1$, let

$u_j, v_j \in [0,\infty)^{|\Sigma_j|}$ (identified with column vectors as in §2), and

$\rho_j \in [0,\infty)^{|\Sigma_j|}$ (identified with diagonal matrices) satisfy

$$\langle \rho_1 u_1, v_1 \rangle = 1 \tag{3.1}$$

$$A_j \rho_{j+1} u_{j+1} = u_j \tag{3.2}$$

$$A_j^{tr} \rho_j v_j = v_{j+1}. \tag{3.3}$$

A family (A,u,v,ρ) satisfying (3.1) - (3.3) is said to be a *quasi-product system* nd often abbreviated as (u,v,ρ). Note that it is sufficient to be given A, $u_j)_{j=1}^{\infty}$, $(\rho_j)_{j=1}^{\infty}$ and v_1 satisfying (3.1) and (3.2), and then use (3.3) to define v_j or $j > 1$.

OTATION 3.1

If $\rho_j(i) \in \mathbb{C}$, $i \in \Sigma_j$, $j = m,\ldots,n$ and $a = (a_m,\ldots,a_n) \in \Sigma_{[m,n]}$, let

$$\rho(a) = \rho_m(a_m)\ldots\rho_n(a_n)A_m(a_m,a_{m+1})\ldots A_{n-1}(a_{n-1},a_n) \tag{3.4}$$

If $a = (a_m,\ldots,a_n) \in \Sigma_{[m,n]}$, the cylinder $[a]$ is defined as

$$[a] = \{x = (x_i) \in \Omega : x_i = a_i,\ i = m,\ldots,n\}. \tag{3.5}$$

rom the data (A,u,v,ρ) we define a Borel probability measure μ on Ω by defining it n the cylinders:

$$\mu([a]) = \rho(a)\ v_m(a_m)u_n(a_n) \tag{3.6}$$

le call μ a quasi-product measure, and also denote it by $\mu(A,u,v,\rho)$ or $\mu(u,v,\rho)$.

EMARK 3.2

If $u_j > 0$ for all j, then $\mu(A,u,v,\rho)$ is a Markov measure. Let $p_1 = (\rho_1(i)u_1(i)v_1(i):i \in \Sigma_1)$ denote the initial distribution, and P_n denote the transition probabilities

$$P_n(i,j) = A_n(i,j)\rho_{n+1}(i)u_{n+1}(j)/u_n(i) \tag{3.7}$$

for $i \in \Sigma_n$, $j \in \Sigma_{n+1}$. Then for $a = (a_1,\ldots,a_n)$,

$$\mu([a]) = p_1(a_1)P_1(a_1,a_2)\ldots P_{n-1}(a_{n-1},a_n) \tag{3.8}$$

EXAMPLE 3.3

Suppose $A_j = u_j v_{j+1}^{tr}$, and ρ_j satisfies $\langle v_j,\rho_j u_j\rangle = 1$. Then (A,u,v,ρ) is a quasi-product system. In particular, if

$$u_j = (1,\ldots,1)^{tr} = v_j \quad \text{(of length } q_j) \tag{3.9}$$

then μ on $\Omega = \Pi\ \Sigma_j$ is the product measure $\Pi\hat{\rho}_j$, where $\hat{\rho}_j$ is the measure $\hat{\rho}_j(i) = \rho_j(i)$ on Σ_j. Thus on product spaces, the class of quasi-product measures are precisely all product measures. No more, no less.

We now describe how to construct the unital AF algebra from the sequence A, and how quasi-product systems gives rise to a class of states on these AF algebras. Let $A = \{A_j\}_{j=1}^{\infty}$ be a sequence of $\{q_j \times q_{j+1}\}$ matrices, with zero-one entries. If $\alpha = (\alpha_m,\ldots,\alpha_n) \in \Sigma_{[m,n]}$, $\beta = (\beta_{n+1},\ldots,\beta_\ell) \in \Sigma_{[n+1,\ell]}$, we let $\alpha\beta$ denote the sequence $(\alpha_m,\ldots,\alpha_n,\beta_{n+1},\ldots,\beta_\ell)$ in $\Sigma_{[m,\ell]}$. If $j \in \Sigma_n$, let

$$\Omega_n^j = \{\mu = (\mu_i)_{i=1}^n : \mu_n = j\}$$

so that $|\Omega_n^j| = \sum_{i\in\Sigma_1} (A_1A_2\ldots A_{n-1})(i,j)$

and $\Omega_n = \bigcup_j \Omega_n^j$.

Let $F_n = \bigoplus_j M(|\Omega_n^j|)$

where $M(p)$, for $p \in \mathbb{N}$ denotes the full $p \times p$ complex matrix algebra. Then $F_n^j = M(|\Omega_n^j|)$ is generated by matrix units $\{e_{\mu\nu} : \mu,\nu \in \Omega_n^j\}$. Define a homomorphism $\phi_n : F_n \to F_{n+1}$ by

$$\phi_n(e_{\mu\nu}) = \sum_{q\in\Sigma_{n+1}} A_n(j,q)\, e_{\mu q,\nu q} \tag{3.10}$$

for $\mu,\nu \in \Omega_n^j$, i.e. ϕ_n is given by the matrix A_n^{tr}. Hence

$$F_A = \lim(F_n, \phi_n)$$

is an AF algebra with dimension group

$$\lim_{\to} (\mathbf{Z}^{|\Sigma_n|}, A_n^{tr})$$

and dimension range $\lim_{\to} D_n$, if

$$D_n = \{x \in \mathbf{Z}^{|\Sigma_n|} : 0 \le x \le (A_{n-1}^{tr}\ldots A_1^{tr})(u)\}$$

where $u = (1,\ldots,1)^{tr}$ (of length $q_1 = |\Sigma_1|$) and $\mathbf{Z}^q$ is given the simplical ordering:

$$\mathbf{Z}_+^q = \{(x_i)_{i=1}^q \in \mathbf{Z}^q : x_i \ge 0, \ \forall i\} .$$

Let $\quad e_\mu = e_{\mu\mu}, \quad \mu \in \Omega_n$

and $\quad C_n = C^*(e_\mu : \mu \in \Omega_n)$

$$\cong C(\Omega_n).$$

hen $C_n \subset F_n \cap C_{n+1}$, and if we let

$$C_A = \lim (C_n, \phi_n|C_n)$$

e have $C_A \simeq C(\Omega_A)$, and is a regular masa in F_A. In fact C_A gives a diagonalisation f F_A in the sense of Stratila and Voiculescu [13]. (See [8] for details.)

It is convenient to make sense of $e_{\mu\nu}$ for $\mu,\nu \in \Sigma_{[m,n]}$. First we define $e_{\mu\nu} = 0$ for $\mu = (\mu_m,\ldots,\mu_n)$, $\nu = (\nu_m,\ldots,\nu_n) \in \Sigma_{[m,n]}$ if either $\mu_n \neq \nu_n$, or $\mu \notin \Omega_{[m,n]}$ r $\nu \notin \Omega_{[m,n]}$. Then for $\mu\nu \in \Sigma_{[m,n]}$ put

$$e_{\mu\nu} = \Sigma\, e_{\theta\mu,\theta\nu}\ , \quad e_\mu = e_{\mu\mu}$$

here the summation is over all θ in $\Sigma_{[1,m-1]}$. There is a conditional expectation P rom F_A onto C_A such that

$$P(e_{\nu\nu'}) = e_\nu\ \delta_{\nu\nu'}\ , \quad \nu,\nu' \in \Sigma_{[m,n]}\ . \tag{3.11}$$

ence a quasi-product system (A,u,v,ρ) determines a state on F_A, which we call a uasi-product state and denote by $\phi = \phi(A,u,v,\rho)$ or $\phi(u,v,\rho)$, so that $\phi = \mu\circ P$. Then is given by

$$\phi(e_{\nu\nu'}) = \rho(\nu)v_m(\nu_m)v_n(\nu_n)\delta_{\nu\nu'} \tag{3.12}$$

or $\nu = (\nu_m,\ldots,\nu_n)$, $\nu' \in \Sigma_{[m,n]}$.

Now let A be a $q \times q$ zero-one matrix, and F_A, C_A the AF algebra and masa associated ith the constant sequence $\{A\}_{j=1}^{\infty}$. We extend the one-sided shift σ_A on Ω_A, or C_A, to he AF algebra F_A as follows. If $\mu \in \Sigma_{[m,n]}$, let $\sigma(\mu)$ denote the same sequence regarded s an element of $\Sigma_{[m+1,n+1]}$. We define a completely positive unital ap $\sigma = \sigma_A$ on F_A by

$$\sigma(e_{\mu\nu}) = e_{\sigma(\mu),\sigma(\nu)} \quad \mu,\ \nu \in \Sigma_{[m,n]}. \tag{3.13}$$

hen $\sigma(f(x)) = f(\sigma(x))$, for $f \in C(\Omega_A) \subset F_A$, $x \in \Omega_A$, and so σ on F_A extends the shift n Ω_A. Note that if $F_A(i)$ denotes the hereditary C*-subalgebra $e_i F_A e_i$, $i \in \Sigma$, then he restriction of σ^n to $F_A(i)$ is a homomorphism for each $n \geq 0$, $i \in \Sigma$.

We define an action β of the torus $\mathbb{T}^q$ on F_A as follows. If $t = (t_i)_{i=1}^q \in \mathbb{T}^q$, $\mu = (\mu_m,\ldots,\mu_n)$, $\nu = (\nu_m,\ldots,\nu_n) \in \Sigma_{[m,n]}$. Let

$$\beta(t)(e_{\mu\nu}) = t_{\mu_m} \cdots t_{\mu_n}\ \overline{t_{\nu_m} \cdots t_{\nu_n}}\ e_{\mu\nu} \tag{3.14}$$

and let H_A denote the fixed point algebra $F_A^{\mathbb{T}^q}$, which is an AF algebra and shift invariant. For the full 2-shift, $q = 2$, $A(i,j) \equiv 1$, F is Fermion algebra $\bigotimes_1^\infty M_2$, H the current algebra, and it is well known that extermal traces on the current algebra arise as the restrictions of Powers' states, or shift invariant product states on the Fermion algebra. The motivation for studying quasi-product states came from the following generalisation [7] of this to aperiodic matrices:

PROPOSITION 3.4

Suppose A is an aperiodic $q \times q$ matrix, such that for any $i,j \in \Sigma$, there exist k_1, $k_2 \in \Sigma$ with

$$A(i,k_1),\ A(j,k_1),\ A(k_2,i),\ A(k_2,j) > 0. \tag{3.15}$$

Let ψ be an extremal shift invariant faithful tracial state on H_A. Then there exist u in $(0,\infty)^q$ and a strictly positive diagonal $q \times q$ matrix ρ such that

$$A\rho u = u \tag{3.16}$$

$$A^{tr}\rho v = v \tag{3.17}$$

where $v = (1,\ldots,1)^{tr}$, (length q) (3.18)

$$\langle \rho v, u\rangle = 1 \tag{3.19}$$

and ψ is the restriction to H_A of the quasi-product state on F_A determined by the system

$$((A)_{j=1}^\infty,\ (u)_{j=1}^\infty,\ (v)_{j=1}^\infty,\ (\rho)_{j=1}^\infty). \tag{3.20}$$

The proof of this is similar to its two sided version in [7] except that the shift is now merely a completely positive map instead of being an automorphism. Thus one needs an appropriate modification of the theory of the chemical potential which was the first step in [7]. First let ϕ be any extremal shift invariant extension of ψ to a state on F_A. Then one has the clustering property

$$\sum_{m=1}^{n} \frac{1}{n}\phi(a\ \sigma^m(b)) \to \phi(a)\phi(b) \tag{3.21}$$

for all $a,b \in F_A$ and

$$\sum_{m=1}^{n} \frac{1}{n}\phi(a\ \sigma^m(b)\ a'\ \sigma^m(b')) \to \phi(aa')\phi(bb') \tag{3.22}$$

if $a,a' \in F_A$, $b,b' \in F_A(i)$, $i \in \Sigma$.

etting $G_\phi = \{t: \phi\,\beta(t) = \phi\}$, one sees readily from (3.15) and (3.22) that since ψ is aithful that $C_i^2 = \{G_\phi \ni t \to \phi(a\beta(t)(b)) : a,b \in F_A(i)$ is dense in $C(G_\phi)$. We let the eader fill in the remaining details of the modification [2] (or see [3]). Once that as been established, the remainder of the proof is as in the two sided case [7]. ote that (3.17-18) is a consequence of shift invariance.

Now F_A is the AF algebra constructed (essentially as a crossed product) from e dynamical system (Ω_A,Γ_A),(see [5,8] for details). Similarly H_A is the AF algebra etermined from a dynamical system (Ω_A,Θ_A), where Θ_A is the subgroup of Γ_A consisting f homeomorphisms which *permute* finitely many coordinates. That is to say, $\Theta = \bigcup_\ell \Theta_\ell$, here Θ_ℓ consists of those homeomorphisms h in Γ_ℓ such that for each $= (x_1,x_2,\ldots,x_\ell,\ldots)$ in Ω_A, $(h(x)_1,\ldots,h(x)_\ell)$ is a permutation of $(x_1,\ldots,x_\ell)$. learly any quasi-product measure arising from (3.20) which satisfies (3.16-19) is -invariant. Conversely, Proposition 3.4 characterises (certain) Θ-invariant easures on Ω_A as quasi-product measures.

4. ERGODIC QUASI-PRODUCT MEASURES

Let (A,u,v,ρ) be a quasi-product system, and Γ the group of uniformly finite imensional homeomorphisms of the path space Ω. If $u_j,v_j,\rho_j > 0$ for all j, the ssociated quasi-product measure is clearly quasi-invariant under Γ. To show rgodicity, we need the following aperiodic type criterion, which is an asymptotic ank one-type property (c.f. [12, Chapter 3]).

YPOTHESIS 4.1

Given $\varepsilon > 0$, there exists n_0, t_0 such that for all $t \geq t_0$, $i \in \Sigma_t$, $j \in \Sigma_{t+n_0+1}$:

$$|(A_t\,\rho_{t+1}\,A_{t+2}\cdots\rho_{t+n_0}\,A_{t+n_0})(i,j) - u_t(i)v_{t+n_0+1}(j)| < \varepsilon\; u_t(i)v_{t+n_0+1}(j) \tag{4.1}$$

For example in the situation of Example 3.3 one has the equality:

$$A_t\rho_{t+1}A_{t+2}\cdots\rho_{t+n_0}A_{t+n_0} = u_t v^{tr}_{t+n_0+1}.$$

ore interesting examples can be constructed with the aid of the following obser- ation.

EMMA 4.2

Suppose $\{B_j\}_{j=1}^\infty$ is a sequence of matrices in $M_q(R^+)$ such that $\lim_{\to} B_j$ exists and is an aperiodic matrix B. Let $\bar{u}$, $\bar{v}_0$ be the right and left Perron eigenvectors of B, normalised so that $\langle\bar{u},\bar{v}_0\rangle = 1$, and suppose the Perron eigenvalue of B is 1. Then given $\varepsilon > 0$, there exists t_0, m_0 such that for all $t \geq t_0$:

$$\| B_t \cdots B_{t+m_0} - \bar{u}\,\bar{v}_0^{tr} \| < \varepsilon . \tag{4.3}$$

PROOF

This is clear because $\lim_{\to} B^j = \bar{u}\,\bar{v}_0^{tr}$ by §2.

PROPOSITION 4.3

Let (A,u,v,ρ) be a quasi-product system such that $(|\Sigma_j|, A_j)$ is constant, say (q,A) for all large j. Suppose $\lim_{\to} u_j$, $\lim_{\to} v_j$ exist and are strictly positive, say $\bar{u}$, $\bar{v}$ respectively, and $\lim A_j \rho_{j+1}$ exists and is an aperiodic matrix B say, with left Perron eigenvector $\bar{v}_0 > 0$, such that $A^{tr}\,\bar{v}_0 = \bar{v}$, and normalised so that $\langle\bar{v}_0,\bar{u}\rangle = 1$. Then Hypothesis 4.1 holds.

PROOF

We have $B\bar{u} = \bar{u}$, $\bar{u} > 0$, so that $\bar{u}$ is a right Perron eigenvector for B, with Perron eigenvalue 1. By Lemma 4.2, given $\varepsilon' > 0$, there exists t_0, m_0 such that for $t \geq t_0$

$$\| A_t \rho_{t+1} \cdots A_{t+m_0} \rho_{t+m_0+1} - \bar{u}\,\bar{v}_0^{tr} \| < \varepsilon' .$$

Hence

$$\| A_t \rho_{t+1} \cdots A_{t+m_0} \rho_{t+m_0+1} A_{t+m_0+1} - \bar{u}(A^{tr}\bar{v}_0)^{tr} \| < \varepsilon' \quad \|A\|$$

and so the result follows from $A^{tr}\,\bar{v}_0 = \bar{v}$.

Although we have restricted our attention to zero-one matrices, the extension to matrices with positive integral entries is straightforward using symbol splitting, (see [8] for details). We then illustrate the use of Proposition 4.3 with the following example.

EXAMPLE 4.4

Let $q_j \equiv 3$, $A_j \equiv \begin{pmatrix} 2 & 1 & 0 \\ 1 & 1 & 1 \\ 0 & 1 & 2 \end{pmatrix} = A$

$$\bar{u} = u_j = (1,1,1)^{tr}$$
$$= \bar{v} = v_j$$
$$\bar{v}_0 = (1,2,1)^{tr}/3.$$

hen $\langle\bar{v}_0,\bar{u}\rangle = 1$, and $A\bar{u} = 3\bar{u}$.

Let $\underline{X} = \{(x_1,x_2,x_3)^{tr} : x_i \geq 0,\ x_1 = x_3,\ 2x_1 + x_2 = 1\}$. Then $A\rho\bar{u} = \bar{u}$, $A^{tr}\rho\bar{v} = \bar{v}$, $\langle\bar{u},\rho\bar{v}\rangle = 1$ for all ρ in $\underline{X}$, identified as diagonal matrices. Let $\{\rho_j\}_{j=1}^{\infty}$ be a sequence n $\underline{X}$ such that $\lim\limits_{\to j} \rho_j = (1/3,\ 1/3,\ 1/3)^{tr}$. Then $((A_j),\ (u_j),\ (v_j),\ (\rho_j))$ is a quasi-roduct system satisfying (3.1) - (3.3). Moreover $\lim A_j\rho_{j+1} = A/3$, which has $\bar{u}$, $\bar{v}_0$ s correctly normalised right and left Perron eigenvectors. Moreover $A^{tr}\bar{v}_0 = \bar{v}$ so that ll the conditions of Proposition 4.3 hold.

Product measures on $\prod\limits_{j=1}^{\infty} \Sigma_j$ are ergodic under finitely many changes of coordinates see e.g. [11]). More generally:

HEOREM 4.5

If (A,u,v,ρ) is a quasi-product system with $u,v,\rho > 0$ satisfying Hypothesis 4.1, hen the quasi-product measure $\mu(A,u,v,\rho)$ is Γ-ergodic.

ROOF

By [13, Proposition 1.3.14] it is enough to show that if $\varepsilon > 0$, and $f \in C(\Omega)$, here exists $\ell > 0$ such that for all Γ_ℓ-invariant g in $C(\Omega)$:

$$\left|\int fg\,d\mu - \int f\,d\mu \int g\,d\mu\right| \leq \varepsilon \|g\|_{C(\Omega)} \tag{4.4}$$

hoose by Hypothesis 4.1, n_0, t_0 such that (4.1) holds. It is enough to verify (4.4) hen $f \in C(\Omega)$ is of the form:

$$f(\beta) = \begin{cases} 0 & \text{if } (\beta_j)_{j=1}^{t} \neq (\alpha_j)_{j=1}^{t} \\ 1 & \text{otherwise} \end{cases}$$

here $\beta = (\beta_1,\beta_2,\ldots) \in \Omega$, and $\alpha = (\alpha_1,\ldots,\alpha_t) \in \Omega_t$ is fixed for some $t \geq t_0$. Now ake $s > \ell = t + n_0$, and suppose g is Γ_ℓ invariant, and depends only on the first s oordinates in Ω. Then

$$g(\beta) = \phi(\beta_{\ell+1},\ldots,\beta_s)$$

or $\beta = (\beta_1,\beta_2,\ldots) \in \Omega$, and some function ϕ on $\Omega_{[\ell+1,s]}$. Then

$$\int f\,d\mu = \rho(\alpha)v_1(\alpha_1)u_t(\alpha_t)$$

$$\int g\,d\mu = \Sigma\phi(\beta_{\ell+1},\ldots,\beta_s)\rho(\beta_{\ell+1},\ldots,\beta_s)v_{\ell+1}(\beta_{\ell+1})u_s(\beta_s)$$

here the summation is over all $(\beta_{\ell+1},\ldots,\beta_s)$ in $\Omega_{[\ell+1,s]}$, and

$$\int fg\,d\mu = \Sigma\phi(\beta_{\ell+1},\ldots,\beta_s)\rho(\alpha_1,\ldots,\alpha_t,\beta_{t+1},\ldots,\beta_s)v_1(\alpha_1)u_s(\beta_s)$$

where the summation is over all $(\beta_{t+1},\ldots,\beta_s)$ in $\Omega_{[t+1,s]}$. Then

$$\left|\int fg\,d\mu - \int f\,d\mu \int g\,d\mu\right|$$

$$= |\Sigma\phi(\beta_{\ell+1},\ldots,\beta_s)v_1(\alpha_1)u_s(\beta_s)\rho(\alpha)\rho(\beta_{\ell+1},\ldots,\beta_s)$$

$$(\Sigma A_t(\alpha_t,\beta_{t+1})\rho(\beta_{t+1},\ldots,\beta_\ell)A_\ell(\beta_\ell,\beta_{\ell+1})-u_t(\alpha_t)v_{\ell+1}(\beta_{\ell+1}))|$$

$$\leq \Sigma|\phi(\beta_{\ell+1},\ldots,\beta_s)|v_1(\alpha_1)u_s(\beta_s)\rho(\alpha)\rho(\beta_{\ell+1},\ldots,\beta_s)$$

$$.\varepsilon \quad u_t(\alpha_t)v_{\ell+1}(\beta_{\ell+1})$$

$$\leq \varepsilon \quad \rho(\alpha)v_1(\alpha_1)u_t(\alpha_t)\int|\phi|d\mu$$

$$\leq \varepsilon \quad \|g\| .$$

Since ε is arbitrary, it follows that (4.4) holds for all Γ_ℓ-invariant g in $C(\Omega)$.

§5. EQUIVALENCE OF QUASI-PRODUCT MEASURES

We compare two quasi-product measures $\mu^1 = \mu(A,u,v,\rho^1)$ and $\mu^2 = \mu(A,u,v,\rho^2)$, where ρ^1 and ρ^2 are two sequences satisfying (2.1) - (2.3) for the same family (A,u,v). Consider the following sequence

$$\chi_n = \sum_\beta \{\mu^1([\beta])\mu^2([\beta])\}^{\frac{1}{2}} \tag{5.1}$$

where the summation is over all $\beta = (\beta_1,\ldots,\beta_n)$ in Ω_n. Then

$$\chi_n = \sum_\beta[\rho^1(\beta)\rho^2(\beta)]^{\frac{1}{2}}\, v_1(\beta_1)u_n(\beta_n) \tag{5.2}$$

and $0 \leq \chi_{n+1} \leq \chi_n \leq 1$. Hence

$$\chi(\mu^1,\mu^2) = \lim_{n\to\infty} \chi_n \text{ exists} \tag{5.3}$$

and $0 \leq \chi(\mu^1,\mu^2) \leq 1$.

An irreducible type hypothesis is clearly required in order to obtain a zero-one law, as for product measures [9]. So as in Hypothesis 4.1, we introduce an asymptotic rank-one hypothesis:

HYPOTHESIS 5.1

There exists $0 < \gamma_n \le 1$ and $w_n \in [0,\infty)^{|\Sigma_n|}$ such that given $\varepsilon > 0$ there exist n_0, t_0 such that for $t \ge t_0$, $i \in \Sigma_t$, $j \in \Sigma_{t+n_0+1}$:

$$(5.4) \qquad |[A_t(\rho^1_{t+1}\rho^2_{t+1})^{\frac{1}{2}}A_{t+1}\cdots(\rho^1_{t+n_0}\rho^2_{t+n_0})^{\frac{1}{2}}A_{t+n_0}](i,j)/\gamma_t\cdots\gamma_{t+n_0-1}$$

$$-u_t(i)w_{t+n_0+1}(j)| < \varepsilon\, u_t(i)v_{t+n_0+1}(j).$$

EXAMPLE 5.2

Suppose $u_j, v_j \in [0,\infty)^{|\Sigma_j|}$,

$$A_j = u_j\, v^{tr}_{j+1} \tag{5.5}$$

and

$$\langle v_j, \rho^i_j u_j\rangle = 1,\ i = 1,2,\quad j = 1,2,\ldots \tag{5.6}$$

as in Example 3.3. Then if

$$\gamma_j = \langle v_{j+1}, (\rho^1_{j+1}\,\rho^2_{j+1})^{\frac{1}{2}}u_{j+1}\rangle$$

we have

$$\gamma_j \le \{\langle v_{j+1}, \rho^1_{j+1}u_{j+1}\rangle + \langle v_{j+1}, \rho^2_{j+1}u_{j+1}\rangle\}/2$$

$$= 1.$$

Moreover

$$A_t(\rho^1_{t+1}\rho^2_{t+1})^{\frac{1}{2}}A_{t+1}\cdots(\rho^1_{t+n_0}\rho^2_{t+n_0})^{\frac{1}{2}}A_{t+n_0} = u_t v^{tr}_{t+n_0+1}\gamma_t\cdots\gamma_{t+n_0-1}$$

so that one can take $w_n \equiv v_n$.

To obtain more interesting examples, consider:

PROPOSITION 5.3

Let (A,u,v,ρ^i), $i = 1,2$ be two quasi-product systems with $(|\Sigma_j|, A_j)$ constant, say (q,A) for all large j. Suppose $\lim_{\to} u_j$ exists and is strictly positive, $\inf_{n,i} v_n(i) > 0$, and that there exist $\gamma_n > 0$ such that

$$(5.7) \qquad A_n(\rho^1_{n+1}\rho^2_{n+1})^{\frac{1}{2}}u_n = \gamma_n u_n \text{ for } n \text{ large}$$

(5.8) $\lim A_j(\rho^1_{j+1}\rho^2_{j+1})^{\frac{1}{2}}/\gamma_j$ exists and is an aperiodic matrix B.

Then Hypothesis 5.1 holds.

PROOF

We have

$$\gamma_n u_n = A_n(\rho^1_{n+1}\rho^2_{n+1})^{\frac{1}{2}} u_{n+1}$$

$$\leq (A_n\rho^1_{n+1}u_{n+1} + A_n\rho^2_{n+1}u_{n+1})/2$$

$$= u_n,$$

and so $\gamma_n \leq 1$.

Let $B_j = A_j(\rho^1_{j+1}\rho^2_{j+1})^{\frac{1}{2}}/\gamma_j$

$B = \lim\limits_{\to} B_j$, $\bar{u} = \lim\limits_{\to} u_j$. Then $B\bar{u} = \bar{u}$, and so $\bar{u}$ is a right Perron eigenvector for B, and the Perron eigenvalue is 1. Let $\bar{v}_0$ denote the left Perron eigenvector of B, normalised so that $\langle \bar{v}_0, \bar{u} \rangle = 1$. If $\varepsilon' > 0$, choose by Lemma 4.2, t_0, m_0 such that for all $t \geq t_0$:

$$\| A_t(\rho^1_{t+1}\rho^2_{t+1})^{\frac{1}{2}} \cdots A_{t+m_0}(\rho^1_{t+m_0+1}\rho^2_{t+m_0+1})^{\frac{1}{2}}/\gamma_t \cdots \gamma_{t+m_0}$$

$$- \bar{u}\,\bar{v}_0^{tr} \| < \varepsilon'.$$

Thus Hypothesis 5.1 holds if we take

$$w_n = A^{tr}\,\bar{v}_0.$$

THEOREM 5.4

Assume (A,u,v,ρ^i), $i = 1,2$ are two quasi product systems satisfying Hypothesis 5.1. Then

(5.9) $\mu^1 = \mu(A,u,v,\rho^1)$ and $\mu^2 = \mu(A,u,v,\rho^2)$ are mutually absolutely continuous if and only if $\chi(\mu^1,\mu^2) > 0$.

(5.10) μ^1 and μ^2 are singular if and only if $\chi(\mu^1,\mu^2) = 0$.

The proof will be reduced to the following sequence of lemmas. But first some notation. Let

$$\mu_k^i = \omega_*\mu^i \tag{5.11a}$$

e the induced measures on Ω_k, using the chopping maps $\omega:\Omega \to \Omega_k$, and

$$\psi_k = [d\mu_k^1/d\mu_k^2]^{\frac{1}{2}}.$$

EMMA 5.5 [9]

If $\chi = 0$, then μ^1 and μ^2 are singular.

ROOF

$$\int_{\Omega_k} \psi_k d\mu_k^2 = \int_{\Omega_k} [d\mu_k^1/d\mu_k^2]^{\frac{1}{2}} d\mu_k^2 = \chi_k.$$

or any $\varepsilon > 0$, there exists N such that $\chi_N < \varepsilon$. Let

$$B = \{a \in \Omega_k : \psi_N(a) > 1\}.$$

hen

$$\mu_N^2(B) = \int_B 1\, d\mu_N^2 \le \int_B \psi_N\, d\mu_N^2 \le \chi_N < \varepsilon$$

nd

$$\mu_N^1(CB) = \int_{CB} \frac{d\mu_N^1}{d\mu_N^2}\, d\mu_N^2 = \int_{CB} (\psi_N)^2\, d\mu_N^2$$

$$\le \int_{CB} \psi_N\, d\mu_N^2 \le \chi_N < \varepsilon .$$

hus $\mu^2(\omega^{-1}(B))$, $\mu^1(C\omega^{-1}(B)) < \varepsilon$, and so μ^1, μ^2 are singular since ε is arbitrary.

EMMA 5.6

If $\chi > 0$, then given $\varepsilon > 0$, there exist positive integers t_0, n_0 such that for $t \ge t_0$, $s \ge \ell = t + n_0$

$$\Big|\sum_\beta [\rho^1(\beta)\rho^2(\beta)]^{\frac{1}{2}}\gamma_t\cdots\gamma_{\ell-1}w_{\ell+1}(\beta_{\ell+1})u_s(\beta_s) - 1\Big| < \varepsilon$$

where the summation is over all $\beta = (\beta_{\ell+1},\ldots,\beta_s)$ in $\Sigma_{[\ell+1,s]}$.

PROOF

Using Hypothesis 5.1, choose n_0, t_0 such that

$$|[A_t(\rho^1_{t+1}\rho^2_{t+1})^{\frac12}A_{t+1}\cdots(\rho^1_\ell\rho^2_\ell)^{\frac12}A_\ell](i,j)/\gamma_t\cdots\gamma_{\ell-1}$$

$$-u_t(i)w_{\ell+1}(j)| < \varepsilon\, u_t(i)v_{\ell+1}(j)/2$$

for all $i \in \Sigma_t$, $j \in \Sigma_{\ell+1}$, if $t \geq t_0$, $\ell = t+n_0$. Then for $s \geq \ell = t + n_0$:

$$\chi_s = \Sigma[\rho^1(\beta^1)\rho^2(\beta^1)]^{\frac12}A_t(\beta_t,\beta_{t+1})[\rho^1(\beta^2)\rho^2(\beta^2)]^{\frac12}A_\ell(\beta_\ell,\beta_{\ell+1})$$

$$[\rho^1(\beta^3)\rho^2(\beta^3)]^{\frac12}\, v_1(\beta_1)u_s(\beta_s)$$

where the summation is over all

$$\beta^1 = (\beta_1,\ldots,\beta_t),\ \beta^2 = (\beta_{t+1},\ldots,\beta_\ell),\ \beta^3 = (\beta_{\ell+1},\ldots,\beta_s)$$

in $\Sigma_{[1,t]}$, $\Sigma_{[t+1,\ell]}$, $\Sigma_{[\ell+1,s]}$ respectively. Then for all $s \geq \ell = t + n_0$:

$$|\chi_s - \sum_{\beta^1,\beta^3}[\rho^1(\beta^1)\rho^2(\beta^1)]^{\frac12}\, u_t(\beta_t)w_{\ell+1}(\beta_{\ell+1})\gamma_t\cdots\gamma_{\ell-1}$$

$$\cdot[\rho^1(\beta^3)\rho^2(\beta^3)]^{\frac12}\, v_1(\beta_1)u_s(\beta_s)|$$

$$= |\sum_{\beta^1,\beta^3}[\rho^1(\beta^1)\rho^2(\beta^1)]^{\frac12}[\Sigma_{\beta^2} A_t(\beta_t,\beta_{t+1})[\rho^1(\beta^2)\rho^2(\beta^2)]^{\frac12}A_\ell(\beta_\ell,\beta_{\ell+1})/\gamma_t\cdots\gamma_{\ell-1}$$

$$- u_t(\beta_t)w_{\ell+1}(\beta_{\ell+1})].\ \gamma_t\cdots\gamma_{\ell-1}[\rho^1(\beta^3)\rho^2(\beta^3)]^{\frac12}\, v_1(\beta_1)u_s(\beta_s)|$$

$$\leq \sum_{\beta^1,\beta^3}[\rho^1(\beta^1)\rho^2(\beta^1)]^{\frac12}(\varepsilon/2)u_t(\beta_t)v_{\ell+1}(\beta_{\ell+1})[(\rho^1(\beta^3)\rho^2(\beta^3)]^{\frac12}v_1(\beta_1)u_s(\beta_s)$$

$$= (\varepsilon/2)\sum_{\beta^1}[(\rho^1(\beta^1)\rho^2(\beta^1)]^{\frac12}v_1(\beta_1)u_t(\beta_t)\sum_{\beta^3}[\rho^1(\beta^3)\rho^2(\beta^3)]^{\frac12}v_{\ell+1}(\beta_{\ell+1})u_s(\beta_s)$$

$$< \varepsilon/2$$

using Cauchy Schwarz,

i.e.

$$|\chi_s - \chi_t \sum_{\beta^3}[\rho^1(\beta^3)\rho^2(\beta^3)]^{\frac12}w_{\ell+1}(\beta_{\ell+1})u_s(\beta_s)\gamma_t\cdots\gamma_{\ell-1}| < \varepsilon/2.$$

But $\lim \chi_s > 0$, hence the result follows.

<u>LEMMA 5.7</u>

If $\chi > 0$, then given $\varepsilon > 0$, there exist positive integers t_0, n_0 such that

$$|1 - \langle \psi_t, \psi_s \rangle| < \varepsilon$$

if $s \geq t + n_0$, $t \geq t_0$, and where $\{\psi_s\}$ are defined in (5.11) and the inner products are computed in $L^2(\Omega, d\mu^2)$.

PROOF.

Take $s \geq t$, then

$$\langle \psi_s, \psi_t \rangle$$

$$= \sum_{\beta} [\rho^1(\omega\beta)/\rho^2(\omega\beta)]^{\frac{1}{2}} [\rho^1(\beta)\rho^2(\beta)]^{\frac{1}{2}} v_1(\beta_1) u_s(\beta_s)$$

(where the summation is over all $\beta = (\beta_1, \ldots, \beta_s)$ in $\Sigma_{[s]}$, and

$\omega : \Sigma_{[s]} \to \Sigma_{[t]}$ is the chopping map)

$$= \sum_{\beta^1, \beta^2} \rho^1(\beta^1) A_t(\beta_t, \beta_{t+1}) [\rho^1(\beta^2)\rho^2(\beta^2)]^{\frac{1}{2}} v_1(\beta_1) u_s(\beta_s)$$

(where the summation is over all $\beta^1 = (\beta_1, \ldots, \beta_t) \in \Sigma_{[t]}$, $\beta^2 = (\beta_{t+1}, \ldots, \beta_s)$ in $\Sigma_{[t+1,s]}$).

Taking $s \geq \ell \geq t$:

$$\langle \psi_t, \psi_s \rangle$$

$$= \Sigma \rho^1(\beta^1) A_t(\beta_t, \beta_{t+1}) [\rho^1(\beta^2)\rho^2(\beta^2)]^{\frac{1}{2}} A_\ell(\beta_\ell, \beta_{\ell+1})$$

$$[\rho^1(\beta^3)\rho^2(\beta^3)]^{\frac{1}{2}} v_1(\beta_1) u_s(\beta_s)$$

(where the summation is all all $\beta^1 = (\beta_1, \ldots, \beta_t) \in \Sigma_{[t]}$,

$\beta^2 = (\beta_{t+1}, \ldots, \beta_\ell) \in \Sigma_{[t+1,\ell]}$, $\beta^3 = (\beta_{\ell+1}, \ldots, \beta_s) \in \Sigma_{[\ell+1,s]}$).

Then

$$|\langle \psi_t, \psi_s \rangle - \sum_{\beta^1, \beta^3} \rho^1(\beta^1) u_t(\beta_t) w_{\ell+1}(\beta_{\ell+1}) \gamma_t \cdots \gamma_{\ell-1}$$

$$v_1(\beta_1) u_s(\beta_s) [\rho^1(\beta^3)\rho^2(\beta^3)]^{\frac{1}{2}}|$$

$$= |\sum_{\beta^1, \beta^3} (\sum_{\beta^2} A_t(\beta_t, \beta_{t+1}) [\rho^1(\beta^2)\rho^2(\beta^2)]^{\frac{1}{2}} A_\ell(\beta_\ell, \beta_{\ell+1}) / \gamma_t \cdots \gamma_{\ell-1}$$

$$- u_t(\beta_t) w_{\ell+1}(\beta_{\ell+1}))$$

$$\rho^1(\beta^1)[\rho^1(\beta^3)\rho^2(\beta^3)]^{\frac{1}{2}}v_1(\beta_1)u_s(\beta_s)\gamma_t\cdots\gamma_{\ell-1}|$$

$$\leq (\varepsilon/2) \sum_{\beta^1,\beta^3} \rho^1(\beta^1)u_t(\beta_t)v_{\ell+1}(\beta_{\ell+1})[\rho^1(\beta^3)\rho^2(\beta^3)]^{\frac{1}{2}}v_1(\beta_1)u_s(\beta_s)$$

$$= (\varepsilon/2) \sum_{\beta^1} \rho^1(\beta^1)v_1(\beta_1)u_t(\beta_t) \sum_{\beta^3} [\rho^1(\beta^3)\rho^2(\beta^3)]^{\frac{1}{2}}v_{\ell+1}(\beta_{\ell+1})u_s(\beta_s)$$

$$\leq \varepsilon/2$$

if $\ell = t + n_0$, $t \geq t_0$, for some positive integers n_0, t_0 by Hypothesis 5.1. The result now follows from Lemma 5.6.

PROOF OF THEOREM 5.4

Suppose $\chi > 0$. Choose inductively by Lemma 1.10, t_i, n_i, $i = 0,1,2,\ldots$ such that

$$|1-\langle\psi_t,\psi_s\rangle| < 1/2^{i+1}$$

if $t \geq t_i$, $s \geq t + n_i$ with $t_i > t_{i-1}$. Then $f_i = \psi_{t_i}$ is a Cauchy sequence in $L^2(\Omega,d\mu^2)$, because

$$\|f_i - f_j\|^2 = 2(1-\langle f_i,f_j\rangle) \leq 2^{-\min(i,j)}.$$

Let f denote the limit function. If [a] is a finite cylinder, then for large i

$$\mu^1([a]) = \mu^1_{t_i}(a) = \int_{[a]} (\psi_{t_i})^2\, d\mu^2_{t_i} = \int_{[a]} (\psi^2_{t_i})d\mu^2$$

$$= \int_{[a]} (f_i)^2\, d\mu^2 \to \int_{[a]} f^2 d\mu^2 \quad \text{as } i \to \infty .$$

Hence $\mu^1 \prec \mu^2$, and by symmetry, $\mu^2 \prec \mu^1$.

This and Lemma 5.5 completes the proof of Theorem 5.4.

From this we can derive the following practical criterion:

Corollary 5.8.

Suppose (A,u,v,ρ^i) are two quasi-product systems such that

(5.12) $(|\Sigma_j|, A_j)$ are eventually constant, say (q,A), and $\lim_{\to} u_j$, $\lim_{\to} v_j$, $\lim_{\to} \rho^i_j$ exist and are strictly positive, say $\bar{u}$, $\bar{v}$, $\bar{\rho}^i$ respectively.

5.13) There exist $0 < \gamma_j \leq 1$ such that $B_j = A_j(\rho_j^1\rho_j^2)^{\frac{1}{2}}/\gamma_j$ converges to an aperiodic matrix B with Perron eigenvalue 1 and right and left Perron eigenvectors $\bar{u}$ and $\bar{v}_0$ respectively with $<\bar{u},\bar{v}_0> = 1$.

5.14) $A^{tr}\bar{v}_0 = \bar{v}$.

Then $\mu(A,u,v,\rho^1)$ and $\mu(A,u,v,\rho^2)$ are singular unless $\bar{\rho}^1 = \bar{\rho}^2$.

PROOF

The conditions of Theorem 5.4 are satisfied (with $w_n = A^{tr}\bar{v}_0 = \bar{v}$ in the notation of Hypothesis 5.1.) Suppose μ^1 and μ^2 are not singular, then by Theorem 5.4, $\chi > 0$. In which case by Lemma 5.6, given $\varepsilon > 0$, there exist t_0, n_0 such that for $t \geq t_0$, $\ell = t + n_0$, $s = \ell + 1$,

$$\Big|\sum_{\beta}[\rho_s^1(\beta)\rho_s^2(\beta)]^{\frac{1}{2}}\gamma_t\cdots\gamma_{\ell-1}\,\bar{v}(\beta)u_s(\beta) - 1\Big| < \varepsilon$$

where the summation is over all β in Σ_s.

Hence

$$1 - \varepsilon \leq \Sigma\,[\rho_s^1(\beta)\rho_s^2(\beta)]^{\frac{1}{2}}\gamma_t\cdots\gamma_{\ell-1}\bar{v}(\beta)u_s(\beta)$$

$$\leq \Sigma\,[\rho_s^1(\beta)\rho_s^2(\beta)]^{\frac{1}{2}}\,\bar{v}(\beta)u_s(\beta)$$

$$\leq [\Sigma\rho_s^1(\beta)\bar{v}(\beta)u_s(\beta)]^{\frac{1}{2}}[\Sigma\rho_s^2(\beta)\bar{v}(\beta)u_s(\beta)]^{\frac{1}{2}}$$

$$= <\rho_s^1,\bar{v}u_s>^{\frac{1}{2}}<\rho_s^2,\bar{v}u_s>^{\frac{1}{2}}.$$

Thus

$$1 - \varepsilon \leq \Sigma\,[\rho_s^1(\beta)\rho_s^2(\beta)]^{\frac{1}{2}}\bar{v}(\beta)u_s(\beta) \leq <\rho_s^1,\ \bar{v}u_s>^{\frac{1}{2}}<\rho_s^1,\bar{v}u_s>^{\frac{1}{2}}$$

and letting $s \to \infty$,

$$1 - \varepsilon \leq \Sigma\,[\bar{\rho}^1(\beta)\bar{\rho}^2(\beta)]^{\frac{1}{2}}\bar{v}(\beta)\bar{u}(\beta) \leq 1.$$

Now let $\varepsilon \to 0$, to get

$$<(\bar{\rho}^1\bar{\rho}^2)^{\frac{1}{2}},\ \bar{u}\ \bar{v}> = 1.$$

But $<\bar{\rho}^i,\bar{u}\bar{v}) = 1$, and so $\bar{\rho}^1 = \bar{\rho}^2$ by the converse of Cauchy Schwarz.

EXAMPLE 5.9.

If $q_j = 2$, then either A_j is invertible and $\rho_j^1 = \rho_j^2$ or A_j is rank one. To get more interesting examples, consider $q_j \equiv 5$. Let

$$A = \begin{pmatrix} 1 & 1 & 1 & 1 & 0 \\ 1 & 1 & 1 & 1 & 0 \\ 1 & 1 & 1 & 1 & 0 \\ 0 & 1 & 1 & 1 & 1 \\ 0 & 1 & 1 & 1 & 1 \end{pmatrix}$$

which is aperiodic and not rank one. Let

$$X = \{(x_1,x_2,x_3,x_4,x_5)^{tr} \ : \ x_i \geq 0,\ x_1 = 1/4 = x_5, x_2+x_3+x_4 = 3/4\}.$$

Then $A\rho\bar{u} = \bar{u}$, for all $\rho \in X$ if

$$\bar{u} = (1, 1, 1, 1, 1)^{tr}.$$

Let $A_j \equiv A$, $u_j = \bar{u}$, $v_j = 4\bar{u}/5 = \bar{v}$.

If $\rho_j^i \in X$, $i = 1,2$, $j = 1,2,\ldots$, then (A,u,v,ρ^i) are quasi-product systems such that

$$A(\rho_{j+1}^1\ \rho_{j+1}^2)^{\frac{1}{2}}\bar{u} = \gamma_j\ \bar{u}$$

if $\gamma_j = \frac{1}{4} + \sum_{r=2}^{4} [\rho_{j+1}^1(r)\rho_{j+1}^2(r)]^{\frac{1}{2}}$.

Suppose $\rho_j^i \to \bar{\rho}^i$ as $j \to \infty$ for $i = 1,2$. Then $\gamma_j \to \gamma > 0$, and $A[\rho_{j+1}^1\ \rho_{j+1}^2]^{\frac{1}{2}}/\gamma_j$ converges to an aperiodic matrix B with right Perron eigenvector $\bar{u}$. If

$$\bar{\rho}^1(r)\bar{\rho}^2(r) = 1/16$$

for $r \in \{2,3,4\}$, then $B = A/4$, and so a left Perron eigenvector of B is $\bar{v}_0 = \bar{u}/5$, with $\langle\bar{u},\bar{v}_0\rangle = 1$. But then $A\bar{v}_0 = 4\bar{u}/5 = \bar{v}$, and so the conditions of Corollary 5.8 hold . Thus the associated quasi-product measures are singular if $\bar{\rho}_1 \neq \bar{\rho}_2$.

§6. WEAK IMPLEMENTABILITY OF GAUGE AUTOMORPHISMS

Let $A = [A(i,j)]$ be a $q \times q$ aperiodic zero-one matrix. Then $\mathcal{O}_A$[5] is the C*-algebra generated by q non-zero partial isometries $\{S_i : i \in \Sigma\}$ satisfying

$$S_i^*S_i = \sum_{j=1}^{q} A(i,j)S_jS_j^*\ , \quad i \in \Sigma. \tag{6.1}$$

The gauge group $\{\alpha(t):t \in \mathbb{T}\}$ is the group of automorphisms such that

$$\alpha(t)S_i = tS_i \qquad t \in \mathbb{T}, \quad i \in \Sigma. \tag{6.2}$$

The conditional expectation $P = \int \alpha(t)dt$ takes the C*-algebra $\mathcal{O}_A$ onto the fixed point algebra which can be identified with the AF algebra F_A. Hence given a quasi-product state $\phi(u,v,\rho)$ on F_A, one can consider its canonical extension

$$\omega(u,v,\rho) = \phi(u,v,\rho) \circ P \tag{6.3}$$

to $\mathcal{O}_A$. Here we consider the question of weak implementability of the stabilised gauge automorphism in associated quasi-free or quasi-product weights.

Let $v = (1,\ldots,1)^{tr}$ (of length q) and $u, \rho_j \in (0,\infty)^q$, $j = 1,2,\ldots$, satisfy

$$\langle \rho_1 u, v\rangle = 1 \tag{6.4}$$

$$A\rho_j u = u \tag{6.5}$$

$$A^{tr}\rho_j v = v \tag{6.6}$$

for all j (c.f. Proposition 3.4). Let $\mu = \mu(u,v,\rho)$ denote the associated quasi-product measure on Ω_A. Let ρ^σ denote the shifted sequence

$$\rho_j^\sigma = \rho_{j+1}, j = 1,2,\ldots, \tag{6.7}$$

so that (u,v,ρ^σ) is also a quasi-product system, and let

$$\mu^\sigma = (u,v,\rho^\sigma) \tag{6.8}$$

denote the associated quasi-product measure so that

$$\mu^\sigma = \mu \circ \sigma^{-1}. \tag{6.9}$$

Consider the two sided shift space

$$\Omega^A = \{x = (x_i) \in \Sigma^{\mathbb{Z}} : A(x_i,x_{i+1}) = 1, \ i \in \mathbb{Z}\}$$

and fix any $x \in \Omega^A$. The unstable manifold of x is

$$W^u = W^u(x) = \bigcup_j W_j^u(x)$$

endowed with the inductive limit topology, if

$$W_j^u = W_j^u(x) = \{y = (y_i) \in \Omega^A : y_i = x_i, \ i \leq j\}.$$

Define an (infinite) measure $\bar{\mu}$ on $W^u(x)$ by

$$\bar{\mu}(W_j^u \cap [a]) = \begin{cases} \rho(x_1,\ldots,x_j,a_{j+1},\ldots,a_n)u(a_n) & j \geq 0 \\ A(x_0,a_1)\rho(a_1,\ldots,a_n)\,u(a_n) & j < 0 \end{cases}$$

if $a = (a_{j+1},\ldots,a_n) \in \Sigma_{[j+1,n]}$, and [a] denotes the cylinder of a as usual.

Let $G = G(W^u(x))$ denote the group of uniformly finite dimensional homeomorphisms of $W^u(x)$, i.e. $G = U\Phi_\ell$ where Φ_ℓ denotes the homeomorphisms g of W^u such that

$$g(y)_i = y_i \qquad i > \ell\,. \tag{6.10}$$

Let $\bar{C}_A = C_0(W^u)$

and $\bar{F}_A = C^*(\bar{C}_A, G)/J$

where J denotes the ideal in the C*-crossed product $C^*(\bar{C}_A,G)$ generated by $P(K)U(g)-P(K)U(g')$, where P(K) is the characteristic function of a compact open subset K of W^u, $g,g' \in G$ agree on K, and U is the canonical representation of G in the multiplier agebra of the crossed product. The shift induces an automorphism r of $\bar{F}_A$ [5] and let $\bar{O}_A$ denote the corresponding crossed product. Letting $\mathcal{D}$ denote a maximal commutative subalgebra of K, the compact operators on a separable Hilbert space, one has:

$$(\bar{O}_A,\bar{C}_A,\hat{r}) \simeq (K \otimes O_A,\ \mathcal{D} \otimes C_A,\ 1 \otimes \alpha) \tag{6.11}$$

where $\hat{r}$ denotes the dual action of r, [5].

The measure $\bar{\mu}$ on W^u induces canonical weights $\bar{\phi}$, $\bar{\omega}$ on $\bar{F}_A$, $\bar{O}_A$ respectively.

<u>THEOREM 6.1</u>

Assume that

(i) (u,v,ρ) satisfies Hypothesis 4.1.

(ii) The pair (u,v,ρ) and (u,v,ρ^σ) satisfies Hypothesis 5.1.

(iii) $\chi(\mu,\mu^\sigma) > 0$.

(iv) $\bar{\phi}(u,v,\rho)$ on $\bar{F}_A$ is not type I.

Then $\bar{\phi}(u,v,\rho)$ and $\bar{\omega}(u,v,\rho)$ are primary, and $\{1 \otimes \alpha_t : t \in \mathbb{T}\backslash\{1\}\}$ are not weakly inner in $\bar{\omega}(u,v,\rho)$.

ROOF

Let $\bar{\pi}^{\rho}$, $\bar{\pi}_{\rho}$ denote the representations of $\bar{\omega}$, $\bar{\phi}$ respectively. The following observations are enough to ensure that the strategy of [1] for the full shift carry over to our aperiodic situation.

a) : (i) ensures by Theorem 4.5 that μ is Γ-ergodic and hence $\bar{\mu}$ is G-ergodic, and is primary.

b) : (ii) and (iii) ensure by Theorem 5.4 that $\bar{\mu}$ and $\bar{\mu}^{\sigma}$ are mutually absolutely continuous. Hence the automorphism r on $\bar{F}_A$ extends to $\tilde{r}$ on the weak closure $\bar{N}$ in . Thus the W*-crossed product $W^*(\bar{\pi}_{\rho}(\bar{F}_A)'',\mathbb{Z},\tilde{r})$ exists and can be identified with $\bar{\pi}^{\rho}(\bar{\mathcal{O}}_A)''$ as in [4,1].

c) : That $\bar{\omega}$ is primary can be argued as follows. Let $\mathbf{Z}$ act on G by conjugation by r, and form the semidirect product $G \times \mathbf{Z}$. Then by (b), $\bar{\mu}$ is clearly $G \times \mathbb{Z}$ quasi-invariant, and by (a) it is certainly $G \times \mathbb{Z}$-ergodic. We can thus form the Krieger crossed product $W^*_{Kr}(W^u,G \times \mathbb{Z},\bar{\mu})$ which is a factor [10]. Now the Krieger product $W^*_{Kr}(W^u,G,\bar{\mu})$ can be identified with $\bar{\pi}_{\rho}(\bar{F}_A)''$ (see [13, pp. 55-56]). Moreover we claim that condition (iv) implies

$$W^*_{Kr}(W^u,G \times \mathbf{Z},\bar{\mu}) \simeq W^*(W^*_{Kr}(W^u,G,\bar{\mu}),\mathbb{Z}). \tag{6.12}$$

Then by (c), the latter is identified with $\bar{\pi}^{\rho}(\bar{\mathcal{O}}_A)''$, which is then seen to be primary. It just remains to show claim (6.12). Let $G = G \times \mathbf{Z}$, and G_y, the stabilizer $\{y \in G: gy = y\}$ for $y = (y_i)_{-\infty}^{\infty} \in W^u$. Then G_y is trivial unless the sequence $\{y_i\}_{i=1}^{\infty}$ is eventually periodic. Thus the set $Y = \{y \in W^u: G_y$ nontrivial$\}$ is countable, hence by (iv), is of $\bar{\mu}$-measure zero. Thus the action is free, up to a set of measure zero. Then for $y \in W^u\backslash Y$, let $H_y = \ell^2(G)$, $K_y = H_y \otimes \ell^2(\mathbf{Z}) \simeq \ell^2(G)$, and set $H = \int_{W^u} H_y\, d\bar{\mu}$, $K = \int K_y d\bar{\mu} = H \otimes \ell^2(\mathbf{Z})$. On K, $W^*_{Kr}(W^u,G \times \mathbf{Z},\bar{\mu})$ is generated by $L^{\infty}(W^u)$ ⊕ and $\mathbf{Z}$ on $H \otimes \ell^2(\mathbf{Z})$. This can also be generated by $W^*_{Kr}(W^u,G)$ and $\mathbf{Z}$ i.e. $W^*(W^*_{Kr}(W^u,G),\mathbf{Z})$. Details are left to the reader.

d): By (b) and (c), $W^*(\bar{\pi}(\bar{F}_A)'',\mathbb{Z})$ is a factor, and so the automorphism $\tilde{r}$ on $\bar{F}_A''$ cannot possibly be inner. In fact no non trivial power of $\tilde{r}$ can be inner, as can be seen by the following argument for $\tilde{r}^2$. Form the associated strong shift equivalent zero-one matrix A' by symbol splitting, and then form the sequences, u', v', ρ' so that (u',v',ρ') is a quasi-product system with $(\Omega_A,\sigma_A^2,\mu(u,v,\rho))$ and $(\Omega_{A'},\sigma_{A'},\mu(u',v',\rho'))$ homeomorphic (see [8, Remark 3.3] for details). Then the previous argument applied to A' shows that $\tilde{r}_{A'}$ (and hence $\tilde{r}_A^2$) is not inner. Details are left to the reader.

⊕ and translation by $G \times \mathbb{Z}$, hence by ($L^{\infty}(W^u)$ and G)

EXAMPLE 6.2

Although we stated and proved Theorem 6.1 for zero-one matrices, it holds of course for matrices with positive integers, using symbol splitting. To provide one situation where the conditions of Theorem 6.1 hold, let A_j, u_j, v_j, $\bar{u}$, $\bar{v}$, $\bar{v}_0$, X be as in Example 4.4, and ρ_j a sequence in X converging to $\bar{u}/3$. Then as shown in Example 4.4, Hypothesis 4.1 holds, and so $\mu(u,v,\rho)$ on Ω_A is Γ-ergodic or $\bar{\phi}(u,v,\rho)$ on $\bar{F}_A$ is primary.

Moreover,

$$A(\rho_{j+1}\rho^{\sigma}_{j+1})^{\frac{1}{2}}u_j = \gamma_j u_j \tag{6.13}$$

if $$\gamma_j = 2[\rho_{j+1}(1)\rho_{j+2}(1)]^{\frac{1}{2}} + [\rho_{j+1}(2)\rho_{j+2}(2)]^{\frac{1}{2}}, \tag{6.14}$$

and $$\lim A(\rho_{j+1}\rho^{\sigma}_{j+1})^{\frac{1}{2}}/\gamma_j = A/3.$$

Hence by Proposition 5.3, the pair (ρ,ρ^{σ}) satisfies Hypothesis 5.1. Thus if

$$\chi(\mu,\mu^{\sigma}) > 0 \tag{6.15}$$

$\bar{\mu}$ and $\bar{\mu}^{\sigma}$ are mutually absolutely continuous.

Thus the only remaining condition required is that $\bar{\phi}(u,v,\rho)$ is not type I. We will not enter into a detailed discussion of this here.

ACKNOWLEDGEMENTS

This work was essentially completed during visits to the ANU, Canberra and RIMS, Kyoto. I am grateful to D.W. Robinson and H. Araki respectively for their hospitality and the Australian National University and the Royal Soceity respectively for their financial support.

REFERENCES

[1] H. Araki, A.L. Carey, D.E. Evans. On O_{n+1}. J. Operator Theory (in press).

[2] H. Araki, R. Haag, D. Kastler, M. Takesaki. Extension of states and chemical potential. Commun. math. Phys. 53 (1977), 97-134.

[3] O. Bratteli, D.W. Robinson, Operator Algebras and Quantum Statistical Mechanics II. Springer Verlag. Berlin, Heidelberg, New York 1981.

[4] J. Cuntz. Simple C*-algebras generated by isometries. Commun. math. Phys. 57 (1977), 173-185.

[5] J. Cuntz, W. Krieger. A class of C*-algebras and topological Markov chains. Inventiones Math. 56 (1980), 251-258.

[6] D.E. Evans. On O_n. Publ.RIMS Kyoto Univ. 16 (1980), 915-927.

[7] D.E. Evans. Entropy of automorphisms of AF algebras. Publ. RIMS Kyoto Univ. 18 (1982), 1045-1051.

[8] D.E. Evans. The C*-algebras of topological Markov chains. Lecture notes. Tokyo Metropolitan University, 1983.

[9] S. Kakutani. On equivalence of infinite product measures, Ann. of Math. 49 (1948), 214-222.

[10] W. Krieger. On constructing non-isomorphic hyperfinite factors of type III. J. Func. Analysis. 6 (1970), 97-109.

[11] C.C. Moore. Invariant measures on product spaces. Proc. of the Fifth Berkeley Symposium on Math. Stat. and Probability. Vol. II, part II, 447-459 (1967).

[12] E. Seneta. Non negative matrices and Markov chains. Springer-Verlag. Berlin, Heidelberg and New York. (2nd edition), 1981.

[13] S. Stratila, D. Voiculescu. Representations of AF algebras and of the group $U(\infty)$. Lecture notes in Mathematics. Springer-Verlag, vol. 486. Berlin, Heidelberg and New York, 1975.

ABELIAN GROUP ACTIONS ON TYPE I C*-ALGEBRAS

by

Elliot C. Gootman*
Department of Mathematics
University of Georgia
Athens, Georgia 30602

ABSTRACT

Let (G,A,α) be a separable C*-dynamical system, with G abelian and A type I. We prove that all points in a quasi-orbit in $\hat{A}$ have the same isotropy subgroup and determine cocycles of this subgroup of the same class. These results are then used to prove that if $\hat{A}$ has a non-transitive quasi-orbit and either the (common) dimension of all representations in this quasi-orbit is finite or the (common) isotropy group is discrete, then the crossed product algebra is non-type I. While this latter result has long been known, we present a new proof using Takai duality. An example is also given of a non-smooth action of $\mathbb{R}^2$ on A for which the crossed product algebra is nevertheless type I. Finally, we characterize the Connes spectrum in terms of the separated primitive ideals of the crossed product algebra. When $A = C_0(X)$ is commutative, we determine which primitive ideals of the crossed product algebra are separated in terms of the behaviour of isotropy groups and orbit closures.

1. Introduction

Let G be a separable locally compact group, A a separable C*-algebra, and $\alpha:G \to \mathrm{Aut}\, A$ a point-norm continuous homomorphism of G into the automorphism group of A. The triple (G,A,α) is called a C*-dynamical system, and determines a C*-algebra $G \times_\alpha A$, the crossed product algebra, whose representations are in one-to-one correspondence with covariant pairs of representations $\langle V,\pi\rangle$ of (G,A,α) [4,5,14,16]. G also acts as a topological transformation group, in a natural manner, on PR A, the primitive ideal space of A endowed with the hull-kernel topology. For $P \in \mathrm{PR}\, A$, let $G_P = \{g \in G: gP = P\}$ denote the isotropy group of P and

---

*Partially supported by a grant from the National Science Foundation.

$\tilde{G}P = \{Q \in PR\ A: \overline{GP} = \overline{GQ}\}$ the quasi-orbit of P. If G is amenable, it follows from the positive solution of the Effros-Hahn conjecture [9] that every primitive ideal of $G \times_\alpha A$ arises as the kernel of an irreducible induced representation. More specifically, let L be an irreducible representation of $G \times_\alpha A$ corresponding to the covariant pair $\langle V, \pi\rangle$. The direct integral decomposition of π into homogeneous representations can be based on a quasi-orbit, say $\tilde{G}P$, in $PR\ A$, and L is weakly equivalent to the representation induced from an irreducible representation $R = \langle W, \tau\rangle$ of $G_P \times_\alpha A$ with kernel $\tau = P$. (See [9] for further details and references).

As a consequence of the above, information concerning the action of G on $PR\ A$, such as the behavior of isotropy groups and of quasi-orbits, should be relevant in determining the ideal structure and representation theory of $G \times_\alpha A$. If G is abelian and $PR\ A$ is Hausdorff, it is obvious that every point in a given quasi-orbit in $PR\ A$ has the same isotropy subgroup. That this is not the case in general, even for G abelian, can be seen from Example 2.3 of [7]. There $A = T \times_\alpha F$ is the crossed product of the fermion algebra by the gauge action and there is only one quasi-orbit for the dual action of the integers $\mathbb{Z}$ on A, but (0) is a primitive ideal of A left fixed by $\mathbb{Z}$ while A has non-zero primitive ideals with trivial isotropy groups. We show in §2 that the above situation is typical, in the following sense: If G is abelian, every quasi-orbit in $PR\ A$ contains a dense subset all of whose points have the same isotropy group, and this isotropy group contains the isotropy group of any other point in the quasi-orbit.

If A is type I, $PR\ A$ can be identified with $\hat{A}$, the space of unitary equivalence classes of irreducible representations of A. As in Mackey's little-group method, each π in $\hat{A}$ determines a multiplier $\overline{a_\pi}$ of G_π and an $\overline{a_\pi}$-representation U_π of G_π such that all irreducible representations of $G_\pi \times_\alpha A$ "lying over" the orbit of π in $\hat{A}$ are induced from representations of $G_\pi \times_\alpha A$ corresponding to pairs $\langle V \otimes U_\pi, I \otimes \pi\rangle$ for irreducible a_π-representations V of G_π. Furthermore, a_π determines a subgroup S_π of G_π, defined by $S_\pi = \{s \in G_\pi : a_\pi(s,t) = a_\pi(t,s) \text{ for all } t \in G_\pi\}$. For G_π abelian, a_π is called totally skew if S_π is trivial. If a_π is totally skew, all irreducible a_π-representations of G_π are weakly equivalent ([2], [10]), and from this fact it follows that S_π is more relevant than G_π for the ideal structure of $G \times_\alpha A$.

For G abelian and A type I, we prove in §2 that all π in a given quasi-orbit have the same isotropy group G_π and the same S_π, and determine multipliers of the same cocycle class. The results on common G_π and S_π extend Proposition 2.2 and Theorem 3.1 of [7] where it is assumed that A is G-simple (i.e., has no non-trivial G-invariant ideals) and, for the S_π result, that A is also of continuous trace. That the continuous trace hypothesis could be dropped was announced, without proof, on page 348 of [7] and page 324 of [8]. We show in §2 that the basic ideas in [7] can be made to work without the hypothesis of G-simplicity, and that they also yield the result on common cocycle classes. Our results on isotropy groups and cocycle classes also extend measure-theoretic versions proven in [1, Proposition 2.4, page 70 and Theorem 4, page 95], where, in a slightly different context, it is shown that for a quasi-invariant ergodic measure α on $\hat{A}$, almost all ($d\alpha$) points have the same isotropy group and determine the same cocycle class.

In §3 we consider the question of whether a non-smooth action of G on $\hat{A}$ forces the crossed product algebra to be non-type I, again for G abelian and A type I. That this need not be true for the more complicated construction of "twisted" crossed product algebras [10, p. 196] follows from pages 88-90 of [1], where examples are given of groups G with closed normal type I subgroups N such that G/N is abelian and acts non-smoothly on $\hat{N}$, but G is type I. We consider whether a similar "unwinding" phenomenon can occur in the simpler case of a crossed product algebra.

We first prove that it cannot occur if, for all points in a given non-transitive quasi-orbit in $\hat{A}$, either the (common) isotropy group is discrete, or the (common) dimension of the Hilbert space upon which the representations act is finite. The first result is known, even for twisted crossed products [1, Chapter II, Theorem 9]. The proof given in [1], however, involves a rather intricate cross-section argument, while our proof involves a relatively straight-forward application of Takai duality. Our proof also extends directly to the twisted crossed product case, by Corollary 31 of [10]. We mention that Kallman has extended Theorem 9 of [1, Chapter II] to the case of non-abelian group actions [11],[19],[20].

It would be interesting to try to apply some version of "non-commutative duality theory" to obtain a proof of Kallman's result, analogous to our proof (Theorem 3.2) of the abelian case.

The second result mentioned above is not true for twisted crossed products, since in the examples in [1], the closed normal subgroup N is actually abelian. For crossed product algebras, however, it has long been known [6] that if A is a commutative C*-algebra and G(not necessarily abelian) acts non-smoothly on $\hat{A}$, then $G \times_\alpha A$ is non-type I. Thus Theorem 3.5 extends Theorem 3.3 of [6], in the case of G abelian. Finally, we close §3 with an example of the "unwinding phenomenon" for crossed product algebras. In our example, G is the group $\mathbb{R}^2$, and we note that the "unwinding phenomenon" cannot occur in the case of $G = \mathbb{R}$.

The notion of a separated point of $\hat{A}$ or of PR A enters critically in the considerations of §2 (see §2 for a definition). In §4 we continue to exploit this notion, by characterizing the Connes spectrum $\Gamma(\alpha)$ of the system (G,A,α) [12, §3] in terms of the isotropy groups of the separated points of $PR(G \times_\alpha A)$ under the dual action of the dual group $\hat{G}$. Although our result follows immediately from Corollary 5.4 of [12], it might nonetheless prove useful to have a characterization of $\Gamma(\alpha)$ involving only (certain) primitive ideals, and not arbitrary ideals. For example, since every primitive ideal of $G \times_\alpha A$ is induced, a characterization of which induced primitive ideals are separated points of $PR(G \times_\alpha A)$ would then lead to a characterization of $\Gamma(\alpha)$ in terms of properties of the system (G,A,α). Although in general a description of the topology of $PR(G \times_\alpha A)$ is quite hard to come by, perhaps a description of the separated points of $PR(G \times_\alpha A)$ might be more accessible. In any case, we conclude by considering the case of abelian C*-algebra $A = C_0(X)$, where the topology on the primitive ideal space has been worked out by Dana Williams [17]. We use Williams' results to characterize the separated points of $PR(G \times_\alpha C_0(X))$ and as a corollary obtain simple necessary and sufficient conditions for the primitive ideal space to be Hausdorff, thus extending the second theorem [18] to the not necessarily type I case.

2. Isotropy groups for quasi-orbits

Recall [3, Exercise 3.9.4] that a point x of a topological space X is said to be separated if for any point y of X not in the closure of {x}, the points y and x admit a pair of disjoint neighborhoods. For a separable C*-algebra A, the set of separated points in PR A form a dense G_δ, and each separated point is a minimal primitive ideal in A.

Lemma 2.1 Let (G,A,α) be a separable C*-dynamical system, with G abelian. For P in PR A, the quasi-orbit $\widetilde{GP}$ contains a dense G_δ subset X such that all points of X have the same isotropy subgroup H and such that $H \supseteq G_Q$ for all Q in $\widetilde{GP}$.

Proof By letting $I = \bigcap_{g \in G} gP$ and replacing the original dynamical system with the natural action of G on A/I, we may assume without loss of generality that the orbit of P is dense in PR A. By Lemma 1.1 of [5], $\widetilde{GP}$ can be written as a countable intersection of open sets, necessarily dense. As PR A is a Baire space [3, proof of Corollary 3.4.13], the set X of separated points of PR A which are contained in $\widetilde{GP}$ is a dense G_δ subset of $\widetilde{GP}$ which is clearly G-invariant. Let $Q \in X$, $R \in \widetilde{GP}$, let g_n be a sequence in G with $g_n R \to Q$ and let $s \in G_R$. It suffices to show $G_R \subseteq G_Q$. Now $g_n R = g_n sR = sg_n R \to sQ$ also. If $sQ \in \overline{Q}$ then $sQ \supseteq Q$ and $sQ = Q$ as separated points of PR A are minimal ideals. If $sQ \notin \overline{Q}$, then Q and sQ can be separated by disjoint open sets, which contradicts $g_n R$ converging to both Q and sQ.

If A is type I, we can improve on Lemma 2.1 as follows:

Lemma 2.2 Let (G,A,α) be a separable C*-dynamical system, with G abelian and A type I. Then all points in a given quasi-orbit in $\hat{A}$ have the same isotropy subgroup.

Proof Let $\pi \in \hat{A}$. As in Lemma 2.1, we may assume without loss of generality that the orbit of π is dense in $\hat{A}$. If τ in $\hat{A}$ lies in the quasi-orbit of π, then the orbit of τ is dense in $\hat{A}$ also. The proof follows exactly as in Proposition 2.2 of [7], since any non-empty open subset V of $\hat{A}$ intersects $G\tau$.

In addition, employing the notation in the Introduction, we have

Theorem 2.3 Let (G,A,α) be a separable C*-dynamical system, with G abelian and A type I. Let π and τ in $\hat{A}$ have the same quasi-orbit. Then the cocycle class of a_π equals the cocycle class of a_τ, and $S_\pi = S_\tau$.

Proof The second statement follows from the first as S_π clearly depends only on the class of a_π. For the first, we may again assume that π and τ have dense orbits in $\hat{A}$. A type I

C*-algebra contains a non-zero ideal I of continuous trace [3, Theorem 4.5.5], and $\hat{I}$ intersects the dense orbit $G\tau$ in $\hat{A}$. As we may choose U_τ equal to $U_{g\tau}$, for any $g \in G$, we may assume $\tau \in \hat{I}$. Let g_n be a sequence in G with $g_n\pi \to \tau$ in $\hat{A}$, so that $g_n\pi \to \tau$ in $\hat{I}$. Now $d(\pi)$, the dimension of the Hilbert space $\mathcal{H}_\pi$ of π, is a lower semi-continuous, G-equivariant function on $\hat{A}$, so $d(\pi) = d(\tau)$ and we now consider π, τ as concrete irreducible representations of A acting on a given Hilbert space $\mathcal{H}$. There exists a sequence of unitaries V_n on $\mathcal{H}$ with $V_n(g_n\pi)(a)V_n^* \to \tau(a)$ strongly, for all $a \in A$. It follows from Proposition 4.5.3 and the proof of Lemma 4.4.2 of [3] that for any rank one projection P on $\mathcal{H}$, there exists $i \in I$ with $\tau(i) = P$ and $\rho(i)$ a rank one projection for all ρ in a suitable neighborhood of the equivalence class of τ in $\hat{I}$. Exactly as in the proof of Theorem 3.1 of [7], we have, for s, t in G_π (which equals G_τ be Lemma 2.2) sequences c_n and d_n of complex numbers of modulus one, such that $c_nV_nU^\pi(s)V_n^*$ converges to $U^\tau(s)$ strongly, and $d_nV_nU^\pi(t)V_n^*$ converges to $U^\tau(t)$ strongly. Upon passing to a subsequence, we may assume c_nd_n converges to z in $\mathbb{C}$, $|z| = 1$, and it follows from strong continuity of multiplication (on bounded sequences) that

$$V_nU^\pi(s)U^\pi(t)V_n^* \to \bar{z}U^\tau(s)U^\tau(t) \quad \text{strongly}$$

and

$$V_nU^\pi(t)U^\pi(s)V_n^* \to \bar{z}U^\tau(t)U^\tau(s) \quad \text{strongly,}$$

or equivalently, that

$$a_\pi(s,t)V_nU^\pi(st)V_n^* \to a_\tau(s,t)\bar{z}U^\tau(st) \quad \text{strongly}$$

and

$$a_\pi(t,s)V_nU^\pi(ts)V_n^* \to a_\tau(t,s)\bar{z}U^\tau(ts) \quad \text{strongly.}$$

Clearly then, since G is abelian,

$$\frac{a_\pi(s,t)}{a_\tau(s,t)} = \frac{a_\pi(t,s)}{a_\tau(t,s)} ,$$

the 2-cocycle $\frac{a_\pi}{a_\tau}$ is symmetric, thus trivial, and a_π is cohomologous to a_τ, as was to be proven.

3. On the type of the crossed product algebra

Throughout this section, we let (G,A,α) be a separable C^*-dynamical system, with G abelian and A type I. Let Q be a quasi-orbit in $\hat{A}$ under the action of G, and let H denote the common isotropy group (cf. Lemma 2.2) for points of Q. We present below a new proof, using Takai duality, of the fact (see the Introduction) that if Q is non-transitive and H is discrete then the crossed product algebra $G \times_\alpha A$ is non-type I. For background information concerning the dual action $\hat{\alpha}$ of the dual group $\hat{G}$ on $G \times_\alpha A$, and on Takai duality, we refer the reader to §§7.8-7.9 of [14] and to [15]. We let $\mathcal{K}(\mathcal{H})$ denote the compact operators on a Hilbert space $\mathcal{H}$, and $\mathcal{K}(X)$ the continuous functions of compact support on a locally compact Hausdorff space X. As a preliminary, we need the following

<u>Lemma 3.1</u> Let L be a representation of $G \times_\alpha A$ corresponding to the covariant pair $\langle V,\pi\rangle$. Under the isomorphism between the dual crossed product algebra $\hat{G} \times_{\hat{\alpha}} (G \times_\alpha A)$ and $\mathcal{K}(L^2(G)) \otimes A$, the representation of $\hat{G} \times_{\hat{\alpha}} (G \times_\alpha A)$ induced from the representation L of $\{e\} \times_{\hat{\alpha}} (G \times_\alpha A)$ corresponds to a representation of $\mathcal{K}(L^2(G)) \otimes A$ which is unitarily equivalent to $I \otimes \pi$.

<u>Proof</u> The proof simply involves computing explicitly the isomorphism between $\hat{G} \times_{\hat{\alpha}} (G \times_\alpha A)$ and $\mathcal{K}(L^2(G)) \otimes A$ established in [14] and [15], and seeing what it does to the representations involved. For the convenience of the reader, we present the formulas. Let ψ and f lie in, respectively, $\mathcal{K}(\hat{G})$ and $\mathcal{K}(G)$, and let a be an element of A. Assume A is faithfully represented as bounded linear operators on $\mathcal{H}$. For $w \in L^2(G)$ and $v \in \mathcal{H}$, let $\theta(\psi \otimes f \otimes a)$ denote the operator on $L^2(G) \otimes \mathcal{H}$ given by

$$\theta(\psi \otimes f \otimes a)(w \otimes v)|_t = (\hat{\psi}w*f)(t)(\alpha_{t^{-1}}a)(v), \; t \in G.$$

Then θ extends to a $*$-isomorphism of $\hat{G} \times_{\hat{\alpha}} (G \times_\alpha A)$ on $\mathcal{K}(L^2(G)) \otimes A$. For $v,w \in \mathcal{K}(G)$, let $P_{v,w}$ denote the rank one projection on $L^2(G)$ defined by $P_{v,w}(f) = \langle f,v\rangle w$, $f \in L^2(G)$. Then for $a \in A$,

$$(\theta^{-1}(P_{v,w} \otimes a))(\gamma,s) = \int_G w(t)\bar{\gamma}(s^{-1}t)\bar{v}(s^{-1}t)\alpha_t a \, dt.$$

If L acts on the Hilbert space $\mathcal{H}_L$, $h \in \mathcal{K}(\hat{G} \times G, A)$ and $\psi \in L^2(\hat{G},\mathcal{H}_L)$, then the induced representation IND L of $\hat{G} \times_{\hat{\alpha}} (G \times_{\alpha} A)$ acts on $L^2(\hat{G},\mathcal{H}_L)$ as follows:

$$(\mathrm{IND}(L))(h)\psi|_{\sigma} = \int_{\hat{G}}\int_{G}\overline{\sigma}(s)\pi(h(\chi,s))V(s)\psi(\chi^{-1}\sigma)ds\, d\chi,$$
$$\text{for } \chi,\sigma \in \hat{G} \text{ and } s \in G.$$

Letting $\mathcal{F}$ denote the Fourier transform $L^2(\hat{G},\mathcal{H}_L) \to L^2(G,\mathcal{H}_L)$ and letting $f \in L^2(G,\mathcal{H}_L)$, we obtain

$$\mathcal{F}(\mathrm{IND}\ L)(h)\mathcal{F}^{-1}f|_t = \int_{\hat{G}}\int_{G}\chi(s^{-1}t)\pi(h(\chi,s))V(s)f(s^{-1}t)ds\, d\chi,$$
$$\text{for } \chi \in \hat{G},\ s,\ t \in G.$$

Letting U denote the unitary operator on $L^2(G,\mathcal{H}_L)$ defined by $(Uf)(t) = V(t)f(t)$, we obtain

$$U(I \otimes \pi)(\theta h)U^{-1}f|_t = \mathcal{F}(\mathrm{IND}\ L)(h)\mathcal{F}^{-1}f|_t,$$ and we are done.

<u>Theorem 3.2</u> Let (G,A,α) be a separable C*-dynamical system, with G abelian and A type I. If G acts non-smoothly on $\hat{A}$ and the (common) isotropy group for the points in some non-transitive quasi-orbit is discrete, then the crossed product algebra $G \times_{\alpha} A$ is non-type I.

<u>Proof</u> Let π, τ be points in $\hat{A}$ with distinct orbits but the same orbit closure, and assume the common isotropy group H is discrete. Let $L_{\pi} = \langle V,\tilde{\pi}\rangle$ and $L_{\tau} = \langle W,\tilde{\tau}\rangle$ denote irreducible representations of $G \times_{\alpha} A$ lying over the orbits of, respectively, π and τ. Then $\tilde{\pi}$ and $\tilde{\tau}$ are unitarily equivalent to multiples of, respectively, the direct integral representations $\int_{G/H} g\pi d\overline{g}$ and $\int_{G/H} g\tau\, d\overline{g}$, and thus are weakly equivalent. It follows from Lemma 3.1 that the representations of $\hat{G} \times_{\hat{\alpha}} (G \times_{\alpha} A)$ induced from the representations L_{π} and L_{τ} of $G \times_{\alpha} A$ are weakly equivalent. From the induction-restriction formula and the continuity of restriction, the $\hat{G}$-orbits of L_{π} and L_{τ} in $(G \times_{\alpha} A)^{\wedge}$ have the same closure. By the construction, discussed briefly in the Introduction, of irreducible representations of $G \times_{\alpha} A$ lying over an orbit in $\hat{A}$, it follows that $H^{\perp}$, the annihilator in $\hat{G}$ of the isotropy group H, leaves L_{π} and L_{τ} fixed. Assuming $G \times_{\alpha} A$ is type I, every point in the $\hat{G}$ quasi-orbit of L_{π} is left fixed by $H^{\perp}$ by Lemma 2.2.

As H is discrete, every point in this quasi-orbit has co-compact isotropy group and the quasi-orbit is an orbit [13, Lemma 2.1]. Thus for some $\chi \in \hat{G}$, χL_π is unitarily equivalent to L_τ, so the systems $\langle \chi V, \tilde{\pi} \rangle$ and $\langle W, \tilde{\tau} \rangle$ are unitarily equivalent, contradicting the hypothesis that π and τ have distinct G-orbits.

<u>Remark</u> In Corollary 31 of [10] and the subsequent remarks, Phil Green proves that if $(G,A,\mathcal{T})$ is a twisted covariant system with $\tilde{G} = G/N_{\mathcal{T}}$ abelian, then the crossed product algebra $C^*(\tilde{G}^\wedge, C^*(G,A,\mathcal{T}))$ is isomorphic to $\mathcal{K}(L^2(G/N_{\mathcal{T}})) \otimes A$. It is clear that our proof goes through in this situation also, and we have an alternate proof of [1, Chapter II, Theorem 9].

There is one other general situation in which we can prove that non-smooth action of G on $\hat{A}$ implies $G \times_\alpha A$ is non-type I, namely, when the (common) dimension of all representations in a given non-transitive quasi-orbit in $\hat{A}$ is finite. In addition to using compactness of the unitary group on a finite-dimensional space, the proof also uses the fact that if $\langle V, \pi \rangle$ is a covariant pair of representations for (H,A,α), H any subgroup of G, then V intertwines correctly with all the representations $g\pi$, $g \in G$. This latter fact does not hold for twisted crossed products in general, which explains why the proof does not apply to the example on pp. 88-90 of [1] even though in these examples $A = C^*(N)$ is a commutative C*-algebra.

Accordingly, let Q denote a quasi-orbit in $\hat{A}$ such that the (common) dimension of all representations in Q is finite. By Lemma 2.2 and Theorem 2.3, all points in Q have the same isotropy group H and determine cocycles of H of the same class b. As in [1, Chapter II] we may assume H has only type I b-representations, since otherwise $G \times_\alpha A$ would already be non-type I by the Mackey machine. Let $S = \{x \in H: b(x,y) = b(y,x)\ \forall y \in H\}$. For $\pi \in Q$, let U_π be a $\bar{b}$-representation of H on $\mathcal{H}_\pi$ which intertwines with π, as in the Introduction.

<u>Lemma 3.3</u> With notation and hypotheses as above, the subset of $(H \times_\alpha A)^\wedge$ lying over π is parametrized by $\hat{S}$. Furthermore, H/S is a finite group and has a totally skew cocycle c and a unique (finite-dimensional) irreducible c-representation W such that, if W' denotes the lift of W to H, every element of $(H \times_\alpha A)^\wedge$ lying over π is given by a pair $(\chi W' \otimes U_\pi, I \otimes \pi)$, $\chi \in \hat{H}$.

Two such pairs $(\chi W' \otimes U_\pi, I \otimes \pi)$ and $(\sigma W' \otimes U_\pi, I \otimes \pi)$ are equivalent if and only if $\chi\sigma^{-1} \in S^\perp$.

Proof By Theorem 3.1 of [2] we may assume b is the lift to H of a totally skew cocycle c on H/S. As H has finite dimensional b-representations (for example, the contragredient of U_π) and only type I b-representations, the remaining statements all follow from Theorems 3.1 and 3.3 of [2].

Remark As we are not assuming S is discrete, we cannot use the result on "covering quasi-orbits" given in [1, Section II, Theorem 10]. As in [1], though, we can prove $G \times_\alpha A$ is non-type I by proving G does not act smoothly (in the topological sense) on that part of $(H \times_\alpha A)^\wedge$ lying over Q, when Q is non-transitive. Our strategy will be to consider only a small G-invariant subset F of the part of $(H \times_\alpha A)^\wedge$ lying over Q, and to check that F has only a finite number of points lying over each $\pi \in Q$. Then non-smoothness of the action of G on F will follow by a completely elementary argument.

Lemma 3.4 Fix notation and hypotheses as in Lemma 3.3, choose $\pi \in Q$ and an intertwining $\bar{b}$-representation U_π of H, and let W' be as in Lemma 3.3. Let F be the closure in $(H \times_\alpha A)^\wedge$ of the representations determined by the set of pairs $\{\langle W' \otimes U_\pi, I \otimes g\pi\rangle : g \in G\}$. For $\tau \in Q$, F contains an element lying over τ, and only finitely many such elements.

Proof Let π_0 and τ_0 denote concrete irreducible representations of A on $\mathcal{K}$ (finite-dimensional) with equivalence classes of, respectively, π and τ. As there exists a sequence $g_n \in G$ with $g_n\pi \to \tau$, there exists a sequence of unitaries U_n on $\mathcal{K}$ with $U_n(g_n\pi_0)(a)\, U_n^{-1} \to \tau_0(a)$, $\forall a \in A$ [3, Theorem 3.5.8]. As $\mathcal{K}$ is finite-dimensional, we may pass to a subsequence and assume U_n converges to a unitary U, such that $U(g_n \pi_0)(a)\, U^{-1} \to \tau_0(a)$, $\forall a \in A$. As U_π intertwines correctly with all the representations $g\pi_0$, $g \in G$, as well as with π_0, we have

$$(UU_\pi(h)U^{-1})\tau_0(a)(UU_\pi(h^{-1})U^{-1}) =$$

$$\lim_n (UU_\pi(h)U^{-1})(U(g_n\pi_0)(a)U^{-1})(UU_\pi(h^{-1})U^{-1})$$

$$= \lim_n U(U_\pi(h)(g_n\pi_0)(a)U_\pi(h^{-1}))U^{-1}$$

$$= \lim_n U(g_n\pi_0)(ha)U^{-1} = \tau_0(ha), \ \forall a \in A, \ h \in H.$$

Thus $UU_\pi U^{-1}$ intertwines with τ_0 and it follows that, as concrete irreducible representations of $H \times_\alpha A$, $(I \otimes U) < W' \otimes U_\pi$, $I \otimes g_n\pi >(I \otimes U^{-1}) \to <W' \otimes UU_\pi U^{-1}, I \otimes \tau_0>$ and F contains a point lying over τ. Any other element of $(H \times_\alpha A)^\wedge$ lying over τ is of the form (up to equivalence) $L = <\chi W' \otimes UU_\pi U^{-1}, I \otimes \tau_0>$ for $\chi \in \hat{H}$. Recall by Lemma 3.3 that W' is also a finite-dimensional representation, so $\mathcal{H}_L$ is finite-dimensional. As $L \in F$, we have a sequence $a_n \in G$ and (by compactness of the unitary group on $\mathcal{H}_L$) a unitary operator ψ such that $\psi<W' \otimes U_\pi, I \otimes a_n\pi>\psi^{-1} \to <\chi W' \otimes UU_\pi U^{-1}, I \otimes \tau_0>$. It follows that as unitary representations of H, $W' \otimes U_\pi$ and $\chi W' \otimes U_\pi$ are unitarily equivalent. Upon recalling that $W'|_S = ID$ and that $U_\pi|_S$ is a unitary representation of S, we see that a multiple of $(U_\pi|_S)$ and $(\chi U_\pi|_S)$ are unitarily equivalent. Thus $\chi|_S$ maps the finite spectrum A of $(U_\pi|_S)$ into itself, and $\chi|_S$ must lie in the finite set $AA^{-1} \subseteq \hat{S}$.

<u>Theorem 3.5</u> Let (G,A,α) be a separable C*-dynamical system with G abelian and A type I. Let Q be a non-transitive quasi-orbit in $\hat{A}$ such that the common dimension of all the representations in Q is finite. Then the crossed product algebra $G \times_\alpha A$ is non-type I.

<u>Proof</u> We fix notation as in Lemmas 3.3 and 3.4, and choose π,τ in distinct orbits in Q. Again, let π_0 and τ_0 denote concrete irreducible representations of A with unitary equivalence classes, respectively, of π and τ. As in the proof of Lemma 3.4, we can find sequences s_n, t_n in G and unitary operators θ,ψ such that

$$\theta(s_n\pi_0)(a)\theta^{-1} \to \tau_0(a) \text{ and}$$

$$\psi(t_n\tau_0)(a)\psi^{-1} \to \tau_0(a), \ \forall a \in A.$$

If $<\chi W' \otimes U_\pi, I \otimes \pi_0> \in F$, then as in the proof of Lemma 3.4 $(I \otimes \theta)<\chi W' \otimes U_\pi, I \otimes s_n\pi_0> (I \otimes \theta^{-1}) \to <\chi W' \otimes \theta U_\pi\theta^{-1}, I \otimes \tau_0>$ which lies in F as F is closed and G-invariant. Likewise, if

$<\chi W' \otimes UU_\pi U^{-1}, I \otimes \tau_0> \in F$, then so does

$$\lim_n (I \otimes \psi)<\chi W' \otimes UU_\pi U^{-1}, I \otimes t_n \tau_0>(I \otimes \psi^{-1})$$
$$= <\chi W' \otimes \psi UU_\pi U^{-1}\psi^{-1}, I \otimes \pi_0>.$$

In other words, if we denote by A and B the finite subsets of F lying over, respectively, π and τ, we have well-defined maps L of A into B and R of B into A given by $L(a) = \lim_n s_n a$ and $R(b) = \lim_n t_n b$. Note that under either map, the image of a point lies in the G-orbit closure in $(H \times_\alpha A)^\wedge$ of the point itself, so that if there exists $a \in A$ such that for some positive integer n, $(RL)^n(a) = a$, then a and $L(a)$ lie in the same G-orbit closure in $(H \times_\alpha A)^\wedge$, but in distinct orbits, and we are done. However, as A is finite, such an a exists.

We now present an example of a separable C*-dynamical system (G,A,α), with G abelian and A type I, such that the action of G on $\hat{A}$ is not smooth but the crossed product algebra $G \times_\alpha A$ is nevertheless type I. Note that by Theorem 3.2, such an example cannot occur for $G = \mathbb{Z}$ or $\mathbb{R}$, since in these cases every subgroup is discrete and/or co-compact.

Theorem 3.6 There is a non-smooth $\mathbb{R}^2$-action on a type I separable C*-algebra A for which the crossed product algebra is type I.

Proof Let $X = T^2 \times R$ be the direct product space of the two-dimensional torus T^2 and the real line. We define two actions α and β of $\mathbb{R}$ on X as follows:

$$\alpha_t(z,w,s) = (e^{it}z, e^{i\theta t}w, t + s) \text{ and}$$

$$\beta_t(z,w,s) = (z,w,t + s), \text{ for } (z,w) \in T^2,\ t,\ s \in \mathbb{R}$$

and θ an irrational multiple of 2π. It is clear that the actions α and β commute and define an action of $\mathbb{R}^2$ on X, and that the $\mathbb{R}^2$-action on X is not smooth although α and β are each separately smooth actions on X. As α determines a smooth, free action of $\mathbb{R}$ on X the C*-algebra $A = \mathbb{R} \times_\alpha C_0(X)$ is type I, and $\hat{A}$ is homeomorphic to $X/\mathbb{R}$ [17, Theorem 5.3] which can be identified with T^2. Since the α and β actions commute, the β-action can easily be lifted to an action (also called β) of $\mathbb{R}$ on A, as follows: $(\beta_t f)(s) = \beta_t(f(s), s, t \in \mathbb{R}, f \in \mathcal{K}(R, C_0(X))$. It is trivial to check that the corresponding β-action of $\mathbb{R}$ on $\hat{A} = T^2$ is irrational rotation. Let $\hat{\alpha}$ denote the action of $\mathbb{R}$ on

A dual to the α-action of $\mathbb{R}$ on $C_0(X)$. Loosely speaking, as the β-action of $\mathbb{R}$ is on the $C_0(X)$ part of A while the $\hat{\alpha}$-action of $\mathbb{R}$ is on the $\mathbb{R}$ part of A, these two actions commute and define an action γ of $\mathbb{R}^2$ on A. It is trivial to verify that $\mathbb{R}^2 \times_\gamma A \cong \mathbb{R} \times_\beta (R \times_{\hat{\alpha}} A)$. Here the β-action is "lifted" one more time in an obvious manner to an action on $R \times_{\hat{\alpha}} A$. To check that the γ-action is non-smooth, note that since the α-action of $\mathbb{R}$ on X is free, all irreducible representations of A are induced from (1-dimensional) representations of $C_0(X)$. It follows that the $\hat{\alpha}$-action of $\mathbb{R}$ on A leaves each point of $\hat{A}$ fixed, so that the orbit structure of $\hat{A}$ modulo the γ-action of $\mathbb{R}^2$ is identical to the orbit structure of $\hat{A}$ modulo the β-action of $\mathbb{R}$. Finally, to see that $\mathbb{R}^2 \times_\gamma A$ is type I, note that $\mathbb{R} \times_{\hat{\alpha}} A = \mathbb{R} \times_{\hat{\alpha}} (\mathbb{R} \times_\alpha C_0(X)) \cong \mathcal{K}(L^2(\mathbb{R})) \otimes C_0(X)$ by the Takai duality theorem. By using the explicit isomorphism given in the proof of Lemma 3.1, one can easily see that the β-action of $\mathbb{R}$ on $\mathbb{R} \times_{\hat{\alpha}} A$ is transformed into $ID \otimes \beta$, where ID is the identity action on $\mathcal{K}(L^2(\mathbb{R}))$ and β is the original action of $\mathbb{R}$ on $C_0(X)$. As this latter action is smooth, the crossed product algebra is type I.

4. The Connes spectrum and separated primitive ideals

For background information concerning the Connes spectrum $\Gamma(\alpha)$, we refer the reader to [12]. For Q in $PR(G \times_\alpha A)$, $\hat{G}_Q$ denotes the isotropy subgroup of Q under the above-mentioned dual action.

Lemma 4.1 Let (G,A,α) be a separable C*-dynamical system, with G abelian. Then the Connes spectrum $\Gamma(\alpha)$ equals $\cap \hat{G}_Q$, where the intersection is taken over all primitive ideals Q of $G \times_\alpha A$ which are separated points in $PR(G \times_\alpha A)$.

Proof The proof follows easily from Corollary 5.4 of [12], which characterizes $\Gamma(\alpha)$ as $\{\gamma \in \hat{G}: \gamma I \cap I \neq (0)$ for all non-zero ideals I of $G \times_\alpha A\}$. This is clearly equivalent to characterizing $\Gamma(\alpha)$ as $\{\gamma \in \hat{G}: \gamma\mathcal{O} \cap \mathcal{O} \neq \emptyset$ for all non-empty open subsets $\mathcal{O}$ of $PR(G \times_\alpha A)\}$. Let $\gamma \in \Gamma(\alpha)$ and let Q be a separated point of $PR(G \times_\alpha A)$. If $\gamma Q \neq Q$, either $\gamma Q \notin \overline{Q}$ or $Q \notin \overline{\gamma Q}$. In the latter case $\gamma^{-1}Q \notin \overline{Q}$ and as $\Gamma(\alpha)$ is a group, we may as well suppose $\gamma Q \notin \overline{Q}$. Then γQ and Q have disjoint neighborhoods, and by joint continuity of the action of $\hat{G}$ on $PR(G \times_\alpha A)$, we have a contradiction. The converse follows by the density of the set of separated points of $PR(G \times_\alpha A)$.

Let $A = C_0(X)$ be a commutative C*-algebra, so that (G,X) is a topological transformation group. To each x in X and $\gamma \in \hat{G}$, one can associate an irreducible representation, hence primitive ideal, of $G \times_\alpha C_0(X)$ as follows: the irreducible representation is induced from the one-dimensional representation of $G_x \times_\alpha C_0(X)$ corresponding to the covariant pair $\langle \gamma|_{G_x}, \delta_x \rangle$ with $\delta_x(h) = h(x)$ for $h \in C_0(X)$. This association defines a map $\psi: X \times \hat{G}$ into $PR(G \times_\alpha C_0(X))$, which is onto (in the separable case) by the Effros-Hahn conjecture. For more details on the above, see [17].

<u>Proposition 4.2</u> Let Λ be the quotient topological space obtained from $X \times \hat{G}$ by identifying (x,γ) and (y,χ) if x and y lie in the same quasi-orbit and $\gamma|_{G_x} = \chi|_{G_x}$ (note that $G_x = G_y$). Then ψ factors through Λ and defines a homeomorphism of Λ onto $PR(G \times_\alpha C_0(X))$.

<u>Proof</u> Theorem 5.3 of [17].

Henceforth we shall use Proposition 4.2 to identify $PR(G \times_\alpha C_0(X))$ with Λ, and we shall denote by ψ the map of $X \times \hat{G}$ onto Λ. Essentially under the hypothesis that G acts smoothly on X, so that $(X/G)^\sim$ equals the orbit space X/G and the crossed product algebra is type I, Williams used the above result to determine when $G \times_\alpha C_0(X)$ has a Hausdorff dual [18]. Our goal is to consider the general (separable) case and to specify the separated points of Λ. As a corollary, we determine when $PR(G \times_\alpha C_0(X))$ is Hausdorff. As a first step, we specify the separated points of the quasi-orbit space $(X/G)^\sim$. Let θ denote the canonical map of X onto $(X/G)^\sim$, and let $S(X)$ denote the space of closed subsets of X, endowed with the Fell topology (see [1, Chapter II, §2]). The following proposition is reminiscent of the identification of the separated points of $\hat{A}$ (A a separable C*-algebra) with the points of continuity of the maps $\pi \to \|\pi(a)\|$, for $\pi \in \hat{A}$ and $a \in A$ [3, Exercise 3.9.4].

Proposition 4.3 With notation as above, $\theta(x_0)$ is a separated point of $(X/G)^{\sim}$ if and only if x_0 is a point of continuity of the map $X \to S(X)$ sending x into its orbit closure $\overline{Gx}$.

Proof ($\Rightarrow$) Let $\mathcal{O}_1,\ldots,\mathcal{O}_n$ be non-empty open subsets of X, and let C be a compact subset of X, with $\overline{Gx_0} \cap \mathcal{O}_i \neq \emptyset$, $1 \le i \le n$, and $\overline{Gx_0} \cap C = \emptyset$. If $x_k \to x_0$ in X, it is clear that eventually $\overline{Gx_k} \cap \mathcal{O}_i \neq \emptyset$, $1 \le i \le n$, and we need only check that $\overline{Gx_k} \cap C = \emptyset$ eventually. If not, by passing to subsequences, we can find $c_k \in C \cap \overline{Gx_k}$ with $c_k \to c_0$ in C. As $c_0 \notin \overline{Gx_0}$, $\overline{Gc_0} \not\subseteq \overline{Gx_0}$, $\theta(c_0) \notin \overline{\theta(x_0)}$ in $(X/G)^{\sim}$ and by hypothesis $\theta(c_0)$ and $\theta(x_0)$ can be separated by disjoint open sets U, V in $(X/G)^{\sim}$. As eventually U is an open neighborhood of $\theta(c_k)$ not containing $\theta(x_k)$, we have $\theta(c_k) \notin \overline{\theta(x_k)}$ or $c_k \notin \overline{Gx_k}$, a contradiction.

($\Leftarrow$). Let $y \in X$ with $\theta(y) \notin \overline{\theta(x_0)}$, so that $y \notin \overline{Gx_0}$ and we can find a compact neighborhood C of y with $C \cap \overline{Gx_0} = \emptyset$. By hypothesis, there exists a neighborhood U of x_0 with $C \cap \overline{Gx} = \emptyset$ for all $x \in U$. Denoting by C^0 the interior of U, and using the fact that θ is open, we have that $\theta(C^0)$ and $\theta(U)$ are disjoint open neighborhoods of y and x_0, respectively.

Before turning to the next result, we remark that for (x,ω) and $(y,\gamma) \in X \times \hat{G}$, $\psi(x,\omega) \in \overline{\psi(y,\gamma)}$ if and only if $\overline{Gx} \subseteq \overline{Gy}$ (which implies $G_y \subseteq G_x$) and $\omega|_{G_y} = \gamma|_{G_y}$. Also, we let $S(G)$ denote the space of all closed subgroups of G, endowed with the Fell topology.

Theorem 4.4 $\psi(x,\omega)$ is a separated point of Λ if and only if x is a point of continuity of the maps $c\colon X \to S(G)$ and $d\colon X \to S(X)$ sending x into, respectively, its isotropy subgroup G_x and its orbit closure $\overline{Gx}$.

Remarks a) As $\psi(x,\omega) = \psi(y,\omega)$ whenever y lies in the quasi-orbit of x, it is clear from the above that whether or not an irreducible representation L of $G \times_{\alpha} C_0(X)$ corresponding to a covariant pair <V,M> determines a separated point of $PR(G \times_{\alpha} C_0(X))$ depends only on the quasi-orbit on which M is based, and not at all on V. Indeed, one can see directly that if x is a point of continuity for both maps in the statement of Theorem 4.4, then so is any point in the quasi-orbit of x.

b) As $PR(A)$ has a dense set of separated points, for A a separable C*-algebra, it follows from Theorem 4.4 that the maps c and d have a dense set of joint continuity points. Presumably this can be proven directly by using the "upper semi-continuity" of d, as was done for the "lower semi-continuous" map c in Lemma 1.1 of [9].

c) It follows from Lemma 4.1 that $\Gamma(\alpha) = \cap G_x^{\perp}$ where the intersection is taken over points of continuity of both maps c and d. In [7, Remark 1.7] it is shown by quite different methods that $\Gamma(\alpha) = \cap G_x^{\perp}$ where the intersection is taken over points of continuity of the map c alone. A moment's reflection shows that, in light of remark (b) above, if $\gamma \in \hat{G}$ and $\gamma \equiv 1$ on G_x, for all x which are points of continuity of both c and d, then $\gamma \equiv 1$ on G_x for all x which are points of continuity of c alone. Hopefully, however, the additional insight concerning the role of points of continuity of the map d will prove useful in the study of actions on non-commutative C*-algebras.

d) The maps c and d do not have identical sets of points of continuity, as can be seen by considering the integers acting by powers of an irrational rotation on the complex plane. The point 0 is a point of continuity of d, but not of c.

<u>Proof</u> The ideas in the proof of the second Theorem in [18] all carry over with our weakened hypotheses and in our somewhat different context. In light of the differences, however, we present the details for the convenience of the reader.

($\Rightarrow$) Let $\psi(x,\omega)$ be a separated point of Λ. If x is not a point of continuity of the map c, we can find a sequence $x_n \to x$ in X with $G_{x_n} \to K$ in $S(G)$ and $K \subsetneq G_x$. Let $\chi \in \hat{G}$ such that $\chi|_{G_x}$ is non-trivial but $\chi|_K$ is trivial. Then $\psi(x,\omega\chi) \notin \overline{\psi(x,\omega)}$ as $\omega \neq \omega\chi$ on G_x, but I claim $\psi(x,\omega\chi)$ and $\psi(x,\omega)$ cannot be separated by disjoint open sets. To see this, note that by Lemma 2 of [18] we may, after passing to a subsequence, find $\chi_n \in \hat{G}$ such that $\chi_n \to \chi$ in $\hat{G}$ and $\chi_n \in G_{x_n}^{\perp}$. Let $\theta_n = \chi_n^{-1}\chi\omega$. Then $(x_n,\theta_n) \to (x,\omega)$ in $X \times \hat{G}$, while $\psi(x_n,\theta_n) = \psi(x_n,\chi\omega) \to \psi(x,\chi\omega)$ in Λ.

If x is not a point of continuity of the map d, we can find a sequence $x_n \to x$ with $\overline{Gx_n} \to C$ in $S(X)$ and $C \supsetneq \overline{Gx}$. Let $c \in C$, $c \notin \overline{Gx}$. Then $\psi(c,\omega) \notin \overline{\psi(x,\omega)}$, but $\psi(c,\omega)$ and $\psi(x,\omega)$ cannot be separated by disjoint open sets since we can find g_n in G with $g_n x_n \to c$, so $\psi(g_n x_n,\omega) \to \psi(c,\omega)$ while $\psi(g_n x_n,\omega) = \psi(x_n,\omega) \to \psi(x,\omega)$.

($\Leftarrow$) Let x be a point of continuity for the maps c and d, and let $\psi(y,\gamma) \notin \overline{\psi(x,\omega)}$. Then either $\overline{Gy} \not\subseteq \overline{Gx}$ or $\overline{Gy} \subseteq \overline{Gx}$ (so $G_x \subseteq G_y$) but $\omega|G_x \neq \gamma|_{G_x}$. In the former case $\theta(y) \notin \overline{\theta(x)}$ in $(X/G)^{\sim}$, and since $\theta(x)$ is a separated point in $(X/G)^{\sim}$ by Proposition 4.3 there exist open neighborhoods U and V of $\theta(y)$ and $\theta(x)$, respectively, with $U \cap V = \emptyset$. Let $A = \theta^{-1}(U)$ and $B = \theta^{-1}(V)$ in X. Then $(y,\gamma) \in A \times \hat{G}$, $(x,\omega) \in B \times \hat{G}$ but $\psi(A \times \hat{G})$ and $\psi(B \times \hat{G})$ are disjoint open sets in Λ. In the latter case, $\overline{Gy} \subseteq \overline{Gx}$ and $\omega\gamma^{-1} \neq 1$ on G_x. We can find $\varepsilon > 0$, a point k in G_x, and open neighborhoods A and B, respectively, of $\omega\gamma^{-1}$ and of k, such that $|\tau(g) - 1| > \varepsilon$ for all $(\tau,g) \in A \times B$. As x is a point of continuity of the map c, there exists a neighborhood N of x such that $B \cap G_p \neq \emptyset$ for all p in N. Choosing a symmetric neighborhood U of e in $\hat{G}$ with $\omega\gamma^{-1}U^2 \subseteq A$, we have $(x,\omega) \in N \times \omega U$ and $(y,\gamma) \in X \times \gamma U$, but $\psi(N \times \omega U)$ and $\psi(X \times \gamma U)$ are disjoint.

<u>Corollary 4.5</u> Let (G,X) be a separable topological transformation group, with G abelian. The C*-algebra $G \times_{\alpha} C_0(X)$ has a Hausdorff primitive ideal space if and only if $(X/G)^{\sim}$ is Hausdorff and the map of $X \to S(G)$ sending x into G_x is continuous.

REFERENCES

1. L. Auslander and C. C. Moore, Unitary representations of solvable Lie groups, Mem. Amer. Math. Soc. 62(1966).
2. L. Baggett and A. Kleppner, Multiplier representations of abelian groups, J. Functional Analysis 14(1973), 299-324.
3. J. Dixmier, <u>C*-algebras</u>, North-Holland Mathematical Library, volume 15, Amsterdam, 1977.
4. S. Doplicher, D. Kastler and D. W. Robinson, Covariance algebras in field theory and statistical mechanics, Comm. Math. Phys. 3(1966), 1-28.
5. E. G. Effros and F. Hahn, Locally compact transformation groups and C*-algebras, Mem. Amer. Math. Soc. 75(1967).
6. E. C. Gootman, The type of some C* and W*-algebras associated with transformation groups, Pacific J. Math. 48(1973), 93-106.
7. E. C. Gootman and D. Olesen, Spectra of actions on type I C*-algebras, Math. Scand. 47(1980), 329-349.

8. ___________, Minimal abelian group actions on type I C*-algebras, in Operator Algebras and Applications, R. V. Kadison, ed., Proc. Symp. Pure Math., vol. 38, part I, pp. 323-325, Amer. Math. Soc., Providence, R.I., 1982.

9. E. C. Gootman and J. Rosenberg, The structure of crossed product C*-algebras: A proof of the generalized Effros-Hahn conjecture, Invent. Math. 52(1979), 283-298.

10. P. Green, The local structure of twisted covariance algebras, Acta Math. 140(1978), 191-250.

11. R. R. Kallman, Certain quotient spaces are countably separated, Illinois J. Math. 19(1975), 378-388.

12. D. Olesen and G. K. Pedersen, Applications of the Connes spectrum to C*-dynamical systems, J. Functional Anal. 30(1978), 179-197.

13. ____________________________, Applications of the Connes spectrum to C*-dynamical systems, III, J. Functional Analysis 45(1982), 357-390.

14. G. K. Pedersen, C*-algebras and their automorphism groups, London Math. Soc. Monographs 14, Academic Press, London/New York, 1979.

15. H. Takai, On a duality for crossed products of C*-algebras, J. Functional Analysis 19(1975), 25-39.

16. M. Takesaki, Covariant representations of C*-algebras and their locally compact automorphism groups, Acta Math. 119(1967), 273-303.

17. D. Williams, The topology on the primitive ideal space of transformation group C*-algebras and C.C.R. transformation group C*-algebras, Trans. Amer. Math. Soc. 226(1981), 335-359.

18. __________, Transformation group C*-algebras with Hausdorff spectrum, Illinois J. Math. 26(1982), 317-321.

19. R.R. Kallman, Certain quotient spaces are countably separated, II, J. Functional Analysis 21(1976), 52-62.

20. R.R. Kallman, Certain quotient spaces are countably separated, III, J. Functional Analysis 22(1976), 225-241.

Injectivity and decomposition of completely bounded maps

Uffe Haagerup

Introduction

A linear map S from a C*-algebra A into a C*-algebra B is completely positive if

$$S \otimes i_m : A \otimes M_m \to B \otimes M_m$$

is positive for all m. Here M_m is the algebra of complex $m \times m$ matrices and i_m is the identity on M_m. Moreover a linear map T from A to B is completely bounded if

$$\sup_{m \in \mathbb{N}} \| T \otimes i_m \| < \infty \quad .$$

The supremum is called the completely bounded norm of T and is denoted $\|T\|_{cb}$.

In 1979 Wittstock proved the striking result that any completely bounded map from a C*-algebra A into an injective C*-algebra B is a linear combination of completely positive maps from A to B. More specificly he proved that if $T : A \to B$ is a completely bounded selfadjoint map (i.e. $T(x^*) = T(x)^*$, $x \in A$), then there exist completely positive maps T_1, T_2 from A to B, such that

$$T = T_1 - T_2 \quad \text{and} \quad \|T_1 + T_2\| \leq \|T\|_{cb}$$

(cf. [27, Satz 4.5]). Later Paulsen found a simpler proof of Wittstock's result based on Arveson's extension theorem (cf. [15, Cor. 2.6] and [2, Thm. 1.2.9]). He also proved that for any (not necessarily selfadjoint) completely bounded linear map T fro a C*-algebra A into an injective C*-algebra B, there exist completely positive maps S_1, S_2 from A to B, such that $\|S_i\| \leq \|T\|_{cb}$ i=1,2, and such that

$$x \rightarrow \begin{pmatrix} S_1(x) & T(x^*)^* \\ T(x) & S_2(x) \end{pmatrix}$$

is a completely positive map from A to $B \otimes M_2$. (This follows from [16, thm. 2.5]).

In the following we let CP(A,B) (resp. CB(A,B)) denote the set of completely positive (resp. completely bounded) maps from a C*-algebra A to a C*-algebra B. The main result of this paper is the following converse to Wittstock's theorem:

Let N be a non-injective von Neumann algebra, then for every infinite dimensional C*-algebra A, there exists a completely bounded map $T : A \rightarrow N$, which is not a linear combination of completely positive maps. In particular a von Neumann algebra N is injective if and only if CB(N,N) = span CP(N,N). (cf. Theorem 2.6 and corollary 2.8).

It is essential that N is a von Neumann algebra, because Huruya has recently given an example of a non-injective C*-algebra B, such that CB(A,B) = span CP(A,B) for all C*-algebras A (cf. [10]). Smith proved in [20, example 2.1] that for the abelian C*-algebra A = C([0,1]), one has

$$\text{span } CP(A,A) \subsetneq CB(A,A).$$

The first example of a von Neumann algebra N for which

$$\text{span } CP(A,N) \subsetneq CB(A,N)$$

for some C*-algebra A was given by Huruya and Tomiyama (cf. [11, example 12]).

We apply our result to show that for every infinite dimensional C^*-algebra A, there exists a completely bounded map T of A into some quotient C^*-algebra B/J, which has <u>no</u> completely bounded lifting $\tilde{T}$ from A to B

$$\begin{array}{ccc} & \tilde{T} & B \\ & \nearrow & \downarrow \\ A & \xrightarrow[T]{} & B/J \end{array}$$

(cf. corollary 2.9). Hence the Choi-Effros lifting theorem for completely positive maps [4] fails for completely bounded maps, even if A is abelian. If $\dim(A) < \infty$, T has of course always a linear lifting. However, we show that for a particular choice of B and J, we can find completely bounded maps T_n from $M_n = M_n(\mathbb{C})$, $n \geqq 3$ to B/J, such that

$$\|\tilde{T}_n\|_{cb} \geqq \frac{n}{2\sqrt{n-1}} \|T_n\|_{cb}$$

for any linear lifting $\tilde{T}_n$ of T_n. (cf. prop. 3.2). This gives the negative answer to a problem posed by Paulsen [17].

To prove the above mentioned results, it is convenient to introduce a norm $\| \ \|_{dec}$ on span $CP(A,B)$ for arbitrary C^*-algebras A and B. For $T \in \text{span}\, CP(A,B)$, we let $\|T\|_{dec}$ denote the infimum of those $\lambda \geqq 0$, for which there exist $S_1, S_2 \in CP(A,B)$ such that

$$x \to \begin{pmatrix} S_1(x) & T(x^*)^* \\ T(x) & S_2(x) \end{pmatrix}$$

is a completely positive map from A to $B \otimes M_2$. If T is self-adjoint, $\|T\|_{dec}$ is simply

$$\|T\|_{dec} = \inf \{ \|T_1 + T_2\| \mid T = T_1 - T_2 \ , T_1 \ , T_2 \in CP(A,B) \}$$

(cf. def. 1.1. and prop. 1.3). We show that the inequality

$$\|T\|_{cb} \leqq \|T\|_{dec}$$

always holds, so by Wittstock's and Paulsen's results

$$\|T\|_{cb} = \|T\|_{dec}$$

whenever B is injective. Our main result (theorem 2.6) is a relative easy consequence of the following characterization of injective von Neumann algebras, which we prove in theorem 2.1:

A von Neumann algebra N is injective if and only if there exists $c \in \mathbb{R}_+$, such that for all linear maps T from ℓ_n^∞ to N,

$$\|T\|_{dec} \leqq c\|T\|_{cb} .$$

Here ℓ_n^∞ denotes n-dimensional abelian C*-algebra $\ell^\infty\{1,\dots,n\}$. The starting point in the proof of theorem 2.1 is that the hyperfinite II_1-factor R can be characterized among all factors on a separable Hilbert space by the property that

$$\|\sum_{i=1}^{n} u_i \otimes u_i^c\|_{H\otimes H^c} = n$$

for any finite set $u_1,\dots,u_n$ of unitaries in R. This was proved by Connes as an offshoot of his work on injective factors (cf. [6, Remark 5.29]). Thus if N is a non-injective finite factor (on a separable Hilbert space) one can choose unitaries $u_1,\dots,u_n \in N$ such that

$$\frac{1}{n}\|\sum_{i=1}^{n} u_i \otimes u_i^c\|_{H\otimes H^c} < 1 .$$

By considering the r'th power of $\sum_{i=1}^{n} u_i \otimes u_i^c$, we can obtain $m = n^r$

unitaries $v_1, \dots, v_m \in N$, such that

$$\frac{1}{m} \| \sum_{i=1}^{m} v_i \otimes v_i^c \|_{H \otimes H^c}$$

is smaller than any given constant γ. Now if one define $T : \ell_m^\infty \to N$ by

$$T(c_1, \dots, c_m) = \sum_{i=1}^{m} c_i v_i$$

it turns out that $\|T\|_{dec} > \gamma^{-\frac{1}{2}} \|T\|_{cb}$, which proves theorem 2.1 in the case of II_1-factors on a separable Hilbert space. The general case is obtained by extending Connes' result to finite von Neumann algebras with a non-trivial center (lemma 2.2) and by using Takesaki's decomposition of a type III von Neumann algebra as a crossed product of a semifinite algebra with a one-parameter group of automorphisms.

In section 3 we give concrete examples of linear maps T_n from ℓ_n^∞ to the von Neumann algebra $\mathcal{M}(\mathbb{F}_2)$ associated with the regular representation of the free group on two generators, such that $\|T_n\|_{dec} > \|T_n\|_{cb}$ for $n \geqq 3$, and

$$\|T_n\|_{dec} / \|T_n\|_{cb} \to \infty \quad \text{for} \quad n \to \infty$$

(cf. example 3.1). On the other hand, we prove in prop. 3.4 that for any linear map T from ℓ_2^∞ to a von Neumann algebra N,

$$\|T\| = \|T\|_{cb} = \|T\|_{dec} .$$

§1.

Decomposable linear maps between C*-algebras.

Let A,B be C*-algebras. We will call a bounded linear map from A to B decomposable if it is a linear combination of completely positive maps from A to B. Note first that a bounded linear map T from A to B is decomposable if and only if there exist $S_1, S_2 \in CP(A,B)$, such that

$$(*) \qquad R(x) = \begin{pmatrix} S_1(x) & T(x^*)^* \\ T(x) & S_2(x) \end{pmatrix}$$

defines a completely positive map from A to $B \otimes M_2$. Assume namely that $T = \sum_{i=1}^{n} c_i T_i$, $c_i \in \mathbb{C}$ and $T_i \in CP(A,B)$. Then clearly $S_1 = S_2 = \sum_{i=1}^{n} |c_i| T_i$ can be used. Conversely if $T \in B(A,B)$ and there exist $S_1, S_2 \in CP(A,B)$ such that (*) defines a completely positive map R from A to $B \otimes M_2$, one checks easily that

$$T = (T_1 - T_2) + i(T_3 - T_4)$$

where

$$T_1 = \tfrac{1}{4}(S_1 + S_2 + T + T^*) \ , \quad T_2 = \tfrac{1}{4}(S_1 + S_2 - T - T^*) \ ,$$

$$T_3 = \tfrac{1}{4}(S_1 + S_2 - iT + iT^*) \ , \quad T_4 = \tfrac{1}{4}(S_1 + S_2 + iT - iT^*)$$

are four completely positive maps from A to B. (T^* is the linear map given by $T^*(x) = T(x^*)^*$, $x \in A$).

For two linear maps R_1, R_2 from A to B we write

$$R_1 \underset{cp}{\leq} R_2$$

if $R_2 - R_1$ is completely positive.

<u>Definition 1.1</u>

Let A and B be C*-algebras and let $T : A \to B$ be a bounded linear map. If T is decomposable we let $\|T\|_{dec}$ denote the infimum of those $\lambda \geqq 0$ for which there exist $S_1, S_2 \in CP(A,B)$, such that $\|S_i\| \leqq \lambda$, $i = 1,2$, and

$$R(x) = \begin{pmatrix} S_1(x) & T(x^*)^* \\ T(x) & S_2(x) \end{pmatrix}$$

is a completely positive map from A to $B \otimes M_2$. If T is not decomposable, we put $\|T\|_{dec} = +\infty$.

<u>Remark 1.2</u>

We could equivalently have defined $\|T\|_{dec}$ as the infimum of those $\lambda \geqq 0$ for which there exist $S_1, S_2 \in CP(A,B)$, such that $\|S_i\| \leqq \lambda$, $i = 1,2$, and

$$\tilde{R}\begin{pmatrix} x_{11} & x_{12} \\ x_{21} & x_{22} \end{pmatrix} = \begin{pmatrix} S_1(x_{11}) & T^*(x_{12}) \\ T(x_{21}) & S_2(x_{22}) \end{pmatrix}$$

is a completely bounded map from $A \otimes M_2$ to $B \otimes M_2$. Indeed if $\tilde{R}$ is completely positive, so is R, because

$$\tilde{R} = R \circ P$$

where P is the completely positive map from A to $A \otimes M_2$ given by

$$P(x) = \begin{pmatrix} x & x \\ x & x \end{pmatrix} .$$

To prove the converse, let $(e_{ij})_{i=1,2}$ be the matrix units of M_2, and let $Q : M_2 \otimes M_2 \to M_2$ be the linear map defined by

$$Q(e_{ij} \otimes e_{k\ell}) = \begin{cases} e_{ij} & \text{for } i=k \text{ and } j=\ell \\ 0 & \text{otherwise.} \end{cases}$$

One checks easily that Q is completely positive (Q can be written as $Q = Q_2 \circ Q_1$ where $Q_1(x) = exe$, $e = e_{11} \otimes e_{11} + e_{22} \otimes e_{22}$, and Q_2 is a $*$-isomorphism of $e(M_2 \otimes M_2)e$ onto M_2). Since

$$\tilde{R} = (i_B \otimes Q) \circ (R \otimes i_2)$$

it follows that $\tilde{R}$ is completely positive whenever R is.

<u>Proposition 1.3</u>

Let A and B be C*-algebras.

(1) If $T \in B(A,B)$ is a selfadjoint decomposable linear map, then

$$\|T\|_{dec} = \inf \{ \|S\| \mid S \in CP(A,B) \; , \; -S \underset{cp}{\leqq} T \underset{cp}{\leqq} S \}$$

$$= \inf \{ \|T_1 + T_2\| \mid T_1, T_2 \in CP(A,B), \; T = T_1 - T_2 \} .$$

(2) Let $T \in B(A,B)$ and let $\tilde{T} \in B(A, B \otimes M_2)$ be the selfadjoint linear map given by

$$\tilde{T}(x) = \begin{pmatrix} 0 & T(x^*)^* \\ T(x) & 0 \end{pmatrix}$$

then

$$\|T\|_{dec} = \|\tilde{T}\|_{dec} .$$

(3) Any decomposable map T from A to B is completely bounded and

$$\|T\|_{cb} \leqq \|T\|_{dec} .$$

(4) If T is a completely positive map from A to B , then

$$\|T\|_{dec} = \|T\|_{cb} = \|T\| .$$

(5) If C is a third C*-algebra, and $T_1 \in B(A,B)$, $T_2 \in B(B,C)$ are two decomposable linear maps, then $T_2 \circ T_1$ is a decomposable map from A to C , and

$$\|T_2 \circ T_1\|_{dec} \leqq \|T_2\|_{dec} \|T_1\|_{dec}$$

<u>proof</u>

(1) If x,y are selfadjoint elements in a C*-algebra D, then

$$-y \leqq x \leqq y \Rightarrow \begin{pmatrix} y & x \\ x & y \end{pmatrix} \geqq 0 .$$

Moreover, if x,y,z are selfadjoint elements in D, then

$$\begin{pmatrix} y_1 & x \\ x & y_2 \end{pmatrix} \geqq 0 \Rightarrow -\tfrac{1}{2}(y_1 + y_2) \leqq x \leqq \tfrac{1}{2}(y_1 + y_2) .$$

Applying this to elements in $B \otimes M_m$, it follows that if $T, S \in B(A,B)$ are selfadjoint maps, then

$$-S \underset{cp}{\leqq} T \underset{cp}{\leqq} S \Rightarrow \begin{pmatrix} S & T \\ T & S \end{pmatrix} \in CP(A, B \otimes M_2)$$

and if $T, S_1, S_2 \in B(A,B)$ are selfadjoint maps, then

$$\begin{pmatrix} S_1 & T \\ T & S_2 \end{pmatrix} \in CP(A, B \otimes M_2) \Rightarrow -\tfrac{1}{2}(S_1 + S_2) \underset{cp}{\leqq} T \underset{cp}{\leqq} \tfrac{1}{2}(S_1 + S_2).$$

This proves the first equality in (1). To prove the second equality in (1), assume that $T \in B(A,B)$, $S \in CP(A,B)$ and

$$-S \underset{cp}{\leqq} T \underset{cp}{\leqq} S$$

Then $T_1 - T_2$ where $T_1 = \tfrac{1}{2}(S + T)$, $T_2 = \tfrac{1}{2}(S - T)$ are completely positive and $T_1 + T_2 = S$. Conversely if $T = T_1 - T_2$, where $T_1, T_2 \in CP(A,B)$, then

$$-(T_1 + T_2) \underset{cp}{\leqq} T \underset{cp}{\leqq} (T_1 + T_2)$$

This proves the second equality.

(2) We prove first that $\|\tilde{T}\|_{dec} \leqq \|T\|_{dec}$. Clearly we can assume that $\|T\|_{dec} < \infty$. Let $\varepsilon > 0$. There exist $S_1, S_2 \in CP(A,B)$ such that

$$R(x) = \begin{pmatrix} S_1(x) & T(x^*)^* \\ T(x) & S_2(x) \end{pmatrix}, \quad x \in A$$

is completely positive, and $\|S_i\| \leqq \|T\|_{dec} + \varepsilon$, $i = 1,2$.
We put

$$\tilde{S}(x) = \begin{pmatrix} S_1(x) & 0 \\ 0 & S_2(x) \end{pmatrix}, \quad x \in A .$$

Then clearly $\tilde{S} \in CP(A, B \otimes M_2)$, $\|\tilde{S}\| \leqq \|T\|_{dec} + \varepsilon$ and

$$-\tilde{S} \underset{cp}{\leqq} \tilde{T} \underset{cp}{\leqq} \tilde{S} .$$

Since ε is arbitrary we have $\|\tilde{T}\|_{dec} \leqq \|T\|_{dec}$. We prove next that $\|T\|_{dec} \leqq \|\tilde{T}\|_{dec}$. We can assume that $\|\tilde{T}\|_{dec} < \infty$. Let $\varepsilon > 0$. By (1) there exists $\tilde{S} \in CP(A, B \otimes M_2)$, such that

$$-\tilde{S} \underset{cp}{\leqq} \tilde{T} \underset{cp}{\leqq} \tilde{S} .$$

and $\|\tilde{S}\| \leqq \|T\|$.
We have

$$\tilde{S}(x) = \begin{pmatrix} S_{11}(x) & S_{12}(x) \\ S_{21}(x) & S_{22}(x) \end{pmatrix}, \quad x \in A$$

where $S_{11}, S_{22} \in CP(A,B)$, $S_{21}, S_{12} \in B(A,B)$ and $S_{12} = S_{21}^*$.
Let $u \in B \otimes M_2$ be the unitary

$$u = \begin{pmatrix} 1 & 0 \\ 0 & -1 \end{pmatrix} .$$

Then

$$u\,\tilde{S}(x)u^* = \begin{pmatrix} S_{11}(x) & -S_{12}(x) \\ -S_{21}(x) & S_{22}(x) \end{pmatrix}, \quad x \in A$$

and

$$u\,\tilde{T}(x)u^* = -\tilde{T}(x) \qquad x \in A .$$

Therefore

$$-\mathrm{ad}(u) \circ \tilde{S} \underset{cp}{\leqq} -\tilde{T} \underset{cp}{\leqq} \mathrm{ad}(u) \circ \tilde{S} .$$

In particular

$$\mathrm{ad}(u)\circ\tilde{S} + \tilde{T} \underset{cp}{\geqq} 0 \quad .$$

Put

$$R(x) = \begin{pmatrix} S_{11}(x) & T(x^*)^* \\ T(x) & S_{22}(x) \end{pmatrix} \qquad x \in A \; .$$

Then R is completely positive, because

$$R(x) = \tfrac{1}{2}(\tilde{S} + \tilde{T}) + \tfrac{1}{2}(\mathrm{ad}(u)\circ\tilde{S} + \tilde{T}) \; .$$

Moreover

$$\max\{\|S_{11}\|, \|S_{22}\|\} = \|\tilde{S}\| < \|\tilde{T}\|_{dec} + \varepsilon \; .$$

This proves that $\|T\|_{dec} \leqq \|\tilde{T}\|_{dec}$.

(3) It is clear that any linear combination of completely positive maps is completely bounded. Let $T \in B(A,B)$ be a decomposable map, and assume first that $T = T^*$. Let $\varepsilon > 0$. By (1) there exist $T_1, T_2 \in CP(A,B)$, such that $T = T_1 - T_2$ and

$$\|T_1 + T_2\| < \|T\|_{dec} + \varepsilon \; .$$

For $R \in B(A,B)$, be put $R^{(m)} = R \otimes i_m$, where i_m is the identity on the m×m-matrices M_m . For $x \in (A \otimes M_m)_{s.a.}$ we have

$$\begin{aligned} T^{(m)}(x) &= T_1^{(m)}(x) - T_2^{(m)}(x) \\ &\leqq T_1^{(m)}(|x|) + T_2^{(m)}(|x|) \\ &= (T_1 + T_2)^{(m)}(|x|) \end{aligned}$$

and similarly

$$-T^{(m)}(x) \leqq (T_1 + T_2)^{(m)}(|x|) \; .$$

Since $T_1 + T_2$ is completely positive,

$$\|T_1 + T_2\|_{cb} = \|T_1 + T_2\| \; .$$

Thus

$$\|T^{(m)}(x)\| \leq \|T_1 + T_2\| \; \|x\| \; .$$

If $x \in A \otimes M_m$ is not selfadjoint, then

$$y = \begin{pmatrix} 0 & x^* \\ x & 0 \end{pmatrix} \in (A \otimes M_{2m})_{s.a.}$$

Since $(T^{(m)})^* = T^{(m)}$ we have

$$T^{(2m)}(y) = \begin{pmatrix} 0 & T^{(m)}(x)^* \\ T^{(m)}(x) & 0 \end{pmatrix} \in (B \otimes M_{2m})_{s.a.}$$

Hence

$$\|T^{(m)}(x)\| = \|T^{(2m)}(y)\| \leq \|T_1 + T_2\| \; \|y\| = \|T_1 + T_2\| \; \|x\| \; .$$

This shows that $\|T\|_{cb} \leq \|T\|_{dec} + \varepsilon$.

(4) It is well known that $\|T\|_{cb} = \|T\|$ for any completely positive map. The equality $\|T\|_{dec} = \|T\|_{cb}$ follows from (1) and (3).

(5) It is clear that $T_2 \circ T_1 \in \text{span}\, CP(A,C)$. Choose

$$S_1^{(1)}, S_1^{(2)} \in CP(A,B) \quad \text{and} \quad S_2^{(1)}, S_2^{(2)} \in CP(B,C)$$

such that

$$R_i(x) = \begin{pmatrix} S_i^{(1)}(x) & T_i^*(x) \\ T_i(x) & S_i^{(2)}(x) \end{pmatrix} , \quad i = 1,2$$

defines completely positive maps $R_1 \in CP(A, B \otimes M_2)$ and $R_2 \in CP(B, C \otimes M_2)$, such that

$$\max \{\|S_i^{(1)}\| , \|S_i^{(2)}\|\} \leq \|T_i\|_{dec} + \varepsilon \; .$$

By remark 1.2 the map $\tilde{R}_2 \in B(B \otimes M_2 , C \otimes M_2)$ given by

$$\tilde{R}_2 \begin{pmatrix} x_{11} & x_{12} \\ x_{21} & x_{22} \end{pmatrix} = \begin{pmatrix} S_2^{(1)}(x_{11}) & T_2^{\,*}(x_{12}) \\ T_2(x_{21}) & S_2^{(2)}(x_{22}) \end{pmatrix}$$

is completely positive. Hence $\tilde{R}_2 \circ R_1 \in CP(A, C \otimes M_2)$.

For $x \in A$,

$$\tilde{R}_2 \circ R_1(x) = \begin{pmatrix} S_2^{(1)} \circ S_1^{(1)}(x) & T_2^* \circ T_1^*(x) \\ T_2 \circ T_1(x) & S_2 \circ S_1(x) \end{pmatrix} .$$

Therefore

$$\|T_2 \circ T_1\|_{dec} \leqq \max\{\|S_2^{(1)} \circ S_1^{(1)}\|, \|S_2^{(2)} \circ S_1^{(2)}\|\}$$

$$\leqq (\|T_2\|_{dec} + \varepsilon)(\|T_1\|_{dec} + \varepsilon)$$

This proves (5).

Proposition 1.4

Let A and B be C*-algebras.

(1) The decomposable maps from A to B form a Banach space with norm $\| \ \|_{dec}$.

(2) If every completely bounded map from A to B is decomposable then there exists a constant $c < \infty$, such that

$$\|T\|_{dec} \leqq c\|T\|_{cb}$$

for all $T \in CB(A,B)$.

proof

(1) Put $V(A,B) = \text{span}\, CP(A,B)$. It is clear that $\| \ \|_{dec}$ is a norm on $V(A,B)$. Since $\|T^*\|_{dec} = \|T\|_{dec}$ for all $T \in V(A,B)$ it is sufficient to prove that the selfadjoint part of $(V(A,B), \| \ \|_{dec})$ is complete. This follows in fact from [20, Remark p. 159], but since no proof is given there, we will include a proof: Let $(T_n)_{n\in\mathbb{N}}$ be a sequence of selfadjoint linear maps from A to B , such that

$$\sum_{n=1}^{\infty} \|T_n\|_{dec} < +\infty .$$

Since $B(A,B)$ is a Banach space, there exists an operator $T \in B(A,B)$ such that

$$\lim_{p\to\infty} \|\sum_{n=1}^{p} T_n - T\| = 0 .$$

By prop. 1.3(2), there exists $S_n \in CP(A,B)$, such that

$$-S_n \underset{cp}{\leqq} T_n \underset{cp}{\leqq} S_n$$

and $\|S_n\| \leqq 2\|T_n\|_{dec}$. In particular

$$\sum_{n=1}^{\infty} \|S_n\| < \infty .$$

Therefore we can define $R_p \in B(A,B)$, by

$$R_p = \sum_{n=p+1}^{\infty} S_n , \quad p = 1,2,3,\dots$$

Each R_p is completely positive. Since the cone $CP(A,B)$ is closed in $B(A,B)$ one gets

$$-R_1 \underset{cp}{\leqq} T \underset{cp}{\leqq} R_1 .$$

Thus $T \in V(A,B)$. Moreover for all $p \in \mathbb{N}$,

$$-R_p \underset{cp}{\leqq} T - \sum_{n=1}^{p} T_n \leqq R_p .$$

This implies that

$$\|T - \sum_{n=1}^{p} T_n\|_{dec} \leqq \|R_p\| \leqq 2 \sum_{n=p+1}^{\infty} \|T\|_{dec} .$$

Therefore

$$\lim_{p\to\infty} \|T - \sum_{n=1}^{p} T_n\|_{dec} = 0 .$$

This proves that the selfadjoint part of $V(A,B)$ is complete in the $\| \ \|_{dec}$-norm (cf. f.inst. [12, lemma 1.5.2]).

(2) Follows from (1) by applying the open mapping theorem to the identity map from

$$(V(A,B) , \| \ \|_{dec}) \text{ to } (CB(A,B) , \| \ \|_{cb}).$$

Remark 1.5

We do not know whether the infimum in the definition of $\|T\|_{dec}$

is actually a minimum i.e. whether S_1, S_2 in definition 1.1 can be chosen such that

$$\max\{\|S_1\|, \|S_2\|\} = \|T\|_{dec} .$$

However, this is true in two important cases, namely if B is a von Neumann algebra or if B is an injective C*-algebra. More generally it is true whenever there exists a conditional expectation ε from B^{**} to B : Assume namely that $T \in B(A,B)$ is decomposable. By a simple compactness argument one can find $S_1, S_2 \in CP(A,B^{**})$, such that

$$R(x) = \begin{pmatrix} S_1(x) & T(x^*)^* \\ T(x) & S_2(x) \end{pmatrix}$$

is a completely positive map from A to $B^{**} \otimes M_2$ and $\max\{\|S_1\|, \|S_2\|\} \leqq \|T\|_{dec}$.

Then

$$R'(x) = \begin{pmatrix} \varepsilon\circ S_1(x) & T(x^*)^* \\ T(x) & \varepsilon\circ S_2(x) \end{pmatrix}$$

defines a completely positive map from A to $B \otimes M_2$, and

$$\max\{\|\varepsilon\circ S_1\|, \|\varepsilon\circ S_2\|\} \leqq \|T\|_{dec} .$$

The converse inequality is trivial.

Clearly, under the same condition on B , one gets also that the two infima in Prop. 1.2(1) are actually minima.

Having remark 1.2 and remark 1.5 in mind Wittstock's and Paulsen's theorems [27, Satz 4.5] and [16, theorem 2.5] can be reformulated in the following way:

<u>Theorem 1.6</u> (Wittstock, Paulsen).

Let T be a completely bounded linear map from a C*-algebra A into an injective C*-algebra B , then T is decomposable and

$$\|T\|_{dec} = \|T\|_{cb} .$$

§2.

The main results.

For $n \in \mathbb{N}$, we let ℓ_n^∞ denote the n-dimensional abelian C*-algebra $\ell^\infty\{1,\dots,n\}$.

Theorem 2.1

Let N be a von Neumann algebra. Then the following four conditions are equivalent

(1) N is injective .

(2) For every C*-algebra A and every completely bounded map T from A to N , $\|T\|_{dec} = \|T\|_{cb}$.

(3) For every $n \in \mathbb{N}$, and for every linear map T from ℓ_n^∞ to N , $\|T\|_{dec} = \|T\|_{cb}$.

(4) There exists a constant $c \in \mathbb{R}_+$, such that for every $n \in \mathbb{N}$ and for every linear map T from ℓ_n^∞ to N , $\|T\|_{dec} \leqq c\|T\|_{cb}$.

Note that (1) => (2) is Wittstock's and Paulsen's result, and that (2) => (3) => (4) is trivial, so we have to prove (4) => (1).

For any complex linear space E we let E^c denote the conjugate space i.e. the set E equipped with the same addition as before, but where the scalar multiplication is given by

$$(c,x) \to \bar{c}x \quad , \quad c \in \mathbb{C} \; , \; x \in E.$$

For $x \in E$, we let x^c denote the corresponding element in E^c. If A is an algebra, we consider A^c as an algebra with unchanged multiplication i.e.

$$(ab)^c = a^c b^c \ , \quad a,b \in A \ .$$

In [6, Remark 5.29] Connes proved that for a factor N op type II_1 acting on a separable Hilbert space H, the following two conditions are equivalent

(i) N is injective .

(ii) For any finite set $u_1,\dots,u_n$ of unitaries in N

$$\|\sum_{i=1}^{n} u_i \otimes u_i^c\|_{H \otimes H^c} = n \ .$$

The key step in the proof of (4) => (1) is the following extension of Connes' result:

<u>Lemma 2.2</u>

Let N be a von Neumann algebra acting on a Hilbert space H. The following two conditions are equivalent:

(i) N is finite and injective .

(ii) For any finite set $u_1,\dots,u_n$ of unitaries in N and for any non-zero central projection p in N,

$$\|\sum_{i=1}^{n} p u_i \otimes (pu_i)^c\|_{H \otimes H^c} = n \ .$$

<u>proof</u>

(i) => (ii) : Assume that N is finite and injective. Since any non-zero central projection in N dominates a σ-finite non-zero central projection it is sufficient to prove (2) when p is σ-finite. By passing to the reduced algebra pN, it is sufficient to consider the case, where N itself is σ-finite and $p = 1$. Let τ be a normal faithful tracial state on N. For $a \in N$ we let L_a (resp. R_a) denote the multiplication with a from left (resp. from right) on $L^2(N,\tau)$. Since any injective von Neumann algebra is semidiscrete (cf. [26] and [7]),

$$\|\sum_{i=1}^{m} L_{a_i} R_{b_i^*}\| \leq \|\sum_{i=1}^{m} a_i \otimes b_i^c\|_{H \otimes H^c}$$

for every $m \in \mathbb{N}$ and every $a_1, \dots, a_m$, $b_1, \dots, b_m \in N$. In particular, for any finite set of unitaries $u_1, \dots, u_n$ in N

$$\|\sum_{i=1}^{n} u_i \otimes u_i^c\|_{H \otimes H^c} \geq \|\sum_{i=1}^{n} L_{u_i} R_{u_i^*}\|$$

$$\geq \|\sum_{i=1}^{n} u_i \, 1 \circ u_i^* \|_2 = n .$$

This proves that (ii) => (i). For the proof of (ii) => (i) we shall need the notion of <u>hypertraces</u> introduced by Connes [6, Remark 5.34]. A state ω on $B(H)$ is called a hypertrace for N if for all $x \in B(H)$ and all $a \in N$,

$$\omega(ax) = \omega(xa).$$

Consider now the following two conditions on a von Neumann algebra N:

(iii) For every non-zero central projection p in N, there exists a hypertrace ω for N, such that $\omega(1-p) = 0$.

(iv) For every state ω_o on $Z(N)$ (the center of N), there exists a hypertrace ω for N, such that $\omega(z) = \omega_o(z)$ for all $z \in Z(N)$.

We will prove that (ii) => (iii) => (iv) => (i). Assume that N satisfies (ii). Let $HS(H)$ denote the space of Hilbert-Schmidt operators on H and let $\| \ \|_{HS}$ be the Hilbert-Schmidt norm. Since $HS(H)$ can be identified in a natural way with $H \otimes H^c$, one gets that for $a_1, \dots, a_n$, $b_1, \dots, b_n \in B(H)$,

$$\|\sum_{i=1}^{n} a_i \otimes b_i^c\|_{H \otimes H^c} = \sup\{\|\Sigma a_i x b_i^*\|_{HS} \mid x \in HS(H), \|x\|_{HS} \leq 1\}.$$

Let p be a non-zero central projection in N. Let $\mathcal{F}$ be the family of sets

$$F = (u_1, u_2, \ldots, u_n, \varepsilon)$$

where $n \in \mathbb{N}$, $u_1, \ldots, u_n$ are distinct unitaries in N, and $\varepsilon > 0$. Let $F = (u_1, \ldots, u_n, \varepsilon) \in \mathcal{F}$. By (ii)

$$\| p \otimes p^c + \sum_{i=1}^{n} (pu_i) \otimes (pu_i)^c \| = n + 1 .$$

Therefore we can choose $x_F \in HS(H)$, such that $\|x_F\|_{HS} \leqq 1$, and

$$\| p\, x_F\, p + \sum_{i=1}^{n} pu_i\, x_F\, pu_i^* \| > (n+1) - \varepsilon .$$

By exchanging x_F with px_Fp, we have still $\|x_F\|_{HS} \leqq 1$. Moreover

$$px_F = x_F p = x_F$$

and

$$\|x_F + \sum_{i=1}^{n} u_i\, x_F\, u_i^*\|_{HS} > (n+1) - \varepsilon .$$

Since for $k=1,\ldots,n$ we have

$$\| \sum_{i \neq k} u_i x_F u_i^* \|_{HS} \leqq n - 1$$

it follows that

$$\|x_F + u_k x_F u_k^*\|_{HS} > 2 - \varepsilon \quad , \quad k=1,\ldots,n .$$

So, by the parallelogramidentity

$$\begin{aligned} \|x_F - u_k\, x_F\, u_k^*\|_{HS}^2 &\leqq 2\|x_F\|_{HS}^2 + 2\|u_k\, x_F\, u_k^*\|_{HS}^2 - (2-\varepsilon)^2 \\ &\leqq 4 - (2-\varepsilon)^2 \\ &< 4\varepsilon . \end{aligned}$$

Since $\|x_F\|_{HS} = \|u_k\, x_F\, u_k^*\|_{HS}$ we have also

$$\|x_F\|_{HS} > 1 - \tfrac{1}{2}\varepsilon .$$

Define a positive functional ω_F on N by

$$\omega_F(a) = (a\,x_F\,,\,x_F)_{HS} = T_r(a\,x_F\,x_F^*)\,.$$

For $a \in N$, and $x,y \in HS(H)$,

$$|(ax,x)_{HS} - (ay,y)_{HS}| = \tfrac{1}{2}|(a(x+y),\,(x-y))_{HS} + (a(x-y),\,(x+y))_{HS}|$$

$$\leq \|a\|\;\|x-y\|_{HS}\,\|x+y\|_{HS}\,.$$

Hence for $a \in N$ and $i=1,\ldots,n$.

$$\omega_F(a - u_i a u^*_i) \leq \|a\|\;\|x_F - u_i x_F u_i^*\|_{HS}\|x_F + u_i x_F u_i^*\|_{HS}$$

$$\leq 4\varepsilon^{\frac{1}{2}}\|a\|\,.$$

Also $\omega_F(1-p) = 0$, and $\omega_F(1) = \|x_F\|_{HS}^2 > 1-\varepsilon$.

The set $\mathfrak{F}$ is directed with the ordering $\leq$ given by

$$(u_1,\ldots,u_n,\varepsilon) \leq (v_1,\ldots,v_m,\delta)$$

if $\{v_1,\ldots,v_m\}$ contains the set $\{u_1,\ldots,u_n\}$ and $\delta \leq \varepsilon$. Let $\omega \in B(H)^*$ be a $\sigma(B(H)^*, B(H))$ cluster point for the net $(\omega_F)_{F\in\mathfrak{F}}$. Clearly ω is a state on $B(H)$,

$$\omega(uxu^*) = \omega(x)\,, \quad x \in B(H)\,, \quad u \in U(N)$$

i.e. ω is a hypertrace for N. Moreover $\omega(1-p) = 0$.
Hence we have proved that (ii) => (iii).

(iii) => (iv): Let ω_o be a state on $Z(N)$, and let

$$P = \{p_1,\ldots,p_r\}$$

be a "partition of the unity" in $Z(N)$, i.e. $r \in \mathbb{N}$ and $p_1,\ldots,p_r$ are non-zero pairwise orthogonal projections in $Z(N)$ with sum 1. If N satisfies (iii) we can choose hypertraces $\omega_1,\ldots,\omega_r \in B(H)^*$ for N, such that $\omega_k(1-p_k) = 0$. Put now

$$\omega_P = \sum_{k=1}^{r} \omega_o(p_k)\omega_k .$$

Then ω_P is a hypertrace on N, and

$$\omega_P(p_k) = \omega_o(p_k).$$

The set $\mathcal{P}$ of partition of the unity in $Z(N)$ is directed by the ordering $\leqq$, where $P \leqq Q$ means that each projection in P can be written as a sum of projections in Q. Let now ω be a $\sigma(B(H)^*, B(H))$-cluster point for the net $(\omega_P)_{P\in\mathcal{P}}$. Then ω is a hypertrace for N, and ω coincides with ω_o on every central projection. Hence

$$\omega(x) = \omega_o(x) \quad , \quad x \in Z(N).$$

(iv) => (i) : Assume that N satisfies (iv). We prove first that N is finite: Let $e \in Z(N)$ be the largest finite projection in $Z(N)$. If $1-e \neq 0$, we can choose a state ω_o on $Z(N)$, such that $\omega_o(1-e) = 1$. By (iv) there exists a hypertrace $\omega \in B(H)^*$ for N such that $\omega(1-e) = 1$. The restriction of ω to $(1-e)N$ is a tracial state. This gives a contradiction, because $(1-e)N$ is properly infinite. Hence $e = 1$ and N is finite. Since any finite von Neumann algebra is a direct sum of σ-finite, finite algebras, we can in the rest of the proof of (3) => (1) assume that N itself is σ-finite and finite. Let ω_o be a normal faithful state on $Z(N)$ and let $\omega \in B(H)^*$ be a hypertrace for N that extends ω_o. The restriction τ of ω to N is a trace on N. Let T be the central-valued trace on N, then

$$\tau = \tau \circ T = \omega_o \circ T .$$

This shows that τ is a normal, faithful tracial state on N.

For $x \in B(H)$, we let φ_x be the functional on N given by

$$\varphi_x(a) = \omega(ax) = \omega(xa) \quad , \quad a \in N$$

In particular $\varphi_1(a) = \tau(a)$.

If $0 \leqq x \leqq 1$, then for all $a \in N_+$,

$$\varphi_x(a) = \omega(ax) = \omega(a^{\frac{1}{2}}xa^{\frac{1}{2}}) \geqq 0$$

and

$$\varphi_x(a) = \tau(a) - \omega(a^{\frac{1}{2}}(1-x)a^{\frac{1}{2}}) \leqq \tau(a).$$

Therefore $0 \leqq \varphi_x \leqq \tau$. Hence there is a unique $b_x \in N_+$, $0 \leqq b_x \leqq 1$, such that

$$\varphi_x(a) = \tau(b_x a) .$$

Since N is spanned by the positive elements in N of norm $\leqq 1$, the map $x \to b_x$ can be extended to a linear map $E : B(H) \to N$, such that

$$\tau(E(x)a) = \varphi_x(a) = \omega(xa) \quad , \quad x \in B(H) \quad , \quad a \in N.$$

Clearly, E is positive, $E(1) = 1$. Moreover for $a_1, a_2 \in N$

$$\tau(E(a_1xa_2)b) = \omega(a_1xa_2b) = \omega(xa_2ba_1)$$
$$= \tau(E(x)a_2ba_1) = \tau(a_1E(x)a_2b)$$

for every $b \in N$. This shows that $E(a_1xa_2) = a_1E(x)a_2$ i.e. E is a conditional expectation of $B(H)$ onto N. Hence N is injective. This completes the proof of lemma 2.2.

Lemma 2.3

Let N be a von Neumann algebra on a Hilbert space H. The following two conditions are equivalent

(i) N is finite and injective .

(ii') There exists a constant $\gamma > 0$, such that for every finite set $u_1,\dots,u_n$ of unitaries in N and any non-zero central projection p in N,

$$\left\| \sum_{i=1}^{n} pu_i \otimes (pu_i)^c \right\|_{H \otimes H^c} \geqq \gamma\, n .$$

<u>proof</u>

(i) => (ii') follows from lemma 2.2. To prove (ii') => (i) assume that N satisfies (ii') with $\gamma = \gamma_o > 0$, but that N does not satisfy (i). By lemma 2.2 we can choose a central projection p and unitaries $u_1,\dots,u_n$ in N, such that

$$\left\| \sum_{i=1}^{n} pu_i \otimes (pu_i)^c \right\|_{H \otimes H^c} < n .$$

Put

$$\alpha = \frac{1}{n} \left\| \sum_{i=1}^{n} pu_i \otimes (pu_i)^c \right\|_{H \otimes H^c} .$$

Since $\alpha < 1$, we can choose $r \in \mathbb{N}$, such that $\alpha^r < \gamma_o$. Put $\Lambda = \{1,\dots,n\}^r$. Note that Λ is a finite set with n^r elements. For $\lambda = (i_1,\dots,i_r) \in \Lambda$, put

$$v_\lambda = u_{i_1} u_{i_2} \dots u_{i_r} .$$

Then

$$\sum_{\lambda \in \Lambda} pv_\lambda \otimes (pv_\lambda)^c = \left(\sum_{i=1}^{p} pu_i \otimes (pu_i)^c \right)^r$$

and therefore

$$\left\| \sum_{\lambda \in \Lambda} pv_\lambda \otimes (pv_\lambda)^c \right\| \leqq (\alpha n)^r < \gamma_o n^r .$$

This contradicts that N satisfies (ii') with $\gamma = \gamma_o$. Hence (ii') => (i).

mma 2.4

t H and K be Hilbert spaces and let $a_1,\dots,a_n \in B(H)$,
$,\dots,b_n \in B(K)$. Then

$$\|\sum_{i=1}^n a_i \otimes b_i^c\|_{H\otimes K^c} \leqq \|\sum_{i=1}^n a_i \otimes a_i^c\|^{\frac12}_{H\otimes H^c} \cdot \|\sum_{i=1}^n b_i \otimes b_i^c\|^{\frac12}_{K\otimes K^c} .$$

roof

ssume first that $H = K$. By the usual identification of
$\otimes H^c$ with the Hilbert-Schmidt operators $HS(H)$ on H, we have

$$\|\sum_{i=1}^n a_i \otimes b_i^c\|_{H\otimes H^c} = \sup\{\|\sum_{i=1}^n a_i x b_i^*\|_{HS} \mid \|x\|_{HS} \leqq 1\}$$

$$= \sup\{\mathrm{Tr}(\sum_{i=1}^n a_i x b_i^* y^*) \mid \|x\|_{HS} \leqq 1,\ \|y\|_{HS} \leqq 1\}.$$

et $x,y \in HS(H)$, $\|x\|_{HS} \leqq 1$, $\|y\|_{HS} \leqq 1$, and let $x = u|x|$,
$= v|y|$ be the polardecompositions of x and y. Put

$$x_1 = u|x|^{\frac12}, \quad x_2 = |x|^{\frac12}$$
$$y_1 = v|y|^{\frac12}, \quad y_2 = |y|^{\frac12} .$$

Then

$$x = x_1x_2, \quad y = y_1y_2$$
$$|x| = x_2^*x_2, \quad |y| = y_2^*y_2$$
$$|x^*| = x_1x_1^*, \quad |y^*| = y_1y_1^* .$$

Therefore

$$\sum_{k=1}^n \mathrm{Tr}(y^*a_kxb_k^*) = \sum_{k=1}^n \mathrm{Tr}(y_1^* a_k x_1x_2b_k^*y_2^*)$$

$$\leqq \sum_{k=1}^n \mathrm{Tr}(y_1^* a_k x_1 (y_1^* a_k x_1)^*)^{\frac12}\, \mathrm{Tr}((x_2b_k^*y_2^*)^*(x_2b_k^*y_2^*))^{\frac12}$$

$$\leqq \Big(\sum_{k=1}^n \mathrm{Tr}(y_1^*a_kx_1x_1^*a_k^*y_1)\Big)^{\frac12}\Big(\sum_{k=1}^n \mathrm{Tr}(y_2b_kx_2^*x_2b_k^*y_2^*)\Big)^{\frac12}$$

$$= \left(\sum_{k=1}^{n} \mathrm{Tr}(|y^*|a_k|x^*|a_k^*)\right)^{\frac{1}{2}}\left(\sum_{k=1}^{n} \mathrm{Tr}(|y|b_k|x|b_k^*)\right)^{\frac{1}{2}}$$

$$\leq \left\| \sum_{k=1}^{n} a_k \otimes a_k^c \right\| \left\| \sum_{k=1}^{n} b_k \otimes b_k^c \right\| .$$

Here we have used that

$$\| |x| \|_{HS} = \| |x^*| \|_{HS} = \|x\|_{HS} \leq 1 \text{ and } \| |y| \|_{HS} = \| |y^*| \|_{HS} = \|y\|_{HS} \leq 1 .$$

This completes the proof in the case $H = K$. The general case can be reduced to this case if one puts

$$\tilde{H} = H \oplus K$$

and considers the operators $\tilde{a}_1,\ldots,\tilde{a}_n,\ \tilde{b}_1,\ldots,\tilde{b}_n \in B(\tilde{H})$ given by

$$\tilde{a}_k(\xi,\eta) = (a_k\xi, 0)$$

$$\tilde{b}_k(\xi,\eta) = (0, b_k\eta)$$

for $\xi \in H$ and $\eta \in K$.

<u>Lemma 2.5</u>

Let $u_1,\ldots,u_n$ be n unitaries in a finite von Neumann algebra N, let p be a non-zero central projection in N, and let T be the linear map from ℓ_n^∞ to N given by

$$T(c_1,\ldots,c_n) = p\left(\sum_{i=1}^{n} c_i u_i\right).$$

Then

a) $\|T\|_{cb} \leq n^{\frac{1}{2}} \left\| \sum_{i=1}^{n} pu_i \otimes (pu_i)^c \right\|^{\frac{1}{2}}$

b) $\|T\|_{dec} = n$.

<u>proof</u>

a) Let $m \in \mathbb{N}$, and put $T^{(m)} = T \otimes i_m$, where i_m is the identity on M_m. An element x in the unitball of $\ell_n^\infty \otimes M_m$ is given by a set $(x_1,\ldots,x_n)$ of n elements in the unitball of M_m. We have

$$T^{(m)}(x) = \sum_{k=1}^{n} pu_k \otimes x_k .$$

We have $M_m \cong B(K)$, where dim $K = n$. Hence by lemma 2.4:

$$\|T^{(m)}(x)\| \leq \|\sum_{i=1}^{n} pu_k \otimes (pu_k)^c\|^{\frac{1}{2}}_{H\otimes H^c} \|\sum_{k=1}^{n} x_k^c \otimes x_k\|_{K^c \otimes K} .$$

$$\leq \|\sum_{i=1}^{n} pu_k \otimes (pu_k)^c\|^{\frac{1}{2}}_{H\otimes H^c} \cdot n^{\frac{1}{2}} .$$

This proves a).

b) Since $\|T\|_{dec} = \|\tilde{T}\|_{dec}$, where $\tilde{T} : A \to B \otimes M_2$ is defined by,

$$\tilde{T}(c_1,\dots,c_n) = \begin{pmatrix} p & 0 \\ 0 & p \end{pmatrix} \cdot \sum_{k=1}^{n} c_k \begin{pmatrix} 0 & u_k^* \\ u_k & 0 \end{pmatrix} \in N \otimes M_2 ,$$

(cf. prop. 1.3(2)) it is sufficient to consider the case where $u_1,\dots,u_n$ are selfadjoint unitaries. Put

$$S(c_1,\dots,c_n) = \left(\sum_{k=1}^{n} c_k\right) p .$$

Then S is a positive map freom ℓ_n^∞ to N and

$$-S \leq T \leq S .$$

However, by [21, thm. 4] a positive map from ℓ_n^∞ to N is automaticly completely positive.

Therefore

$$\|T\|_{dec} \leq \|S(1)\| = n\|p\| = n.$$

Let now τ be a normal tracial state on N for which $\tau(1-p) = 0$, and let $e \leq p$ be the support projection of τ. It is well known that

$$\|x\|_1 = \tau(|x|), \quad x \in M$$

is a norm on eN , and since

$$\|x\|_1 = \|ex\|_1$$

for all $x \in N$, $\|\cdot\|_1$ is a seminorm on N. Assume that $R : \ell_n^\infty \to N$ is a completely positive map, such that

$$-R \underset{cp}{\leq} T \underset{cp}{\leq} R .$$

Put $x_k = R(p_k)$, where $p_1,\dots,p_n$ are the minimal projections in ℓ_n^∞. Then

$$-x_k \leq pu_k \leq x_k \qquad k = 1,\dots,n .$$

Therefore

$$\tau(x_k) = \|\tfrac{1}{2}(x_k + pu_k)\|_1 + \|\tfrac{1}{2}(x_k - pu_k)\|_1$$

$$\geq \|\tfrac{1}{2}(x_k + pu_k) - \tfrac{1}{2}(x_k - pu_k)\|_1 = \|pu_k\|_1 = \tau(p) = 1.$$

Hence

$$\|R(1)\| = \|\sum_{k=1}^{n} x_k\| \geq \sum_{k=1}^{n} \tau(x_k) \geq n .$$

This shows that $\|T\|_{dec} \geq n$.

<u>proof of theorem 2.1</u>

It remains to be proved that (4) => (1). Assume first that N is finite. Let $u_1,\dots,u_n$ be n unitaries in N, let p be a non-zero central projection in N, and let $T : \ell_n^\infty \to N$ be the linear map

$$T(c_1,\dots,c_n) = p(\sum_{i=1}^{n} c_i u_i) .$$

By lemma 2.5

$$\|T\|_{cb} \leq n^{\frac{1}{2}} \|\sum_{i=1}^{n} pu_i \otimes (pu_i)^c\|^{\frac{1}{2}}$$

and

$$\|T\|_{dec} = n.$$

Thus, if $\|T\|_{dec} \leq c\|T\|_{cb}$, we get that

$$\|\sum_{i=1}^{n} pu_i \otimes (pu_i)^c\| \geq n/c^2 .$$

Hence, if N satisfies condition (4), it follows from lemma 2.3 that N is injective. This proves (4) => (1) for N finite.

To prove (4) => (1) for a general von Neumamn algebra, we show first that if a von Neumann algebra M satisfies condition (4) in theorem 2.1, then

(a) Any reduced algebra $N = pMp$ of M satisfies condition (4) in theorem 2.1 .

(b) Any sub von Neumann algebra of N which is the range of a conditional expectation $\varepsilon : M \to N$ satisfies condition (4) in theorem 2.1.

Let namely $T : \ell_n^\infty \to N$ be a linear map. Since in both cases (a) and (b) , $N \subseteq M$, where M satisfies (4) with $c = c_o$, there exist completely positive maps S_1, S_2 from ℓ_n^∞ to M , such that $\|S_i\| \leq c_o\|T\|_{cb}$, $i = 1,2$ and

$$R(x) = \begin{pmatrix} S_1(x) & T(x^*)^* \\ T(x) & S_2(x) \end{pmatrix} , \quad x \in \ell_n^\infty$$

is a completely positive map from ℓ_n^∞ to $M \otimes M_2$ (cf. def. 1.1 and remark 1.4). By putting $S_i' = pS_i(\cdot)p$ in case a) and $S_i' = \varepsilon \circ S$ in case b) one gets completely positive maps S_1', S_2' from ℓ_n^∞ to N , such that

$$R'(x) = \begin{pmatrix} S_1'(x) & T(x^*)^* \\ T(x) & S_2'(x) \end{pmatrix} , \quad x \in \ell_n^\infty$$

defines a completely positive map from ℓ_n^∞ to $N \otimes M_2$. Hence

$$\|T\|_{dec} \leq \max\{\|S_1'\|, \|S_2'\|\} \leq c_o\|T\|_{cb} .$$

This proves (a) and (b) above.

Let now N be a semifinite von Neumann algebra that satisfies condition (4) in theorem 2.1. By [23, Chap. 5, prop. 1.40], N can be written in the form

$$N = \bigoplus_{i\in I} (N_i \hat{\otimes} B(H_i))$$

where $(H_i)_{i\in I}$ is a family of Hilbert spaces and $(N_i)_{i\in I}$ is a family of finite von Neumann algebras. By (a) above each N_i satisfies condition (4). Thus by the first part of the proof each N_i is injective, which implies that N itself is injective completing the proof of (4) => (1) for semifinite algebras.

Assume next that N is a von Neumann algebra of type III that satisfies condition (4). By [22] N is the crossed product of a semifinite Neumann algebra M and a one-parameter group of automorphisms (θ_s) on M

$$N = M \times_\theta \mathbb{R} .$$

Let $\hat{\theta}$ be the dual action of $\mathbb{R}$ on N (cf. [22, Def. 4.1]), and let m be a left invariant mean on $\mathbb{R}$. Then

$$x \to \int_{-\infty}^{\infty} \hat{\theta}_s(x)\, dm(s).$$

defines a conditional expectation ε of N onto the fixed point algebra $N_{\hat{\theta}}$ for $\hat{\theta}$. By [22, thm. 6.1] $N_{\hat{\theta}}$ is isomorphic to M. Thus by (b) above, M also satisfies condition (4), and hence M is injective by the first part of the proof. But the crossed product of an injective von Neumann algebra by an abelian group is again injective (cf. [6, prop. 6.8]). Hence (4) => (1) for von Neumann algebras of type III. Since a general von Neumann algebra is the direct sum of a semifinite algebra and a type III-algebra, we are don

Theorem 2.6

Let N be a non-injective von Neumann algebra.

a) For every infinite dimensional C*-algebra A, there exists a map $T \in CB(A,N)$, which is not a linear combination of completely positive maps from A to N.

b) For every infinite dimensional von Neumann algebra M, there exists a normal map $T \in CB(M,N)$ which is not a linear combination of completely positive maps from M to N.

For the proof of theorem 2.6 we shall need

Lemma 2.7

Let A be an infinite dimensional C*-algebra. For each $n \in \mathbb{N}$, there exist completely positive maps

$$R_n : \ell_n^\infty \to A \quad , \quad S_n : A \to \ell_n^\infty$$

such that $\|R_n\| \leqq 1$ $\|S_n\| \leqq 1$, and

$$S_n \circ R_n(x) = x \, , \quad x \in \ell_n^\infty \, .$$

If A is a von Neumann algebra R_n and S_n can be chosen normal and unitpreserving.

proof

Let B be a maximal abelian *-subalgebra of A. Since B is infinite dimensional (cf. : [12, exercise 4.6.12]), the spectrum $\hat{B}$ of B is infinite. Let $n \in \mathbb{N}$. We can choose n distinct characters

$$\omega_1, \ldots, \omega_n \in \hat{B} \, .$$

Moreover, since B is isomorphic to $C_o(\hat{B})$, we can choose n positive selfadjoint elements

$$b_1,\dots,b_n \in B$$

such that $\|b_i\| \leqq 1$, $\omega_i(b_i) = 1$ for $i=1,\dots,n$ and such that the corresponding functions on $C_o(\hat{B})$ have disjoint supports. Let $\varphi_1,\dots,\varphi_n$ be extensions of $\omega_1,\dots,\omega_n$ to states on A. Put

$$R_n(c_1,\dots,c_n) = \sum_{i=1}^{n} c_i b_i \qquad c_i \in \mathbb{C}$$

and

$$S_n(a) = (\varphi_1(a),\dots,\varphi_n(a)) \qquad a \in A .$$

Since a positive map from a C*-algebra to another C*-algebra is automaticly completely positive if one of the algebras is abelian (cf. [21, thm. 4] and [2, prop. 1.2.2]), R_n and S_n are completely positive. Moreover one gets easily that $\|R_n\| \leqq 1$, $\|S_n\| \leqq 1$ and $S_n \circ R_n(x) = x$ for $x \in \ell_n^\infty$.

If A is an infinite dimensional von Neumann algebra, let instead $c_1,\dots,c_n$ be n non-zero orthogonal projections with sum 1, let $\varphi_1,\dots,\varphi_n$ be normal states on A, such that the support projection of φ_i is less or equal to c_i, $i=1,\dots,n$, and define R_n and S_n by the above formulas. Then R_n, S_n satisfy all the conditions stated in the second part of lemma 2.7.

<u>proof of theorem 2.6</u>

a) Let N be a von Neumann algebra, and let A be any infinite dimensional C*-algebra. Assume that every completely bounded map from A to N is decomposable. By prop. 1.5, there exists a constant $c \in \mathbb{R}_+$, such that

$$\|T'\| \leqq c\|T'\|_{cb}$$

for all $T' \in CB(A,N)$. For every $n \in \mathbb{N}$ we can choose completely bounded maps $R_n : \ell_n^\infty \to A$ and $T_n : A \to \ell_n^\infty$ which satisfy the conditions of lemma 2.7. Let T be a linear map from ℓ_n^∞ to N.

Since

$$T = (T \circ S_n) \circ R_n$$

we get from prop. 1.3(4)(5) that

$$\|T\|_{dec} \leqq \|T \circ S_n\|_{dec} .$$

Therefore

$$\|T\|_{dec} \leqq c\|T \circ S_n\|_{cb} \leqq c\|T\|_{cb} .$$

Hence N satisfies the condition (4) in theorem 2.1, i.e. N is injective.

b) Let M,N be von Neumann algebras, $\dim M = +\infty$, and assume that any normal map $T \in CB(M,N)$ is decomposable. Since

$$V_n(M,N) = \{T \in \text{span}\, CP(M,N) \mid T \text{ normal}\}$$

is a closed subspace of the Banach space

$$(\text{span}\, CP(M,N) \,,\, \| \ \|_{dec})$$

it follows as in the proof of prop. 1.4 that there exists $c \in \mathbb{R}_+$, such that

$$\|T'\|_{dec} \leqq c\|T'\|_{cb}$$

for all normal maps $T' \in CB(M,N)$. Hence, as in the proof of a) we can conclude that N is injective. This proves theorem 2.6.

If M and N are two von Neumann algebras, we let $CP_n(M,N)$ (resp. $CB_n(M,N)$) denote the set of normal completely positive (resp. normal completely bounded) maps from M to N.

<u>Corollary 2.8</u>

Let N be a von Neumann algebra. The following three conditions are equivalent

(1) N is injective.

(2) CB(N,N) = span CP(N,N).

(3) $CB_n(N,N)$ = span $CP_n(N,N)$.

proof

From theorem 2.6 it follows that (1) <=> (2) <=> (3'), where (3') is the condition

$$(3') \quad CB_n(N,N) \subseteq \text{span } CP(N,N) \ .$$

However, if a normal map T from N to N is a linear combination of completely positive maps $T_1,\dots,T_n$ from N to N

$$T = \sum_{i=1}^{n} c_i T_i$$

then also

$$T = \sum_{i=1}^{n} c_i T_i^{(n)}$$

where $T_i^{(n)},\dots,T_n^{(n)}$ are the normal parts of $T_1,\dots,T_n$ (cf. [23, def. 2.15]). Therefore (3) <=> (3').

Corollary 2.9

Let R be the hyperfinite II_1-factor with tracial state τ, and let ω be a free ultrafilter on R,

$$R^\omega = \ell^\infty(\mathbb{N},R)/I_\omega$$

where I_ω is the ideal in $\ell^\infty(\mathbb{N},R)$ consisting of those bounded sequences (x_n) in R for which

$$\lim_{n\to\omega} \tau(x_n^* x_n) = 0.$$

Then for every infinite dimensional C*-algebra A, there exists a completely bounded map T from A to R^ω, such that T has no completely bounded lifting $\tilde{T} : A \to \ell^\infty(\mathbb{N},R)$.

proof

It is well known that R^ω is a II_1-factor with tracial state

τ_ω given by

$$\tau_\omega(x) = \lim_{n\to\omega} \tau(x_n) ,$$

where $(x_n)_{n\in\mathbb{N}}$ is a representing sequence for $x \in R^\omega$ (cf. [19, Chap. II, sects. 6,7] and [14, p. 451]. Moreover by an argument due to Wassermann R^ω is not injective: Let $\mathbb{F}_2$ be the free group on two generators, then by [25, p. 244], there exists a sequence of representations $(\pi_n)_{n\in\mathbb{N}}$ of finite $\mathbb{F}_2$ into finite dimensional subfactors F_n of R such that

$$\lim_{n\to\infty} \tau(\pi_n(g)) = \begin{cases} 1 & g = e \\ 0 & g \neq e \end{cases}$$

where τ is the normalized trace. Hence as in [25, page 245] one sees that R^ω contains a subfactor isomorphic to $\mathcal{M}(\mathbb{F}_2)$, the von Neumann algebra associated with the regular representation of $\mathbb{F}_2$, which implies that R^ω is not injective (cf. proof of [25, prop. 1.7]).

Let now A be any infinite dimensional C*-algebra. By theorem 2.6 there exists a completely bounded map $T : A \to R^\omega$, which is not decomposable. Assume that $\tilde{T} : A \to \ell^\infty(\mathbb{N},R)$ is a completely bounded lifting of T. Since R is injective, $\ell^\infty(\mathbb{N},R)$ is also an injective von Neumann algebra. Thus by prop. 1.6, $\tilde{T}$ is a linear combination of completely positive maps. But since, $T = \rho \circ \tilde{T}$, where $\rho : \ell^\infty(\mathbb{N},R) \to R^\omega$ is the quotient map, T is also a linear combination of completely positive maps, which gives a contradiction. Hence T has no completely bounded lifting.

§3.

Examples and complements.

Example 3.1

Let $\mathbb{F}_2$ be the free group on two generators a and b, and let λ be the left regular representation of $\mathbb{F}_2$. Choose a free, infinite set $\{x_1, x_2, \ldots\}$ in $\mathbb{F}_2$, f.inst.

$$x_n = b^n a b^{-n}, \quad n \in \mathbb{N}$$

and define a linear map T_n from ℓ_n^∞ to $\mathfrak{M}(\mathbb{F}_2) = \lambda(\mathbb{F}_2)''$ by

$$T_n(c_1, \ldots, c_n) = \frac{1}{2\sqrt{n-1}} \sum_{i=1}^{n} c_i \lambda(x_i) \qquad (n \geqq 2) .$$

We will show that

$$\|T_n\| = \|T_n\|_{cb} = 1$$

while

$$\|T_n\|_{dec} = \frac{n}{2\sqrt{n-1}} .$$

In [1], Akemann and Ostrand proved that

$$\| \sum_{i=1}^{n} \lambda(x_i) \| = 2\sqrt{n-1} , \quad n \geqq 2 .$$

They also proved ([1], Theorem III F) that, for $c_1, \ldots, c_n \in \mathbb{C}$,

$$\| \sum_{i=1}^{n} c_i \lambda(x_i) \| = \| \sum_{i=1}^{n} |c_i| \lambda(x_i) \| .$$

In particular,

$$\| \sum_{i=1}^{n} c_i \lambda(x_i) \| = 2\sqrt{n-1}$$

for $n \geqq 2$ and $|c_1| = |c_2| = \cdots = |c_n| = 1$.

Hence $\|T_n(u)\| = 1$ for every unitary operator $u \in \ell_n^\infty$, and since the unit ball in any finite dimensional C*-algebra is the convex hull of the unitary operators, we conclude that $\|T_n\| = 1$.

Let $m \in \mathbb{N}$, and put $T^{(m)} = T\otimes i_m$, where i_m is the identity on M_m. Every unitary operator $u \in \ell_n^\infty \otimes M_m$ is of the form

$$u = (u_1, \ldots, u_n)$$

where $u_1, \ldots, u_n$ are unitary $m\times m$-matrices. Clearly,

$$T_n^{(m)}(u) = \frac{1}{2\sqrt{n-1}} \sum_{i=1}^{n} \lambda(x_i)\otimes u_i .$$

We can identify the subgroup of $\mathbb{F}_2$ generated by $\{x_1, x_2, \ldots\}$ with the free group $\mathbb{F}_\infty$ on infinite (countable) many generators. The restriction λ' of λ to $\mathbb{F}_\infty$ is just a multiple of the left regular representation λ_∞ of $\mathbb{F}_\infty$. Therefore,

$$\|T_n^{(m)}(u)\| = \frac{1}{2\sqrt{n-1}} \left\| \sum_{i=1}^{n} \lambda_\infty(x_i)\otimes u_i \right\| .$$

Let π be the unitary representation of $\mathbb{F}_\infty$ on the m-dimensional Hilbert space $\mathbb{C}^m$ for which

$$\pi(x_i) = u_i , \quad i \in \mathbb{N} .$$

Then, by [8, Addendum 13.11.3], $\lambda\otimes\pi$ is unitary equivalent to $\lambda\otimes\tau_o$, where τ_o is the trivial representation of $\mathbb{F}_\infty$ on $\mathbb{C}_m$.

Hence,

$$\|T_n^{(m)}(u)\| = \frac{1}{2\sqrt{n-1}} \left\| \sum_{i=1}^{n} \lambda_\infty(x_i) \right\|$$

$$= \frac{1}{2\sqrt{n-1}} \left\| \sum_{i=1}^{n} \lambda(x_i) \right\| = 1 ,$$

which proves that $\|T_n^{(m)}\| = 1$ for all m. Hence $\|T_n\|_{cb} = 1$. Finally, by Lemma 2.5 (b), we have

$$\|T_n\|_{dec} = \frac{n}{2\sqrt{n-1}} .$$

From Example 3.1 and the proof of Corollary 2.8, we get:

<u>Proposition 3.2</u>

Let R be the hyperfinite factor, let ω be a free ultrafilter on $\mathbb{N}$, and let

$$R^{\omega} = \ell^{\infty}(\mathbb{N},R)/I_{\omega}$$

as in Corollary 2.8.

(1) For $n \in \mathbb{N}$, $n \geq 3$, there exists a linear map,

$$T : \ell_n^{\infty} \to R^{\omega} ,$$

such that, for any lifting of T to a linear map $\tilde{T}$ from ℓ_n^{∞} to $\ell^{\infty}(\mathbb{N},R)$,

$$\|\tilde{T}\|_{cb} \geq \frac{n}{2\sqrt{n-1}} \|T\|_{cb} .$$

(2) For $n \in \mathbb{N}$, $n \geq 3$, there exists a linear map,

$$T : M_n \to R^{\omega} ,$$

such that, for every linear lifting of T to a map $\tilde{T}$ from M_n to $\ell^{\infty}(\mathbb{N},R)$,

$$\|\tilde{T}\|_{cb} \geq \frac{n}{2\sqrt{n-1}} \|T\|_{cb} .$$

Proof

(1) By the proof of Corollary 2.8 we can identify $\mathcal{M}(\mathbb{F}_2)$ with a subfactor of R^ω. Let $n \geq 3$, and let $T : \ell_n^\infty \to R^\omega$ be the map obtained by composing T_n from Example 3.1 with the inclusion map. Then $\|T\|_{cb} = 1$, and by Lemma 2.5 (b), we have still

$$\|T\|_{dec} = \frac{n}{2\sqrt{n-1}} .$$

Let $\rho : \ell^\infty(\mathbb{N},R) \to R^\omega$ be the quotient map. If $\tilde{T}$ is a linear lifting of T, then clearly

$$\|\tilde{T}\|_{dec} \geq \|\rho\circ\tilde{T}\|_{dec} = n/2\sqrt{n-1} ,$$

and since $\ell^\infty(\mathbb{N},R)$ is injective, we have $\|\tilde{T}\|_{cb} = \|\tilde{T}\|_{dec}$. This proves (1).

(2) Let $n \geq 3$, and let $(e_{ij})_{i,j=1,\dots,n}$ be the matrix units in M_n. Define a linear map R from ℓ_n^∞ to M_n and a linear map S from M_n to ℓ_n^∞ by

$$R(c_1,\dots,c_n) = \sum_{i=1}^{n} c_i e_{ii}$$

$$S(\Sigma\, a_{ij}\, e_{ij}) = (a_{11},\dots,a_{nn}) .$$

Then R,S are completely positive,

$$R(1) = 1 , \quad S(1) = 1$$

and

$$(S\circ R)(x) = x , \quad x \in \ell_n^\infty .$$

Let $T : \ell_n^\infty \to R^\omega$ be chosen as in (1) and define $T' \in B(M_n, R^\omega)$ by

$$T' = T\circ S .$$

Then

$$T = T' \circ R \ .$$

From these two equalities we get

$$\|T'\|_{cb} = \|T\|_{cb} \quad \text{and} \quad \|T'\|_{dec} = \|T\|_{dec}$$

(cf. Proposition 1.3 (4) and (5)). If $\tilde{T}'$ is any linear lifting of T', then, as in (1), we get

$$\|\tilde{T}'\|_{cb} = \|\tilde{T}'\|_{dec} \geqq \|T'\|_{dec} = \|T\|_{dec} = \frac{n}{2\sqrt{n-1}}$$

while $\|T'\|_{cb} = \|T\|_{cb} = 1$. This proves (2).

It is worthwhile to compare Example 3.1 with an example due to Landford, which has been discussed in papers of Loebl [13, Lemma 2.1], Tsui [24, Lemma 3.2], and Huruya and Tomiyama [11, Lemma 1]. We present the example in an updated version:

Example 3.3 (Landford)

Let B be the C*-algebra generated by a sequence $(u_n)_{n\in\mathbb{N}}$ of selfadjoint anticommuting operators:

$$u_k = u_k^* \ , \quad u_k^2 = 1 \ , \quad u_k u_\ell + u_\ell u_k = 0 \ , \quad k \neq \ell \ .$$

From the theory of Clifford algebras it follows that $u_1, u_2, \ldots, u_{2n}$ generates a finite dimensional factor of type $I_{(2^n)}$. Therefore B is isomorphic on the infinite tensorproduct of a sequence of 2×2-matrices. In particular, B has a unique tracial state τ. We will consider B in the representation induced by τ. Thus the weak closure of B is the hyperfinite II_1-factor R.

Consider now the linear map T from ℓ_n^∞ to R given by

$$T_n(c_1, \ldots, c_n) = \frac{1}{\sqrt{2n}} \sum_{k=1}^{n} c_k u_k \ .$$

Based on computations made in [13] and [24], it was showed in [11, Lemma 1] that $\|T\| \leqq 1$ and $\|T\|_{cb} \geqq \sqrt{n/2}$. In fact, it is not hard to show that

$$\|T_n\| = 1 \quad \text{and} \quad \|T_n\|_{dec} = \|T_n\|_{cb} = \sqrt{n/2} \quad .$$

To prove the first equality, put

$$c_k = e^{ik\pi/n} \ , \quad k=1,\dots,n \quad ,$$

and let a_k and b_k be the real and imaginary parts of c_k . Since $|c_k| = 1$ and since

$$\sum_{k=1}^{n} c_k^2 = 0$$

we have

$$\sum_{k=1}^{n} a_k^2 = \sum_{k=1}^{n} b_k^2 = \frac{n}{2} \quad \text{and} \quad \sum_{k=1}^{n} a_k b_k = 0 \ .$$

Let A and B be the self-adjoint operators defined by

$$A = \sqrt{\frac{2}{n}} \sum_{k=1}^{n} a_k u_k \ , \quad B = \sqrt{\frac{2}{n}} \sum_{k=1}^{n} b_k u_k \quad .$$

A straightforward computation shows that

$$A^2 = B^2 = 1 \quad \text{and} \quad AB+BA = 0 \ ,$$

from which it follows that

$$(A+iB)(A+iB)^*(A+iB) = 4(A+iB) \ .$$

Therefore $\frac{1}{2}(A+iB)$ is a partial isometry, and since $\frac{1}{2}(A+iB) \neq 0$, we get $\|\frac{1}{2}(A+iB)\| = 1$. Using that

$$T_n(c_1,\dots,c_n) = \tfrac{1}{2}(A+iB) \ ,$$

we conclude that $\|T_n\| \geqq 1$. Hence $\|T_n\| = 1$. From Lemma 2.5 (b) we have $\|T_n\|_{dec} = \sqrt{n/2}$, and since R is injective, also $\|T_n\|_{cb} = \sqrt{n/2}$.

In Example 3.1, $\|T_n\|_{cb} < \|T_n\|_{dec}$ for $n \geq 3$ and in Example 3.3, $\|T_n\| < \|T_n\|_{cb}$ for $n \geq 3$. However, in both cases

$$\|T_2\| = \|T_2\|_{cb} = \|T_2\|_{dec} .$$

This turns out to be true in general:

Proposition 3.4

For every von Neumann algebra N and every linear map T from ℓ_2^∞ to N,

$$\|T\| = \|T\|_{cb} = \|T\|_{dec} .$$

The proof of Proposition 3.4 is based on the following lemma:

Lemma 3.5

Let N be a von Neumann algebra with a separating vector. Let $x_1,\dots,x_n \in N$ and let $T : \ell_n^\infty \to N$ be given by

$$T(c_1,\dots,c_n) = \sum_{i=1}^{n} c_i x_i , \quad c_i \in \mathbb{C} .$$

Then

$$\|T\|_{dec} = \sup\{ \|\sum_{i=1}^{n} x_i v_i\| \mid v_i \in N', \|v_i\| \leq 1\} ,$$

where N' is the commutant of N.

Proof

We prove first the inequality $\geq$. We may assume that $\|T\|_{dec} = 1$. Using Remark 1.3, we can choose completely positive maps S_1, S_2 from ℓ_n^∞ to N, such that $\|S_i\| \leq 1$, $i=1,2$, and such that

$$R(x) = \begin{pmatrix} S_1(x) & T(x^*)^* \\ T(x) & S_2(x) \end{pmatrix}, \quad x \in \ell_n^\infty$$

defines a completely positive map from ℓ_n^∞ to $N \otimes M_2$. Let $p_1, \dots, p_n$ be the minimal projections in ℓ_n^∞, and put

$$y_i = S_1(p_i), \quad z_i = S_2(p_i), \quad i=1,\dots,n.$$

Then $y_i \geqq 0$, $z_i \geqq 0$, $\sum_{i=1}^n y_i \leqq 1$, $\sum_{i=1}^n z_i \leqq 1$, and

$$\begin{pmatrix} y_i & x_i^* \\ x_i & z_i \end{pmatrix} \geqq 0, \quad i=1,\dots,n.$$

Let $u_1, \dots, u_n$ be unitaries in N', and put

$$a = \sum_{i=1}^n x_i u_i .$$

Then

$$\begin{pmatrix} 1 & a^* \\ a & 1 \end{pmatrix} \geqq \sum_{i=1}^n \begin{pmatrix} 1 & 0 \\ 0 & u_i \end{pmatrix} \begin{pmatrix} y_i & x_i^* \\ x_i & z_i \end{pmatrix} \begin{pmatrix} 1 & 0 \\ 0 & u_i^* \end{pmatrix} \geqq 0$$

which implies that $\|a\| \leqq 1$. By the Russo-Dye Theorem [18, Thm. 1], the unit ball of N' is the norm closed convex hull of the unitary operators in N'. Hence

$$\sup\{\|\sum_{i=1}^n x_i v_i\| \mid v_i \in N', \|v_i\| \leqq 1\} \leqq 1 = \|T\|_{dec} .$$

To prove next the inequality $\leqq$ in Lemma 3.5, we can assume that

$$(*) \qquad \sup\{\|\sum_{i=1}^n x_i v_i\| \mid v_i \in N', \|v_i\| \leqq 1\} = 1 .$$

Let E be the subspace of $(N' \otimes \ell_n^\infty) \otimes M_2$ of operators of the form

$$\begin{pmatrix} a \otimes 1 & w \\ v & b \otimes 1 \end{pmatrix}$$

where $a,b \in N'$ and $v,w \in N'\otimes\ell_n^\infty$. Then E is a self-adjoint set of operators and $1 \in E$, i.e. E is an operator-system in the sense of Choi and Effros [4, p. 162]. Let ξ_o be a separating unit vector for N and let ω be the linear functional on E given by

$$\omega\begin{pmatrix} a\otimes 1 & w \\ v & b\otimes 1\end{pmatrix} = ((a + b + \sum_{i=1}^{n} (x_i v_i + x_i^* v_i^*))\xi_o,\xi_o)$$

where

$$v = (v_1,\dots,v_n) \ , \quad w = (w_1,\dots,w_n) \ , \quad v_i,w_i \in N' \ .$$

We will prove that ω is a positive functional on E . Assume that

$$x = \begin{pmatrix} a\otimes 1 & w \\ v & b\otimes 1\end{pmatrix} \in E_+ \quad .$$

Then clearly $w = v^*$ and $a,b \in N'_+$. For $\varepsilon>0$, put $a_\varepsilon = a+\varepsilon 1$ and $b_\varepsilon = b+\varepsilon 1$. Then

$$\begin{pmatrix} 1 & (a_\varepsilon\otimes 1)^{-\frac{1}{2}}v^*(b_\varepsilon\otimes 1)^{-\frac{1}{2}} \\ (b_\varepsilon\otimes 1)^{-\frac{1}{2}}v(a_\varepsilon\otimes 1)^{-\frac{1}{2}} & 1\end{pmatrix}$$

is a positive operator, because it is equal to

$$\begin{pmatrix} a_\varepsilon\otimes 1 & 0 \\ 0 & b_\varepsilon\otimes 1\end{pmatrix}^{-\frac{1}{2}} (x+\varepsilon 1) \begin{pmatrix} a_\varepsilon\otimes 1 & 0 \\ 0 & b_\varepsilon\otimes 1\end{pmatrix}^{-\frac{1}{2}} .$$

Hence $\|(b_\varepsilon\otimes 1)^{-\frac{1}{2}}v(a_\varepsilon\otimes 1)^{-\frac{1}{2}}\| \leqq 1$, or equivalently

$$\|b_\varepsilon^{-\frac{1}{2}}v_i a_\varepsilon^{-\frac{1}{2}}\| \leqq 1 \ , \quad i=1,\dots,n \ .$$

Therefore, by the assumption (*)

$$\|\sum_{i=1}^{n} x_i b_\varepsilon^{-\frac{1}{2}} v_i a_\varepsilon^{-\frac{1}{2}}\| \leqq 1 \ .$$

Since $x_i \in N$ and $v_i, a_\varepsilon, b_\varepsilon \in N'$, we get that

$$-\sum_{i=1}^{n} ((x_i v_i + x_i^* v_i^*)\xi_o, \xi_o) = -2\,\mathrm{Re}((\sum_{i=1}^{n} x_i v_i)\xi_o, \xi_o)$$

$$= -2\,\mathrm{Re}((\sum_{i=1}^{n} x_i b_\varepsilon^{-\frac{1}{2}} v_i a_\varepsilon^{-\frac{1}{2}}) a_\varepsilon^{\frac{1}{2}}\xi_o, b_\varepsilon^{\frac{1}{2}}\xi_o)$$

$$\leqq 2\,\|a_\varepsilon^{\frac{1}{2}}\xi_o\|\ \|b_\varepsilon^{\frac{1}{2}}\xi_o\|$$

$$\leqq (a_\varepsilon\xi_o, \xi_o) + (b_\varepsilon\xi_o, \xi_o)$$

$$= ((a+b)\xi_o, \xi_o) + 2\varepsilon \quad .$$

Since ε was arbitrary, we conclude that ω is positive. Hence

$$\|\omega\| = \omega(1) = 2 \ .$$

(The fact that $\|\omega\| = \omega(1)$ for positive functionals on operator systems can be proved as for C*-algebras, cf. proof of [12, Theorem 4.3.2].) Let $\tilde{\omega}$ be a Hahn-Banach extension of ω to $N' \otimes \ell_n^\infty \otimes M_2$. Then

$$\|\tilde{\omega}\| = \tilde{\omega}(1) = 2$$

so $\tilde{\omega}$ is a positive functional on $N' \otimes \ell_n^\infty \otimes M_2$.

Let $p_1, \ldots, p_n$ be the minimal projections in ℓ_n^∞. Put

$$\varphi_i(a) = \tilde{\omega}\begin{pmatrix} a \otimes p_i & 0 \\ 0 & 0 \end{pmatrix}$$

$$\psi_i(b) = \tilde{\omega}\begin{pmatrix} 0 & 0 \\ 0 & b \otimes p_i \end{pmatrix}$$

for $a,b \in N'$ and $i=1,\dots,n$. By the definition of ω

$$\sum_{i=1}^{n} \varphi_i(a) = \omega\begin{pmatrix} a\otimes 1 & 0 \\ 0 & 0 \end{pmatrix} = (a\xi_o,\xi_o)$$

and

$$\sum_{i=1}^{n} \psi_i(b) = \omega\begin{pmatrix} 0 & 0 \\ 0 & b\otimes 1 \end{pmatrix} = (b\xi_o,\xi_o)$$

for $a,b \in N'$. From [9, Part I, Chap. 4, Lemma 1] there exist positive operators $y_1,\dots,y_n,z_1,\dots,z_n \in N$, such that

$$\varphi_i(a) = (ay_i\xi_o,\xi_o) \ , \quad a \in N'$$

$$\psi_i(b) = (bz_i\xi_o,\xi_o) \ , \quad b \in N' \ .$$

Note that $\sum_{i=1}^{n} y_i = \sum_{i=1}^{n} z_i = 1$, because ξ_o is cyclic for N' and for all $a,b \in N'$:

$$\sum_{i=1}^{n} (y_i a\xi_o,b\xi_o) = \sum_{i=1}^{n} \varphi_i(b^*a) = (a\xi_o,b\xi_o)$$

$$\sum_{i=1}^{n} (z_i a\xi_o,b\xi_o) = \sum_{i=1}^{n} \psi_i(b^*a) = (a\xi_o,b\xi_o) \quad .$$

Let $a,b \in N'$. By the Cauchy-Schwartz inequality for positive functionals, we have

$$\begin{aligned}(x_i a\xi_o,b\xi_o) &= (x_i b^*a\xi_o,\xi_o) \\ &= \omega\begin{pmatrix} 0 & 0 \\ b^*a\otimes p_i & 0 \end{pmatrix} \\ &= \omega\left(\begin{pmatrix} 0 & 0 \\ 0 & b\otimes p_i \end{pmatrix}^* \begin{pmatrix} 0 & 0 \\ a\otimes p_i & 0 \end{pmatrix}\right) \\ &\leq \tilde{\omega}\begin{pmatrix} a^*a\otimes p_i & 0 \\ 0 & 0 \end{pmatrix}^{\frac{1}{2}} \tilde{\omega}\begin{pmatrix} 0 & 0 \\ 0 & b^*b\otimes p_i \end{pmatrix}^{\frac{1}{2}} \\ &= (y_i a\xi_o,a\xi_o)^{\frac{1}{2}}(z_i b\xi_o,b\xi_o)^{\frac{1}{2}} \quad .\end{aligned}$$

Since ξ_o is cyclic for N' , we conclude that

$$\begin{pmatrix} y_i & x_i^* \\ x_i & z_i \end{pmatrix} \geq 0 , \quad i=1,\dots,n .$$

Define now $S_1, S_2 : \ell_n^\infty \to N$ by

$$S_1(c_1,\dots,c_n) = \sum_{i=1}^{n} c_i y_i$$

$$S_2(c_1,\dots,c_n) = \sum_{i=1}^{n} c_i z_i \quad .$$

Then

$$R(x) = \begin{pmatrix} S_1(x) & T(x^*)^* \\ T(x) & S_2(x) \end{pmatrix} , \quad x \in \ell_n^\infty$$

is clearly a positive map from ℓ_n^∞ to $N \otimes M_2$, and since ℓ_n^∞ is abelian, it is also completely positive. Since $S_1(1) = S_2(1) = 1$ we have

$$\|T\|_{dec} \leq 1 = \sup \{ \| \sum_{i=1}^{n} x_i v_i \| \mid v_i \in N', \|v_i\| \leq 1 \} .$$

This completes the proof of Lemma 3.5.

<u>Proof of Proposition 3.4</u>

Let T be a linear map from ℓ_2^∞ into a von Neumann algebra N . Since

$$\|T\| \leq \|T\|_{cb} \leq \|T\|_{dec} \quad ,$$

it is sufficient to prove that $\|T\|_{dec} \leq \|T\|$. Let p_1, p_2 be the two minimal projections in ℓ_2^∞ and put $x_i = T(p_i)$, $i=1,2$. Since the extreme points of the unit ball of ℓ_n^∞ are of the form

$$(c_1, c_2) , \quad c_1, c_2 \in \mathbb{C} , \quad |c_1| = |c_2| = 1 ,$$

we have

$$\|T\| = \sup\{\|x_1+cx_2\| \mid c\in\mathbb{C},\ |c|=1\} .$$

Assume first that N is σ-finite. Then, via the G.N.S.-representation, we can obtain that N acts on a Hilbert space H with a cyclic and separating vector ξ_o. By Lemma 3.5,

$$\|T\|_{dec} = \sup\{\|x_1v_1+x_2v_2\| \mid v_i\in N',\ \|v_i\| \leqq 1\} .$$

By the Russo-Dye theorem, it is sufficient to consider unitary operators v_1, v_2 in N'. In this case,

$$\|x_1v_1+x_2v_2\| = \|x_1+x_2v_2v_1^*\| .$$

Therefore

$$\|T\|_{dec} = \sup\{\|x_1+x_2u\| \mid u\in N',\ u \text{ unitary}\} .$$

If u has finite spectrum, then

$$u = \sum_{i=1}^{r} \lambda_i p_i ,$$

where $\lambda_i \in sp(u)$ and p_i are orthogonal projections in N' with sum 1. Since the subspaces $p_i(H)$, $i=1,\ldots,r$ are invariant under x_1 and x_2, we get in this case

$$\|x_1+x_2u\| = \sup\{\|x_1+\lambda x_2\| \mid \lambda\in sp(u)\} .$$

Since every unitary in N' can be approximated in norm by unitaries with finite spectrum,

$$\|T\|_{dec} \leqq \sup\{\|x_1+cx_2\| \mid c\in\mathbb{C},\ |c|=1\} = \|T\| .$$

If N is not σ-finite, we can choose a net (p_λ) of σ-finite projections in N, such that $p_\lambda \to 1$ strongly. Using the first part of the proof on the map $T_\lambda : \ell_2^\infty \to p_\lambda M p_\lambda$ given by

$$T_\lambda(x) = p_\lambda x p_\lambda \ , \quad x \in \ell_2^\infty \ ,$$

we find completely positive maps $S_\lambda^{(1)}$, $S_\lambda^{(2)}$ from ℓ_2^∞ to $p_\lambda M p_\lambda \subseteq M$, such that $\|S_\lambda^{(i)}\| \leqq \|T\|$, $i=1,2$, and such that

$$R_\lambda(x) = \begin{pmatrix} S_\lambda^{(1)}(x) & T_\lambda(x^*)^* \\ T_\lambda(x) & S_\lambda^{(2)}(x) \end{pmatrix} , \quad x \in \ell_2^\infty$$

is a completely positive map from ℓ_2^∞ to $N \otimes M_2$. Let $R : \ell_2^\infty \to N \otimes M_2$ be a clusterpoint for the net (R_λ) in the topology of pointwise σ-weak convergence on $B(\ell_2^\infty, N \otimes M_2)$. Then R is a completely positive map of the form

$$R(x) = \begin{pmatrix} S^{(1)}(x) & T(x^*)^* \\ T(x) & S^{(2)}(x) \end{pmatrix} , \quad x \in \ell_2^\infty \ ,$$

where $S^{(1)}, S^{(2)} : \ell_2^\infty \to N$ are completely positive and $\|S^{(i)}\| \leqq \|T\|$. Hence $\|T\|_{dec} \leqq \|T\|$.

<u>Corollary 3.6</u>

Let N be a von Neumann algebra, and let $x \in N$. The following two conditions are equivalent

(i) There exists $a \in N_{s.a.}$, $0 \leqq a \leqq 1$, such that

$$\begin{pmatrix} a & x^* \\ x & 1-a \end{pmatrix} \geqq 0$$

(ii) $w(x) \leqq \frac{1}{2}$, where $w(x)$ is the numerical radius of x.

<u>Proof</u>

Recall that the numerical range $W(x)$ of an operator $x \in B(H)$ is

$$\{ (x\xi,\xi) \mid \xi \in H, \ \|\xi\| = 1 \} \ ,$$

and the numerical radius $w(x)$ of x is

$$w(x) = \sup\{|\lambda| \mid \lambda \in W(x)\}$$
$$= \sup\{|(x\xi,\xi)| \mid \xi \in H, \|\xi\| = 1\}$$

(cf. [3, pp. 1-2]). To prove (i) ⇒ (ii), let $\xi \in H$ be a unit vector and let $c \in \mathbb{C}$, $|c|=1$. Put $\xi' = (\xi, c\xi) \in H \oplus H$ and put

$$b = \begin{pmatrix} a & x^* \\ x & 1-a \end{pmatrix} .$$

If $b \geq 0$, then $(b\xi',\xi') \geq 0$. Thus

$$1 + 2\,\mathrm{Re}(c(x\xi,\xi)) \geq 0 ,$$

so by choosing c , such that $c(x\xi,\xi) = -|(x\xi,\xi)|$, we get

$$|(x\xi,\xi)| \leq \tfrac{1}{2} .$$

Conversely, if $w(x) \leq \frac{1}{2}$, then for $c \in \mathbb{C}$, $|c|=1$,

$$\begin{aligned} \|cx+\bar{c}x^*\| &= \sup\{|((cx+\bar{c}x^*)\xi,\xi)| \mid \xi \in H, \|\xi\| = 1\} \\ &= 2\sup\{|\mathrm{Re}(c(x\xi,\xi))| \mid \xi \in H, \|\xi\| = 1\} \\ &\leq 2w(x) \\ &\leq 1 . \end{aligned}$$

Hence also $\|x+\bar{c}^2x^*\| \leq 1$. Consider now the map $T : \ell_2^\infty \to N$ given by

$$T(c_1,c_2) = c_1x+c_2x^* .$$

Clearly,

$$\|T\| = \sup\{\|x+\gamma x^*\| \mid \gamma \in \mathbb{C}, |\gamma|=1\} \leq 1 .$$

Hence, by Prop. 3.4, $\|T\|_{dec} \leqq 1$. Thus there exist $y_1, y_2, z_1, z_2 \in N_+$, such that $y_1+y_2 \leqq 1$, $z_1+z_2 \leqq 1$, and

$$\begin{pmatrix} y_1 & x^* \\ x & y_2 \end{pmatrix} \geqq 0 , \quad \begin{pmatrix} z_1 & x \\ x^* & z_2 \end{pmatrix} \geqq 0 .$$

Hence also

$$\begin{pmatrix} y_1+z_2 & x^* \\ x & y_2+z_1 \end{pmatrix} \geqq 0 .$$

Put $a = y_1+z_2$. Then $1-a \geqq y_2+z_1$. This proves (i).

<u>Remark 3.7</u>

In [4, Thm. 3.4], Choi and Effros proved that a von Neumann algebra N is injective if and only if for $n \in \mathbb{N}$, $n \geqq 2$, any unit preserving, completely positive map T from an operator system $E \subseteqq M_n$ of codimension 1 into N can be extended to a completely positive map $\tilde{T}$ from M_n to N . It is somewhat surprising that for $n=2$ such an extension exists, even if N is not injective. This follows easily from Corollary 3.6:

Let E be any three-dimensional operator system in M_2 , then, by a change of basis, we can obtain that

$$E = \left\{ \begin{pmatrix} c_{11} & c_{12} \\ c_{21} & c_{22} \end{pmatrix} \middle| \; c_{11} = c_{22} \right\} .$$

Let $T : E \to N$ be completely positive and unit preserving, and put

$$x = T\begin{pmatrix} 0 & 0 \\ 1 & 0 \end{pmatrix} .$$

Since

$$1+cx+\bar{c}x^* = T\begin{pmatrix} 1 & \bar{c} \\ c & 1 \end{pmatrix} \geqq 0$$

whenever $|c| = 1$, it follows that $w(x) \leqq \frac{1}{2}$. Hence, by Cor. 3.6, there exists $a \in N_+$, such that

$$\begin{pmatrix} a & x^* \\ x & 1-a \end{pmatrix} \geqq 0 .$$

Therefore,

$$\tilde{T}\begin{pmatrix} c_{11} & c_{12} \\ c_{21} & c_{22} \end{pmatrix} = c_{11}a + c_{22}(1-a) + c_{21}x + c_{12}x^*$$

defines a complete positive extension $\tilde{T} : M_2 \to N$ of T (use [4, Lemma 2.1]).

Problem 3.8

Let N be a von Neumann algebra, such that

$$\|T\|_{cb} = \|T\|_{dec}$$

for every linear map T from ℓ_3^∞ to N. Is N injective?

Acknowledgement.

The main part of the present research was carried out while the author was visiting the United States during the academic year 1982/83, supported by the National Science Foundations of Denmark and of U.S.A. I will like to thank my colleagues at U.C.L.A. and University of Pennsylvania for their warm hospitality during the visit. A special thank to Ed Effros and Vern Paulsen for a number of fruitful conversations concerning the subject of this paper.

References.

[1] C.A. Akemann and P.A. Ostand, *Computing norm in group C*-algebras. Amer. J. Math. 98 (1976), 1015-1047.*

[2] W. Arveson, *Subalgebras of C*-algebras, Acta Math. 123 (1969), 141-224.*

[3] F.F. Bonsall and J. Duncan, *Numerical ranges of operators on normed spaces and of elements of normed algebras, London Math. Soc. Lecture Note Series 2 (1971).*

[4] M.D. Choi and E.G. Effros, *Injectivity and matrix ordered spaces, J. Funct. Analysis 24 (1977), 156-209.*

[5] M.D. Choi and E.G. Effros, *The completely positive lifting problem for C*-algebras, Annals of Math. 104 (1976), 585-609.*

[6] A. Connes, *Classification of injective factors, Annals of Math. 104 (1976), 73-115.*

[7] A. Connes, *On the equivalence between injectivity and semidiscreteness for operator algebras, pp. 107-112 in "Algebres d'operateurs et leurs application en physique theorique", Edition C.N.R.S. 1979.*

[8] J. Dixmier, *C*-algebras, North Holland Mathematical Library 15, North Holland 1972.*

[9] J. Dixmier, *Von Neumann algebras, North Holland Mathematical Library 27, North Holland 1981.*

[10] T. Huruya, *Linear maps between certain non separable C*-algebras, preprint 1983.*

[11] T, Huruya and J. Tomiyama, *Completely bounded maps between C*-algebras, J. Operator Theory 10 (1983), 141-152.*

[12] R.V. Kadison and J. Ringrose, *Fundamentals of the theory of operator algebras I, Academic Press 1983.*

[13] R.I. Loebl, *A Hahn decomposition for linear maps, Pacific J. Math. 65 (1976), 119-133.*

[14] D. McDuff, *Central sequences and the hyperfinite factor*, *Proc. London Math. Soc. 21 (1970), 443-461.*

[15] V.I. Paulsen, *Completely bounded maps on C*-algebras and invariant operator ranges*, *Proc. Amer. Math. Soc. 86 (1982), 91-96.*

[16] V.I. Paulsen, *Every completely polynomial bounded operator is similar to a contraction. To appear in J. Funct. Analysis.*

[17] V.I. Paulsen, *Lecture at the AMS-meeting in Denver, January 1983.*

[18] B. Russo *and* H.A. Dye, *A note on unitary operators in C*-algebras, Duke Math. J. 33 (1966), 413-416.*

[19] S. Sakai, *The theory of W*-algebras, Lecture Notes, Yale University 1962.*

[20] R.R. Smith, *Completely bounded maps between C*-algebras, J. London Math. Soc. 27 (1983), 157-166.*

[21] W.F. Stinespring, *Positive maps on C*-algebras, Proc. Amer. Math. Soc. 6 (1955), 211-216.*

[22] M. Takesaki, *Duality for crossed products and the structure of von Neumann algebras of type III, Acta Math. 131 (1973), 249-310.*

[23] M. Takesaki, *Theory of operator algebras I, Springer-Verlag 1979.*

[24] S.J. Tsui, *Decomposition of linear maps, Trans. Amer. Math. Soc. 230 (1977), 87-112.*

[25] S. Wassermann, *On the tensor product of certain group C*-algebras, J. Funct. Analysis 23 (1976), 239-254.*

[26] S. Wassermann, *Injective W*-algebras, Math. Proc. Cambridge Phil. Soc. 82 (1977), 39-47.*

[27] G. Wittstock, *Ein operatorvertiger Hahn-Banach Satz, J. Funct. Analysis 40 (1981), 127-150.*

JB-ALGEBRAS WITH TENSOR PRODUCTS ARE C^*-ALGEBRAS

Harald Hanche-Olsen

1. In their celebrated 1934 paper Jordan, Von Neumann, and Wigner [5] pointed out that Jordan algebras, rather than C^*-algebras, seemed to provide the "right" framework for quantum mechanics. However, quantum mechanics still seems to happen in C^*-algebras. This may be because C^*-algebras are much better understood than Jordan operator algebras, but the general feeling seems to be that there are deeper reasons for this phenomenon - for example, the need to aassociate a one parameter group of automorphisms with the Hamiltonian operator.

In this paper we investigate another possible reason for this phenomenon: The need for tensor products. If $\mathcal{A}$ and $\mathcal{B}$ are the C^*-algebras of observables associated with two physical systems, their tensor product $\mathcal{A} \otimes \mathcal{B}$ is generally assumed to be the algera of observables for the union of the two systems. Thus we would like to generalize the notion of tensor product to Jordan algebras.

In [2] the author carried this plan out for JC-algebras, i.e. norm closed Jordan algebras of self-adjoint operators on a Hilbert space. There was shown the existence of a "universal" tensor product $A \tilde{\otimes} B$ of two unital JC-algebras A and B. This is defined as containing mutually commuting copies of A and B, and being universal with this property. $A \tilde{\otimes} B$ contains the linear tensor product $A \otimes B$ as a subspace. However, $A \otimes B$ is usually not dense in $A \tilde{\otimes} B$. In particular, it is shown that if $A \otimes H_2(\mathbb{C})$ is a JC-algebra with some product satisfying some reasonable identities, then A is the self-adjoint part of a C^*-algebra. Here, $H_2(\mathbb{C})$ is the Jordan algebra of hermitian 2 by 2 matrices over the complex numbers, i.e. the Jordan algebra of observables associated with the spin of an electron.

In this paper we extend the above result to JB-algebras, the abstract counterparts of JC-algebras. For the general theory of JB-algebras the reader is referred to [1], [3].

2. Consider a C^*-algebra $\mathcal{A}$ and the algebra M_2 of 2 by 2 matrices. Then $\mathcal{A} \otimes M_2$ is a C^*-algebra with the product $(x\otimes\alpha)(y\otimes\beta) = xy \otimes \alpha\beta$. Considering the induced Jordan products on $\mathcal{A}$, M_2, and $\mathcal{A} \otimes M_2$ respectively, we immediately find

(1) $\quad (x\otimes 1)\circ(y\otimes\beta) = x\circ y \otimes \beta,$

(2) $\quad (1\otimes\alpha)\circ(y\otimes\beta) = y \otimes \alpha\circ\beta.$

It is easily verified that these two formulas are equivalent to the following four:

(3) $\quad (x\otimes 1)\circ(y\otimes 1) = x\circ y \otimes 1,$

(4) $\quad (1\otimes\alpha)\circ(1\otimes\beta) = 1 \otimes \alpha\circ\beta,$

(5) $\quad (1\otimes\alpha)\circ(x\otimes 1) = x \otimes \alpha,$

(6) $\quad [T_{1\otimes\alpha}, T_{x\otimes 1}] = 0,$

where T_a is the operator $b \to a\circ b$.

<u>3. Theorem</u>. <u>Assume A is a unital JB-algebra and that $A\otimes H_2(\mathbb{C})$ is equipped with some Jordan product such that (1) and (2) hold, and $A\otimes H_2(\mathbb{C})$ is a JB-algebra in a suitable norm. Then A is the self-adjoint part of a C^*-algebra</u>.

Before we start working on the proof, we note that the theorem is valid if the word "JC-algebra" is substituted for "JB-algebra". This is the content of [2, Theorem 5.5]. However, we cannot use the techniques of [2] in the present setting.

Note that the complexification $\mathcal{A}$ of A is a complex Jordan algebra, that the complexification of $H_2(\mathbb{C})$ is $M_2(\mathbb{C})$ and the complexification of $A \otimes H_2(\mathbb{C})$ is $\mathcal{A} \otimes M_2(\mathbb{C})$, and that the formulas (1), (2) hold in the complex algebras as well. We can now state the purely algebraic version of our theorem, where the field $\mathbb{C}$ has been generalized:

4. Theorem. Let $\mathcal{A}$ be a Jordan algebra with identity over a field K of characteristic different from two. Assume that there exists a Jordan product in $\mathcal{A} \otimes M_2(K)$ satisfying (1) and (2). Then there exists a unique associative product in $\mathcal{A}$ inducing the given Jordan products both in $\mathcal{A}$ and in $\mathcal{A} \otimes M_2(K)$.

The remainder of this section is devoted to the proof of Theorem 4. Thus, the assumptions of the theorem will be kept throughout. As noted above, the formulas (3)-(6) are also valid in the present setting.

We will make extensive use of the "linearised Jordan identity" [4; p.34]:

$$(7) \qquad [T_{a\circ b}, T_c] + [T_{b\circ c}, T_a] + [T_{c\circ a}, T_b] = 0,$$

which is valid for a,b,c in any Jordan algebra. Putting $a = c = 1 \otimes \alpha$ and $b = x \otimes 1$ in (7) and using (6) yields:

$$(8) \qquad [T_{x\otimes\alpha}, T_{1\otimes\alpha}] = 0.$$

Using this together with (1) and (2) we may compute:

$(x\otimes\alpha)\circ(y\otimes\alpha) = T_{x\otimes\alpha}T_{1\otimes\alpha}(y\otimes 1) = T_{1\otimes\alpha}T_{x\otimes\alpha}(y\otimes 1) = (x\circ y)\otimes\alpha^2$, which applied to the matrix units $\alpha = \varepsilon_{ij}$, yields the formulas:

(9) $(x\otimes\varepsilon_{11})\circ(y\otimes\varepsilon_{11}) = (x\circ y)\otimes\varepsilon_{11}$ and similarly for ε_{22},

(10) $(x\otimes\varepsilon_{21})\circ(y\otimes\varepsilon_{21}) = 0$ and similarly for ε_{12},

(11) $(x\otimes\varepsilon_{11})\circ(y\otimes\varepsilon_{22}) = 0$.

The last formula comes from (2.9), (2.1) and $\varepsilon_{22} = 1-\varepsilon_{11}$.

The following lemma is the main step in the proof of the theorem.

5. Lemma. If $x,y \in \mathcal{A}$, there exists $z \in \mathcal{A}$ such that $(x\otimes\varepsilon_{21})\circ(y\otimes\varepsilon_{11}) = z\otimes\varepsilon_{21}$.

Proof: Writing

(12) $(x\otimes\varepsilon_{21})\circ(y\otimes\varepsilon_{11}) = \sum_{i,j} z_{ij}\otimes\varepsilon_{ij}$,

we have to show $z_{11} = z_{12} = z_{22} = 0$.

First, we apply $T_{1\otimes\varepsilon_{11}}$ to both sides of (12). On the left-hand side we find, using (8) and (2):

$$\begin{aligned} T_{1\otimes\varepsilon_{11}}((x\otimes\varepsilon_{21})\circ(y\otimes\varepsilon_{11})) &= T_{1\otimes\varepsilon_{11}}T_{y\otimes\varepsilon_{11}}(x\otimes\varepsilon_{21}) = \\ &= T_{y\otimes\varepsilon_{11}}T_{1\otimes\varepsilon_{11}}(x\otimes\varepsilon_{21}) = \\ &= T_{y\otimes\varepsilon_{11}}(\tfrac{1}{2}x\otimes\varepsilon_{21}) = \\ &= \tfrac{1}{2}(x\otimes\varepsilon_{21})\circ(y\otimes\varepsilon_{11}), \end{aligned}$$

i.e., $(x\otimes\varepsilon_{21})\circ(y\otimes\varepsilon_{11})$ is an eigenvector of $T_{1\otimes\varepsilon_{11}}$ with eigenvalue $\tfrac{1}{2}$.

On the right-hand side we find

$$T_{1\otimes\varepsilon_{11}}(\sum_{i,j} z_{ij}\otimes\varepsilon_{ij}) = z_{11}\otimes\varepsilon_{11}+\tfrac{1}{2}(z_{12}\otimes\varepsilon_{12}+z_{21}\otimes\varepsilon_{21}).$$

Since the ε_{ij}'s are linearly independent, $z_{11} = z_{22} = 0$ follows.

Next, applying $T_{1\otimes\varepsilon_{21}}$ and using the same technique, we find $z_{12} = 0$; thus proving the lemma.

6. Note that Lemma 5 is also valid when ε_{22} is sbstituted for ε_{11}. Thus, we may define maps $R_x : A \to A$ and $L_x : A \to A$ for given $x \in A$ by

(13) $$R_x(y)\otimes\varepsilon_{21} = 2(x\otimes\varepsilon_{11})\circ(y\otimes\varepsilon_{21}),$$

(14) $$L_x(y)\otimes\varepsilon_{21} = 2(x\otimes\varepsilon_{22})\circ(y\otimes\varepsilon_{21}).$$

7. Lemma. $[R_y, L_x] = 0$ for all $x, y \in \mathcal{A}$.

Proof: This will follow from the identity $[T_{y\otimes\varepsilon_{11}}, T_{x\otimes\varepsilon_{22}}] = 0$, which is proved as follows: Let $a = y\otimes\varepsilon_{11}$, $b = x\otimes\varepsilon_{22}$, $c = 1\otimes\varepsilon_{22}$. Then $a\circ b = a\circ c = 0$ and $b\circ c = x\otimes\varepsilon_{22}$. Applying (7) yields the desired equaity.

8. Corollary. There exists an associative product in $\mathcal{A}$ with 1 as an identity and such that $xy = L_x(y) = R_y(x)$ and $x\circ y = \frac{1}{2}(xy+yx)$ for all $x, y \in \mathcal{A}$.

Proof: By (2), (13) and (14), $L_x(1) = R_x(1) = x$. Lemma 7 now guarantees existence and associativity of the product defined by $xy = L_x(y) = R_y(x)$. Also, $T_{x\otimes\varepsilon_{11}} + T_{x\otimes\varepsilon_{22}} = T_{x\otimes 1}$, which together with (2.1) proves that $x\circ y = \frac{1}{2}(R_x+L_x)(y) = \frac{1}{2}(yx+xy)$.

9. Concluding proof of Theorem 4. Now, $\mathcal{A}$ is shown to be an associative algebra, and so is $\mathcal{A}\otimes M_2$. What remains to prove is that $a\circ b = \frac{1}{2}(ab+ba)$ for all a,b in $\mathcal{A}\otimes M_2$. First, rewrite (13) and (14) as follows:

$$(15)\qquad (x\otimes\varepsilon_{11})\circ(y\otimes\varepsilon_{21}) = \tfrac{1}{2}yx\otimes\varepsilon_{21}\ ,$$

$$(16)\qquad (x\otimes\varepsilon_{22})\circ(y\otimes\varepsilon_{21}) = \tfrac{1}{2}xy\otimes\varepsilon_{21}\ .$$

Next, put $a = x\otimes\varepsilon_{21}$, $b = y\otimes\varepsilon_{11}$, $c = 1\otimes\varepsilon_{11}$, so that $a\circ b = \frac{1}{2}xy\otimes\varepsilon_{21}$, $b\circ c = y\otimes\varepsilon_{11}$, $a\circ c = \frac{1}{2}x\otimes\varepsilon_{21}$. Applying (7) we get $[T_{xy\otimes\varepsilon_{21}},T_{1\otimes\varepsilon_{11}}]+[T_{y\otimes_{11}},T_{x\otimes\varepsilon_{21}}] = 0$. Applying this operator identity to $1\otimes\varepsilon_{12}$ and computing by means of previous formulas, we find

$$(17)\qquad (x\otimes\varepsilon_{21})\circ(y\otimes\varepsilon_{12}) = \tfrac{1}{2}yx\otimes\varepsilon_{11}+\tfrac{1}{2}xy\otimes\varepsilon_{22}.$$

Now, repeat the foregoing discussion with the indices 1,2 interchanged. The result will be another associative product on A such that the analogues of the previous formulas hold. But (17) is invariant under this reversal of indices, thus proving that the two products are equal. Now the formulas (9), (10), (11), (15), (16), (17) together with the "reversed" analogues of (15), (16) show that the formula $a\circ b = \frac{1}{2}(ab+ba)$ holds for all a,b of the form $x\otimes\varepsilon_{ij}$, thus for all $a,b\in\mathcal{A}\otimes M_2$. This completes the proof of Theorem 4.

10. Proof of Theorem 3.

Let A be a JB-algebra satisfying the requirements of Theorem 3. Then its complexification $\mathcal{A}$ satisfies the conditions of Theorem 4, and it follows that A and $A\otimes H_2(\mathbb{C})$ are special, and therefore JC-algebras; see [1;Lemma 9.4]. By appeal to [2, Theorem 5.5] the proof is finished.

11. Remark.

The above shortcut, reducing the problem to the JC-algebra case, was not necessary. Indeed, it is not difficult to prove directly that any complex associative $\star$-algebra $\mathcal{A}$ whose hermitian part is a JB-algebra in some norm, can itself be normed so as to become a C^*-algebra. Applying this to the complexification of the JB-algebra A in Theorem 3, we get a direct proof. Moreover, this shows that we do not need any norm on $A \otimes H_2(\mathbb{C})$ as stated in Theorem 3.

BIBLIOGRAPHY

1. E.M. Alfsen, F.W. Shultz, E. Størmer, A Gelfand-Neumark theorem for Jordan algebras. Adv. Math. 28 (1978), 11-56.

2. H. Hanche-Olsen, On the structure and tensor products of JC-algebras. Can. J. Math. (to appear).

3. H. Hanche-Olsen, E. Størmer, "Jordan operator algebras". London, 1984.

4. N. Jacobson, "Structure and representation of Jordan algebras", Providence 1968.

5. P. Jordan, J. von Neumann, E. Wigner, On an algebraic generalization of the quantum mechanical formalism. Ann. of Math. 35 (1934), 29-64.

Institute for Energy Technology
Box 40
N-2007 KJELLER, Norway

REDUCED C*-ALGEBRAS OF DISCRETE GROUPS WHICH ARE SIMPLE WITH A UNIQUE TRACE

Pierre de la Harpe

Let Γ be a (discrete) group, not reduced to $\{1\}$; Let $C^*_r(\Gamma)$ be its reduced C*-algebra. My interest is to know when $C^*_r(\Gamma)$ is simple with a unique (positive and normalized) trace.

This is of course not always the case. For example, if Γ is amenable, then the trivial representation of Γ in $\mathbb{C}$ provides a morphism $C^*_r(\Gamma) \longrightarrow \mathbb{C}$ which is onto and $C^*_r(\Gamma)$ cannot be simple (theorems 7.6.6 and 7.7.7 in [Ped]). More generally, if Γ has an amenable non trivial normal subgroup (e.g. if Γ has a centre), then $C^*_r(\Gamma)$ is not simple and has more than one trace (proposition 1.6 in [PS]).

But this seems to be very often the case, as the following sample indicates.

<u>Theorem.</u> For any group Γ in one of the classes below, $C^*_r(\Gamma)$ is simple and has a unique trace.

(a) Γ is a subgroup of $PSL(2,\mathbb{R})$ which is not solvable.

(b) Γ is a subgroup of $PSL(2,\mathbb{C})$ which is not almost solvable.

(c) Γ is a subgroup of G which contains a lattice, where $G = PSL(d,\mathbb{R})$ or $G = PSL(d,\mathbb{C})$ for some integer $d \geqslant 2$.

(d) Γ is a free product $H*K$, where H has at least two and K at least three elements.

(e) $\Gamma = H \underset{A}{*} K$ is a free product with amalgamation over a group $A \neq \{1\}$ such that, given any finite subset $F \subset \Gamma - \{1\}$, there exists $g \in \Gamma$ with $g^{-1} F g \cap A = \emptyset$.

(f) $\Gamma = HNN(H, A, \theta)$ is an extension à la G. Higman, B.H. Neumann and H. Neumann with $|H/A| \geqslant 3$ such that, given any finite subset $F \subset \Gamma - \{1\}$, there exists $g \in \Gamma$ with $g^{-1} F g \cap A = \emptyset$.

(g) $\Gamma = PSL(d, \mathbb{K})$ for some integer $d \geqslant 2$ and for some field $\mathbb{K}$ of characteristic zero.

Recall that a group Γ is almost solvable if it contains a solvable subgroup of finite index; a subgroup Γ of a locally compact topological group G is a lattice if it is discrete and if G/Γ carries a G-invariant finite measure; in the conjecture below, a subgroup Γ of $PSL(d, \mathbb{C})$ is strongly irreducible if the inverse image $\tilde{\Gamma}$ of Γ in $SL(d, \mathbb{C})$ as well as any subgroup of finite index in $\tilde{\Gamma}$ act irreducibly on $\mathbb{C}^d$. There are partial results about statements (a) and (b) in [HJ]; statement (d) is due to Paschke and Salinas, and (e), (f) are due to Bédos, but the proofs below are different.

The theorem describes work in progress and looks far from a definitive formulation. For example, I have little doubt that a lattice in any non compact simple connected real Lie group without centre qualifies (the case of lattices in semi-simple groups is slightly less clear); moreover the proof is probably nothing more than an exercise about minimal parabolic subgroups (see section 4). But this generalization seems out of order, because the correct hypothesis for the theorem do probably not involve discreteness. As a possible step for further work, we state :

<u>Conjecture.</u> If Γ is a strongly irreducible subgroup of $PSL(d,\mathbb{C})$ for some integer $d \geqslant 2$, then $C_r^*(\Gamma)$ is simple and has a unique trace.

The proof of the theorem above has the following structure : From the geometrical context in which the group Γ (or some of its finitely generated subgroups) is given (cases (a) to (c), as well as (g)) or led to (cases (d) to (f)), it can be shown that Γ has a combinatorial property making it in some sense "like" the non abelian free group F_2 on two generators (sections 3 and 4). Then the proof devised by Powers [P] for $C_r^*(F_2)$ extends (section 2). Section 1 contains some generalities.

I am grateful to M. Boileau, Y. Guivarc'h, G. Schibler, R. Spatzier and T. Vüst for a number of discussions and letters about what is reported herebelow.

1.- COMBINATORICS

Let Γ be an infinite group. In loose words, a subset A of Γ is "small" if there are "many" distinct translated sets γA disjoint from A. My purpose is to study groups which are made up of two small subsets, in the following very strong sense.

<u>Definition.</u> A <u>Powers' group</u> is a group Γ having the following property : given any non empty finite subset $F \subset \Gamma - \{1\}$ and any integer $N \geqslant 1$, there exist a partition $\Gamma = A \amalg B$ and elements $g_1, \ldots, g_N$ in Γ such that

(i) $fA \cap A = \emptyset$ for all $f \in F$

(ii) $g_j B \cap g_k B = \emptyset$ for $j,k = 1,\ldots,N$ with $j \neq k$.

Here are some elementary observations.

Proposition 1. Let Γ be a Powers' group.

(a) Any conjugacy class in Γ other than $\{1\}$ is infinite.

(b) The group Γ is not amenable.

(c) Any subgroup Γ' of Γ of finite index is a Powers' group.

Proof. (a) Assume that there exists in Γ an element $f \neq 1$ with finite conjugacy class F. Set $N = 3$, and let A, B, g_1, g_2, g_3 be as in the definition. For $j \in \{2,3\}$ one has $g_1 B \cap g_j B = \emptyset$, so that

$$fg_1 B \subset fg_j A = g_j f' A \subset g_j B$$

with $f' = g_j^{-1} f g_j \in F$. This implies the impossible relation $g_2 B \cap g_3 B \neq \emptyset$.

(b) Assume that Γ has an invariant mean m. With notations as in the definition and $N \geqslant 3$, one has $m(A) \leqslant \frac{1}{2}$ by (i) and $m(B) \leqslant \frac{1}{3}$ by (ii). This implies $m(\Gamma) < 1$, an absurdity.

(c) Let T be a finite complete transversal to Γ' in Γ: any $\gamma \in \Gamma$ can be written $\gamma = t\gamma'$ with $t \in T$ and $\gamma' \in \Gamma'$; let c be the cardinality of T. Let a finite subset $F' \subset \Gamma' - \{1\}$ and an integer $N' \geqslant 1$ be given; set $N = cN'$. There exist a partition $\Gamma = A \amalg B$ and elements $g_1 = t_1 g_1', \ldots, g_N = t_N g_N'$ in Γ with $fA \cap A = \emptyset$ for $f \in F'$ and $t_j g_j' B \cap t_k g_k' B = \emptyset$ for $j \neq k$. One may assume without loss of generality that $t_1 = \ldots = t_{N'}$. Then $\Gamma' = (\Gamma' \cap A) \amalg (\Gamma' \cap B)$ and $g_1', \ldots, g_{N'}'$ have the desired properties. ■

Let Γ be a Powers' group.

It follows first from (a) that Γ has no centre, and that it can thus be neither abelian nor indeed nilpotent. If Γ is moreover finitely generated, a theorem of Gromov [G] and (c) imply that Γ cannot have polynomial growth. Observe also that Γ has not any non trivial finite normal subgroup by (a). In particular $SL(2,\mathbb{Z})$ or the direct product of the free group F_2 by a finite group are not Powers' groups.

It follows also from (b) that Γ is not almost solvable. If Γ is moreover isomorphic to a subgroup of $GL(d,\mathbb{C})$ for some $d \geq 1$, a theorem of Tits [T] implies that Γ contains a subgroup isomorphic to F_2.

It can be shown that a Powers' group does not have any non trivial amenable normal subgroup (proposition 1.6 in [PS]).

<u>Example.</u> The non abelian free group F_2 on two generators s and t is a Powers' group.

<u>Proof.</u> Let F and N be as in the definition. There exists an integer $m \geq 0$ such that the reduced word defined by $t^m f t^{-m}$ begins and ends with a non-zero power of t for each $f \in F$. Define A to be the set of reduced words in Γ beginning with t^{-m} (followed by a non-zero power of s or by nothing at all) and $g_k = s^k t^m$ for $k = 1,\dots,N$. ∎

The following questions are open (some could be easy).

(1) Let Γ' be a subgroup of a group Γ. Assume that Γ' is a Powers' group and that its centralizer in Γ is trivial. Is Γ a Powers' group ? Interesting particular cases : Γ' is of finite index in Γ (see proposition 1(c)), and Γ' is free non abelian (see [AL]).

(2) Let $1 \longrightarrow \Gamma' \longrightarrow \Gamma \longrightarrow \Gamma'' \longrightarrow 1$ be a short exact sequence of groups, with Γ' and Γ'' Powers' groups. Is Γ a Powers' group ? This is unknown to me even when $\Gamma \approx \Gamma' \times \Gamma''$, and indeed when moreover $\Gamma' \approx \Gamma'' \approx F_2$ (although it is easy to check that

$^*_r(F_2 \times F_2)$ is simple and has a unique trace because it is a
*-tensor product of $C^*_r(F_2)$ by itself, as was observed to me by
avid Handelman).

(3) Let $\Gamma = H * K$ be a free product where H has at least two
nd K at least three elements. Then Γ is a Powers' group. This
arries over to the cases of a free product with amalgamation and of
HNN-extension as in our main theorem. See [B1], [B2] and section 3
elow. E. Bédos has also observed that, consequently, Powers' groups
nclude groups with at least three generators and a single defining
elation.

One could probably obtain other results about fundamental groups
f graphs of groups [Se], and it would be nice to deduce them from
eometric properties of appropriate actions.

(4) Given a Powers' group Γ, does the group algebra $\mathbb{C}[\Gamma]$
ave any remarkable property ?

Let me quote a few known facts. The first is Rickart's theorem :
or any group Γ, the algebra $\mathbb{C}[\Gamma]$ is semi-simple (i.e. the
ntersection of the kernels of the irreducible representations of
$[\Gamma]$ is reduced to zero); see chapter 7 in [Pa]. If Γ has no
inite normal subgroup besides $\{1\}$, then $\mathbb{C}[\Gamma]$ is also prime (i.e.
f J and J' are two two-sided ideals in $\mathbb{C}[\Gamma]$ with $JJ' = 0$,
hen $J = 0$ or $J' = 0$); in particular $\mathbb{C}[\Gamma]$ is without
entral idempotent besides 0 and 1 ; this is theorem 4.2.10 in
Pa]. It may actually happen that $\mathbb{C}[\Gamma]$ has no (two-sided) ideal at
ll besides its augmentation ideal [BHPS].

If Γ is finitely generated torsion free and has some extra
roperty implied by being linear, then $\mathbb{C}[\Gamma]$ has no idempotent at
ll besides 0 and 1 ; see [Ba2] and [Fo]; see also after
roposition 7 below.

(5) A Powers' group has clearly property (po) of [HJ]. In fact (following a suggestion of Guivarc'h) we have changed from the definition of (po) to the definition above just because the latter is easier to remember. It could be worth to find simpler combinatorial properties of a group Γ which would still imply that $C_r^*(\Gamma)$ is simple with a unique trace, or indeed any statement which could be an entry in a dictionary translating from properties of Γ to properties of $C_r^*(\Gamma)$.

2.- FUNCTIONAL ANALYSIS

This section is a repetition of part of [P] and [HJ]. Let Γ be a Powers' group, let H be the Hilbert space $l^2(\Gamma)$ of square summable functions from Γ to $\mathbb{C}$, and let $\mathbb{C}[\Gamma]$ be the group algebra of Γ viewed as an algebra of operators on H. The reduced C*-algebra $C_r^*(\Gamma)$ is the uniform closure of $\mathbb{C}[\Gamma]$. The trace $\tau : \mathbb{C}[\Gamma] \longrightarrow \mathbb{C}$ defined by

$$\tau\left(\sum_{\text{finite}} \lambda_g U(g)\right) = \lambda_1$$

extends uniquely to a positive trace $C_r^*(\Gamma) \longrightarrow \mathbb{C}$, again denoted by τ.

In lemma 2, the constant c is $\frac{1}{5}\left\{\frac{2}{\sqrt{5}} + 4\right\}$. One has $c = 0{,}97885...$, and the important estimate is $0 < c < 1$.

<u>Lemma 2.</u> Let Y be an hermitian operator in $\mathbb{C}[\Gamma]$ with $\tau(Y) = 0$. There exist $g_1, \dots, g_5$ in with

$$\left\| \frac{1}{5} \sum_{j=1}^{5} U(g_j) Y U(g_j)^* \right\| \leq c\|Y\| .$$

Proof. One may write

$$Y = \sum_{s=1}^{n} \lambda_s U(f_s) + \bar{\lambda}_s U(f_s)^*$$

for some finite subset $F_o = \{f_1, \ldots, f_n\} \subset \Gamma - \{1\}$ and for complex numbers $\lambda_1, \ldots, \lambda_n$. Set $F = F_o \cup (F_o)^{-1}$ and $N = 5$. Let $\Gamma = A \perp B$ and $g_1, \ldots, g_5$ be as in the definition of section 1. For each $j \in \{1, \ldots, 5\}$, identify $l^2(g_jA)$ with the subspace M_j of H of functions which vanish outside g_jA, and $l^2(g_jB)$ with $M_j^{\perp}$. Denote by E_j the orthogonal projection of H onto M_j. Then $U(g_j)YU(g_j)^*M_j \subset M_j^{\perp}$. It follows that one has for all $\varphi \in H$

$$|\langle U(g_j)YU(g_j)^*\varphi|\varphi\rangle| \leq |\langle U(g_j)YU(g_j)^*\varphi|E_j\varphi\rangle| +$$
$$+ |\langle U(g_j)YU(g_j)^*\varphi|(1-E_j)\varphi\rangle| \leq 2\|Y\| \|(1-E_j)\varphi\| \|\varphi\| .$$

Now define

$$Z = \frac{1}{5} \sum_{j=1}^{5} U(g_j)YU(g_j)^* .$$

I claim that $\|Z\| \leq c\|Y\|$. Indeed, let $\varphi \in H$ with $\|\varphi\| \leq 1$. As $1-E_1, \ldots, 1-E_5$ are pairwise orthogonal, there exists $j \in \{1, \ldots, 5\}$ with $\|(1-E_j)\varphi\|^2 \leq \frac{1}{5}$. Then

$$|\langle Z\varphi|\varphi\rangle| \leq$$
$$\leq \frac{1}{5}\left\{|\langle U(g_j)YU(g_j)^*\varphi|\varphi\rangle| + \sum_{k\neq j} |\langle U(g_k)YU(g_k)^*\varphi|\varphi\rangle|\right\}$$
$$\leq \frac{\|Y\|}{5}\left\{\frac{2}{\sqrt{5}} + 4\right\} .$$

It follows that $\|Z\| \leq c\|Y\|$. ■

Proposition 3. For any Powers' group Γ, the reduced C*-algebra $C_r^*(\Gamma)$ is simple and has a unique trace.

Proof. Let J be a non-zero two-sided ideal in $C_r^*(\Gamma)$ and let $X \in J$ with $X \neq 0$. As τ is faithful, $t = \tau(X^*X) \neq 0$ and $t^{-1}X^*X \in J$. One may thus assume to start with that X is self-adjoint with $\tau(X) = 1$. Let $\varepsilon \in \mathbf{R}$ with $0 < \varepsilon < 1$.

Choose a self-adjoint $Y \in \mathbf{C}[\Gamma]$ with $\tau(Y) = 1$ and $\| Y-X \| \leq \frac{\varepsilon}{2}$. By lemma 2 applied several times, there exist an integer $m \geq 1$ and elements $g_1, \ldots, g_m$ in Γ with

$$\left\| \frac{1}{m} \sum_{1 \leq j \leq m} U(g_j)(Y - 1)U(g_j)^* \right\| \leq \frac{\varepsilon}{2} \qquad (*)$$

Define

$$Z = \frac{1}{m} \sum_{1 \leq j \leq m} U(g_j)XU(g_j)^* \in J \ .$$

Then

$$\| Z-1 \| \leq \| X-Y \| + \left\| \frac{1}{m} \sum_{1 \leq j \leq m} U(g_j)(Y - 1)U(g_j)^* \right\| \leq \varepsilon$$

and Z is invertible. Hence $J = C_r^*(\Gamma)$.

Let σ be a positive normalized trace on $C_r^*(\Gamma)$. If Y is as above, it follows from (*) and from the continuity of σ that $\sigma(Y) = 1 = \tau(Y)$. One has therefore $\sigma = \tau$. ■

Observe that proposition 1(a) means precisely that the centre of $C_r^*(\Gamma)$ is trivial, and also that the von Neumann algebra $W^*(\Gamma)$ of Γ is a factor (continuous and finite). It follows from theorem 5.3 in [KLR] that inner derivations of $C_r^*(\Gamma)$ are closed in the Banach space of all its derivations; it also follows that inner *-automorphisms of $C_r^*(\Gamma)$ are closed in the complete metric group of all its *-automorphisms (with the topology from the usual norm on linear transformations of the Banach space $C_r^*(\Gamma)$). Lemma 2 above shows that the algebra $C_r^*(\Gamma)$ has the Dixmier property in the sense of [Ar]. All this suggests the following questions (the first one being due to U. Haagerup).

(1) Given a group Γ , is it equivalent for $C_r^*(\Gamma)$ to be simple with unique trace and for $W^*(\Gamma)$ to be a full factor ?

(2) Does there exist a group Γ such that $C^*_r(\Gamma)$ is simple and has several traces ? or such that $C^*_r(\Gamma)$ has a unique trace and is not simple ?

(3) Let Γ' be a subgroup of Γ such that the centralizer of Γ' in Γ is trivial. Does the simplicity of $C^*_r(\Gamma')$ imply that of $C^*_r(\Gamma)$? (Compare with question 1 of the previous section.)

3.- GEOMETRY : THE CASE OF RANK ONE

One efficient tool for detecting Powers' groups is the following lemma from topological dynamics. Let us agree that a homeomorphism φ of a Hausdorff topological space L is <u>hyperbolic</u> if it has two distinct fixed points s_φ and r_φ with the following properties : for any neighbourhoods S and R of s_φ and r_φ in L, there exists an integer $n_o > 0$ such that $\varphi^n(L-S) \subset R$ for all $n \geq n_o$. Then s_φ is the <u>source</u> and r_φ the <u>range</u> of φ, and φ has clearly no other fixed point. Two hyperbolic homeomorphisms of L are <u>transversal</u> if they do not have any common fixed point.

<u>Lemma 4.</u> Let Γ be a group acting by homeomorphisms on a Hausdorff topological space L. Assume that the action is

<u>minimal</u> : for any $y \in L$, the orbit Γy is dense in L ;

<u>strongly faithful</u> : for any finite subset $F \subset \Gamma - \{1\}$, there exists $y \in L$ with $fy \neq y$ for all $f \in F$;

<u>strongly hyperbolic</u> : for any integer $N \geq 0$, there exist $\gamma_o, \ldots, \gamma_N$ in Γ which are pairwise transversal hyperbolic homeomorphisms of L.

Then Γ is a Powers' group.

Proof. Consider a finite subset $F \subset \Gamma - \{1\}$ and an integer $N \geq 1$. Let $y \in L$ be such that $fy \neq y$ for all $f \in F$. Let $\chi : \Gamma \longrightarrow L$ be the map $\gamma \longmapsto \gamma y$. Let A_L be a neighbourhood of y in L such that $f(A_L) \cap A_L = \emptyset$ for all $f \in F$. Define $A = \chi^{-1}(A_L)$, $B = \Gamma - A$, and observe that $fA \cap A = \emptyset$ for all $f \in F$.

Let S be the set of those y in L such that there exist pairwise transversal hyperbolic homeomorphisms $\gamma_o, \ldots, \gamma_N$ in Γ with y the source of γ_o. Then S is non empty by hypothesis, and clearly Γ-invariant. By minimality, S is dense in L. Hence there exist $g_o, \ldots, g_N$ in Γ which are pairwise transversal hyperbolic, and with $s_o \in A_L$ (where $s_o, \ldots, r_N$ hold for $s_{g_o}, \ldots, r_{g_N}$). Upon conjugating g_j by an appropriate power of g_o, one may assume furthermore $s_1, r_1, \ldots, s_N, r_N$ in A_L.

Let $R_1, \ldots, R_N$ be disjoint neighbourhoods of $r_1, \ldots, r_N$ respectively. One may raise if necessary g_j to a large enough power and consequently assume that $g_j(L-A_L) \cap g_k(L-A_L) = \emptyset$, and thus also $g_jB \cap g_kB = \emptyset$ for $j,k \in \{1,\ldots,N\}$ with $j \neq k$. ∎

One of the easiest consequences is the following

Proposition 5. Let Γ be a subgroup of $PSL(2,\mathbb{R})$ which is not solvable. Then Γ is a Powers' group.

Proof. Consider Γ as acting by fractional linear transformations on the real projective line (which may be thought of as the boundary $\mathbb{R} \cup \{\infty\}$ of the Poincaré half-plane). As Γ is not solvable, it contains hyperbolic elements by corollary 1.3(c) of [Ba1]. Let L be the closure in $P^1(\mathbb{R})$ of the set of those points which are fixed by some hyperbolic element of Γ. Then L is not reduced to one or two points, otherwise Γ would be solvable (compare with section 4 in [H]). It follows that L is a perfect set, so that L is infinite and indeed not countable !).

The action of Γ on L is strongly hyperbolic : given $N \geq 0$, check first that Γ contains two transversal hyperbolic elements γ_o, γ, and choose then for $\gamma_1, \dots, \gamma_N$ conjugates of γ by convenient powers of γ_o. The other hypothesis of lemma 4 are easily checked, and Γ is a Powers' group. ■

For example, any non elementary fuchsian group is a Powers' group. Also any group containing such a group, such as $PSL(2,\mathbb{K})$ for any subfield $\mathbb{K}$ of $\mathbb{R}$, is a Powers' group. All this carries over to groups of isometries of the Poincaré half-plane which do not respect the orientation.

To replace $\mathbb{R}$ by $\mathbb{C}$ in proposition 5, we need so far a finiteness hypothesis. (Late news : no more; see after corollary 15.)

<u>Proposition 6.</u> Let Γ be a finitely generated subgroup of $PSL(2,\mathbb{C})$ which is not almost solvable. Then Γ is a Powers' group.

<u>Proof.</u> Consider the natural projection $\pi : SL(2,\mathbb{C}) \longrightarrow PSL(2,\mathbb{C})$. There exists a normal subgroup Γ' of finite index in Γ such that any eigenvalue of any $\gamma \in \pi^{-1}(\Gamma')$ is not a non trivial root of unity, by theorem 6.11 of [Ra].

If Γ contains an hyperbolic element γ_o, consider the closure L in $P^1(\mathbb{C}) \approx S^2$ of the set of those points which are fixed by some hyperbolic element of Γ. As Γ is not almost solvable, the set L is perfect and the previous proof carries over to the present case.

In all cases, there exists a locally compact field $\mathbb{K}$ given together with an absolute value $\omega : \mathbb{K} \longrightarrow \mathbb{R}_+$ such that Γ can be identified with a subgroup of $PSL(2,\mathbb{K})$, with Γ' containing a $\mathbb{K}$-hyperbolic element. Here, we say that γ in $PSL(2,\mathbb{K})$ is

$\mathbb{K}$-hyperbolic if it is the image of a matrix in $SL(2,\mathbb{K})$ with eigenvalues λ_1, λ_2 satisfying $\omega(\lambda_1) \neq \omega(\lambda_2)$. (See lemma 4.1 and the proof of proposition 4.3 in [T].) Let $L_{\mathbb{K}}$ be the closure in the compact space $P^1(\mathbb{K})$ of the set of those points which are fixed by some $\mathbb{K}$-hyperbolic element of Γ. It is again clear that $L_{\mathbb{K}}$ is perfect, and one may conclude as above. ∎

In particular, any finitely generated non elementary discrete group of isometries of the hyperbolic space of dimension 3 is a Powers' group.

Let Γ be a subgroup of $PSL(2,\mathbb{C})$ which is not almost solvable, but possibly not finitely generated. Then any finite subset of Γ is contained in a finitely generated subgroup of $PSL(2,\mathbb{C})$ which is not almost solvable (indeed : Γ is almost solvable if and only if any 2-generator subgroup of Γ is so by [W]). It follows from the proofs of section 2 that $C^*_r(\Gamma)$ is also simple with a unique trace.

The argument of the two proofs above applies to the sphere at infinity of any visibility manifold [EO] as well as to a projective line P^1. For example, let G be a non compact simple connected real Lie group of real rank one without centre, let K be a maximal compact subgroup of G, let $\mathcal{H} = G/K$ be the associated symmetric space and let Y be the space of points at infinity of $\mathcal{H}$. Then $\mathcal{H} \cup Y$ has a natural topology making it homeomorphic to a closed disc with interior $\mathcal{H}$ and with boundary Y.

Let Γ be a discrete subgroup of G, let L be its limit set in Y, and assume that Γ is non elementary. (This means that Γ is not almost abelian, or equivalently that L is an infinite set.) The action of Γ on L is strongly faithful in each of the following cases :

(a) Γ is torsion free (because any γ in $\Gamma-\{1\}$ has at most two fixed points in the infinite set L)

(b) $\mathcal{H}$ is of dimension 2 or 3 (because any g in $G-\{1\}$ has at most two fixed points in Y)

(c) Γ is a lattice (because L = Y by lemma 8.5 of [Mo]).

<u>roposition 7.</u> Let G be a non compact simple connected real Lie roup of real rank one without centre, let Γ be a non elementary iscrete subgroup of G , and assume that Γ is strongly faithful on ts limit set. Then Γ is a Powers' group.

<u>orrections to previous work.</u> Let $\mathcal{H}^d \approx SO(1,d)^o/SO(d)$ denote the -dimensional hyperbolic space. For $d' < d''$, there is a standard nclusion of the group of isometries of $\mathcal{H}^{d'}$ into that of $\mathcal{H}^{d''}$. onsider in particular PSL(2,$\mathbb{Z}$) acting as usual on the Poincaré alf-plane $\mathcal{H}^2 = \{ u \in \mathbb{C} \mid \mathrm{Im} u > 0 \}$, and thus also on $\mathcal{H}^4 = \{ (u,v) \in \mathbb{C}^2 \mid \mathrm{Im} u > 0 \}$; consider also Z/2Z acting on $\mathcal{H}^4$ y $(u,v) \longmapsto (u,\pm v)$. These actions define an action of the group $\Gamma = PSL(2,\mathbb{Z}) \times \mathbb{Z}/2\mathbb{Z}$ on $\mathcal{H}^4$, for which the limit set L is the ircle

$$\{ (u,v) \in \mathbb{C}^2 \mid u \in \mathbb{R} \ , \ v = 0 \} \cup \{\infty\}$$

n

$$Y = \{ (u,v) \in \mathbb{C}^2 \mid u \in \mathbb{R} \} \cup \{\infty\} \approx S^3 \ ,$$

o that $\mathbb{Z}/2\mathbb{Z}$ acts trivially on L . In particular this Γ is not (strongly) faithful on its limit set. We also know from proposition 1 hat this Γ is not a Powers' group. Thus the hypothesis of strong aithfulness cannot be deleted in proposition 7, and [HJ] has to be orrected accordingly.

The following corrections should also be made to [HJ]. Firstly, here is a conjecture about $C^*_r(\Gamma)$ having no non trivial projections hich is stated in [HJ] for discontinuous groups of hyperbolic isometries, but which clearly makes sense for torsion-free groups only ! In fact, given any torsion-free group Γ, it would follow from a conjecture of P. Baum and A. Connes [BC] that $C_r^*(\Gamma)$ has no non-trivial projection; compare with (4) at the end of section 1. Secondly, there are printing mistakes : $\frac{1}{s}$ should be deleted on top of page 51 of [HJ], and $\Gamma\circ\phi$ is Γ and not ϕ. Finally [Lan] is missing in the list at the end; it should refer to C. Lance : "On nuclear C*-algebras", J. Functional Analysis 12 (1973) 157-176.

* * * * * * *

One may also translate the proof of proposition 5 from the Poincaré half-plane to relevant graphs. Consider a free product with amalgamation $\Gamma = H \underset{A}{*} K$. Let X be the oriented graph defined as follows (where $\amalg$ denotes a disjoint union)

the set of vertices is $\mathrm{Ver}X = \Gamma/H \amalg \Gamma/K$

the set of oriented edges is $\mathrm{Edg}_+X = \Gamma/A$

the origin of an edge γA is the vertex γH

the end of γA is the vertex γK.

Then X is a tree by n° I.4.1 of [Se]. Let Y be the space of ends of X. (See n° I.2.2 in [Se]; I use the word "end" even though X is not necessarily locally finite.) The group Γ acts naturally on X by orientation-preserving morphisms, and on Y by homeomorphisms.

We deal first with the easiest case, when A is trivial.

Proposition 8. Let $\Gamma = H*K$ be a free product, where H has at least two and K at least three elements. Then Γ is a Powers' group.

Proof. The group Γ acts freely on Edg_+X. Any $\gamma \in \Gamma - \{1\}$ has either 0 or 1 fixed vertex. If γ has 0 fixed vertex, γ is hyperbolic both as automorphism of the tree X and as homeomorphism f Y by proposition 25 of n° I.6.4 in [Se]. If γ has 1 fixed ertex, then Y has no fixed point in Y by proposition 23 of the ame. Now Y is infinite by hypothesis on the cardinalities of H nd K. Thus Γ acts strongly faithfully on Y.

Given a pair y, y' of distinct points in Y and neighbour-oods U, U' of these in Y, there exists an hyperbolic element n Γ with fixed points in U, U' by proposition 25(b) of n° I.6.4 n [Se]. Hence the action of Γ on Y is strongly hyperbolic, and .lso minimal.

The hypothesis of lemma 4 are verified. ∎

We deal now with the case of a non trivial group A. Assume :hat $\bigcap \gamma A \gamma^{-1} = \{1\}$ (intersection over all γ in Γ), so that :he action of Γ on Edg_+X is faithful. On one hand, this implies .hat A is not of index two in both H and K, so that Y is .nfinite. On the other hand, this is implied by the condition

$$\text{(SF)} \quad \begin{cases} \text{given any finite subset } F \subset \Gamma - \{1\}, \\ \text{there exists } g \in \Gamma \text{ with } g^{-1}Fg \cap A = \emptyset \end{cases}$$

vhich expresses that the action of Γ on Edg_+X is strongly faithful.

Lemma 9. Condition (SF) implies that the action of Γ on Y is strongly faithful.

Proof. Let $F \subset \Gamma - \{1\}$ be a finite set. Denote by X^f the subtree of X which is fixed by some $f \in F$, by Y the union of the X^f's and by Z the union of Y with all geodesics connecting two trees of Y (see n° I.6.4 in [Se]); then Z is a tree and Z-Y is finite.

It is clear from (SF) that X-Y is non empty. We claim that it is moreover an infinite graph, so that X-Z itself is non empty.

Indeed, let $g \in \Gamma$ be a cyclically reduced element (n^o I.1.3 in [Se]); then g cannot be in any conjugate in Γ of either H or K, so that g has no fixed vertex in X (and a fortiori no fixed edge), namely g is hyperbolic. Let n be an integer, $n \geqslant 1$, and let $F_+ = \bigcup_{1 \leqslant k \leqslant n} g^{-n} F g^n$. By (SF) again, there exists $\mathcal{E}_o \in Edg_+ X$ such that $f_+ \mathcal{E}_o \neq \mathcal{E}_o$ for all $f_+ \in F_+$, namely such that $fg^k \mathcal{E}_o \neq g^k \mathcal{E}_o$ for all $f \in F$ and for all $k \in \{1, \ldots, n\}$. This shows the claim.

As the tree X has no terminal vertex and as the subtree Z is proper, the complement $X-Z$ contains a half-infinite geodesic δ such that $f\mathcal{E} \neq \mathcal{E}$ for all $f \in F$ and for all edges $\mathcal{E}$ in δ; we denote by y the infinite endpoint of δ.

Let $F' = \{ f \in F \mid fy \neq y \}$. As F' is finite, there exists a neighbourhood U of y in Y such that $fU \cap U = \emptyset$ whenever $f \in F'$. By proposition 25 in n^o I.6.4 of [Se] again, any $f \in F-F'$ is hyperbolic. Upon restricting U, one may assume that y is the only fixed point of f in U for any $f \in F-F'$. If $z \in U-\{y\}$, one has therefore $fz \neq z$ for all $f \in F$. ■

It is straightforward that the converse of lemma 9 holds.

<u>Proposition 10.</u> Let $\Gamma = H \underset{A}{*} K$ be a free product with amalgamation over a group $A \neq \{1\}$ fulfilling condition (SF) above. Then Γ is a Powers' group.

<u>Proof</u> : see lemma 9 and the end of the proof of proposition 8. ■

This and proposition 11 below are essentially due to E. Bédos, who gives a different proof. He uses a condition slightly more restrictive than (SF), namely that one may find a "blocking pair" for A in K as defined for theorem V.11.3 of [LS].

The obliged variation for HNN-extensions can be played as follows. Consider a group H, two isomorphic subgroups A, B of H, an isomorphism $\theta : A \longrightarrow B$, and the resulting HNN-extension $\Gamma = HNN(H,A,\theta)$: it is the group generated by H and a so-called stable letter s, with the defining relations $sas^{-1} = \theta(a)$ for all $a \in A$. Let X be the oriented graph defined by

$$\mathrm{Ver}X = \Gamma/H \qquad o(\gamma A) = \gamma H$$
$$\mathrm{Edg}_+X = \Gamma/A \qquad e(\gamma A) = \gamma sH \quad .$$

Then X is a tree by n° I.5.3 of [Se]. Let Y be the space of ends of X. The group Γ acts naturally on both X and Y.

We assume that A has at least two elements (otherwise Γ would be a free product of H by an infinite cyclic group) and that the index of A in H is at least three, so that Y is infinite. We assume moreover

$$(SF') \quad \begin{cases} \text{given any finite subset} \quad F \subset \Gamma - \{1\} \ , \\ \text{there exists} \quad g \in \Gamma \quad \text{with} \quad g^{-1}Fg \cap A = \emptyset \end{cases}$$

so that Γ acts in a strongly faithful way on Y (lemma 9).

<u>Proposition 11.</u> Under the assumptions just made, the HNN-extension $\Gamma = HNN(H,A,\theta)$ is a Powers' group.

<u>Proof</u> : see proposition 10. ■

Suppose that Γ is a free product with amalgamation or a HNN-extension, but that Γ does not fulfil the conditions required by proposition 10 and 11. (Examples : $SL(2,\mathbb{Z})$, groups of torus knots, ...). It should not be difficult to find results about two-sided ideals in and traces on $C^*_r(\Gamma)$, at least in some cases.

4.- MORE GEOMETRY : HIGHER RANK EXAMPLES

Consider an integer $d \geqslant 2$. The purpose of this section is to show that any lattice Γ in the simple connected real Lie group $G = PSL(d,\mathbb{C})$ is a Powers' group. (For possible other G 's, see the comment in the introduction.)

Let Y be the space of (complete) flags in $\mathbb{C}^d$. Recall that a <u>flag</u> is a nested sequence $y = (y^1, \ldots, y^{d-1})$ of linear subspaces of $\mathbb{C}^d$ with $\dim(y^j) = j$ for $j = 1,\ldots,d-1$. Two flags x and y are <u>transversal</u> if $x^j \cap y^{d-j} = 0$ for $j = 1,\ldots,d-1$. There is an obvious action of G on Y which is faithful, transitive, and continuous with respect to both the transcendental topology and the Zariski topology on Y . It may be that one can apply lemma 4 to the action of Γ on an appropriate space of subsets of Y , but this would be far-fetched and I rather proceed as follows. (I am grateful again to Y. Guivarch' for a good hint at this point.)

For any $y \in A$, let $C(y)$ be the closed subset of Y defined by

$$C(y) = \left\{ (z^1, \ldots, z^{2m-1}) \in Y \mid z^1 \subset y^m \right\}$$

if $d = 2m$ is even and by

$$C(y) = \left\{ (z^1, \ldots, z^{2m}) \in Y \;\middle|\; \begin{array}{l} z^1 \subset y^{m+1} \quad \text{and} \\ \dim(z^2 \cap y^m) \geqslant 1 \end{array} \right\}$$

if $d = 2m+1$ is odd. Observe that $C(y)$ depends continuously on y in the following sense : let A_Y be a neighbourhood of $C(y)$ in Y ; then there exists a neighbourhood U of y in Y such that $C(x) \subset A_Y$ whenever $x \in U$. It is easily checked that $C(x) \cap C(y) = \emptyset$ for transversal flags $x, y \in Y$.

An element $g \in G$ is polar regular if g has a representati-
e $\bar{g} \in SL(d,\mathbb{C})$ with eigenvalues having pairwise distinct moduli
see § 2 in [Mo]). Let $e_1, \dots, e_d$ be eigenvectors of $\bar{g}$ with
$e_j = \lambda_j e_j$ for $j = 1,\dots,d$ and $|\lambda_1| < |\lambda_2| < \dots < |\lambda_d|$.
he sourceof g is the flag

$$s(g) = (s(g)^1, \dots, s(g)^{d-1})$$
$$s(g)^j = \text{linear span } (e_1, \dots, e_j)$$

nd the range of g is

$$r(g) = (r(g)^1, \dots, r(g)^{d-1})$$
$$r(g)^j = \text{linear span } (e_d, \dots, e_{d-j+1}) \qquad r(g) = s(g^{-1}) .$$

wo polar regular elements g , h in G are transversal if the
lags $s(g)$, $r(g)$, $s(h)$, $r(h)$ are pairwise transversal.

(Digression : a polar regular element g in G has exactly $d!$
ixed points on Y . One is a source which is $s(g)$, one is a sink
hich is $r(g)$. All the others are saddles, exactly half of them
eing in $C(s(g))$ and the others in $C(r(g))$. For $y \in Y$, the
rbit $(y, g(y), g^2(y), \dots)$ may stay a long time near a saddle
efore wandering away. However, if this orbit approaches near enough
saddle in $C(r(g))$, then it cannot wander away from $C(r(g))$ any
ore.)

Let g be a polar regular element and let U [respectively V]
e a neighbourhood of $C(s(g))$ [resp. $C(r(g))$] in Y ; then there
xists an integer $n_o \geq 1$ such that $g^n(Y-U) \subset V$ for all
$n \geq n_o$.

emma 12. Let Γ be a lattice in G . For any integer $N \geq 0$
here exist polar regular elements $\gamma_o, \dots, \gamma_N$ in Γ which are
airwise transversal.

Proof. Lemma 12 for $N = 0$ is contained in theorem 2.5 of [PR] (see also the remark after lemma 8.2 in [Mo]). We may thus proceed by induction, so that we assume that $N \geq 1$ and that the lemma holds for $N-1$.

Let $\gamma_0, \ldots, \gamma_{N-1}$ be pairwise transversal polar regular elements in Γ. Let $\tilde{\gamma}_N$ be a polar regular element in Γ. Then

$$W = \left\{ g \in G \;\middle|\; \begin{array}{l} s(\gamma_0),\ r(\gamma_0), \ldots, r(\gamma_{N-1}),\ s(g\tilde{\gamma}_N g^{-1}), \\ r(g\tilde{\gamma}_N g^{-1}) \quad \text{are pairwise transversal} \end{array} \right\}$$

is a Zariski-open set in G; and W is not empty because G is transitive on Y. As Γ is Zariski-dense in G (corollary 5.16 of [Ra]), there exists $\gamma \in W \cap \Gamma$ and one may define $\gamma_N = \gamma\tilde{\gamma}_N\gamma^{-1}$. ∎

Proposition 13. Any lattice Γ in $G = PSL(d,\mathbb{C})$ is a Powers' group.

Proof. Let $F \subset \Gamma - \{1\}$ be a finite subset and let $N \geq 1$ be an integer.

For any $g \in G - \{1\}$, the set

$$\{ y \in Y \mid g(y) \text{ is transversal to } y \}$$

is Zariski-open in Y. Hence there exists $y_0 \in Y$ with fy_0 and y_0 transversal for all $f \in F$, so that $C(fy_0) = fC(y_0)$ and $C(y_0)$ have empty intersection for all $f \in F$. One may thus choose a (transcendental) neighbourhood A_Y of $C(y_0)$ in Y such that $f(A_Y) \cap A_Y = \emptyset$ for all $f \in F$. Let $\chi : \Gamma \longrightarrow Y$ be defined by $\chi(\gamma) = \gamma y_0$. Set $A = \chi^{-1}(A_Y)$ and $B = \Gamma - A$; one has clearly $fA \cap A = \emptyset$ for all $f \in F$.

Consider the set of those $y \in Y$ such that there exist $(N+1)$-uples $\gamma_0, \ldots, \gamma_N$ of pairwise transversal polar regular elements in Γ with $y = s(\gamma_0)$. This set is Γ-invariant, and non empty by lemma 12. As Γ is minimal on Y by lemma 8.5 of [Mo], this set is also dense. Hence there exist pairwise transversal polar regular elements $g_o, \ldots, g_N$ in Γ with $r(g_o)$ very near y_o, so that $C(r(g_o)) \subset A_Y$. One may if necessary conjugate g_j by a large enough power of g_o, and thus assume that the N disjoint sets $C(r(g_1)), \ldots, C(r(g_N))$ are moreover in A_Y.

Let $R_1, \ldots, R_N$ be disjoint neighbourhoods of $C(r(g_1))$, $\ldots, C(r(g_N))$ respectively. One may if necessary raise g_j to a large enough power, and thus assume that $g_j(Y-A_Y) \subset R_j$ for $j = 1, \ldots, N$. In particular $g_j(Y-A_Y) \cap g_k(Y-A_Y) = \emptyset$, and thus also $g_j B \cap g_k B = \emptyset$ for $j,k = 1, \ldots, N$ with $j \neq k$. ∎

Proposition 13 and its proof work just as well for $G = PSL(d, \mathbb{R})$. Though theorem 2.5 of [PR] quoted above is rather technical, its conclusion is quite obvious for familiar groups such as $\Gamma = PSL(d, \mathbb{Z})$.

Proposition 13 holds also for a group Γ containing a lattice in $PSL(d, \mathbb{R})$ or $PSL(d, \mathbb{C})$. In particular :

Corollary 14. For any subfield $\mathbb{K}$ of $\mathbb{C}$, the group $PSL(d, \mathbb{K})$ is a Powers' group.

We may push this a bit further, as we did after proposition 6. Say that a group Γ is locally a Powers' group if any finite subset of Γ is contained in a Powers' subgroup of Γ. On one hand, proposition 3 and its proof carry over to locally Powers' groups; on the other hand, any field of characteristic 0 is locally a subfield of $\mathbb{C}$ (see e.g. § 4 in [Ma]). Consequently :

Corollary 15. For any field $\mathbb{K}$ of characteristic 0 and for any integer $d \geq 2$, the reduced C*-algebra of the group $PSL(d, \mathbb{K})$ is simple and has a unique trace.

I wrote this too fast, because it is clear that a group which is locally Powers' is a Powers group !! Thus proposition 6 holds without "finitely generated", and any $PSL(d, \mathbb{K})$ as above is a Powers' group.

REFERENCES

AL C.A. Akemann and T.Y. Lee : "Some simple C*-algebras associated with free groups". Indiana Math. J. 29 (1980) 505-511.

Ar R.J. Archbold : "On the Dixmier property of certain algebras". Math. Proc. Camb. Phil. Soc. 86 (1979) 251-259.

BC P. Baum and A. Connes : "Geometric K theory for Lie groups and foliations". Preprint, November 1982.

Ba1 H. Bass : "Groups of integral representation type". Pacific J. Math. 86 (1980) 15-51.

Ba2 H. Bass : "Euler characteristics and characters of discrete groups". Invent. Math. 35 (1976) 155-196.

B1 E. Bédos : "Operator algebras associated with free products of groups with amalgamation". Preprint, Oslo 1982.

B2 E. Bédos : "Operator algebras associated with HNN-extensions". Preprint, Oslo 1983.

BHPS K. Bonvallet, B. Hartley, D.S. Passman and M.K. Smith : "Group rings with simple augmentation ideals". Proc. Amer. Math. Soc. 56 (1976) 79-82.

EO P. Eberlein and B. O'Neill : "Visibility manifolds". Pacific J. Math. 46 (1973) 45-109.

Fo E. Formanek : "Idempotents in noetherian group rings". Can. J. Math. 25 (1973) 366-369.

G M. Gromov : "Groups of polynomial growth and expanding maps". Publ. Math. IHES 53 (1981) 53-78.

HJ P. de la Harpe and K. Jhabvala : "Quelques propriétés des algèbres d'un groupe discontinu d'isométries hyperboliques". Monographie 29 in l'Enseignement math. (1981) 47-55.

H P. de la Harpe : "Free groups in linear groups". L'Enseignement math. 29 (1983) 129-144.

KLR R.V. Kadison, E.C. Lance and J.R. Ringrose : "Derivations and automorphisms of operator algebras II". J. Functional anal. 1 (1967) 204-221.

LS R.C. Lyndon and P.E. Schupp : "Combinatorial group theory". Springer 1977.

[Ma] A.I. Mal'cev : "On the faithful representations of infinite groups by matrices". Amer. Math. Soc. Transl. (2) 45 (1965) 1-18.

[Mo] G.D. Mostow : "Strong rigidity of locally symmetric spaces". Princeton Univ. Press 1973.

[PS] W.L. Paschke and N. Salinas : "C*-algebras associated with free products of groups". Pacific J. Math. 82 (1979) 211-221.

[Pa] D.S. Passmann : "The algebraic structure of group rings". Wiley Interscience 1977.

[Ped] G.K. Pedersen : "C*-algebras and their automorphism groups". Academic Press 1979.

[P] R.T. Powers : "Simplicity of the C*-algebra associated with the free group on two generators". Duke Math. J. 42 (1975) 151-156.

[PR] G. Prasad and M.S. Raghunathan : "Cartan subgroups and lattices in semi-simple groups". Ann. of Math. 96 (1972) 296-317.

[Ra] M.S. Raghunathan : "Discrete subgroups of Lie groups". Springer 1972.

[Se] J.P. Serre : "Arbres, amalgames, SL_2". Astérisque 46 (Soc. Math. France 1977).

[T] J. Tits : "Free subgroups in linear groups". J. of Algebra 20 (1972) 250-270.

[W] B.A.F. Wehrfritz : "2-generator conditions in linear groups". Archiv. Math. 22 (1971) 237-240.

P.H.

Section de mathématiques

C.P. 240

1211 Genève 24.

Added in proof: Claims (c) and (g) of our main theorem are only proved for $d \leqslant 3$. Indeed, the open set $\{y \in Y \mid g(y)$ is transversal to $y\}$ in the proof of proposition 13 may be empty when $d \geqslant 4$.

SIGNATURE OPERATOR ON LIPSCHITZ MANIFOLDS AND UNBOUNDED KASPAROV BIMODULES.

- Michel HILSUM -

Laboratoire de Mathématiques Fondamentales
U.E.R. 48 - Université P.M. CURIE
75230 PARIS CEDEX 05
FRANCE

INTRODUCTION

Novikov's theorem states that the rational Pontryagin classes of a smooth manifold are topological invariants ([N]).

Recently, D. Sullivan and N. Teleman have given an analytical proof of Novikov's theorem, based on the theory of Lipschitz manifolds; and thus answering to a question raised by I.M. Singer. ([S-T] , [T_1])

As was announced in [H], we explain a different and shorter proof of the results of [T_1],[T_2] by putting the construction of N. Teleman into the framework of the KK-theory of G.G-Kasparov.

The Atiyah-Singer theorem shows us that the rational Pontryagin classes of a smooth manifold are determined by the knowledge of the homomorphism $K^o(M) \longrightarrow \mathbb{Z}$ given by :

$$E \longrightarrow \text{Index } D_E$$

where D_E is the signature operator with coefficients in E ([A - S] , [A - B - P]).

On the other hand, in 1977, D. Sullivan shows the existence and quasi-uniqueness of a Lipschitz structure on any topological manifold of dimension $\neq 4$ ([5]).

Later, N. Teleman exhibit on a Lipschitz Riemannian manifold (M,g) the construction of a signature operator D_E acting in the graded Hilbert space $L^2(M,E)$ of square integrable differential forms with values in E.

ne shows then that the map :

$$E \longrightarrow \text{Index } (D_E)$$

goes to a map $K^o(M) \longrightarrow \mathbb{Z}$ which only depends on the Lipschitz structure on M.

As a consequence of this, N. Teleman ([T_2]) associates to the signature operator D_g an element in the K-homology group of M, $K_o(M)$.
But, as the self-adjointness of D_g is not put in evidence, one must define a new additive functor, the group $K_o^{(1)}(M)$ where one deals with module over some dense subalgebra of C(M) (the "C^1"-functions).

One has to show then that $K_o^{(1)}(M) \cong K_o(M)$, and that the triple $(W, L^2(M), D_g)$ defines a class in $K_o^{(1)}(M)$ (where $W = \text{Domain } (D_g)$ is not a C(M)-module).

In this article, we propose the inverse way :
More precisely, when M is a smooth manifold, we have then the formula :

$$\text{Index } D_E = [E] \otimes_M [D]$$

where the right-hand side is the cup-product in KK-theory, $KK(\mathbb{C},M) \times KK(M,\mathbb{C}) \to \mathbb{Z}$ of the class of E in $K^o(M)$ by the class of [D] in $K_o(M)$ ([K]).

This, with the Atiyah-Singer Index theorem, shows that Novikov's theorem is equivalent to the unicity in $K_o(M)$ of the K- homology class defined by D. (Theorem 1-1).

We are now reduced to prove that on a Riemannian Lipschitz manifold, the signature operator D_g of N. Teleman defines a class of $K_o(M)$ which so depends of the Lipschitz structure.

We show this directly, (theorem 3-1) by showing that $(L^2(M), D_g)$ satisfies the axioms of "Unbounded Kasparov bimodules". ([B-J]).

We mention two points in our proof :

1) We prove the self-adjointness of D (Section 4)

2) We establish a very simple relation (valid in a more abstract setting)

between the signature operators D_o, D_1 associated with two Riemannian metrics g_o, g_1 on M. (Section 5).

As a consequence we get :

1) $(1+D^2)^{-p/2}$ is of trace class, $p > \dim M$

(as N. Teleman's result gives just that it is compact, [T_1]).

In particular the "heat kernel" satisfies (Section 5)

$$\mathrm{Trace}\,(e^{-tD^2}) < +\infty$$

2) We obtain also in an abstract setting an operatorial homotopy between D_o, D_1 which proves uniqueness of the K-homology class (Section 6).

For the sake of completness and as this conference is not intended for topologists, we shall expose first some classical results.

Section 1 is devoted to the signature operator theory on a smooth manifold and the connection with K-homology.

In section 2, we recall the basic results on Lipschitz manifolds and the construction of N. Teleman.

In section 3, we state our principal result and show on the lines of [S-T] how it implies Novikov's theorem.

1. PONTRYAGIN CLASSES AND SIGNATURE OPERATOR ON A SMOOTH MANIFOLD

1.1. Pontryagin classes.

On every smooth manifold M one can compute Pontryagin classes $p_i(M) \in H^{4i}(M,\mathbb{Z})$ which are invariants of the smooth structure : if $f : M \to N$ is a diffeomorphism between two smooth manifolds, then

$$f^*(p_i(N)) = p_i(M)$$

where $f^* : H^*(N,\mathbb{Z}) \longrightarrow H^*(M,\mathbb{Z})$ is the associated map in cohomology (Milnor, [M]).

However these classes are not topological invariants ([M-S],[M]) but an important theorem of S.P. Norikov ([N]) states that they are rational invariants of the topology : the equality above is true if f is just a homeomorphism and if we restrict to the group $H^*(M,\mathbb{Q})$, $H^*(W,\mathbb{Q})$ of cohomology with rational coefficients.

Novikov's proof is based on cobordism and surgery theories.

However the Atiyah-Singer theorem shows us that the Pontryagin classes are related to the signature operator on the manifold.

1.2. Signature Operator ([A-S],[A-B-P]).

In this section we fix M to be a smooth manifold, oriented, of even dimension $m = 2n$, and g a smooth Riemannian metric on M.

We recall the definition of the $\star$-operator of Hodge : this is the endomorphism of the bundle $\Lambda_{\mathbb{C}}(T^*M)$ of the exterior algebra of T^*M defined at $x \in M$ as follows :

Let $e_1,\dots,e_m$ be an orthonormal basis of T^*_xM for g_x; then the $e_I = e_{i_1} \wedge \dots \wedge e_{i_p}$ form an orthonormal basis of $\Lambda_{\mathbb{C}}(T^*_xM)$ where $I = \{i_1,\dots,i_p\}$ is a p-uple of integers such that $1 \le i_1 < i_2 \dots < i_p \le m$.

Then we put

$$* e_I = \delta(I)\, e_{\hat{I}}$$

where $\hat{I}$ is the strictly increasing $m-p$—uple complementary to I in $\{1,2,\dots,m\}$ and $\delta(I)$ is the signature of the permutation $(1,2,\dots,m) \to (I,\hat{I})$.

This operator maps $C^\infty(\Lambda^p_{\mathbb{C}}(T^*M))$ onto $C^\infty(\Lambda^{m-p}_{\mathbb{C}}(T^*M))$ and satisfies on that space :

$$*^2 = (-1)^p$$

The operator $\tau = i^{p(p-1)+m} *$ satisfies then $\tau^2 = 1$ and defines $\mathbb{Z}_2$-grading on the linear space of smooth differential forms by :

$$C^\infty(\Lambda_{\mathbb{C}}(T^*M))^{\pm} = \ker(\tau \mp 1)$$

For $\xi \in C^\infty(T^*M)$ we note $\mathrm{ext}(\xi)$ the endomorphism of $C^\infty(\Lambda_{\mathbb{C}}(T^*M))$.

$$\mathrm{ext}(\xi)\alpha = \xi \wedge \alpha$$

We put $\mathrm{int}(\xi) = * \,\mathrm{ext}(\xi)\, *$ which is the adjoint of $\mathrm{ext}(\xi)$ with respect to the hermitian structure defined by g.

Then the operator

$$C(\xi) = \mathrm{ext}(\xi) + \mathrm{int}(\xi)$$

is a degree one invertible linear map of $C^\infty(\Lambda_{\mathbb{C}}(T^*M))$ whenever $\xi \neq 0$.

Let $d : C^\infty(\Lambda_{\mathbb{C}}(T^*M)) \longrightarrow C^\infty(\Lambda_{\mathbb{C}}(T^*M))$ be the exterior derivative. The <u>signature operator</u> is the differential operator depending of the Riemannian metric g :

$$D_g = d - * d *$$

Usually we put $\delta = - * d *$ and we recall that $D^2 = d\delta + \delta d$ the Laplacian operator associated with g.

The signature operator with coefficients in a bundle E is analogously defined by

$$D^E_g = \nabla - * \nabla *$$

where ∇ is a linear connection on E, acting by extension on $C^{\infty}(\Lambda_{\mathbb{C}}(T^{*}M \otimes E))$, the smooth differential forms with values in E.

The principal symbol of D is

$$\sigma_D(x,\xi) = C(\xi)$$

so that D, and D_E are elliptic differential operators.

1.3. Index theorem.

As D_E is elliptic, it has finite index and the Atiyah-Singer Index theorem gives the formula :

$$\mathrm{Index}(D_E) = \langle \mathrm{ch}(E) \cup L(M) , [M] \rangle$$

where $\mathrm{ch}(E)$ and $L(M)$ are cohomology classes defined as follows :

1) $\mathrm{ch} : K^{o}(M) \longrightarrow H^{*}(M,\mathbb{Q})$ is the chern character (cf.[M]).

2) $L(M)$ is the L-polynomial of Hirzebruch : it is an universal polynomial in the Pontryagin classes of M such that if $L_k(M)$ is the composent of L in $H^{4k}(M,\mathbb{Q})$ then we can write :

$$L_k(M) = \alpha_k p_k + L'_k(p_1,\dots,p_{k-1})$$

where L'_k is a polynomial which depends only on $p_1,\dots,p_{k-1}$.

Thus the Pontryagin classes of M are inductively determined by $L_k(M)$, $k = 0,1,2,\dots$

On the other hand, as the chern character induces an isomorphism

$$\mathrm{ch} : K^{o}(M) \otimes \mathbb{Q} \xrightarrow{\simeq} H^{*}_{\mathrm{even}}(M,\mathbb{Q})$$

it follows that the Pontryagin classes are determined by the knowledge of the map :

$$E \longrightarrow \mathrm{Index}\, D_E$$

from $K^o(M)$ to $\mathbb{Z}$.

1.4. K-homology of M.

We recall that $K_o(M) = KK(M,\mathbb{C})$ is defined as the group of stably homotopy classes of couples (ε,F) where ε is a graded Hilbert space, C(M)-left-module, and F is a bounded linear operator in H such that $F-F^*$, F^2-1 and all the commutators $[F,a]$, $a \in C(M)$ are compact operators. Such a couple (ε,F) is called a Kasparov bimodule ([K]).

The signature operator of the Riemannian smooth manifold (M,g) determines an element $[D] \in K_o(M)$, independant of the choice smooth Riemannian metric, as follows :

We take $\varepsilon = L^2(M,\Lambda_{\mathbb{C}}(T^*M))$ with the obvious grading coming from τ.

The signature operator D_g is essentially self-adjoint on $C^\infty(M,\Lambda_{\mathbb{C}}(T^*M))$ and we take $F = \bar{D}_g(1+\bar{D}_g^2)^{-1/2}$ which is bounded self adjoint.

By the pseudo-differential calculus, we know that F + (compact operator) is a pseudo differential operator of order 0 on M whose principal symbol is

$$\sigma_F(x,\xi) = (1 + C(\xi)^2)^{-1/2} C(\xi)$$

It follows then, modulo compact operators, that F^2-1 and $[F,a]$ are pseudo-differential operators of strictly negative orders, and so are compact operators.

By the same reasoning, the signature operator with coefficients in the vector bundle E determines a class $[D_E] \in KK(\mathbb{C},\mathbb{C})$; if we identify $KK(\mathbb{C},\mathbb{C})$ with $\mathbb{Z}$ by the map $(\varepsilon,F) \longrightarrow \mathrm{Index}(F)$, then the class $[D_E]$ becomes Index $D_E \in \mathbb{Z}$.

We now have the formula :

(1) $$\text{Index } D_E = [E] \otimes_M [D]$$

where the right hand side is the cup product in KK-theory of G.G. Kasparov ([K]) :

$$K^{o}(M) \times K_{o}(M) \longrightarrow KK(\mathbb{C},\mathbb{C}) \simeq \mathbb{Z}$$

Let $ch : K_o(M) \longrightarrow H_*(M,\mathbb{Q})$ denotes the chern character in homology, and P the isomorphism given by the Poincaré duality.

The Atiyah-Singer theorem and (1) show then :

(2) $$P(ch([D])) = L(M)$$

It means that we can focus on the K-homology class defined by D and that Novikov's theorem follows from

<u>Theorem 1.1.</u> : The class $[D] \in K_o(M)$ depends only of the topology.

This theorem will be a consequence of theorem 3.1. (cf. Sections 3.3). It proves Novikov's theorem at least when dimension (M) is even. If dimension $M = 2k+1$, then we can form the even dimensional manifold $M \times S^1$, and the result follows from the equality $L(M \times S^1) = p_*(L(M))$.

2. LIPSCHITZ MANIFOLDS AND N. TELEMAN'S CONSTRUCTION

2.1. Lipschitz maps.

Let $U \subset \mathbb{R}^m$ be an open subset. We recall that a map $\phi : U \longrightarrow \mathbb{R}^k$ is called Lipschitz if one of the following equivalent two conditions are satisfied :

1) There exists $k > 0$ such that for all $x,y \in U$,

$$\|\phi(x) - \phi(y)\| \leq k\,\|x - y\|$$

2) ϕ possesses partial derivatives almost everywhere and the maps

$x \longrightarrow \frac{\partial \phi_i}{\partial x_j}(x)$ belongs to $L^\infty(U)$.

We can define the Jacobian matrix of ϕ as the function $d\phi \in L^\infty(U, M_{m,k}(\mathbb{R}))$

given by $d\phi(x) = (\frac{\partial \phi_i}{\partial x_j}(x))$.

Let $U \subset \mathbb{R}^m$, $V \subset \mathbb{R}^k$, $W \subset \mathbb{R}^e$ be open subsets and $\phi : U \to V$, $\psi : V \to W$ Lipschitz maps, then $\phi \circ \psi$ is Lipschitz and we have for almost every $x \in U$:

(1) $$d(\phi \circ \psi)(x) = d\phi(\psi(x)) \circ d\psi(x)$$

(cf. $[T_1]$).

If ϕ, ϕ^{-1} are Lipschitz homeomorphism between two open subsets $U, V \subset \mathbb{R}^m$, then the class of Lebesgue measure is conserved by ϕ (this follows from condition 1)).

2.2. Measurable differential forms.

Let ϕ a Lipschitz map $U \to \mathbb{R}^k$ and ω a measurable map $\omega : \mathbb{R}^k \to \Lambda_{\mathbb{C}}(\mathbb{R}^k)$. By condition 2) we can form the pull back $\phi^*(\omega)$ on U as follows :

We can suppose that $\omega(y) = a(y)\, dy_{i_1} \wedge dy_{i_2} \wedge \ldots . \wedge dy_{i_p}$:

$$\phi^*(\omega)(x) = a(\phi(x))\, \phi^*(dy_{i_1}) \wedge \ldots . \wedge \phi^*(dy_{i_p})$$

where $\phi^*(dy_i) = \Sigma \frac{\partial \phi_i}{\partial x_j} dx_j$ is in $L^\infty(U, \Lambda^1_{\mathbb{C}}(\mathbb{R}))$.

In particular we get a continuous linear map :

$$\phi^* : L^2(V, \Lambda_{\mathbb{C}}(\mathbb{R}^k)) \longrightarrow L^2(U, \Lambda_{\mathbb{C}}(\mathbb{R}^m))$$

2.3. Lipschitz maps and exterior derivative.

Let $\omega \in L^2(U, \Lambda_{\mathbb{C}}(\mathbb{R}^m))$ considered as a current on U, by the formula

$$< \omega , \alpha > = \int \omega \wedge \alpha$$

where $\alpha \in C_c^\infty(U, \Lambda_{\mathbb{C}}(\mathbb{R}^m))$.

The exterior derivative of ω is the current defined by :

$$< d\omega , \alpha > = \varepsilon \int_U \omega \wedge d\alpha$$

where $\varepsilon = \pm 1$, $\alpha \in C_c^\infty(U, \Lambda_{\mathbb{C}}(\mathbb{R}^m))$, α is homogenous.

We define $\Omega_d(U)$ to be the subspace of $L^2(U, \Lambda_{\mathbb{C}}(\mathbb{R}^m))$ of the ω for which the current $d\omega$ is again a square-integrable differential form; $\Omega_d(U)$ is the maximal domain of d.

<u>Lemma 2.1.</u> ($[T_1]$) : We have $\phi^*(\Omega_d(V)) \subset \Omega_d(U)$ and the following diagram is commutative :

$$\begin{array}{ccc} \Omega_d(V) & \xrightarrow{\phi^*} & \Omega_d(U) \\ \downarrow d & & \downarrow d \\ L^2(V, \Lambda(\mathbb{R}^m)) & \xrightarrow{\phi^*} & L^2(U, \Lambda(\mathbb{R}^m)) \end{array}$$

In other words, for any $\alpha \in \Omega_d(U)$:

$$\phi^*(d\alpha) = d(\phi^*(\alpha))$$

2.4. Lipschitz manifolds.

A Lipschitz manifold M is a topological manifold provided with an atlas $(\mathcal{O}_\alpha, \phi_\alpha)_{\alpha \in A}$ such that for any $\alpha, \beta \in A$, the homeomorphism :

$$\phi_\beta \circ \phi_\alpha^{-1} : \phi_\alpha(\mathcal{O}_\alpha \cap \mathcal{O}_\beta) \longrightarrow \phi_\beta(\mathcal{O}_\alpha \cap \mathcal{O}_\beta)$$

is a Lipschitz map.

It follows from 2.1. that a Lipschitz manifold possesses a well defined Lebesgue class of measure : this is the class of measure μ on M such that $\phi_\alpha^*(\mu)$ is equivalent to the Lebesgue measure on $\phi_\alpha(\mathcal{O}_\alpha) \subset \mathbb{R}^m$ $(m = \dim M)$.

The Lipschitz structure on M determines the dense $\star$-subalgebra $\mathcal{L}$ of $C(M)$ of Lipschitz functions : namely $f \in \mathcal{L}$ if and only if $f \circ \phi_\alpha^{-1}$ is a Lipschitz function on $\phi_\alpha(\mathcal{O}_\alpha)$, for all $\alpha \in A$.

Conversely a sub-algebra of $C(M)$ satisfying this type of conditions will determine a unique Lipschitz structure.

The importance of this notion comes from the crucial theorem of D. Sullivan [S]:

<u>Theorem 2.2.</u> : Any topological manifold of dimension $\neq 4$ possesses a Lipschitz structure which is unique up to a homeomorphism close and isotopic to the identity.

We recall that the analogous statement for smooth structure, or even PL-structure is false ([M],[M-S]) : either there exist topological manifolds without smooth structure, either there exist topological manifolds with many non-equivalent smooth structures.

The second part of the theorem can be reformulated as follows :

Let N a topological manifold with two Lipschitz structures on it, and $\mathcal{L}_1$, $\mathcal{L}_2$ the corresponding algebras of Lipschitz functions on N : then there exists a homeomorphism ϕ of N such that :

$$\phi^*(\mathcal{L}_2) = \mathcal{L}_1$$

where $\phi^* : C(N) \longrightarrow C(N)$ is the induced map (and ϕ^* is isotopic to the identity).

2.5. Differential forms on Lipschitz manifold.

Let M be a Lipschitz manifold, oriented, of dimension m and $(\mathcal{O}_\alpha, \phi_\alpha)_{\alpha \in A}$ an atlas of M.

We shall denote by T^*M and $\Lambda_{\mathbb{C}}(M)$ the measurable fields over M obtained by patching together the local trivial measurable fields $T^*\mathcal{O}_\alpha = \mathcal{O}_\alpha \times \mathbb{R}^m$, $\Lambda_{\mathbb{C}}(\mathcal{O}_\alpha) = \mathcal{O}_\alpha \times \Lambda_{\mathbb{C}}(\mathbb{R}^m)$.

The sections of $\Lambda_{\mathbb{C}}(M)$ are families $(\omega_\alpha)_{\alpha \in A}$ where $\omega_\alpha : \phi_\alpha(\mathcal{O}_\alpha) \longrightarrow \Lambda_{\mathbb{C}}(\mathbb{R}^m)$ is measurable and $(\phi_\beta \circ \phi_\alpha^{-1})^*(\omega_\alpha) = \omega_\beta$ on $\mathcal{O}_\alpha \cap \mathcal{O}_\beta$ (Section 2.2).

Let $L^2(M, \Lambda_{\mathbb{C}}(M))$ be the topological vector space of square integrable differential forms, i.e. $\omega_\alpha \in L^2(\mathcal{O}_\alpha, \Lambda_{\mathbb{C}}(M))$ for $\alpha \in A$.

If $\xi, \eta \in L^2(M, \Lambda_{\mathbb{C}}(M))$ and $d°\xi + d°\eta = m$, then the m-form $\xi \wedge \eta$ is integrable (M being orientable) and we get a bilinear pairing :

$$(\alpha , \beta) = \int_M \alpha \wedge \beta$$

on $L^2_p(M, \Lambda_{\mathbb{C}}(M) \times L^2_{m-p}(M, \Lambda_{\mathbb{C}}(M))$, $p = d°\xi$, which extends to a bilinear pairing on $L^2(M, \Lambda_{\mathbb{C}}(M))$.

The map t from $L^2(M, \Lambda_{\mathbb{C}}(M))$ into its topological dual $L^2(M, \Lambda_{\mathbb{C}}(M))$ defined by

$$< \alpha , t(\beta) > = \int_M \alpha \wedge \beta$$

is an isomorphism.

To simplify, we shall call differential forms on M the elements of $L^2(M, \Lambda_{\mathbb{C}}(M))$.

2.6. Riemannian metrics.

Let $U \subset \mathbb{R}^m$ an open set and $Q(U)$ the space of all measurable rieman-nian metric on U equivalent to the standard one : then $Q(U)$ is the space of measurable map $x \longrightarrow g(x)$ where $g(x)$ is a positive definite quadratic form on $\mathbb{R}^m$ for which there exists $\alpha, \beta > 0$ such that for almost all $\xi \in T^*U$ we have :

$$\alpha \|\xi\|^2 \leq g_x(\xi,\xi) \leq \beta \|\xi\|^2$$

A riemannian metric on M is a collection $(g_\alpha)_{\alpha \in A}$ where $g_\alpha \in Q(\phi_\alpha(\mathcal{O}_\alpha))$ and such that

$$(\phi_\beta \circ \phi_\alpha^{-1})^*(g_\alpha) = g_\beta$$

on $\mathcal{O}_\alpha \cap \mathcal{O}_\beta$.

As M is oriented, a riemannian metric g determines a $*$ operator defined analogously with Sec. 1.2 : in any local chart $\phi : \mathcal{O} \longrightarrow U \subset \mathbb{R}^m$ of M, we define the measurable field of operators :

$$* \in L^\infty(U_\alpha, \mathrm{End}\,(\Lambda_{\mathbb{C}}(\mathbb{R}^m))$$

by the same formula as in Sec. 1.2.

These operators patch together to give an inversible map of $L^2(M, \Lambda_{\mathbb{C}}(T^*M))$, which satisfy

$$*^2 = (-1)^p$$

on $L^2_p(M, \Lambda_{\mathbb{C}}(M))$.

In particular the operator τ of $L^2(M, \Lambda_{\mathbb{C}}(M))$ defined by

$$\tau = i^{p(p-1) + \frac{m}{2}} *$$

on $L^2_p(M, \Lambda_{\mathbb{C}}(M))$ is an involution, $\tau^2 = 1$, and we get a grading on $L^2(M, \Lambda_{\mathbb{C}}(M))$ by taking

$$L^2(M, \Lambda_{\mathbb{C}}(M))^\pm = \ker(\tau \mp 1)$$

The metric g on M gives a hermitian bilinear form on $L^2(M, \Lambda_{\mathbb{C}}(M))$:

$$(\alpha, \beta)_g = \int \alpha \wedge * \bar{\beta}$$

α,β being two differential forms.

Let K_g be the space $L^2(M,\Lambda_{\mathbb{C}}(M))$ equipped with this hermitian form : it is a Hilbert space, for which $*$ and τ are isometrics.

2.6. Deformation of the Riemannian metric.

Let as before (M,g) a Riemannian Lipschitz manifold, oriented, of even dimension.

We can define the hermitian scalar product of the preceding section in a different fashion :

Let ω_g be the L^∞- n-differential form such that for almost all $x \in M$ we have

$$\omega_g(\xi_1 \wedge \ldots \wedge \xi_n) = 1$$

whenever $\xi_1,\ldots,\xi_n$ is on orthonormal basis of T_xM.

We lift g to a scalar product on $L^2(M,\Lambda_{\mathbb{C}}(M))$ by putting :

$$\Lambda_g(\xi_1 \wedge \ldots \wedge \xi_p , \eta_1 \wedge \eta_2 \wedge \ldots \wedge \eta_p) = \det((g(\xi_i,\eta_j))_{i \leq i,j \leq p})$$

whenever $\xi_1 \wedge \ldots \wedge \xi_p$, $\eta_1 \wedge \ldots \wedge \eta_p$ are in $L^2_p(M,\Lambda_{\mathbb{C}}(M))$.

We then have the formula :

$$(1) \qquad \int_M \alpha \wedge * \bar{\beta} = (\alpha,\beta)_g = \int_M \Lambda_g(\alpha,\bar{\beta})\,\omega_g$$

Let g_o, g_1 be two riemannian metrics on M.

We shall now exhibit a canonical isomorphism of K_{g_o} with K_{g_1}.

There exists a measurable field $A \in L^\infty(M,\mathrm{End}(T^*M))$ such that for almost all $x \in M$:

1) $A_x \geq 0$ with respect to $g_o(x)$

2) $g_1(\xi,\xi) = g_o(A_x\xi , A_x\xi)$

for all $\xi \in T^*_xM$.

3) A is invertible, i.e. $A^{-1} \in L^{\infty}(M, \mathrm{End}(T^*M))$.

Then we have

$$\omega_{g_1} = \det(A)\, \omega_{g_o}$$

Let $\Lambda(A) \in L^{\infty}(M, \mathrm{End}(\Lambda_{\mathbb{C}}(M))$ be the exterior power of A, and

$$C = \Lambda(A) \det(A)^{-1/2}$$

Then it follows from formula (1) :

$$(\alpha, \beta)_{g_1} = (C\alpha, C\beta)_{g_o}$$

for $\alpha, \beta \in L^2$.

Thus the map $\alpha \longrightarrow C^{-1}\alpha$ is an isometry of K_{g_o} with K_{g_1}.

2.7. Signature operator on Lipschitz manifold.

Let (M,g) a Lipschitz riemannian manifold, oriented, of even dimension.

Let $\Omega_d \subset L^2(M, \Lambda_{\mathbb{C}}(M))$ be the dense linear space of these ω such that for any local chart $\phi : \mathcal{O} \to U \subset \mathbb{R}^m$ of M, we have

$$\phi^*(\omega) \in \Omega_d(U)$$

(cf. Section 2.2).

By the lemma of sec. 2.2, we see that for $\omega \in \Omega_d$, the differential form $d\omega$ such that $\phi^{*-1}(d\omega) = d\phi^{*-1}(\omega)$ is well defined.

The map $\omega \to d\omega$ of L^2 with domain Ω_d is a closed operator on M and it satisfies :

$$d^2 = 0$$

Remark : If M is a smooth manifold, then the operator d just defined is the closure of usual exterior derivative acting on smooth differential forms.

As in the classical theory, put $\delta = - * d *$. Then

Definition : The signature operator on M is the couple (K_g, D_g) where

1) K_g is $\mathbb{Z}_2$- graded by τ.

2) $D_g = d + \delta$ is a degree one unbounded operator.

By section 1.4, the problem now is to define from (K_g, D_g) a class in $K_o(M)$.

This is explaind in the next section.

Remark : N. Teleman defines also a signature operator D_E with coefficients in a vector bundle E over M. ([11]).

As we shall see later, we will have

$$[D_E] = E \otimes [D_g]$$

i.e. D_E is the cup product of E with $[D_g]$, $K^o(M) \times K_o(M) \longrightarrow KK(\mathbb{C}, \mathbb{C})$.

3. SIGNATURE OPERATOR AS AN UNBOUNDED KASPAROV BIMODULE

3.1. Unbounded Kasparov bimodules.

We shall describe in two particular cases the results of S. Baaj and P. Julg ([3]) :

Let X be a topological compact space.

A) Let $\mathcal{E}$ be a $\mathbb{Z}_2$-graded Hilbert space, $C(X)$ left module, and T a closed operator on $\mathcal{E}$ of degree 1 such that

i) T is self-adjoint.

ii) $(1 + T^2)^{-1/2}$ is a compact operator.

iii) There is a dense $*$-subalgebra $\mathfrak{M} \subset C(X)$ such that for all $a \in \mathfrak{M}$, the commutators $[T,a]$ are bounded operators.

Then $(\mathcal{E}, T(1+T^2)^{-1/2})$ is a Kasparov bimodule (cf. 1.4) and defines a class in $K_o(X)$.

Such a couple $(\mathcal{E},T)$ is called an Unbounded Kasparov bimodule.

B) Operatorial homotopy : Let $\mathcal{E} = (\mathcal{E}_t)_{t \in [0,1]}$ be a Hilbert $C*$module over $[0,1]$, $\mathbb{Z}_2$-graded, left module over $C(X)$ and $T = (T_t)$ a field of unbounded operators on $\mathcal{E}_t$ such that :

i) Each T_t is closed and self adjoint on $\mathcal{E}_t$.

ii) The field $[1+T_t^2]^{-1/2}$ define a compact endomorphism of $\mathcal{E}$.

iii) There exists a dense $*$-subalgebra $\mathfrak{M}$ of $C(X)$ such that $[T,a]$ are bounded endomorphisms of $\mathcal{E}$, for $a \in \mathfrak{M}$.

Then $(\mathcal{E}, T(I+T^2)^{-1/2}$ is a Kasparov-bimodule, and defines a class in $KK(X,[0,1])$.

In particular we have an operational homotopy between $(\mathcal{E}_o,T_o)$ and $(\mathcal{E}_1,T_1)$ and their classes in $K_o(X)$ are equal ([5]) :

$$[(\mathcal{E}_o, T_o)] = [(\mathcal{E}_1, T_1)].$$

3.2. Statement of the theorem.

Let M be a Lipschitz manifold, oriented, of even dimension, g a Riemannian metric on M, (K_g,D_g) the signature operator define in 2.7.

Theorem 3.1. :

1) (K_g, D_g) is a unbounded Kasparov $C(M) \times \mathbb{C}$ - bimodule, and defines a class $[D_g]$ in $K_o(M)$.

2) The class $[D_g]$ in $K_o(M)$ does not depend of the choice of g, but depends only of the Lipschitz structure.

More precisely, in 2), if g_o, g_1 are two Riemannian metrics on M, then there exists an unbounded operational homotopy from (K_o, D_o) to (K_1, D_1).

The demonstration of the theorem will occupy sections 4), 5), 6).

We shall prove first :

i) D_g is self adjoint in K_g.

ii) $(1 + D_g^2)^{-1/2}$ is compact.

iii) The condition on commutators.

3.3. Proof of theorem 1.1.

By theorem 3.1, on every Lipschitz manifold M oriented, and even dimensional there exists a signature class $S_M \in K_o(M)$ defined by any Riemannian structure on M and the associated signature operator.

Let $f : M \to N$ be a Lipschitz homeomorphism between two such Lipschitz manifold.

As Teleman's construction is clearly functorial, we have

$$f_*(S_M) = S_N$$

where f_* $K_o(M) \longrightarrow K_o(N)$ is the induced map.

Now D. Sullivan's theorem tells us that two homeomorphic smooth manifolds are Lipschitz homeomorphic, when dimension $\neq 4$: this proves theorem 1.1. at least when dimension $M \neq 4$.

If $\dim M = 4$, then we form the 6 dimensional manifold $M \times \mathbb{T}^2$

The projection $f : M \times \mathbb{T}^2 \longrightarrow M$ induces a morphism $p_* : K_o(M \times \mathbb{T}^2) \longrightarrow K_o(M)$ and the class

$$2\,S_M = p_*(S_{M \times \mathbb{T}^2})$$

is well defined and depends only of the topology.

4. SELF ADJOINTNESS

As we have remarked in sec. 2.5. the map t from $L^2(M)$ to its topological dual defined by :

$$< \alpha , t(\beta) > = \int \alpha \wedge \beta$$

is an isomorphism.

Let ε be the endomorphism of "parity" of $L^2(M)$ $\varepsilon(\gamma) = -\gamma$ if γ is a differential form of even degree, $\varepsilon(\gamma) = \gamma$ if γ is of odd degree.

The exterior derivative $d : L^2(M) \longrightarrow L^2(M)$ being a closed operator, it has a topological adjoint : $d' : L^2(M)' \longrightarrow L^2(M)'$.

<u>Lemma 4.1.</u> : We have $d \circ \varepsilon = t^{-1} d' t$

<u>Proof</u> : Let $\alpha, \beta \in \Omega_d$ (= Dom d).
Then, by the lemma 4.2 of [T_1] we have :

$$\int d\alpha \wedge \beta = \int \alpha \wedge d\varepsilon(\beta)$$

By definition of t, we get :

$$< d\alpha , t(\beta) > = < \alpha , t(d\varepsilon(\beta)) >$$

which shows that $t(\beta) \in \mathrm{Dom}\,(d')$ and $d \circ \varepsilon(\beta) = t^{-1} d' t(\beta)$, so that $d \circ \varepsilon \subset t^{-1} d' t$.

Conversely, let $\gamma \in \mathrm{Dom}(d')$, so that for any $\alpha \in \mathrm{Dom}(d)$, we have

$$< d\alpha , \gamma > \; = \; < \alpha , d'\gamma >$$

or :

$$\int d\alpha \wedge t^{-1}(\gamma) \; = \; \int \alpha \wedge t^{-1}(d'\gamma)$$

By the lemma 4.2 of [T_1], again, we have $t^{-1}(\gamma) \in \mathrm{Dom}(d)$ and $t^{-1}(d'\gamma) = d(\varepsilon(t^{-1}(\gamma)))$, which shows that $d \circ \varepsilon \supset t^{-1} d' t$. □

Let s be the antilinear isomorphism of $L^2(M)$ with $L^2(M)'$ defined by the Hilbert space stucture K_g on $L^2(M)$:

$$s(\alpha) \; = \; t\,(* \bar{\alpha})$$

Let T be a closed operator of $L^2(M)$, let T' the topological adjoint of T, acting on $L^2(M)'$, and T^* the adjoint operateur of T on the Hilbert space K_g. We have then the relation :

$$T^* \; = \; s^{-1}\, T' s$$

If we apply this to d, δ we have

<u>Lemma 4.2.</u> : On K_g, $d^* = \delta$

<u>Proof</u> : We have $d^* = s^{-1} d s' = (s^{-1} t) d \cdot \varepsilon (t^{-1} s)$

As $t^{-1} s(\alpha) = * \bar{\alpha}$ and $*^{-1}(\alpha) = \varepsilon(*\alpha) = *(\varepsilon(\alpha))$, the last term is $- *d* = \delta$. □

<u>Lemma 4.3.</u> : Let T be a closed densely defined operator on a Hilbert space H, such that $T^2 = 0$. Then $T + T^*$ is self adjoint on H and $H = \ker(T) \cap \ker(T^*) \oplus \mathrm{support}(T) \oplus \mathrm{support}(T^*)$ (direct orthogonal sum).

Proof : As the last part is evident, we can write, with respect to this decomposition :

$$T + T^* = \begin{bmatrix} 0 & 0 & 0 \\ 0 & 0 & S^* \\ 0 & S & 0 \end{bmatrix}$$

where S is the compression of $T : \mathrm{supp}(T) \longrightarrow \mathrm{supp}(T^*)$, from what it follows that

$$(T + T^*)^* = \begin{bmatrix} 0 & 0 & 0 \\ 0 & 0 & S^* \\ 0 & S & 0 \end{bmatrix} = T + T^*. \quad \square$$

Applying this lemma to $d^2 = 0$, we have :

Proposition 4.4. : 1) $D_g = d + \delta$ is self adjoint.

2) $K_g = H(M) \oplus \overline{\mathrm{Im}\, d} \oplus \overline{\mathrm{Im}\, \delta}$

where $H(M)$ is the space of harmonic forms.

We have the following corollary which was not evident in $[T_1]$:

Corollary 4.5. : Dom $D_g = \{\omega \in L^2(M)$ such that $d\omega, \delta\omega \in L^2(M)\}$ is a dense subspace of $L^2(M)$.

5. COMPACTNESS OF THE RESOLVANT

5.1. Abstract change of the metric.

We shall consider the following situation :

i) Let K be a hilbertian space, $\| \ \|_o$ and $\| \ \|_1$ two norms defining the topology of K coming from hermitian scalar products $(,)_o$ and $(,)_1$.

ii) T a closed operator, densely defined on K, such that $T^2 = 0$.

iii) Let K_i the hilbert space K with $\| \|_i$, and S_o, S_1 the adjoints of T with respect to K_o and K_1.

For $p \geq 1$ we recall that the Schatten ideal $L^p(K_o)$ is the ideal of operators $R \in \mathcal{L}(K_o)$ such that

$$\text{Trace } (|R|^p) < +\infty$$

It is clear that this ideal does not depend of the metric chosen, and we shall denote it more simply $L^p(K)$.

Proposition 5.1. : For any $p \geq 1$, the two assertions are equivalent

i) $(S_o + T + i)^{-1} \in L^p(K)$

ii) $(S_1 + T + i)^{-1} \in L^p(K)$

By lemma 4.3, $(S_i + T + i)^{-1}$ is already bounded. The proposition will be a consequence of some lemmas.

We shall note $I \in \mathcal{L}(K_o, K_1)$ and $J \in \mathcal{L}(K_1, K_o)$ the linear operators determined by the identity mapping $\text{Id} = K \to K$.

For $i = 0$ or 1, let p_i, q_i, r_i the orthogonal projections in K_i on respectively the support of T in K_i, the support of S_i in K_i, and on $\ker T \cap \ker S_i$, so that $p_i + q_i + r_i = 1$.

We define $B \in \mathcal{L}(K_o, K_1)$ by :

$$B = p_1 I p_o + q_1 J^* q_o + r_1 r_o$$

where J^* = adjoint of J from $K_1 \longrightarrow K_o$.

<u>Lemma 5.2.</u> : B is an invertible operator.

<u>Proof</u> : As B is the sum of the three operators $p_o(K_o) \longrightarrow p_1(K_1)$, $q_o(K_o) \longrightarrow q_1(K_1)$, $r_o(K_o) \longrightarrow r_1(K_1)$, it is sufficient to check that each is invertible.

As $\mathrm{Im}(p_o)$ and $\mathrm{Im}(p_1)$ are both topological supplementar subspaces to $\ker(T)$, this is true for $p_1\, p_o$.

The subspace $\mathrm{Im}(q_i)$ is equal to the close subspace $\overline{\mathrm{Im}\, T}$, so that the linear operator $q_o\, I\, q_1 \in \mathcal{L}(\mathrm{Im}\, q_1, \mathrm{Im}\, q_o)$ is the identity; this proves that $q_1\, J^*\, q_o = (q_o\, I\, q_1)^*$ is invertible.

Finally, $\mathrm{Im}(r_1)$ and $\mathrm{Im}(r_o)$ are both topological supplementar subspace of $\overline{\mathrm{Im}\, T}$ in $\ker(T)$, so that $r_1 r_o \in \mathcal{L}(\mathrm{Im}(r_o), \mathrm{Im}(r_1))$ is invertible. □

Let $B^* \in \mathcal{L}(K_1, K_o)$ be the adjoint of $B \in \mathcal{L}(K_o, K_1)$.

<u>Lemma 5.3.</u> : We have

$$B^*(T+S_1)B = (T+S_o)$$

<u>Proof</u> : We have to develop the product :

$$(p_o\, I^*\, p_1 + q_o\, q_1 + r_o\, I^*\, r_1)(T+S_1)(p_1\, p_o + q_1\, J^*\, q_o + r_1\, r_o)$$

But by definition of p_i, q_i, r_i, and as $T^2 = 0$, we have :

$$\begin{cases} r_1(T+S_1) = (T+S_1)r_1 = 0 \\ T = q_i\, T\, p_i \\ S_i = p_i\, S_i\, q_i \end{cases}$$

so that we are reduced to prove :

$$q_o\, q_1\, T\, p_1\, p_o + (p_o\, I^* p_1)(S_1)(q_1\, J^*\, q_o) = T + S_o$$

By the definition,

$$q_o\, q_1\, T\, p_1\, p_o \quad = \quad T$$

and by taking adjoint, we find

$$(p_o\, I^*\, p_1)(S_1)(q_1\, J^*\, q_o) \quad = \quad S_o \;. \;\square$$

For $i = 0,1$, we shall note $W_i = \mathrm{Domain}(T + S_i)$.
As $T^2 = 0$, we have also $W_i = \mathrm{Dom}(T) \cap \mathrm{Dom}\, S_i$, so that :

$$W_i \quad = \quad \{\omega \in K \quad ||T\omega||_i^2 \; + \; ||S_i\, \omega||_i^2 \; < \, + \, \infty\}$$

becomes a Hilbert space with the norm

$$N_i(\omega) \quad = \quad (\, ||\omega||_i^2 \; + \; ||T\omega||_i^2 \; + \; ||S_i\, \omega||_i^2\,)^{1/2}$$

Let $\theta_i : W_i \to K$ the canonical continuous injection.

<u>Lemma 5.4.</u> : We have $\mathrm{Im}((T + S_i + i)^{-1}) = W_i$; the map $L_i = (T + S_i + i)^{-1} \in \mathcal{L}(K_i, W_i)$ is an isomorphism and the following diagram is commutative

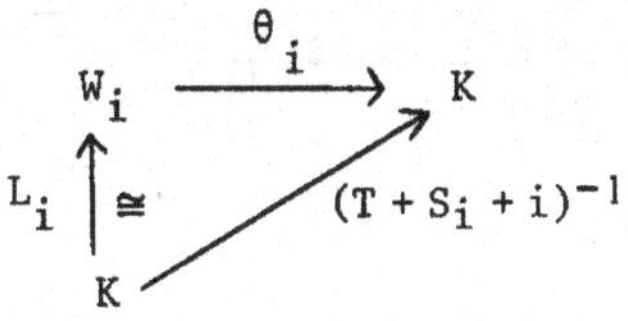

so that $(T + S_i + i)^{-1} \in L^p(K)$ if and only if $|\theta_i| \in L^p(W_i)$.

<u>Proof</u> : Let $\Delta_i = (T + S_i)^2 = T\, S_i + S_i\, T$. Then we have $W_i = \mathrm{Dom}((1 + \Delta_i)^{-1/2})$ and the lemma follows from the fact that $(i + T + S_i)^{-1}(1 + \Delta_i)^{-1/2}$ is bounded and invertible. $\square$

<u>Proof of prop. 5.1.</u> :

As B is invertible, there exists $c>0$ such that $\|B^{*-1}\omega\|_1^2 \leq c_1 \|\omega\|_1^2$. Let $\omega \in W_o$: then $B\omega \in W_1$ and we have :

$$\begin{aligned} N_1(B\omega)^2 &= \|\omega\|_1^2 + \|T\, B\omega\|_1^2 + \|S_1\, B\omega\|_1^2 \\ &= \|\omega\|_1^2 + \|B^{-1}\, T\omega\|_1^2 + \|B^{*-1}\, S_o\omega\|_1^2 \\ &\leq (1+c_1)\,\{\|\omega\|_1^2 + \|T\omega\|_1^2 + \|S_o\omega\|_1^2\} \end{aligned}$$

so that there exists $c>0$ such that

$$N_1(B\omega) \leq c\, N_o(\omega)$$

Let $L \in \mathcal{L}(W_o, W_1)$ the linear operator induced by B :

$L\omega = B\omega$, for $\omega \in W_o$: then L is an isomorphisme of W_o on W_1.

Moreover the following diagram is commutative :

$$\begin{array}{ccc} W_o & \xrightarrow{\theta_o} & K_o \\ L\downarrow & & \downarrow B \\ W_1 & \xrightarrow{\theta_1} & K_1 \end{array}$$

As L and B are isomorphism, it shows that $|\theta_o| \in L^p(W_o)$ if and only if $|\theta_1| \in L^p(W_1)$ and the proposition follows from lemma 8. □

5.2. Resolvant of the signature operator.

We keep in the notations of the theorem 3.1.

<u>Proposition 5.6.</u> : The bounded operator $(D_g+i)^{-1}$ belongs to $L^p(K_g) = L^p(L^2(M))$ for any $p>m$.

Let $\Delta_g = D_g^2$ the Laplacian operator on K_g associated to g. As Δ_g is a positive self adjoint, we can form the bounded operator $e^{-t\Delta_g}$, for $t>0$, the "heat kernel" of g.

Corollary 5.7. : $\mathrm{Trace}(e^{-t\Delta_g}) < +\infty$ for ant $t > 0$.

Proof of the corollary : As $(D_g + i)^{-1} \in L^p(L^2(M))$ for $p > n$, then $(1 + \Delta_g)^{-1} \in L^{2p}(L^2(M))$ which imply that $e^{-t\Delta_g}$ is of trace class. □

We divide the proof of the proposition 5.6. into several lemmas. As a corollary of the proposition 5.1, we get first :

Lemma 5.8. : If M is already a smooth manifold, then the proposition is true : For any Lipschitz Riemannian structure g on M, we have $(i + D_g)^{-1} \in L^p(L^2(M))$.

Proof : We choose a smooth riemannian metric g_o on M and form the signature operator D_{g_o} acting on K_{g_o} : then we know that for any $p > m$, $(i + D_{g_o})^{-1} \in L^p(L^2(M))$ ([B]).
On the other hand, as $D_{g_o} = d + \delta_{g_o}$ and $D_g = d + \delta_g$ we get that δ_{g_o} and δ_g are the adjoint of the closed operator d for respectively K_{g_o} and K_g; so that by proposition 5.1 we have that $(i + D_g)^{-1} \in L^p(L^2(M))$. □

To reduce the proposition 5.9 to the lemma 5.11, we have to "localize" : For any open subset $U \subset M$ let $\mathfrak{M}_o = \{\omega \in \mathrm{Dom}(D_g)$ with $\mathrm{support}(\omega) \subset U\}$.
Then we define the local Sobolev space of g :

$$W_o(U) = \{\text{Closure of } \mathfrak{M}_o \text{ with respect to the norm :}$$
$$N_g(\omega) = \{ \|\omega\|^2 + \|D_g\,\omega\|^2 \}^{1/2}$$

In particular $W(M) = \mathrm{Domain}(D_g)$ with the norm of the graph and $W_o(U)$ is a closed subspace of $W(M)$.
Let ϕ a Lipschitz function on M; then for any $\omega \in W(M)$ we have

$$d(\phi\omega) = d\phi \wedge \omega + \phi\, d\omega$$

and this shows that $W(M)$ is stable under multiplication by Lipschitz function.

Let $\theta_U : W_o(U) \longrightarrow K_g(U)$ be the canonical injection (where $K_g(U) = \{\omega \in K_g,\ \mathrm{supp}(\omega) \subset U\}$). Then as in lemma 5.8, we see that $(i + D_g)^{-1} \in L^p(L^2(M))$ if and only if $|\theta_M| \in L^p(W(M))$.

Lemma 5.9. : 1) If $|\theta_M| \in L^p(W(M))$ then for any open subset $U \subset M$, we have $|\theta_U| \in L^p(W_o(U))$.

2) Let $(U_i)_{i=1,2,..,n}$ be a finite cover of M by open subsets such that $|\theta_{U_i}| \in L^p(W_o(U_i))$.

Then $|\theta_M| \in L^p(W(M))$.

Proof : 1) Let $p_U : K_g \longrightarrow K_g(U)$ be the canonical projection, and $i_U : W_o(U) \longrightarrow W(M)$ the canonical injection.

Then the following diagram is commutative :

$$\begin{array}{ccc} W(M) & \xrightarrow{\ \theta_M\ } & K_g \\ {\scriptstyle i_U}\uparrow & & \uparrow{\scriptstyle p_U} \\ W_o(U) & \xrightarrow[\ \theta_U\ ]{} & L^2(U) \end{array}$$

so that if $|\theta_M| \in L^p$ then so does $\theta_U = p_U\, \theta_M\, i_U$.

2) Conservely let (ϕ_i) $i = 1,\ldots,n$ be a partition of the unity affiliated with $(U_i)_{i=1,..,n}$, each ϕ_i being a Lipschitz function ; we define three maps :

$$\psi_1 \ : \ W(M) \longrightarrow \sum_{i=1}^{n} \oplus\ W_o(U_i) \quad \text{(direct sum)}$$

$$\psi_2 \ : \ \Sigma \oplus W_o(U_i) \longrightarrow \sum_{i=1}^{n} \oplus\ K_g(U_i)$$

$$\psi_3 \ : \ \Sigma \oplus K_g(U_i) \longrightarrow K_g$$

by :

$$\begin{cases} \psi_1(\omega) = (\phi_1\omega, \phi_2\omega, \ldots, \phi_n\omega) \\ \psi_2(\omega_1,\ldots,\omega_n) = (\theta_1\omega_1, \theta_2\omega_2, \ldots, \theta_n\omega_n) \\ \psi_3(\omega_1,\ldots,\omega_n) = \sum_{i=1,n} \omega_i \end{cases}$$

Then we have obviously $\theta_M = \psi_3 \circ \psi_2 \circ \psi_1$ which shows that $|\theta_M| \in L^p(W(M))$. □

By the preceding lemma, it is clear that the proposition 5.6 follows from :

<u>Lemma 5.10.</u> : Let (M,g) as before. Then for any open subset $\mathcal{O} \subset M$ of a local chart, i.e. such that there exists a Lipschitz homeomorphism $\phi : \theta \longrightarrow U \subset \mathbb{R}^m$, we have

$$|\theta_{\mathcal{O}}| \in L^p(W_o(\mathcal{O})).$$

<u>Proof</u> : We can look on U as an open subset of $S^m = \mathbb{R} \cup \{\infty\}$. There exists a Riemannian metric g_1 on the Lipschitz manifold S^m such that $g_{1|U} = \phi^{*-1}(g|_{\mathcal{O}})$. By lemma 5.8, we see that $(i + D_{g_1})^{-1} \in L^p(L^2(S^m))$ and by lemma 5.9, it follows that $|\theta_U| \in L^p(W_o(U))$ which is equivalent to $|\theta_{\mathcal{O}}| \in L^p(W_o(\mathcal{O}))$. □

6. CLASS OF THE SIGNATURE IN THE K-HOMOLOGY OF THE LIPSCHITZ MANIFOLD

6.1. (K_g, D_g) as an unbounded Kasparov bimodule .

We now can check part 1 of the theorem 3.1 :

1) $D_g = d + \delta$ is self adjoint (Sec. 4).

2) $(1 + D_g^2)^{-1/2}$ is a compact operator. This follows from prop. 5.3.

For the last condition, we take $\mathcal{L} \subset C(M)$ the $\star$-algebra of Lipschitz functions $\phi : M \longrightarrow \mathbb{C}$.

We have then, for any $\omega \in W(M)$:

$$d(\phi\omega) = d\phi \wedge \omega + \phi\, d\omega$$

$$\delta(\phi\omega) = *(d\phi \wedge *\omega) + \phi\, \delta\omega$$

so that

$$[d+\delta\, , \phi](\omega) = \mathrm{ext}(d\phi)(\omega) + \mathrm{int}(d\phi)(\omega)$$

where $\mathrm{int}(d\phi) = \mathrm{ext}(d\phi)^*$ in K_g.

As $d\phi \in L^\infty(M,\mathbb{C})$, we find that $[D,\phi]$ is bounded which proves theorem 3.1, 1).

6.2. Abstract operatorial homotopy.

We go back to the notation of Sec. 5.1 : Let K a Hilbertian space, $\| \ \|_o$, $\| \ \|_1$ two hermitian norms defining the topology of K, and T be a closed operator on K such that $T^2 = 0$.

We shall suppose that there is a continuous family $\| \ \|_t$, $t \in [0,1]$ of hermitian norms between $\| \ \|_o$ and $\| \ \|_1$: we mean by that the existence of a norm continuous map

$$[0,1] \longrightarrow \mathcal{L}(K_o)$$

$$t \longrightarrow A_t$$

such that $A_t \geq 0$ in K_o, A_t is invertible and for $\xi \in K$:

$$< \xi\, , \eta >_t \; = \; < A_t\xi\, , A_t\xi > .$$

We note S_t = the adjoint of T in K_t.

The object of this section is to prove

Proposition 6.1. : The map $t \longrightarrow (i + S_t + T)^{-1}$, $[0,1] \longrightarrow \mathcal{L}(K)$ is norm continuous.

We prove first some lemmas :

Lemma 6.2. : Let $H \subset K$ a closed subspace of K and $e_t : K_t \longrightarrow H$ the orthogonal projection in K_t.
Then the map $t \longrightarrow e_t$ is norm continuous.

Proof : Let a_t be the adjoint of the linear operator e_o in K_t, so that for $\xi, \eta \in K$:

$$< a_t \xi , \eta >_t = < \xi , e_o \eta >_t$$

We have then directly :

$$a_t = A_t^{-2} a_o A_t^2$$

so that the map $t \longrightarrow a_t$ is norm continuous.
The operator $a_t e_o$ is the square ot the module of e_o in K_t.
As spectrum$(e_o) = \{0,1\}$, we find then that there exists $a > 0$ such that

$$\text{Spectrum}(a_t e_o) \subset \{0\} \cup [a,1] .$$

Let C be the circle of radius $1 - \frac{a}{2}$, with center 1.
Then we have by the Cauchy formula :

$$e_t = \int_0 (z - a_t e_o)^{-1} dz$$

which shows the norm continuity of e_t. □

Lemma 6.3. : Let P a self-adjoint (unbounded) operator on K_o and $t \longrightarrow B_t$ a norm continuous map from $[0,1]$ to $\mathcal{L}(K)$, such that B_t is invertible and positive in K_o for all t.
Then the map $(i + B_t P B_t)^{-1}$ is norm continuous.

Proof : It suffices to show the continuity of the map

$$f(t) = (P + i\, B_t^{-2})^{-1}$$

Let $s,t \in [0,1]$, we have :

$$f(t) - f(s) = i\, f(t)\, (B_t^{-2} - B_s^{-2})\, f(s)$$

As $t \longrightarrow B_t$ is continuous, there exists $a > 0$ such that

$$\sup_{0 \le t \le 1} \{ \|B_t\|_o + \|B_t^{-1}\|_o \} \le a.$$

It follows then

$$\|f(t) - f(s)\| \le a^4 \|B_t^{-2} - B_s^{-2}\|$$

which tends to 0 as $s,t \longrightarrow 0$. □

Proof of prop. 6.1. : Let B_t the linear operator given by Prop. 5.1 such that

$$B_t^*(T + S_t)B_t = (T + S_o)$$

By the preceding lemmas it is sufficient to prove that B_t, B_t^* are norm continuous. We recall that

$$B_t = p_t\, I_t\, p_o + q_t\, J_t^*\, q_o + r_t\, r_o$$

where p_t, q_t, r_t are orthogonal projection in K_t on support(T) in K_t, on support(S_t) in K_t, and on $\ker T \cap \ker S_t$; the operators $I_t \in \mathcal{L}(K_o, K_t)$, $J_t \in \mathcal{L}(K_t, K_o)$ are just the identity operator $K \to K$.

By lemma 6.2, $t \longrightarrow p_t$ is norm continuous because $\mathrm{Im}(1 - p_t) = \ker T$ does not depend of t.

Again $t \longrightarrow q_t$ is norm continuous because it is the orthogonal projection of $\ker T$ on $\overline{\mathrm{Im}\, T}$ with respect to $\|\ \ \|_t$.

Finally we have the relations :

$$J_t^* = A_t^{-2}$$

$$B_t^* = A_t^{-2} \; C_t \; A_t^2$$

where C_t = adjoint of B_t in K_o, so that B_t, B_t^* are norm continuous. □

6.3. Unicity in $K_o(\underline{M})$.

We prove now theorem 3.1, 2).

Let as before M to be a Lipschitz manifold, oriented of even dimension, and let g_o, g_1 two Riemannian metrics on M.

We define a path g_t of Riemannian metrics by

$$g_t = (1-t)g_o + t\, g_1 \qquad t \in [0,1]$$

and we note K_t the Hilbert space constructed from $(L^2(M), g_t)$ with the scalar product :

$$(\alpha,\beta)_t = \int \alpha \wedge *_t \beta$$

where $*_t$ is the operator of Hodge associated with g_t (cf. Sec. 2.5).

We equip K_t with the $\mathbb{Z}_2$-grading by $\ker(\tau_t \mp 1)$ where $\tau_t = i^{p(p-1)+\frac{m}{2}} *_t$, and we note D_t the signature operator associated with g_t.

Let $\mathcal{E}$ the Hilbert C^*-module over $[0,1]$:

$$\mathcal{E} = C([0,1], L^2(M))$$

with the product : $\mathcal{E} \times \mathcal{E} \longrightarrow C([0,1])$:

$$(\alpha,\beta)(t) = (\alpha,\beta)_t$$

where $\alpha,\beta \in \mathcal{E}$ (so that $\mathcal{E}_t = K_t$) and let D the unbounded endomorphism of $\mathcal{E}$ such that

$$(D\alpha)(t) = D_t\, \alpha(t)$$

for $\alpha \in \mathcal{E}$.

<u>Proposition 6.4.</u> : $(\mathcal{E},D)$ is an unbounded operatorial homotopy from $(\mathcal{E}_o,D_o)$ to $(\mathcal{E}_1,D_1)$.

We have to check that $(\mathcal{E},D)$ satisfy the axioms of unbounded operatorial homotopy (Section 3.1).

Let C_t be the unique family of invertible and positive operators in $\mathcal{L}(K_o)$ such that

$$(\alpha,\beta)_t = (C_t\alpha , C_t\beta)_o$$

for $\alpha,\beta \in L^2(M)$.

<u>Lemma 6.5.</u> : The map $t \longrightarrow C_t$ is norm continuous.

<u>Proof</u> : Let $A \in L^\infty(M,\mathcal{L}(T^*M))$ the measurable field of operators on T^*M such that

$$g_1(\xi,\eta) = g_o(A\xi , A\eta)$$

for ξ,η two measurable sections of cotangent forms on M (cf. Sec. 2.6). We have then

$$g_t(\xi,\eta) = g_o(A_t\xi , A_t\eta)$$

where $A_t = ((1-t)A^2 + t\ I)^{1/2}$

so that we find that A_t is positive almost everywhere with respect to g_o, A_t and A_t^{-1} belong to $L^\infty(M , \mathrm{End}(T^*M))$ and $t \longrightarrow A_t$ is norm continuous.

By section 2.6 again, we find then

$$C_t = \Lambda(A_t)\ \det(A_t)^{-1/2}$$

which proves the lemma. □

The continuity of the graduation follows from the formula :

<u>Lemma 6.5.</u> : $\tau_t = \tau_o C_t^2$

<u>Proof</u> : In fact we have to prove $*_t = *_o C_t^2$

We have

$$\int \alpha \wedge *_t \bar{\beta} = (\alpha,\beta)_t = (\alpha , C_t^2 \beta)_o = \int \alpha \wedge *_o \overline{C_t^2 \beta}$$

which gives $*_t = *_o C_t^2$. □

As spectrum $(\tau_t) = \{-1,+1\}$, we can choose a circle C in $\mathbb{C}$ with center in 1 such that the orthogonal projection :

$$f_t = \int_C (z - \tau_t)^{-1} dz$$

defines the grading and is norm continuous.

Secondly, let $\mathcal{L} \subset C(M)$ the dense $*$-subalgebra of Lipschitz functions on M. Then for $\phi \in \mathcal{L}$, we have :

$$[D_t , \phi] = \mathrm{ext}(d\phi) + *_t(\mathrm{ext}\ d\phi) *_t$$

As $*_t = *_o C_t^2$ and $d\phi \in L^\infty(M,T^*M)$ this is a bounded endomorphism of $\&$.

Finally we have by definition $D_t = d + \delta_t$ where δ_t is the adjoint of d in K_t. Then by propositions 5.3 and 6.1, the map $t \longrightarrow (D_t + i)^{-1}$ is norm continuous with values in the algebra of compact operators.

This shows proposition 6.4. □

In conclusion, by section 3.1 we have proven the equality in $K_o(M)$:

$$[(K_o , D_o)] = [(K_1 , D_1)]$$

which is theorem 3.1, 2).

BIBLIOGRAPHY

[A-S] M.F. Atiyah, I.M. Singer : The Index of Elliptic Operators, I, III. Annals of Maths 87, (1968), p.484.

[A-B-P] M.F. Atiyah, R. Bott, V.K. Patodi : On the Heat Equation and the Index Theorem, Inventiones Maths 19, (1973), p.279.

[B-J] S. Baaj, P. Julg : Theorie bivariante de Kasparov et Multiplicateurs non bornés dans les C^*-modules, Comptes Rendus de l'Académie des Sciences, 296, Serie I, 1983, p.875.

[B] H. Berger, P. Gauduchon, P. Mazet : Le spectre d'une variété Riemannienne, Lecture Notes in Maths n°194, Springer-Verlag.

[H] M. Hilsum : Operateurs de Signature sur une Variété Lipschitzienne et modules de Kasparov non bornés, Comptes Rendus de d'Académie des Sciences, 297, Serie I, 1983, p.49.

[K] G.G. Kasparov : The Operator K-functor and Extensions of C^*-algebras, Math. USSR Izvestija, 16, 1981, n°3.

[M-S] J. Milnor, J. Stasheff : Characteristic classes, Annals of Maths Studies, 76, (1974), Princeton.

[M] J. Milnor : Microbundles, Topology, 3 sup, 1964.

[N] S.P. Novikov : Topological Invariance of rational Pontryagin classes, Doklady, Tome 163, n°2, 1965, p.921.

[S] D. Sullivan : *In* Geometric Topology, Proc. Georgia Conference, Athens, 1977.

[S-T] D. Sullivan, N. Teleman : An Analytical Proof of Novikov's Theorem on Rational Pontryagin classes, Publ. Math. I.H.E.S., 58, 1983.

[T_1] N. Teleman : The Index of the signature Operator on Lipschitz Manifolds, Publ. Math. I.H.E.S., 58, 1983.

[T_2] N. Teleman : The Index theorem for Topological Manifolds, will appear in Actae Mathematicae.

GROUP ACTIONS ON TREES AND K-AMENABILITY

Pierre JULG and Alain VALETTE (*)

1. Introduction

Let F_n be the free group on n generators ($2 \leq n < \infty$). In their paper [12], Pimsner and Voiculescu developed an impressive machinery to show that the reduced C*-algebra $C_r^*(F_n)$ contains no non-trivial idempotent, and to compute the K-theory of $C_r^*(F_n)$: they found $K_0(C_r^*(F_n)) = \mathbb{Z}$ and $K_1(C_r^*(F_n)) = \mathbb{Z}^n$. On the other hand, it is fairly easy to show that the full C*-algebra $C^*(F_n)$ contains no non-trivial idempotent, and that $K_0(C^*(F_n)) = \mathbb{Z}$, $K_1(C^*(F_n)) = \mathbb{Z}^n$. Starting from this remark, Cuntz tried to give an easier proof of the results of Pimsner-Voiculescu and, in [4], he succeeded in proving that, for any C*-dynamical system (A, α, F_n), the canonical map in K-theory $\lambda_{A*}: K_i(A \rtimes_\alpha F_n) \to K_i(A \rtimes_{\alpha,r} F_n)$ $(i=0,1)$ is an isomorphism.

This led Cuntz to introduce the notion of K-amenability. Roughly speaking, a locally compact group G is K-amenable if, for any C*-dynamical system (A, α, G), the map $\lambda_{A*}: K_i(A \rtimes_\alpha G) \to K_i(A \rtimes_{\alpha,r} G)$ $(i=0,1)$ is an isomorphism (the precise definition will be given in § 3). Apart from establishing the K-amenability of the free groups, Cuntz also proved that the class of discrete K-amenable groups is stable under direct and free products.

In the case of a connected Lie group G, the computation of the K-theory groups of $C^*(G)$ and $C_r^*(G)$ is related to problems in the representation theory of G; for instance, any discrete series representation of G contributes a copy of $\mathbb{Z}$ to $K_0(C_r^*(G))$. If G is semi-simple, it is known that the symmetric space G/K (K is a maximal compact subgroup of G) plays an important role in the Atiyah-Schmid construction of the discrete series. More generally, it is possible to

(*) Research assistant at the Belgian National Fund for Scientific Research.

construct geometrically elements of $K_p(C^*(G))$ (or $K_p(C_r^*(G))$, after composition with λ_* ; here p = dim G/K) as analytical indices of G-invariant elliptic operators on G/K; this is done in Connes-Moscovici [3] or Kasparov [10]. Connes and Kasparov conjecture that one gets the whole of $K_p(C_r^*(G))$ by considering the indices of Dirac operators on G/K with coefficients in finite-dimensional representations of K. So it is natural to ask which are the K-amenable Lie groups; but here we meet the well-known obstruction of <u>property (T)</u> (see [5]): if a non-compact group has property (T), then the kernel of λ_* contains at least a copy of $\mathbb{Z}$. It is known from work of Kazhdan and Kostant [5] that an almost simple Lie group G with finite centre has property (T) if and only if G is locally isomorphic neither to SO(n,1) nor SU(n,1). By contrast with this result, it may be conjectured that closed subgroups of connected Lie groups locally isomorphic to SO(n,1) or SU(n,1) are K-amenable. This was proved very recently by Kasparov [11] in the case of SO(n,1). This gives the first non-trivial examples of K-amenable Lie groups.

2. Main result

We tried to understand Cuntz'results for F_n in a geometric way, using the associated tree instead of the space G/K of the Lie group case. We were led to the following result, which is a rather tractable sufficient condition for K-amenability.

<u>Theorem: Let G be a locally compact group acting on a tree, and such that the stabilizer of any vertex is amenable (e.g. compact or abelian). Then G is K-amenable.</u>

This theorem was announced in [7]; we will give the proof in § 3. By action of G on a tree, we mean a simplicial action on a 1-dimensional simply connected simplicial complex.

As a corollary of this result, we see that F_n is K-amenable (of course), but also that free or amalgamated products of discrete amenable groups are k-amenable (that our assumptions are fulfilled is proved in [13, I.4.1]); by a similar argument, HNN-extensions of discrete amenable groups are K-amenable. In this way we get a more conceptual proof of many of Cuntz'results in [4]. Moreover, if G is a discrete non-cocompact subgroup of $SL_2(\mathbb{R})$, then G has a natural action on a tree (if, for example, a fundamental domain for G has just one cusp, the union of the compact edges of the hyperbolic tiling associated to G is a tree). Of course, this result is now superseded by Kasparov's result mentioned above.

However, the main corollary of our theorem is the K-amenability of $SL_2(\mathbb{Q}_p)$, which is the first non-trivial example of a non-discrete totally disconnected K-amenable group. Indeed $SL_2(\mathbb{Q}_p)$ has a natural action with compact stabilizers on a tree (see [13, I.1.1]). More generally, if F is a local field with finite residue field, and if G is the group of F-rational points of some algebraic group, simple, simply connected, of split rank 1 over F, then the Bruhat-Tits building associated to G is a tree (see [14]), hence the K-amenability of G. This is consistent with Tits'philosophy in [14] which says that the building associated to a reductive group over a local field is the analogue of the space G/K (for G a semi-simple connected Lie group, K a maximal compact subgroup), which is a connected, simply connected Riemannian manifold with non-positive curvature. Indeed both objects share many properties, e.g. contractibility, uniqueness of geodesics, fixed point property for compact group actions...

3. Proof of the main result

Let us recall that, in [9], Kasparov associates to any locally compact group G a unital ring $KK_G(\mathbb{C}, \mathbb{C})$ (which, for compact G, co-incides with R(G)), the elements of which being triples $(\mathcal{H}_0, \mathcal{H}_1, F)$ such that $\mathcal{H}_0$, $\mathcal{H}_1$ are Hilbert spaces carrying unitary representations of G, and $F: \mathcal{H}_0 \to \mathcal{H}_1$ is a Fredholm operator commuting modulo compact operators with the action of G. The ring $KK_G(\mathbb{C}, \mathbb{C})$ is the quotient of the set of such triples by the homotopy relation of [9], the ring structure being given by Kasparov's cup-product (see [9]), and the unit being given by the triple $1_G = (\mathbb{C}, 0, 0)$ where $\mathbb{C}$ carries the trivial representation.

Definition: G is said to be K-amenable if 1_G is homotopic to a triple $(\mathcal{H}_0, \mathcal{H}_1, F)$ where the representations of G on $\mathcal{H}_0$ and $\mathcal{H}_1$ are weakly contained in the left regular representation of G.

It is obvious that K-amenability is inherited by closed sub-groups, and preserved under direct products.

If G is K-amenable, then for any C*-dynamical system (A, α, G), the canonical map $\lambda_A: A \rtimes_\alpha G \to A \rtimes_{\alpha,r} G$ defines an invertible element λ_{A*} in $KK(A \rtimes_\alpha G, A \rtimes_{\alpha,r} G)$. This was proved by Cuntz [4] for discrete groups, but his proof goes over to the general case with little modification. Note that since λ_{A*} is invertible, λ_A induces isomorphisms both in K-theory and in K-homology.

Turning to the proof of our theorem, to any action of G on a tree X, we will associate an element γ_0 in $KK_G(\mathbb{C}, \mathbb{C})$.

Remark 1: This element γ_0 is a kind of discrete analogue of the element $\delta \in KK_G(\mathbb{C}, C_0(M))$ associated by Kasparov [9] to any action of G by isometries on a connected, simply connected, complete Riemannian manifold M with non-positive curvature. In the case M = G/K, where G is a connected semi-simple Lie group, δ is a right inverse for the Dirac operator $D \in KK_G(C_0(G/K), \mathbb{C})$, i.e. $D \otimes_{\mathbb{C}} \delta = 1_{C_0(G/K)}$. This shows that the element $\gamma = \delta \otimes_{C_0(G/K)} D$ is an idempotent in $KK_G(\mathbb{C}, \mathbb{C})$; one has $\gamma = 1_G$ if G is amenable, see [9]. Moreover, if $\gamma = 1_G$, then G is K-amenable. This is for example the case for SO(n,1), see [11].

Let Δ^0 (resp. Δ^1) be the set of vertices (resp. edges) of X. The natural metric on Δ^0 will be denoted by d. For any vertices x, y, we denote by [x, y] the set of vertices belonging to the unique geodesic between x and y. Fix some origin x_0 on Δ^0. We define a map $\beta: \Delta^0\backslash\{x_0\} \to \Delta^1$ as follows: for any $x \neq x_0$, $\beta(x)$ will be the edge [x, y], where y is the unique element of $[x_0, x]$ such that $d(x, y) = 1$. The edge $\beta(x)$ can be seen as the "tangent vector at x", pointing to x_0, to the unique geodesic through x_0 and x (just like in the construction of δ in [9]).

Lemma 1: i) $\beta: \Delta^0\backslash\{x_0\} \to \Delta^1$ is a bijection

ii) Fix g in G. The set of x's in Δ^0 such that $g\beta(g^{-1}x) \neq \beta(x)$ is $[x_0, gx_0]$.

Part i) of this lemma is obvious; part ii) is clear if one thinks that $g\beta(g^{-1}x)$ is the tangent vector at x, pointing to gx_0, to the geodesic through x and gx_0. This is a more conceptual form for the map appearing in [2], [4] for F_n.

Define now an operator $F: \ell^2(\Delta^0) \to \ell^2(\Delta^1)$ by $F\delta_{x_0} = 0$ and $F\delta_x = \delta_{\beta(x)}$ for $x \neq x_0$. Then F is a co-isometry of index one, and it follows from Lemma 1 that the triple $(\ell^2(\Delta^0), \ell^2(\Delta^1), F)$ defines an element $\gamma_0 \in KK_G(\mathbb{C}, \mathbb{C})$. By adding midpoints on the edges of X, we may assume that G acts without inversion.

Proposition: $\gamma_0 = 1_G$ in $KK_G(\mathbb{C}, \mathbb{C})$.

The theorem now follows from this proposition, since the representations of G on $\ell^2(\Delta^0)$ and $\ell^2(\Delta^1)$ are weakly contained in the left regular representation of G if the stabilizer of any vertex is amenable.

The proposition is proved by exhibiting an explicit homotopy between $\gamma_0 - 1_G = (\ell^2(\Delta^0), \ell^2(\Delta^1) \oplus \mathbb{C}, \tilde{F})$ (where $\tilde{F}\delta_{x_0} = (0, 1)$ and $\tilde{F}\delta_x = (F\delta_x, 0)$ for $x \neq x_0$) and the triple $(\ell^2(\Delta^0), \ell^2(\Delta^0), 1)$, which is obviously zero in $KK_G(\mathbb{C}, \mathbb{C})$. This homotopy involves a continuous ield of Hilbert spaces connecting $\mathbb{C} \oplus \ell^2(\Delta^1)$ to $\ell^2(\Delta^0)$. The key lemma for the construction of that field is the following:

Lemma 2: The map $(x,y) \to d(x,y)$ is a kernel of negative type on Δ^0.

For the relevant definition, see [1]. To prove this lemma, let us choose an orientation on X, and denote an edge by (x,y), where x is its origin, y its extremity. We have to prove that the scalar product

$$\langle f, g\rangle = - \sum_{x,y \in \Delta^0} d(x,y) f(x)\overline{g(y)}$$

is positive definite on the space V of functions on Δ^0 which are almost everywhere zero and of total mass zero. For that we check that the family $2^{-\frac{1}{2}}(\delta_x - \delta_y)$, $(x,y) \in \Delta^1$, is an orthonormal basis of V.

Remark 2: If G has property (T), then any action without inversion of G on a tree has at least one fixed vertex (i.e. G has property (FA) of Serre [13]). Indeed, it has been proved by Akemann-Walter [1] that any function of negative type on a group G having property (T) is bounded. So that by lemma 2, any orbit of G on Δ^0 is bounded, and by [13, I.2.2.] there is a fixed point (compare with Watatani [15]).

Let us return to the proof of the proposition. Thanks to Lemma 2, we find, for any $\lambda \in]0, \infty[$, a Hilbert space H_λ carrying a unitary representation ρ_λ of G, and an injective map $\Delta^0 \to H_\lambda : x \to \xi_x^\lambda$ such that

$$\langle \xi_x^\lambda, \xi_y^\lambda \rangle = \exp(-\lambda d(x,y))$$

$$\rho_\lambda(g)\xi_x^\lambda = \xi_{gx}^\lambda$$

Let π_0 (resp. π_1) be the representation of G on $\ell^2(\Delta^0)$ (resp. $\ell^2(\Delta^1)$). Now we define $H_0 = \mathbb{C} \oplus \ell^2(\Delta^1)$ carrying the representation $\rho_0 = 1_G \oplus \pi_1$ and $H_\infty = \ell^2(\Delta^0)$ carrying the representation $\rho_\infty = \pi_0$. To construct the continuous field $(H_\lambda)_{\lambda \in [0, \infty]}$, we extend the sections ξ_x^λ $(x \in \Delta^0)$ at 0 and ∞ by $\xi_x^\infty = \delta_x \in \ell^2(\Delta^0)$ and $\xi_x^0 = (1, 0) \in \mathbb{C} \oplus \ell^2(\Delta^1)$, and we also define the sections $\eta_{(x,y)}^\lambda$ $((x,y) \in \Delta^1)$ by

$$\eta_{(x,y)}^\lambda = (2 - 2e^{-\lambda})^{-\frac{1}{2}}(\xi_x^\lambda - \xi_y^\lambda) \quad \text{for } \lambda \in]0, \infty[,$$

$$\eta_{(x,y)}^0 = (0, \delta_{(x,y)}) \in \mathbb{C} \oplus \ell^2(\Delta^1) \text{ and } \eta_{(x,y)}^\infty = 2^{-\frac{1}{2}}(\delta_x - \delta_y).$$

The action of G being without inversion, we may choose a G-invariant orientation on X.

Lemma 3: The family $(H_\lambda, \rho_\lambda)_{\lambda \in [0, \infty]}$ is endowed with a structure of continuous field of G-Hilbert spaces on $[0, \infty]$, with a total space of sections (see [6, § 10.2]) generated by ξ_x^λ $(x \in \Delta^0)$ and $\eta_{(x,y)}^\lambda$ $((x,y) \in \Delta^1)$.

Proof: It is clear that $\lim_{\lambda \to \infty} (\xi_x^\lambda, \xi_y^\lambda) = (\delta_x, \delta_y)$. On the other hand, it is easily verified that: $\lim_{\lambda \to 0} (\eta_{(x,y)}^\lambda, \eta_{(s,t)}^\lambda) = (\delta_{(x,y)}, \delta_{(s,t)})$

and $\lim_{\lambda \to 0} (\xi_x^\lambda, \eta_{(s,t)}^\lambda) = 0$

Moreover, G transforms sections into sections, and the action of G is strongly continuous since, for g in the stabilizer of x (which is an open subgroup), $\rho_\lambda(g)\xi_x^\lambda = \xi_x^\lambda$. QED.

Now we want to show that the field $(H_\lambda)_{\lambda \in [0,\infty]}$ is isomorphic, in a quasi-G-invariant way, to the constant field $\ell^2(\Delta^0)$. Let us define for $\lambda \in]0, \infty[$,

$$U_\lambda(\xi_x^\lambda) = e^{-\lambda d(x_0,x)}\delta_{x_0} + (1 - e^{-2\lambda})^{\frac{1}{2}} \sum_{y \in]x_0,x]} e^{-\lambda d(x,y)}\delta_y$$

Then, as easily checked, $(U_\lambda(\xi_x^\lambda), U_\lambda(\xi_y^\lambda)) = \exp(-\lambda d(x,y))$, so U_λ extends to a unitary operator from H_λ to $\ell^2(\Delta^0)$. Now, the operator $U_\lambda\rho_\lambda(g) - \pi_0(g)U_\lambda$ has finite rank for any g in G. Indeed, a straightforward computation shows that for $x \in \Delta^0$, $(U_\lambda\rho_\lambda(g) - \pi_0(g)U_\lambda)(\xi_x^\lambda)$ lies in the linear span of the functions δ_y for $y \in [x_0, gx_0]$. This shows that we have, for each $\lambda \in]0, \infty[$, an element of $KK_G(\mathbb{C}, \mathbb{C})$. In fact, we will prove that this element is independant of λ, equal to $1 - \gamma$ (by taking $\lambda \to 0$) and to 0 (by taking $\lambda \to \infty$), hence the proposition. Lemma 4 will give the limit of U_λ for $\lambda \to 0$ or ∞. For that, we consider $U_\infty = 1_{\ell^2(\Delta^0)}$, and $U_0 = Q^*$ where $Q: \ell^2(\Delta^0) \to \mathbb{C} \oplus \ell^2(\Delta^1)$ is obtained from $\tilde{F}$ by a sign modification: $Q = (1_{\mathbb{C}} \oplus S)\tilde{F}$, where $S: \ell^2(\Delta^1) \to \ell^2(\Delta^1)$ is given by $S\delta_{(x,y)} = \varepsilon(x,y)\delta_{(x,y)}$ where $\varepsilon(x,y) = 1$ (resp. -1) if $d(x,x_0) > d(y,x_0)$ (resp. $d(x,x_0) < d(y,x_0)$). It is clear that Q and $\tilde{F}$ define the same element in $KK_G(\mathbb{C}, \mathbb{C})$.

<u>Lemma 4</u>: $(U_\lambda)_{\lambda \in [0,\infty]}$ <u>defines a unitary isomorphism between the continuous field</u> $(H_\lambda)_{\lambda \in [0,\infty]}$ <u>and the constant field</u> $\ell^2(\Delta^0)$ <u>on</u> $[0, \infty]$.

<u>Proof</u>: The map $\lambda \to U_\lambda(\xi_x^\lambda)$ is clearly continuous on $]0, \infty[$. It is easily verified that for $x \in \Delta^0$,

$$\lim_{\lambda \to \infty} U_\lambda(\xi_x^\lambda) = \delta_x$$

and for $(x,y) \in \Delta^1$,

$$\lim_{\lambda \to 0} U_\lambda(\eta_{(x,y)}^\lambda) = \varepsilon(x,y)\delta_{\beta^{-1}(x,y)} = Q^*\delta_{(x,y)}$$

This shows that $(U_\lambda)_{\lambda \in [0,\infty]}$ maps sections of $(H_\lambda)_{\lambda \in [0,\infty]}$ into sections of $\ell^2(\Delta^0)$, and also that any section of $\ell^2(\Delta^0)$ is locally approximable by sections of the form $U_\lambda(\sigma(\lambda))$ where σ is a section of $(H_\lambda)_{\lambda \in [0,\infty]}$.

To prove the proposition, we just have to show that $(H_\lambda, \ell^2(\Delta^0), U_\lambda)_{\lambda \in [0,\infty]}$ defines a homotopy, i.e. an element of $KK_G(\mathbb{C}, C[0, \infty])$. This is done by the following lemma:

Lemma 5: <u>For any g in G, the family</u> $(U_\lambda\rho_\lambda(g) - \pi_0(g)U_\lambda)_{\lambda \in [0, \infty]}$ <u>defines a compact operator in the sense of C*-modules over</u> $C[0, \infty]$ (see [8]). <u>Moreover, the map</u> $g \to (U_\lambda\rho_\lambda(g) - \pi_0(g)U_\lambda)_{\lambda \in [0, \infty]}$ <u>is locally constant on G</u>.

Proof: Using Lemma 4, it is enough to show that the map $\lambda \to F_\lambda(g) = U_\lambda\rho_\lambda(g)U_\lambda^*\pi_0(g^{-1}) - 1$ is norm continuous. The range of $F_\lambda(g)$ is contained in the finite dimensional subspace E_g spanned by the δ_x $(x \in [x_0, gx_0])$. On the other hand, we have $F_\lambda(g)^* = \pi_0(g)F_\lambda(g^{-1})\pi_0(g^{-1})$ so that the orthogonal subspace to the kernel of $F_\lambda(g)$ is also contained in E_g. It follows that the norm continuity of $\lambda \to F_\lambda(g)$ is equivalent to its strong continuity, which is clear by lemmas 3 and 4. The second part of Lemma 5 follows from the fact that $U_\lambda\rho_\lambda(g) = \pi_0(g)U_\lambda$ if g belongs to the stabilizer of x_0, which is an open subgroup of G.

4. Some problems

We conclude this paper by mentioning briefly some open questions suggested by our study.

i) Let G be a connected Lie group, K a maximal compact subgroup. Kasparov has shown ([9], [11]) that the restriction map $KK_G(\mathbb{C}, \mathbb{C}) \to R(K)$ is a ring isomorphism if G is either amenable or a Lorentz group. Is there a description of $KK_G(\mathbb{C}, \mathbb{C})$ for general G? Note that if G has property (T), the element γ defined in Remark 1 is a non-trivial idempotent in $KK_G(\mathbb{C}, \mathbb{C})$.

ii) What is $KK_G(\mathbb{C}, \mathbb{C})$ looking like, for G a reductive Lie group over a local field (e.g. $SL_2(\mathbb{Q}_p)$)?

iii) Compute $K_*(C_r^*(G))$ for G as in ii). Is it possible to give a "geometric" construction of elements of $K_*(C_r^*(G))$ analogous to what is done for Lie groups using the Bruhat-Tits building instead of the symmetric space G/K?

References

1. C.A. Akemann and M.E. Walter, Unbounded negative definite functions, Can. J. Math. 33 (1981), 862-871.
2. A. Connes, The Chern character in K-homology, Preprint I.H.E.S. (1982).
3. A. Connes and H. Moscovici, The L^2-index theorem for homogeneous spaces of Lie groups, Ann. of Math. 115 (1982), 291-330.
4. J. Cuntz, K-theoretic amenability for discrete groups, Preprint Univ. of Pennsylvania (1982).
5. C. Delaroche and A.A. Kirillov, Sur les relations entre l'espace dual d'un groupe et la structure de ses sous-groupes fermés,

Séminaire Bourbaki, 1967-68, Exposé 343.
6. J. Dixmier, Les C*-algèbres et leurs représentations, 2nde éd., Gauthier-Villars, 1969.
7. P. Julg and A. Valette, K-moyennabilité pour les groupes opérant sur les arbres, to appear in Comptes-Rendus Acad. Sc. (Paris).
8. G.G. Kasparov, Hilbert C*-modules: theorems of Stinespring and Voiculescu, J. Oper. Th. 4 (1980), 133-150.
9. G.G. Kasparov, K-theory, group C*-algebras, and higher signatures (conspectus), Preprint Chernogolovka (1981).
10. G.G. Kasparov, The index of elliptic operators, K-theory and Lie group representations, Doklady Akad. Nauk. SSSR 268 (1983), 533-537
11. G.G. Kasparov, Lorentz groups: K-theory of unitary representations and crossed products, Preprint Chernogolovka (April 1983).
12. M. Pimsner and D. Voiculescu, K-groups of reduced crossed products by free groups, J. Oper. Theory 8 n°1 (1982).
13. J.P. Serre, Arbres, amalgames, SL_2, Astérisque n°46 (1977).
14. J. Tits, Reductive groups over local fields, in Proc. Symp. Pure Math. 33 (1979), 29-69.
15. Y. Watatani, Property (T) of Kazhdan implies property (FA) of Serre, Preprint 1981.

Authors addresses:

Laboratoire de Mathématiques Fondamentales
Université Pierre et Marie Curie
45-46, 3ème étage
4, Place Jussieu 75230 Paris Cedex 05
France.

Département de Mathématiques CP 214
Université Libre de Bruxelles
Boulevard du Triomphe
B-1050 Bruxelles
Belgium.

Diagonals in algebras of continuous trace

by Alex Kumjian*

tilegnet Inge

1 Prolegomena:

n the following the notion of diagonal (cf. [7], [14]) is studied under the estriction that the ambient algebra be of continuous trace. In a sense this tudy may be viewed as a sequel to [6], as local homeomorphisms figure rominently in the exposition. The conspicuous absence there of any discussion f second cohomology of the associated relation $R(\psi)$, is here remedied thru pplication of the results of Raeburn and Taylor (cf. [11]); their construction rovides the means to interpret the cohomology of the relation in terms of that f the base space.

n assortment of useful facts is collected in the preliminaries (§2). It is hown, for example, that the map from the spectrum of the diagonal to that of he ambient algebra, which assigns to a pure state the class of its unique xtension, is a local homeomorphism. This map gives rise to a seven term xact sequence which resembles the long exact sequence of cohomology (though ur sequence is not quite as long). Contenting ourselves with interpretations f a C*-theoretic nature, we leave the matter of this similarity to more subtle inds.

Let $\psi: X \longrightarrow T$ be a local homeomorphism and $\mathcal{S}$ denote the sheaf of germs of continuous circle-valued functions. The sequence of abelian groups:

$$0 \longrightarrow H^1(R(\psi),\mathbb{T}) \xrightarrow{\eta} H^1(T,\mathcal{S}) \xrightarrow{\psi^*} H^1(X,\mathcal{S}) \xrightarrow{\varepsilon} Tw(R(\psi)) \xrightarrow{\delta} H^2(T,\mathcal{S}) \xrightarrow{\psi^*} H^2(X,\mathcal{S})$$

will be shown exact in §3. The map η is a special case of those considere in [4] and [9]. The elements of $Tw(R(\psi))$ are isomorphism classes of pairs (A,B), with A continuous trace and $B \subseteq A$ a diagonal, for which the spectral map is given by ψ (equivalently, it is the group of twists over the relation $R(\psi)$ and thus a replacement for $H^2(R(\psi),\mathbb{T})$ cf. [7] §4iii). The map δ is the Dixmier-Douady invariant.

In the last section (§4) a class of C*-algebras is introduced in the hopes that our methods may be useful in a preliminary classification.

I record here my gratitude to the Alexander-von-Humboldt Stiftung for its support and to the Universities of Tübingen, Copenhagen, and New South Wales for their hospitality. Thanks are due T. Natsume, R. Archbold, and I. Raebur for helpful discussions pertaining to the matter at hand.

All C*-algebras are assumed to be separable (dually, topological spaces are assumed to be locally compact, Hausdorff, and to satisfy the second axiom of countability). The symbol "$\amalg$" denotes disjoint union.

Recall from [7] that an abelian subalgebra B containing the identity of a unital C*-algebra A is called a diagonal in A if there is a faithful expectation P onto B and if the free normalizer $\{a \in A : a^*Ba \subset B,\ aBa^* \subset B \text{ and } a^2=0\}$ is tota in ker P. For the non unital case, adjoin the same identity to A and B.

In [7] a twist over an equivalence relation R is defined as a groupoid extension of R by the circle group. The collection of (equivalence classes of) twis over R endowed with the usual sum of extensions form an abelian group Tw(R).

A preliminary C*-algebra is defined in [6] as a C*-algebra which is Morita equ valent to a commutative C*-algebra.

*Research supported in part by a grant from Australian Grant Scheme.

§2 Preliminaries

A C*-subalgebra B of a C*-algebra A is said to have the extension property if it contains an approximate unit for A and if pure states of B extend uniquely to (pure) states of A (cf. [1],[2]). If the subalgebra is abelian, say $B \cong C_0(X)$, there are a number of equivalent characterizations (cf. [2]). Each implies the existence of a unique conditional expectation, $P: A \longrightarrow B$, so that the unique extension of point evaluation at $x \in X$ is given by : $P(\cdot)(x)$. If A is of continuous trace, the unique extension yields a continuous map:

$$\psi : X \longrightarrow T \quad \text{by} \quad \psi(x) = [P(\cdot)(x)], \text{ (where } T = \hat{A}).$$

We establish that ψ is a local homeomorphism and that B is diagonal. We require a technical lemma.

1) <u>Lemma</u>: Let X be a second countable locally compact Hausdorff space and $\{U_i: i \in I\}$ a covering of open sets. There is a refinement $\mathcal{U}$ with the property that if $U,V \in \mathcal{U}$ and $U \cap V \neq \phi$, there is $i \in I$ with $U \cup V \subseteq U_i$.

<u>proof</u>: It may be assumed that $\{U_i\}$ is a countable locally finite covering consisting of precompact sets for which $\{i \mid K \cap U_i \neq \phi\}$ is finite for each compact $K \subseteq X$. Choose a collection of open sets $\{V_i^j: i \in I, j \geqslant 0\}$ satisfying:

i) $\overline{V_i^j} \subset V_i^{j+1} \subset U_i$ each $i \in I, j \geqslant 0$

ii) $\{V_i^j : i \in I\}$ covers for each $j \geqslant 0$.

Denote the collection of finite subsets of I by $M(I)$;

set $\quad V_\mu^j = \bigcap_{i \in \mu} V_i^j \setminus (\bigcup_{i \notin \mu} \overline{V_i^j}) \quad$ for $\mu \in M(I)$, $j \geqslant 0$ and note that V_μ^j is open. If $j \leq k$ and $V_\mu^j \cap V_\lambda^k \neq \phi$ $(\mu,\lambda \in M(I))$, then $\mu \subseteq \lambda$; thus, $V_\mu^j \cup V_\lambda^k \subseteq \bigcap_{i\in\mu} U_i$. The reader is left to check that the collection, $\{V_\mu^j : j \geq 0, \mu \in M(I)\}$ covers.

(I am grateful to Iain Raeburn for giving me the idea of the proof)

Let $J(A)$ denote the Pedersen ideal of A and let $\tau: J(A) \longrightarrow C_c(T)$ be the continuous trace (cf. [8]).

2) Theorem: With A and B as above, the spectral map, $\psi: X \longrightarrow T$, is a local homeomorphism and B is diagonal in A.

proof: For $t \in T$, let $\pi_t: A \longrightarrow B(\mathcal{H}_t)$ denote the associated irreducible representation. Since $\pi_t(B)$ is maximal abelian (cf [2], cor. 3.2), there is an orthonormal basis, $\{\xi_i\}_{i\geq 0}$, for $\mathcal{H}_t$ such that $\pi_t(B)$ is generated by the associated projections. Further, $\psi^{-1}(t) = \{x_i\}_{i\geq 0} \subseteq X$, and one has:

$$P(a)(x_i) = (\pi_t(a)\xi_i,\xi_i) \quad \text{for} \quad a \in A.$$

Evidently, P is faithful and ψ is surjective. Suppose ψ is not locally injective at x_0 and choose $f \in C_c(X)_+$ so that $\pi_t(f)$ is the one-dimensional projection onto $\mathbb{C}\xi_0$ (so, $f \in J(A)$ and $\tau(f)(t) = 1$). There are by supposition two sequences $y_n, z_n \in X$ with $\psi(y_n) = t_n = \psi(z_n)$ and $\lim_n y_n = x = \lim_n z_n$. Since $f \geq 0$, we have $\tau(f)(t_n) \geq f(y_n) + f(z_n) \longrightarrow 2$; this violates the continuity of $\tau(f)$ $(t_n \longrightarrow t)$. Thus, ψ is locally injective. That ψ is an open map is clear, hence it is a local homeomorphism. Note that $\mathcal{U}(\psi) = \{U \subseteq X \text{ open} : \psi|_U \text{ is injective}\}$ covers, and if $f \in C_c(X)_+$ with $\operatorname{supp} f \subseteq U \in \mathcal{U}(\psi)$, then $fAf \subseteq B$. By the above lemma, we may choose a partition of unity $\{f_i : i \geq 0\} \subseteq C_c(X)_+$ so that:

i) if $f_i f_j \neq 0$ there is $U \in \mathcal{U}(\psi)$ with $\mathrm{supp}(f_i + f_j) \subseteq U$

ii) for each compact $K \subseteq X$, there is $n > 0$ with $\sum_1^n f_i(x) = 1 \quad x \in K$.

n particular, $g_n = \sum_1^n f_i$ is an approximate identity for A.

ote that if $f_i f_j = 0$ then $f_i a f_j$ is a free normalizer of B for any $\in A$. For $n > 0$ set $\lambda_n = \{(i,j): f_i f_j = 0, 0 \leq i,j \leq n\}$. Now, iven $a \in \ker P$ and $\varepsilon > 0$, there is $n > 0$ for which $\|a - g_n a g_n\| < \varepsilon$. routine calculation using the expectation property verifies:

$$g_n a g_n = \sum_{i,j \in \lambda_n} f_i a f_j .$$

hus the span of the free normalizers is dense in $\ker P$ and B is iagonal. □

) <u>Example</u>: Given a Cech two-cocycle $\lambda: U_{ijk} \rightarrow \mathbb{T}$ relative to an open over $\{U_i\}$ of T, one forms the disjoint union, $X_\lambda = \amalg U_i$, and notes hat the reinclusion map, $\psi_\lambda: X_\lambda \rightarrow T$, is a local homeomorphism. Let $_\lambda: \Gamma_\lambda \rightarrow R(\psi_\lambda)$ denote the twist considered in [11] $H^2(T,\{U_i\},\mathcal{S}) \cong H^2(R(\psi_\lambda),\mathbb{T}) \subseteq Tw(R(\psi_\lambda)))$. Then, $\delta(C^*(\sigma_\lambda)) = [\lambda] \in H^2(T,\mathcal{S})$ $\cong H^3(T,\mathbb{Z}))$.

) <u>Corollary</u>: If A is unital then $\psi: X \rightarrow T$ is a covering map.

) <u>Corollary</u>: Let A_1 and A_2 be complementary full corners in a continuous trace algebra A and let $B_i \subseteq A_i$ be diagonals for $i = 1,2$. Then $B_1 \oplus B_2$ is diagonal in A, and thus is diagonal.

5) <u>Remarks</u>: Thus the sum of diagonals in two strong Morita equivalent continuous trace algebras constitutes a diagonal in any linking algebra (cf. [4]). In other words, if $\psi_i: X_i \rightarrow T$ are local homeomorphisms for

$i = 1,2$, then two twists $\sigma_i: \Gamma_i \longrightarrow R(\psi_i)$ give rise to strong Morita equivalent C*-algebras iff they are restrictions of a twist $\sigma: \Gamma \longrightarrow R(\psi_1 \amalg \psi_2)$ where

$$\psi_1 \amalg \psi_2 : X_1 \amalg X_2 \longrightarrow T$$

is the obvious map.

We illustrate this phenomenon. Consider a pair of local homeomorphisms:

$$Z \xrightarrow{\phi} X \xrightarrow{\psi} T.$$

Given a twist $\sigma: \Gamma \longrightarrow R(\psi)$, a twist may be induced on $R(\psi\phi)$. Note that $R(\psi\phi) = Z*R(\psi)*Z = \{(z_1,(x_1,x_2),z_2): \phi(z_i) = x_i,\ i=1,2\}$. Set $\Gamma^\phi = Z*\Gamma*Z = \{(z_1,\gamma,z_2) : \phi(z_1) = r(\gamma),\ \phi(z_2) = s(\gamma)\}$. One checks that $\sigma^\phi: \Gamma^\phi \longrightarrow R(\psi\phi)$ (by $(z_1,\gamma,z_2) \longrightarrow (z_1,\sigma(\gamma),z_2)$) is a twist. The above criterion may be invoked to show that $C^*(\sigma^\phi)$ is strong Morita equivalent to $C^*(\sigma)$. Consider the map $\phi': Z \amalg X \longrightarrow X$ (where $\phi'|_Z = \phi$ and $\phi'|_X = \mathrm{id}$); the induced twist $\sigma^{\phi'}: \Gamma^{\phi'} \longrightarrow R(\psi\phi')$ restricts to Γ^ϕ(on $R(\psi\phi)$) and to Γ (on $R(\psi)$).

We require the notion of pull-back for continuous trace algebras as defined in [12].

7) <u>Definition</u>: Let A be a continuous trace algebra with spectrum T and $f: Y \longrightarrow T$ be a continuous map. The pull-back of A along f, written f^*A, is the balanced tensor product:

$$f^*A = C_0(Y) \underset{C_0(T)}{\otimes} A.$$

It is immediate that the spectrum of f^*A is Y. It follows that $\delta(f^*A) = f^*\delta(A)$ where f^* is the induced map on cohomology (cf.[12] Prop.1.4

ssume now that $B \cong C_0(X)$ is diagonal in A with spectral map $: X \longrightarrow T$ and twist $\sigma\colon \Gamma \longrightarrow R(\psi)$. Then the algebra,

$$*B = C_0(Y) \underset{C_0(T)}{\otimes} B \cong C_0(Y*X) \quad (\text{where } Y*X = \{(y,x)\colon f(y) = \psi(x)\} \subseteq Y \times X),$$

s clearly diagonal in f^*A. Moreover, one has the following commutative quare of continuous maps:

$$\begin{array}{ccc} Y*X & \xrightarrow{\ \pi_X\ } & X \\ {\scriptstyle \pi_y}\downarrow & & \downarrow{\scriptstyle \psi} \\ Y & \xrightarrow[\ f\ ]{} & T \end{array}$$

) <u>Fact</u>: The spectral map for the pair (f^*A, f^*B) is π_y. The pull-back wist $f^*\sigma\colon f^*\Gamma \longrightarrow R(\pi_y)$ is given by:

$$f^*\Gamma = Y*\Gamma \ (= \{(y,\gamma) \mid f(y) = \psi(s(\gamma))\} \subseteq Y\times\Gamma\ ),$$

$$f^*\sigma : Y*\Gamma \longrightarrow Y*R(\psi) \cong R(\pi_y) \quad \text{by} \quad (y,\gamma) \longrightarrow (y,\sigma(\gamma)).$$

) <u>Prop</u>: If $Y = X$ and $f = \psi$, then ψ^*A is preliminary. ,

<u>roof</u>: Let $\Delta\colon X \longrightarrow X*X$ be given by $x \longrightarrow (x,x)$. Then Δ is a continuous ection for the spectral map of the pair, (ψ^*A, ψ^*B). Thus, there is an belian element which is not contained in any proper ideal. □

3 The sequence

Let $\mathcal{S}$ denote the sheaf of germs of continuous circle-valued functions. The first Cech cohomology of a space X with coefficients in $\mathcal{S}$ is here viewed as the group of isomorphism classes of circle-bundles over X. This corresponds in turn to the group of $C_0(X)$ - $C_0(X)$ imprimitivity bimodules leaving the spectrum intact. The Picard group of a C*-algebra A, Pic(A), is defined as the group of (isomorphism classes of) A - A imprimitivity bimodules (cf. [4]); let pic(A) denote the subgroup preserving the lattice of ideals (so $\mathrm{pic}(C_0(X)) \cong H^1(X,\mathcal{S})$). If A is continuous trace, Raeburn has shown $\mathrm{pic}(A) \cong H^1(\hat{A},\mathcal{S})$, [10]. If $B \subseteq A$ is diagonal and $J \in \mathrm{pic}(B)$ then $J \otimes_B A \otimes_B J^*$ is strong Morita equivalent to A with diagonal $J \otimes_B B \otimes_B J^* \cong B$. Moreover, the spectral map remains the same. The map ε should be seen in this light.

<u>Theorem</u>: Let $\psi: X \longrightarrow T$ be a local homeomorphism. One has an exact sequence of abelian groups:

$$0 \longrightarrow H^1(R(\psi),\mathbb{T}) \xrightarrow{\eta} H^1(T,\mathcal{S}) \xrightarrow{\psi^*} H^1(X,\mathcal{S}) \xrightarrow{\varepsilon} \mathrm{Tw}(R(\psi)) \xrightarrow{\delta} H^2(T,\mathcal{S}) \xrightarrow{\psi^*} H^2(X,\mathcal{S}$$

<u>proof</u>: i) The group $H^1(R(\psi),\mathbb{T})$ is the usual first groupoid cohomology with coefficients in $\mathbb{T}$ (cf. [13]). Alternatively, it may be viewed as the image of the diagonal fixing automorphisms in the outer automorphism group. As such, our map η is a special case of the ones considered in [4] and [9].

For $f \in Z^1(R(\psi),\mathbb{T})$ (i.e. $f: R(\psi) \longrightarrow \mathbb{T}$ is a continuous groupoid morphism), a circle-bundle $\tau_f: S_f \longrightarrow T$ may be defined as a quotient of $X \times \mathbb{T}$. Write $(x_1,z_1) \sim (x_2,z_2)$ iff $(x_1,x_2) \in R(\psi)$ and $z_1 = f(x_1,x_2)z_2$; set $S_f = X \times \mathbb{T}/\sim$ and $\tau_f[(x,z)] = \psi(x)$. Suppose S_f is trivial, i.e. there is a continuous section $\sigma: T \longrightarrow S_f$; there is then a continuous function $h: X \longrightarrow \mathbb{T}$ so that $\sigma(\psi(x)) = [(x,h(x))]$; for $(x_1,x_2) \in R(\psi)$ one has $(x_1,h(x_1)) \sim (x_2,h(x_2))$ and thus $f(x_1,x_2) = h(x_1)\overline{h(x_2)}$ (i.e. $f \in B^1(R(\psi),\mathbb{T})$). Hence $\ker \eta = 0$.

ii) The map $\psi^*: H^1(T,\mathcal{S}) \longrightarrow H^1(X,\mathcal{S})$ is the usual pull-back map. Given a circle-bundle, $\tau: S \longrightarrow T$, the pull-back, $\psi^*(S) = \{(s,x): \tau(s) = \psi(x)\}$, is then a circle-bundle over X. Suppose $\psi^*(S)$ is trivial; one obtains a continuous map $\sigma: X \longrightarrow S$ so that $\psi = \tau\sigma$ $(x \longrightarrow (\sigma(x),x)$ is then a rivializing section). For $(x_1,x_2) \in R(\psi)$, there is a unique $(x_1,x_2) \in \mathbb{T}$ with $\sigma(x_1) = f(x_1,x_2)\sigma(x_2)$. Clearly, $f \in Z^1(R(\psi),\mathbb{T})$ nd $S \cong S_f$. Thus $\ker \psi^* \subseteq \operatorname{Im} \eta$; the reverse inclusion is obvious.

ii) The map ε is discussed in [7] §4 iv; to a circle-bundle $\sigma: S \longrightarrow X$, ne associates its external product with its dual: $\tau \times \tau^*: S \times S^* \longrightarrow X \times X$, nd notes that it carries the natural structure of a groupoid with unit pace X; futher, $\tau \times \tau^*$ is a groupoid morphism. The map ε assigns he twist obtained by restriction to $R(\psi)$, denoted $\sigma_\tau: \Gamma_\tau \longrightarrow R(\psi)$. This wist is trivial i.e. $\varepsilon(\tau) = 0$ iff there is a continuous section, $: R(\psi) \longrightarrow \Gamma_\tau$, which is a groupoid morphism. Suppose this to be the case. iewing S as a left $S \times S^*$-space by the formula:

$$(s_1,s_2^*)s_3 = (s_2^* s_3)s_1 \quad (\text{for } \tau^*(s_2^*) = \tau(s_3)),$$

ut $s_1 \sim s_2$ iff $(\tau(s_1),\tau(s_2)) \in R(\psi)$ and $s_1 = j(\tau(s_1),\tau(s_2))s_2$. he equivalence relation is equivariant; the quotient $S/\sim$ is then a ircle-bundle over T and $S = \psi^*(S/\sim)$.
hus $\ker \varepsilon \subseteq \operatorname{Im} \psi^*$; the reverse inclusion is left to the reader.

v) The map $\delta: Tw(R(\psi)) \longrightarrow H^2(T,\mathcal{S})$ assigns the Dixmier-Douady class f the convolution algebra $C^*(\sigma)$ to a twist $\sigma: \Gamma \longrightarrow R(\psi)$. This class s a complete strong Morita equivalence invariant (for continuous trace lgebras with fixed spectrum T); thus, $\ker \delta$ consists of twists yielding reliminary algebras. By the above remarks (§2.6), a twist $\sigma: \Gamma \longrightarrow R(\psi)$ s in $\ker \delta$ iff it is the restriction of a twist $\sigma': \Gamma' \longrightarrow R(\psi')$ where $': X \amalg T \longrightarrow T$ (is the obvious map). Consider the injection $j: X \hookrightarrow R(\psi')$ y $j(x) = (x,\psi(x))$ and set $S = j^*(\Gamma')$ (qua circle-bundle). With $: S \longrightarrow X$, one has $\Gamma \cong \Gamma_\tau$; so $\ker \delta \subseteq \operatorname{Im} \varepsilon$ (showing the reverse nclusion amounts to extending Γ_τ to a Γ').

v) The last map, $\psi^*: H^2(T,\mathcal{S}) \longrightarrow H^2(X,\mathcal{S})$, results from the contravariance of the functor $H^2(\cdot,\mathcal{S})$. We show $\ker \psi^* \subseteq \operatorname{Im} \delta$ (the reverse inclusion follows from §2.9). Fix $g \in \ker \psi^*$; there is then a local homeomorphism $\phi: Y \longrightarrow T$ and a twist $\sigma: \Gamma \longrightarrow R(\phi)$ for which $\delta(\sigma) = g$ (one may for example take $\sigma_\lambda: \Gamma_\lambda \longrightarrow R(\psi_\lambda)$ where λ is a representative for g). Consider the diagram:

$$\begin{array}{ccc} X*Y & \xrightarrow{\pi_y} & Y \\ \downarrow{\scriptstyle \pi_x} & & \downarrow{\scriptstyle \phi} \\ X & \xrightarrow{\psi} & T \end{array}$$

and recall the notion of pull-back twist $(\psi^*\sigma: \psi^*\Gamma \longrightarrow R(\pi_x))$; then $\delta(\psi^*\sigma) = \psi^*\delta(\sigma) = 0$. Whence $\psi^*\sigma \in \operatorname{Im} \varepsilon_{\pi_x}$. There is a commutative diagram of groupoids with common unit space $(= X * Y)$:

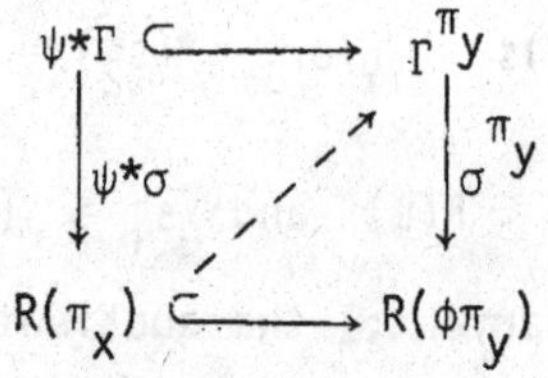

Adjusting, if necessary, by an element of $H^1(X*Y,\mathcal{S})$, there is a groupoid injection $j: R(\pi_x) \hookrightarrow \Gamma^{\pi_y}$. This induces an equivalence relation on Γ^{π_y}, the quotient of which is a twist, $\rho: \Lambda \longrightarrow R(\psi)$. Since $\Lambda^{\pi_x} \simeq \Gamma^{\pi_y}$, it follows that $\delta(\rho) = \delta(\rho^{\pi_x}) = \delta(\sigma^{\pi_y}) = \delta(\sigma) = g$. Thus $\ker \psi^* \subseteq \operatorname{Im} \delta$ as desired.

4 Ultraliminary algebras

iven a diagonal pair (A,B) with A continuous trace, and a subalgebra of continuous trace with $B \subseteq C \subseteq A$, it is easy to show that B is iagonal in C as well. Let $\psi: X \longrightarrow T$ and $\phi: X \longrightarrow Y$ be the spectral aps for the pairs (A,B) and (C,B) with twists $\sigma: \Gamma \longrightarrow R(\psi)$ and : $\Lambda \longrightarrow R(\phi)$. One has the following commutative diagram of groupoids:

$$\begin{array}{ccc} \Lambda & \hookrightarrow & \Gamma \\ \downarrow & & \downarrow \\ R(\phi) & \hookrightarrow & R(\psi) \end{array} \qquad \text{(common unit space } X\text{)}$$

ince $R(\phi) \subseteq R(\psi) \subseteq X \times X$, there is local homeomorphism $\beta: Y \longrightarrow T$ or which $\psi = \beta\phi$.

) Prop: With A,B,C as above, one has $\delta(C) = \beta^*\delta(A)$.

roof: We show that C is strong Morita equivalent to the pull-back *A, and then apply Prop. 1.4 of [12]. Consider the diagram:

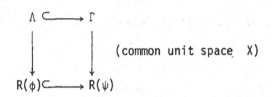

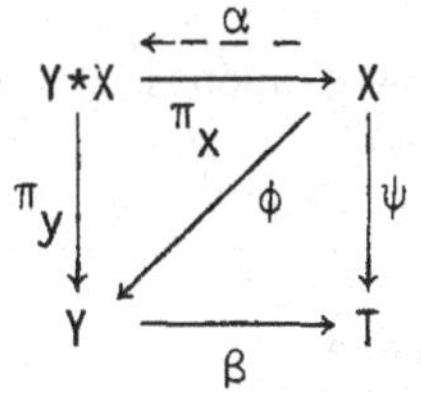

here is a unique $\alpha: X \longrightarrow Y*X$ making the diagram commutative namely: $\alpha(x) = (\phi(x),x))$. Recall the notion of pull-back twist cf. §2.8 ; write : $\beta^*\sigma : \beta^*\Gamma \longrightarrow R(\pi_y))$; note $\delta(\beta^*\sigma) = \beta^*\delta(\sigma)$. aking the reduction of $\beta^*\Gamma$ to $\alpha(X)$, one obtains

$$\Lambda \simeq \beta^*\Gamma_{\alpha(X)}; \quad \text{thus,} \quad \delta(C) = \delta(\rho) = \delta(\beta^*\sigma) = \beta^*\delta(\sigma) = \beta^*\delta(A). \qquad \square$$

his result motivates the following definition:

2) Def: A C*-algebra is said to be ultraliminary if there is an increasing sequence of C*-subalgebras A_i $(i \geq 0)$, such that:

i) A_i is continuous trace

ii) A_i has the extension property in A

iii) $A = \overline{\bigcup_i A_i}$

3) Remarks: If A_0 is abelian, then it is diagonal in each A_i and thus in A (cf. §2.2). Further, one has a sequence of local homeomorphisms:

$$X_0 \xrightarrow{\psi_1} X_1 \xrightarrow{\psi_2} X_2 \xrightarrow{\psi_3} X_3 \xrightarrow{\psi_4} \cdots \quad \text{where } X_i = \hat{A}_i \text{ so that } \delta(A_{i-1}) = \psi_i^* \delta(A_i).$$

Conversely, given such a sequence together with elements $\delta_i \in H^2(X_i, \mathcal{S})$ with $\delta_0 = 0$ and $\delta_{i-1} = \psi_i^* \delta_i$, one may construct an ultraliminary algebra. In a subsequent study, we investigate to what extent the "triple", $\{(X_i, \psi_i, \delta_i)\}$, forms a stable invariant (with some suitable notion of equivalence of triples).

4) Fact: If A is ultraliminary, then so is $A \otimes K$ (each $A_i \otimes K$ has the extension property in $A \otimes K$). A diagonal in $A_0 \otimes K$ is then diagonal in $A \otimes K$.

One would hope for a converse, to wit, hereditary algebras of ultraliminary algebras are ultraliminary.

This class of C*-algebras includes AF algebras, the Bunce-Deddins algebras (as well as their hereditary subalgebras cf. [3]), and a stable simple C*-algebra without idempotent (cf. [3]).

References

[1] J. Anderson, Extensions, restrictions and representations of states on C*-algebras, Trans. Amer. Math. Soc. 249 (1979) 303-329.

[2] R.J. Archbold, J.W. Bunce and K.D. Gregson, Extensions of states of C*-algebras II, Proc. Royal Soc. Edin., 92A (1982) 113-122.

[3] B. Blackadar and A. Kumjian, Skew products of relations and the structure of simple C*-algebras, preprint.

[4] L. Brown, P. Green and M. Rieffel, Stable isomorphism and strong Morita equivalence of C*-algebras, Pac. J. Math., 71(1977) 349-363.

[5] J. Dixmier, C*-algebras, North-Holland, Amsterdam, 1977.

[6] A. Kumjian, Preliminary algebras arising from local homeomorphisms, Math. Scand., 52 (1983) 269-278.

[7] A. Kumjian, On C*-diagonals and twisted relations, Tübinger Semesterbericht (Winter 1983) 179-191.

[8] G.K. Pedersen, C*-algebras and their automorphism groups, Academic Press, London, New York, San Francisco, 1979.

[9] J. Phillips and I. Raeburn, Automorphisms of C*-algebras and second Cech cohomology, Indiana Univ. Math. J., 29 (1980) 799-822.

[10] I. Raeburn, On the Picard group of a continuous trace C*-algebra, Trans. Amer. Math. Soc., 263 (1981) 183-205.

[11] I. Raeburn and J. Taylor, Continuous trace C*-algebras with given Dixmier-Douady class, to appear in J. of Aust. Math. Soc.

[12] I. Raeburn and D. Williams, Pull-backs of C*-algebras and crossed products by certain diagonal actions, to appear in Trans. A.M.S.

[13] J. Renault, A groupoid approach to C*-algebras, Lecture notes in Math., v.793, Springer Verlag, Berlin, Heidelberg, New York, 1980.

[14] J. Renault, Two applications of the dual groupoid of a C*-algebra, in these proceedings.

[15] M.A. Rieffel, Induced representations of C*-algebras, Adv. in Math., 13 (1974) 176-257.

[16] G.K. Pedersen and N.H. Petersen, Ideals in a C*-algebra, Math. Scand. 27 (1970) 193-204.

School of Mathematics
University of New South Wales
Post Office Box 1
Kensington, NSW 2033
AUSTRALIA

P.S. I have invited T. Natsume to write the following appendix in which topological obstructions for the existence of diagonals are considered. Examples of n-homogeneous algebras without diagonal appear in K.D. Gregson's thesis (cf. [2] 3.5).

P.P.S. I wish to thank Jean Renault for sharing his ideas with me.

Appendix

by

Toshikazu NATSUME

In this appendix we present an example of a C*-algebra having no diagonals.

Let H be a hermitian vector bundle on a compact, connected and simply connected space T. The bundle $\mathrm{Hom}(H,H)$ gives rise to a continuous field of elementary C*-algebras on T. The space $A_H = \Gamma(\mathrm{Hom}(H,H))$ of continuous sections is a continuous trace algebra with spectrum T.

Suppose that A_H has a diagonal B, which is identified with $C(X)$ (where X is compact). Since A_H is a continuous trace algebra, there exists a local homeomorphism Ψ from X onto T. By the compactness of X and T, Ψ is a covering map. From the assumption on T, it follows that X is a disjoint union of a finite number of copies of T. Then $C(X)$ is isomorphic to the direct sum of a finite number of copies of $C(T)$. This implies that the vector bundle H is decomposed into a Whitney sum of line bundles over T.

Consider a 4-sphere S^4. There is a one-to-one correspondance between the isomorphism classes of $U(n)$-bundles over S^4 and $\pi_3(U(n))$. Since $\pi_3(U(2)) \simeq \pi_3(S^3) \simeq \mathbb{Z}$, there exists a nontrivial complex vector bundle E of rank 2 on S^4. On the other hand, every complex line bundle on S^4 is trivial, because $\pi_3(U(1)) = 0$. This means that E cannot be decomposed into a Whitney sum of line bundles. Thus we have:

Proposition. There is a hermitian vector bundle E of rank 2 over S^4 such that the associated continuous trace algebra A_E has no diagonals.

Department of Mathematics, Saitama University, Japan. Partially supported by Danish Natural Science Research Council.

MARKOV DILATIONS ON THE 2×2 MATRICES*

Burkhard Kümmerer
Mathematisches Institut
Universität Tübingen
Auf der Morgenstelle 10
D-7400 Tübingen
Germany

Abstract: We present new results concerning Markov dilations on the algebra M_2 of 2×2 matrices. In particular, we show that each identity preserving completely positive operator on M_2 which has an invariant faithful state φ and commutes with the modular automorphism group for φ always possesses a Markov dilation. Moreover we obtain information on the structure of these Markov dilations.

§ 1 Introduction

1.1 Motivation. Dilations of completely positive operators are interesting mainly for three reasons: firstly, the analogy with the theory of unitary dilations (cf. [11]), secondly, the interpretation of a Markov dilation as a Markov process corresponding to a given transition operator (cf.[4]), and thirdly, its application to the quantum theory of irreversible processes (cf.[8],[9]).
While on commutative W^*-algebras probability theory provides a well developed dilation theory, our knowledge about dilations on non commutative W^*-algebras is rather rudimentary. For studying the features coming in by non-commutativity it seems to be a good strategy to investigate dilations on the 2×2 matrices. On the one hand the occuring structures remain manageable while on the other hand many methods and results obtained for 2×2 matrices immediately carry over to more general W^*-algebras.

1.2 Contents. In ([5], 2.1.8) we have found a non trivial necessary condition for a morphism of $(\mathfrak{A},\varphi)$ (for the notation cf. 1.3) to have a dilation: it has to commute with the modular automorphism group σ^{φ}. In the present paper we show that on M_2 this condition is also sufficient (This is no longer true on higher dimensional matrix algebras).

*) This paper is part of a research project supported by the Deutsche Forschungsgemeinschaft.

Moreover, for some cases we obtain more insight into the structure of any possible dilation. The methods developed there can also be applied to more general W^*-algebras.

The rest of this paragraph provides the necessary definitions and states some preliminaries. In §2 and §3 we investigate dilations for morphisms depending on whether or not there is an invariant trace.

1.3 Definitions. As the objects of a category we consider pairs $(\mathfrak{A},\varphi)$ each consisting of a W^*-algebra with a faithful normal state φ on $\mathfrak{A}$. A *morphism* $T:(\mathfrak{A}_1,\varphi_1) \to (\mathfrak{A}_2,\varphi_2)$ is a completely positive operator $T: \mathfrak{A}_1 \to \mathfrak{A}_2$ satisfying $T(\mathbb{1}) = \mathbb{1}$ and $\varphi_2 \circ T = \varphi_1$. In particular, a morphism is a normal operator. If T is a morphism of $(\mathfrak{A},\varphi)$ into itself then we call $(\mathfrak{A},\varphi,T)$ a *dynamical system*. If, moreover, T is a *-automorphism we call it an *automorphism* of $(\mathfrak{A},\varphi)$ and $(\mathfrak{A},\varphi,T)$ a *reversible dynamical system*.

Definition. Let $(\mathfrak{A},\varphi,T)$ be a dynamical system. If there exist a reversible dynamical system $(\hat{\mathfrak{A}},\hat{\varphi},\hat{T})$ and morphisms $i:(\mathfrak{A},\varphi) \to (\hat{\mathfrak{A}},\hat{\varphi})$, $P:(\hat{\mathfrak{A}},\hat{\varphi}) \to (\mathfrak{A},\varphi)$ such that the diagram

$$\begin{array}{ccc} (\mathfrak{A},\varphi) & \xrightarrow{\;T^n\;} & (\mathfrak{A},\varphi) \\ {\scriptstyle i}\big\downarrow & & \big\uparrow{\scriptstyle P} \\ (\hat{\mathfrak{A}},\hat{\varphi}) & \xrightarrow[\;\hat{T}^n\;]{} & (\hat{\mathfrak{A}},\hat{\varphi}) \end{array}$$

commutes for all $n \in \mathbb{N} \cup \{0\}$ then we call $(\hat{\mathfrak{A}},\hat{\varphi},\hat{T};P)$ a *dilation* of $(\mathfrak{A},\varphi,T)$. If the diagram commutes only for $n = 0$ and $n = 1$, then we call it a *dilation of first order*.

Remark. It follows from the above diagram for $n = 0$ that i is an injective *-homomorphism and $i \circ P$ is a normal conditional expectation of $\hat{\mathfrak{A}}$ onto $i(\mathfrak{A})$ leaving $\hat{\varphi}$ invariant. In particular, given P then i is uniquely determined.

Let $(\hat{\mathfrak{A}},\hat{\varphi},\hat{T};P)$ be a dilation (of first order) of $(\mathfrak{A},\varphi,T)$. For $I \subseteq \mathbb{Z}$ we denote by $\mathfrak{A}_I$ the W^*-subalgebra of $\hat{\mathfrak{A}}$ generated by $\bigcup_{k \in I} \hat{T}^k i(\mathfrak{A})$. The subalgebra $\mathfrak{A}_{\{0,1\}}$ will be denoted by $\mathfrak{A}_{0,1}$. From [12] it follows easily (cf. [5],2.1.3) that there exists a conditional expectation P_I of $\hat{\mathfrak{A}}$ onto $\mathfrak{A}_I$ leaving $\hat{\varphi}$ invariant.

Definition. A dilation $(\hat{\mathfrak{A}},\hat{\varphi},\hat{T};P)$ of $(\mathfrak{A},\varphi,T)$ is called *minimal* if $\hat{\mathfrak{A}} = \mathfrak{A}_{\mathbb{Z}}$. It is called a *Markov dilation* if for all $x \in \mathfrak{A}_{[0,\infty)}$: $P_{\{0\}}(x) = P_{(-\infty,0]}(x)$.

Note that from a dilation $(\hat{\mathfrak{A}},\hat{\varphi},\hat{T};P)$ one obtains a minimal dilation by restricting to $\mathfrak{A}_{\mathbb{Z}}$.

For some basic results on dilations we refer to [5] .

1.4 Dilations on M_2. Throughout the rest of this paper $\mathfrak{A}$ denotes the algebra of 2×2 matrices, φ some faithful state on $\mathfrak{A}$, and T a morphism of $(\mathfrak{A},\varphi)$.
Assume now that $(\hat{\mathfrak{A}},\hat{\varphi},\hat{T};P)$ is a dilation (of first order) of $(\mathfrak{A},\varphi,T)$. Since $\mathfrak{A}$ is a type I factor it follows that there exists $(\mathfrak{C},\psi)$, a W^*-algebra $\mathfrak{C}$ with faithful normal state ψ such that $(\hat{\mathfrak{A}},\hat{\varphi})$ is canonically isomorphic to $(\mathfrak{A}\otimes\mathfrak{C},\varphi\otimes\psi)$ with $i(x) = x\otimes\mathbb{1}$, $P(x\otimes y) = \psi(y)\cdot x$ for $x \in \mathfrak{A}$, $y \in \mathfrak{C}$, i.e., in the terminology of [5] $(\hat{\mathfrak{A}},\hat{\varphi},\hat{T};P)$ is necessarily a *tensor dilation*. This isomorphism induces an isomorphism between $(\mathfrak{A}_{0,1}, \hat{\varphi}|\mathfrak{A}_{0,1})$ and $(\mathfrak{A}\otimes\mathfrak{C}_1,\varphi\otimes\psi_1)$ for some W^*- subalgebra $\mathfrak{C}_1$ of $\mathfrak{C}$ and $\psi_1 := \psi|\mathfrak{C}_1$.
In view of this special structure of $(\hat{\mathfrak{A}},\hat{\varphi})$ we will write elements of $\hat{\mathfrak{A}}$ as 2×2 matrices with entries from $\mathfrak{C}$ whenever this is convenient.

1.5 Structure of dilations. In this section we present a result concerning the structure of dilations which may motivate some results in the sequel (cf. 2.5, 3.8).
If $(\mathfrak{B},\chi)$ is a W^*-algebra with faithful normal state then we denote by $\chi\otimes\mathbb{1}$ the morphism of $(\mathfrak{B},\chi)$ given by $x \mapsto \chi(x)\cdot\mathbb{1}$.
(i) Assume that $(\mathfrak{A}\otimes\mathfrak{C}_1,\varphi\otimes\psi_1,\mathrm{Ad}U;P_1)$ is a dilation of first order of $(\mathfrak{A},\varphi,T)$, i.e., this dilation of first order is given by an inner automorphism induced by some unitary $U \in \mathfrak{A}\otimes\mathfrak{C}_1$.
If $(\mathfrak{C},\psi,\sigma;Q)$ is a Markov dilation of $(\mathfrak{C}_1,\psi_1,\psi_1\otimes\mathbb{1})$ then $(\mathfrak{A}\otimes\mathfrak{C},\varphi\otimes\psi,\mathrm{Ad}U\circ(\mathrm{Id}\otimes\sigma);P)$ is a Markov dilation of $(\mathfrak{A},\varphi,T)$ where we consider $\mathfrak{A}\otimes\mathfrak{C}_1$ as a subalgebra of $\mathfrak{A}\otimes\mathfrak{C}$ and P is defined by $P(x\otimes y) = \psi(y)\cdot x$ for $x \in \mathfrak{A}$, $y \in \mathfrak{C}$.
This is a generalization (for a special type of dilations of first order) of the construction of Markov dilations in ([5],4.2.2) where we used the Bernoulli shift on the infinite tensor product of $\mathfrak{C}_1$ with itself as a dilation of $(\mathfrak{C}_1,\psi_1,\psi_1\otimes\mathbb{1})$.
(ii) Conversely, let $(\hat{\mathfrak{A}},\hat{\varphi},\hat{T};P)$ be a (minimal) Markov dilation of $(\mathfrak{A},\varphi,T)$ and assume that there exists an inner automorphism AdU of $(\hat{\mathfrak{A}},\hat{\varphi})$ such that $U \in \mathfrak{A}_{0,1} = \mathfrak{A}\otimes\mathfrak{C}_1$ and $\hat{T}\circ i(x) = U^*\cdot i(x)\cdot U = \mathrm{Ad}U\circ i(x)$ for $x \in \mathfrak{A}$.
Then $\mathrm{Ad}U^*\circ\hat{T}$ leaves $i(\mathfrak{A}) = \mathfrak{A}\otimes\mathbb{1} \subseteq \mathfrak{A}\otimes\mathfrak{C} = \hat{\mathfrak{A}}$ fixed, hence there exists an automorphism σ of $(\mathfrak{C},\psi)$ such that $\mathrm{Ad}U^*\circ\hat{T} = \mathrm{Id}_{\mathfrak{A}}\otimes\sigma$. Denote by Q the conditional expectation of $\mathfrak{C}$ onto $\mathfrak{C}_1$ which leaves ψ invariant (this conditional expectation exists).
Then $(\mathfrak{C},\psi,\sigma;Q)$ is a (minimal) Markov dilation of $(\mathfrak{C}_1,\psi_1,\psi_1\otimes\mathbb{1})$.

his result shows that such a dilation can be decomposed into a dilation f first order and into a part which may be interpreted as a generalized ernoulli shift. This throws a new light on a possible physical interretation of Markov dilations (cf.[9].).
he following results 2.5 and 3.8 show that for certain morphisms each ilation is of the type described above.

.6 Retrospect. Dilations for morphisms on M_2 have been investigated efore in several papers.
arkov dilations for quasifree morphisms have been constructed in [3].
n [15] a dilation has been constructed for every morphism in the convex ull of the quasifree morphisms which, however, lacks the Markov propery (see [8]). Markov dilations for these morphisms have been construced in [8] and for the corresponding one-parameter semigroups in [6].
n ([5], chapter 5), [6], and [9] we have investigated dilations for orphisms with two dimensional fixed space. In particular we arrived at classification of dilations of first order for these morphisms. In 9] we worked out a physical interpretation of these dilations.

he purpose of this paper is the construction and investigation of Marov dilations for morphisms which are not necessarily in the convex hull f the quasifree operators.
rom the results in ([5],2.3) concerning the factor type of a dilation t may be expected that Markov dilations of $(\mathfrak{A},\varphi,T)$ behave quite diferently depending on whether or not T has an invariant trace. Thereore, in the next two paragraphs we investigate these two cases separately.

§ 2 Dilations of Morphisms with Invariant Trace

.1 Throughout this paragraph we assume that T leaves the (normalized) race tr on $\mathfrak{A}$ invariant. In particular, this is the case when the fixed space of T is at least two dimensional. We denote by $\mathcal{M}(\mathfrak{A},tr)$ the losed convex set of all morphisms of $(\mathfrak{A},tr)$.
By [2] a morphism T of $(\mathfrak{A},\varphi)$ has a representation as $T(x) = \sum_{i=1}^{n} a_i^{*}\cdot x\cdot a_i$ with suitably chosen elements $a_i \in \mathfrak{A}$, $1 \leq i \leq n$, such that $\{a_i : 1 \leq i \leq n\}$ forms a set of linearily independent elements in $\mathfrak{A}$ (hence $1 \leq n \leq 4$)

2.2 From the results in [10] , [1], and [3] it is not difficult to deduce the following lemma whose proof will be given in [7].

<u>Lemma</u>. Let $T \in \mathcal{M}(\mathfrak{A},tr)$ be given by $T(x) = \sum a_i^* x a_i$ $(1 \leq i \leq n)$ with $\{a_i : 1 \leq i \leq n\}$ linearily independent in $\mathfrak{A}$. Then T is extreme in $\mathcal{M}(\mathfrak{A},tr)$ if and only if $\sum \lambda_{ij} a_i^* a_j = 0 = \sum \lambda_{ij} a_j a_i^*$ $(1 \leq i,j \leq n)$ implies $\lambda_{ij} = 0$ for $1 \leq i,j \leq n$.

2.3 <u>Theorem</u>. Any morphism $T \in \mathcal{M}(\mathfrak{A},tr)$ is a convex combination of automorphisms.

Proof: We show that the extreme points in $\mathcal{M}(\mathfrak{A},tr)$ are the automorphisms of $\mathfrak{A}$.
Let $T \in \mathcal{M}(\mathfrak{A},tr)$ be given by $T(x) = \sum a_i^* x a_i$ $(1 \leq i \leq n)$ with $\{a_i : 1 \leq i \leq n\}$ linearily independent, hence $1 \leq n \leq 4$.
If n=3 (n=4) then each of the conditions $\sum \lambda_{ij} a_i^* a_j = 0$ and $\sum \lambda_{ij} a_j a_i^* = 0$ for $\{\lambda_{ij} : 1 \leq i,j \leq n\}$ determines a subspace of $\mathbb{C}^9$ ($\mathbb{C}^{16}$) of dimension greater or equal than 5 (12). Since any two 5-dimensional (12-dimensional) subspaces of $\mathbb{C}^9$ ($\mathbb{C}^{16}$) have a non trivial intersection T cannot be extremal in $\mathcal{M}(\mathfrak{A},tr)$ by 2.2.
If n=2 then $T(x) = a_1^* x a_1 + a_2^* x a_2$ with $a_1^* a_1 + a_2^* a_2 = \mathbb{1} = a_1 a_1^* + a_2 a_2^*$ (T leaves $\mathbb{1}$ and tr invariant).
Therefore, if $a_1 = u_1 |a_1|$ is the polar decomposition of a_1 then the polar decomposition of a_2 is given by $a_2 = u_1 v |a_2|$ where v can be chosen to be a unitary commuting with $|a_2|$.
From this it follows easily that $a_j a_i^* = u_1 a_i^* a_j u_1^*$ $(1 \leq i,j \leq 2)$ and $\{a_i^* a_j : 1 \leq i,j \leq 2\}$ forms a commuting self adjoint set of operators in $\mathfrak{A} = M_2$. Hence it spans a two dimensional subspace of $\mathfrak{A}$. Consequently, there exist non trivial $(\lambda_{ij})_{i,j}$ with $\sum \lambda_{ij} a_i^* a_j = 0$ and, moreover, $0 = u_1^* (\sum \lambda_{ij} a_i^* a_j) u_1 = \sum \lambda_{ij} u_1^* a_i^* a_j u_1 = \sum \lambda_{ij} a_j a_i^*$ $(1 \leq i,j \leq 2)$.
Therefore, T is not extremal in $\mathcal{M}(\mathfrak{A},tr)$.
Thus T is extremal in $\mathcal{M}(\mathfrak{A},tr)$ (if and) only if n = 1, i.e., T is an automorphism.

2.4 <u>Corollary</u>. Any morphism $T \in \mathcal{M}(\mathfrak{A},tr)$ has a Markov dilation.

Proof: In ([5],4.3.3) we have shown that a convex combination of automorphisms has a Markov dilation.

2.5 For a motivation of the following result cf. 1.5 (ii).

<u>Theorem</u>. Let $(\hat{\mathfrak{A}},\hat{\varphi},\hat{T};P)$ be a minimal Markov dilation of $(\mathfrak{A},tr,T)$. Then there exists a unitary $U \in \mathfrak{A}_{0,1}$ with $\hat{T} \circ i(x) = U^* i(x) U$ for $x \in \mathfrak{A}$.

roof: By ([5],2.3.3) we may assume that $\hat{\varphi}$ is a trace. In particular, $\mathfrak{A}_{0,1}$ is a finite W^*-algebra with center valued trace $P^\natural$. The algebra $\mathfrak{A}_{0,1}$ is canonically isomorphic to $\mathfrak{A}\otimes\mathfrak{C}_1$ for some W^*-algebra $\mathfrak{C}_1$ cf. 1.4) and the center of $\mathfrak{A}\otimes\mathfrak{C}_1$ is contained in $\mathbb{1}\otimes\mathfrak{C}_1$. Therefore, utting $p := i\left(\begin{pmatrix}1 & 0\\ 0 & 0\end{pmatrix}\right)$, we obtain $P^\natural(p) = 1/2\cdot\mathbb{1}$. Moreover, $P^\natural\hat{T}(p) = 1/2\cdot\mathbb{1}$ since $\hat{T}$ commutes with $P^\natural$ by the uniqueness of the enter valued trace (cf. [14],V.2.6). Hence there exists some unitary $\in \mathfrak{A}_{0,1}$ such that $\hat{T}(p) = V^*\cdot p\cdot V$ (cf. [14],V.2.8).
t follows that $x \mapsto V\cdot\hat{T}(x)\cdot V^*$ is an automorphism of $(\hat{\mathfrak{A}},\hat{\varphi})$ leaving he subalgebra $\{i(x): x = \begin{pmatrix}a & 0\\ 0 & b\end{pmatrix} \in \mathfrak{A}$ with $a,b \in \mathbb{C}\}$ fixed. Hence by [5],5.10 (i)) there exists another unitary $V_1 \in \mathfrak{A}_{0,1}$ such that $_1V(\hat{T}\circ i(x))V^*V_1^* = i(x)$ for $x \in \mathfrak{A}$. Therefore, $\hat{T}\circ i(x) = U^*\cdot i(x)\cdot U$ ith $U := VV_1 \in \mathfrak{A}_{0,1}$ for $x \in \mathfrak{A}$.

3 Dilations of Morphisms without Invariant Trace

.1 <u>The morphisms</u>. Throughout this paragraph we assume that T is a orphism of $(\mathfrak{A},\varphi)$ having no invariant trace. It follows that φ is the nique invariant state of T and we may assume that φ is given by $\varphi\left(\begin{pmatrix}x_{11} & x_{12}\\ x_{21} & x_{22}\end{pmatrix}\right) = \lambda x_{11} + (1-\lambda)x_{22}$ with $0 < \lambda < 1/2$. Then the modular automorphism group σ^φ is given by $\sigma_t^\varphi\colon \begin{pmatrix}x_{11} & x_{12}\\ x_{21} & x_{22}\end{pmatrix} \mapsto \begin{pmatrix}x_{11} & \kappa^{it}x_{12}\\ \kappa^{-it}x_{21} & x_{22}\end{pmatrix}$ ith $\kappa := \lambda/(1-\lambda)$, $t \in \mathbb{R}$.
y $\mathcal{M}(\mathfrak{A},\varphi,\sigma^\varphi)$ we denote the closed convex set of all morphisms of $\mathfrak{A},\varphi)$ which commute with σ^φ.
f the morphism T of $(\mathfrak{A},\varphi)$ has a dilation then $T \in \mathcal{M}(\mathfrak{A},\varphi,\sigma^\varphi)$ y ([5],2.1.8).
n order to obtain a better understanding of this set we represent T s $T(x) = \sum a_k^* x a_k$ $(1 \le k \le n)$ (cf. 2.1). By writing $a_k = \begin{pmatrix}a_{11}^k & a_{12}^k\\ a_{21}^k & a_{22}^k\end{pmatrix}$ e define the vectors $A_{ij} := (a_{ij}^k)_{1\le k\le n} \in \mathbb{C}^n$ $(1\le i,j\le 2)$.
aking into account that $T \in \mathcal{M}(\mathfrak{A},\varphi,\sigma^\varphi)$ leaves the eigenspaces of σ^φ invariant we can write T as

$$T\colon \begin{pmatrix}x_{11} & x_{12}\\ x_{21} & x_{22}\end{pmatrix} \mapsto \begin{pmatrix}\langle A_{11},A_{11}\rangle x_{11} + \langle A_{21},A_{21}\rangle x_{22} & \langle A_{22},A_{11}\rangle x_{12}\\ \langle A_{11},A_{22}\rangle x_{21} & \langle A_{12},A_{12}\rangle x_{11} + \langle A_{22},A_{22}\rangle x_{22}\end{pmatrix}$$

here $\langle\ ,\ \rangle$ denotes the scalar product in $\mathbb{C}^n$.
e introduce a parameter μ by $\langle A_{12},A_{12}\rangle = \lambda\mu$. Then the conditions $T(\mathbb{1}) = \mathbb{1}$ and $\varphi\circ T = \varphi$ lead to $0 < \mu \le 1/(1-\lambda)$, $\langle A_{21},A_{21}\rangle = (1-\lambda)\mu$, $\langle A_{11},A_{11}\rangle = 1-(1-\lambda)/\mu$, $\langle A_{22},A_{22}\rangle = 1-\lambda\mu$ and T reads as

$$T\colon \begin{pmatrix}x_{11} & x_{12}\\ x_{21} & x_{22}\end{pmatrix} \mapsto \begin{pmatrix}(1-(1-\lambda)\mu)x_{11} + (1-\lambda)\mu x_{22} & \delta x_{12}\\ \bar{\delta}x_{21} & (1-\lambda\mu)x_{22} + \lambda\mu x_{11}\end{pmatrix}$$

with $\delta := \langle A_{22}, A_{11}\rangle$, hence $0 \leq |\delta| \leq \sqrt{1-\mu + \lambda(1-\lambda)\mu^2}$.
Thus the set of all morphisms in $\mathcal{M}(\mathfrak{A},\varphi,\sigma^{\varphi})$ with parameter $\delta = |\delta| \geq 0$ is affine isomorphic to the following convex set

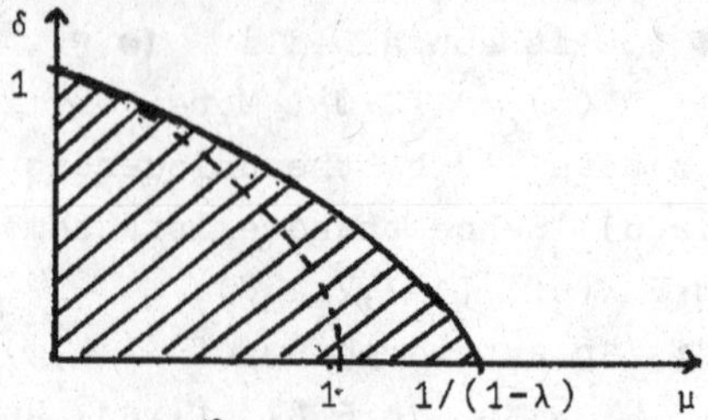

Therefore, $(\mathfrak{A},\varphi,\sigma^{\varphi})$ may be identified with the lying sugar-loaf resulting from turning this figure around the horizontal axis what corresponds to allowing δ to have a phase.
The automorphisms of $(\mathfrak{A},\varphi)$ correspond to the edge of the basis of this sugar-loaf (i.e., $\mu = 0$, $|\delta| = 1$) while the whole basis corresponds to the morphisms of $(\mathfrak{A},\varphi)$ with two dimensional fixed space.(Dilations of these morphisms have been investigated in ([5], chapter 5) and [9].)
The quasifree morphisms (cf.[3]) are obtained for $|\delta| = \sqrt{1-\mu}$ (the morphisms with $\delta = \sqrt{1-\mu}$ are on the dotted line in the above figure) and only for morphisms in the convex hull of the quasifree morphisms dilations have been constructed so far. Moreover, only these morphisms can occur as elements of a continuous one-parameter semigroup of morphisms in $\mathcal{M}(\mathfrak{A},\varphi,\sigma^{\varphi})$ (cf. [15]).

3.2 <u>Theorem</u>. Each morphism $T \in \mathcal{M}(\mathfrak{A},\varphi,\sigma^{\varphi})$ has a Markov dilation.

Proof: By ([5],4.2.3, 4.3.2) it is sufficient to construct a dilation of first order for the extreme points of $\mathcal{M}(\mathfrak{A},\varphi,\sigma^{\varphi})$. Moreover, two such extreme points with the same value of μ differ only by an inner automorphism of $(\mathfrak{A},\varphi)$, hence by ([5],2.2.8) we may restrict ourselves to the case $\delta = \sqrt{(1-\mu) + \lambda(1-\lambda)\mu^2}$, $0 < \mu \leq 1/(1-\lambda)$.
Putting $\mathbb{N}_o := \mathbb{N} \cup \{0\}$ we define the Hilbert space $\mathcal{H} := l^2(\mathbb{N}_o)$ and denote by $(e_i)_{i\in\mathbb{N}_o}$ its canonical basis. The trace class operator $\sum_{i=0}^{\infty}(1-\kappa)^i e_i \otimes e_i$ induces a faithful normal state ψ_1 on $\mathcal{L}(\mathcal{H}) =: \mathcal{C}_1$.
Now put $\mathfrak{A}_1 := \mathfrak{A} \otimes \mathcal{C}_1$ and $\varphi_1 := \varphi \otimes \psi_1$.
In $\mathcal{C}_1$ we define for $\nu \in [0,1] \subseteq \mathbb{R}$ an operator $a := \sum_{i=0}^{\infty} \nu^i e_i \otimes e_i$ and an isometry u by $u(e_i) := e_{i+1}$ $(i \in \mathbb{N}_o)$. Now it is easy to check that $U := \begin{pmatrix} \nu a & -u^*\sqrt{1-a^2} \\ \sqrt{1-a^2}\cdot u & a \end{pmatrix}$ is a unitary in $\mathfrak{A}_1$ which lies in the centralizer of φ_1.
Finally we put $i_1 : \mathfrak{A} \to \mathfrak{A}_1 : x \mapsto x \otimes 1$, and

$$_1: \mathfrak{A}_1 \to \mathfrak{A}: \begin{pmatrix} y_{11} & y_{12} \\ y_{21} & y_{22} \end{pmatrix} \mapsto \begin{pmatrix} \psi_1(y_{11}) & \psi_1(y_{12}) \\ \psi_1(y_{21}) & \psi_1(y_{22}) \end{pmatrix} .$$

hen for $x = \begin{pmatrix} x_{11} & x_{12} \\ x_{21} & x_{22} \end{pmatrix} \in \mathfrak{A}$ we obtain

$$\cdot \mathrm{AdU} \circ i_1(x) = \begin{pmatrix} \nu^2 \beta x_{11} + (1-\nu^2\beta)x_{22} & \nu\beta x_{12} \\ \nu\beta x_{21} & \beta x_{22} + (1-\beta)x_{11} \end{pmatrix}$$

ith $\beta := \psi_1(a^2) = (1-\kappa)/(1-\kappa\nu^2)$ $\quad (\kappa = \lambda/(1-\lambda))$.

hus $(\mathfrak{A}_1, \varphi_1, \mathrm{AdU}; P_1)$ is a dilation of first order for the morphism $\in \mathcal{M}(\mathfrak{A}, \varphi, \sigma^{\varphi})$ with $\mu = (1-\beta)/\lambda = \kappa(1-\nu^2)/\lambda(1-\kappa\nu^2)$ and $= \nu\beta = \nu(1-\kappa)/(1-\kappa\nu^2)$.

oreover, if ν varies between 0 and 1 then μ varies between $/(1-\lambda)$ and 0 and finally we have indeed that $\delta = \sqrt{(1-\mu) + \lambda(1-\lambda)\mu^2}$.

n the rest of this paragraph we investigate the structure of dilations or morphisms in $\mathcal{M}(\mathfrak{A}, \varphi, \sigma^{\varphi})$.

.3 <u>Proposition</u>. If $(\hat{\mathfrak{A}}, \hat{\varphi}, \hat{T}; P)$ is a minimal Markov dilation of $\mathfrak{A}, \varphi, T)$ then the modular automorphism group $\sigma^{\hat{\varphi}}$ is periodic with inimal period $\tau := -2\pi/\log\kappa$ and the fixed space of $\hat{T}$ is one dimensional.

roof: Since $\sigma_t^{\hat{\varphi}} i(x) = i(\sigma_t^{\varphi} x)$ for $x \in \mathfrak{A}$, $t \in \mathbb{R}$, $i(\mathfrak{A})$ is contained n the fixed space of $\sigma_\tau^{\hat{\varphi}}$. Moreover, $\hat{T}$ commutes with σ_t^{φ} $(t \in \mathbb{R})$ ence $\hat{T}^k i(\mathfrak{A})$ $(k \in \mathbb{Z})$ is contained in the fixed space of $\sigma_\tau^{\hat{\varphi}}$ and by inimality of the dilation we see $\sigma_\tau^{\hat{\varphi}} = \mathrm{Id}_{\hat{\mathfrak{A}}}$.
ince by assumption T has one dimensional fixed space, the same is rue for $\hat{T}$ by ([5],3.1.7(i)).

herefore, $\hat{\varphi}$ is a homogeneous periodic state in the terminology of [13] here a rich structure theory for W^*-algebras with such a state is eveloped. This structure theory turns out to be useful for the ollowing.

.4 In the situation of 3.3 we denote by E_n the eigenspace of $\sigma^{\hat{\varphi}}$ corresponding to the eigenvalue $n\kappa$ $(n \in \mathbb{Z})$, i.e., $\sigma_t^{\hat{\varphi}}(x) = \kappa^{n \cdot it} x$ for $\in E_n$. By ([13],1.27) there exists an isometry $v \in E_1$ such that $E_n = E_o v^n$, $E_{-n} = v^{*n} E_o$ for $n \in \mathbb{N}$. Moreover, if $P^{\natural}$ denotes the center alued trace on E_o then there exists an automorphism θ on the center $\mathcal{Z}_o$ of E_o such that $\theta(z)P^{\natural}(vv^*) = P^{\natural}(vzv^*)$ for all $z \in \mathcal{Z}_o$ ([13],1.20) which does not depend on the special choice of the isometry v ([13],1.21).

For the following we put $p := i\left(\begin{pmatrix} 1 & 0 \\ 0 & 0 \end{pmatrix}\right) = \begin{pmatrix} \mathbb{1} & 0 \\ 0 & 0 \end{pmatrix} \in E_o \subseteq \hat{\mathfrak{A}}$ and put $q := \mathbb{1} - p$.

3.5 Lemma. With the above notation we obtain $P^{\natural}(p) = \lambda \mathbb{1}$.

Proof: The element $w := \begin{pmatrix} 0 & \mathbb{1} \\ 0 & 0 \end{pmatrix}$ is a partial isometry in E_1 with $ww^* = p$, $w^* \cdot w = q$, hence by ([13],1.25)

(*) $P^{\natural}(p) = \kappa\theta(P^{\natural}q)$

Now assume that $P^{\natural}p \neq \lambda\mathbb{1}$, hence $P^{\natural}q \neq (1-\lambda)\mathbb{1}$.

Since $\hat{\varphi}(P^{\natural}q) = \hat{\varphi}(q) = 1-\lambda$ there exists a real number $\alpha > 0$ such that $(1-\lambda) + \alpha \in \sigma(P^{\natural}q)$ where $\sigma(x)$ denotes the spectrum of $x \in \hat{\mathfrak{A}}$.

Since $P^{\natural}p + P^{\natural}q = \mathbb{1}$ we obtain $\lambda-\alpha \in \sigma(P^{\natural}p)$ hence, by (*), $(\lambda-\alpha)/\kappa = (1-\lambda) - \alpha/\kappa \in \sigma(\theta(P^{\natural}q)) = \sigma(P^{\natural}q)$. From this we see that $\lambda + \alpha/\kappa \in \sigma(P^{\natural}p)$ and another application of (*) yields $(\lambda + \alpha/\kappa)/\kappa = (1-\lambda) + \alpha/(\kappa^2) \in \sigma(P^{\natural}q)$.

Since $0 < \kappa < 1$ an induction argument leads to a contradiction to $P^{\natural}q \leq \mathbb{1}$.

3.6 Theorem. Let $(\hat{\mathfrak{A}},\hat{\varphi},\hat{T};P)$ be a minimal Markov dilation of $(\mathfrak{A},\varphi,T)$. Then there exists a unitary $U \in E_0$ such that $\hat{T}(x) = U^* \cdot x \cdot U$ for all $x \in i(\mathfrak{A})$.

Proof: The automorphism $\hat{T}$ leaves E_0 globally invariant and by the uniqueness of the center valued trace its restriction to E_0 commutes with $P^{\natural}$. Therefore, by the above lemma we obtain $P^{\natural}p = \lambda\mathbb{1} = \hat{T}(\lambda\mathbb{1}) = \hat{T}(P^{\natural}p) = P^{\natural}(\hat{T}p)$. Now we proceed as in the proof of 2.5

3.7 In general for a morphism $S \in \mathcal{M}(\mathfrak{A},\varphi,\sigma^{\varphi})$ there exist many different dilations of first order. Examples can be constructed by writing S in various ways as a convex combination of other morphisms in $\mathcal{M}(\mathfrak{A},\varphi,\sigma^{\varphi})$ and glueing together their dilations as in ([5],4.3.2) to obtain a dilation for S (cf.[5], chapter 5 and [9]). Therefore, it seems difficult to classify in general all dilations of first order.

If $S \in \mathcal{M}(\mathfrak{A},\varphi,\sigma^{\varphi})$ has two dimensional fixed space (this corresponds to $\mu = 0$) we found such a classification in ([5],5.10) (cf.[9],2.5). The following results show that for a morphism which is extremal in $\mathcal{M}(\mathfrak{A},\varphi,\sigma^{\varphi})$ a minimal dilation of first order is essentially unique. This is trivial for $\mu = 0$, i.e., if S is an automorphism, and we may exclude this case in the following.

3.8 Theorem. Assume that T is extremal in $\mathcal{M}(\mathfrak{A},\varphi,\sigma^{\varphi})$ with $\mu \neq 0$. If $(\hat{\mathfrak{A}},\hat{\varphi},\hat{T};P)$ is a minimal Markov dilation of T then there exists a unitary $U \in \mathfrak{A}_{0,1} \cap E_0$ such that $U^* \cdot x \cdot U = \hat{T}(x)$ for all $x \in i(\mathfrak{A})$ and the resulting dilation of first order $(\mathfrak{A}_{0,1},\hat{\varphi}|\mathfrak{A}_{0,1},\mathrm{Ad}U|\mathfrak{A}_{0,1};P|\mathfrak{A}_{0,1})$ is necessarily isomorphic to the dilation of first order constructed in 3.2. ("Isomorphism" of dilations is defined in the obvious way).

Proof: We restrict ourselves to outline the main steps of our proof omitting some elementary computations.

As in the proof of 3.2 we may assume without loss of generality that $\delta = |\delta|$ so that $\delta = \sqrt{(1-\mu) + \lambda(1-\lambda)\mu^2} = \sqrt{(1-(1-\lambda)\mu)(1-\lambda\mu)}$.

In the following we freely use the notation introduced in 1.4 and 3.4.

By 3.6 there exists a unitary $U = \begin{pmatrix} u_{11} & u_{12} \\ u_{21} & u_{22} \end{pmatrix}$ in the centralizer $E_0 \subset \hat{\mathfrak{A}} = \mathfrak{A} \otimes \mathfrak{C}$ such that $\hat{T}i(x) = U^*\cdot i(x)\cdot U$ for $x \in \mathfrak{A}$. By multiplying U from the left with a suitably chosen unitary in $(\mathbb{1}\otimes\mathfrak{C}) \cap E_0$ we may assume u_{22} to be a positive element in $\mathfrak{C}$ denoted by a.

Now the dilation property leads to

$1-(1-\lambda)\mu) = \psi(u_{11}^* u_{11})$, $(1-\lambda\mu) = \psi(u_{22}^* u_{22})$, $\delta = \psi(u_{22}^* u_{11})$.

Hence $\psi(u_{22}^* u_{11}) = \sqrt{\psi(u_{11}^* u_{11})\ \psi(u_{22}^* u_{22})}$, i.e., the Cauchy-Schwarz inequality becomes an equality. Since ψ is faithful it follows that $u_{11} = \nu u_{22} = \nu a$ for some $\nu \in \mathbb{C}$. One easily deduces that $0 \leq \nu < 1$ and the value of ν is determined by the value of μ.

By means of the polar decomposition of u_{12} and u_{21} we write U as $U = \begin{pmatrix} \nu a & v_{12}a_{12} \\ a_{21}v_{21} & a \end{pmatrix}$ with positive elements a_{12}, a_{21} and partial isometries v_{12}, v_{21} in $\mathfrak{C}$.

Using the fact that U is a unitary in E_0 one easily deduces that $0 \leq a \leq \mathbb{1}$, $a_{12}^2 = a_{21}^2 = \mathbb{1} - a^2$ and $\mathbb{1}\otimes v_{21}$, $\mathbb{1}\otimes v_{12}^*$ are isometries in E_1 with equal final projections in E_0 and $v_{21}^* a^2 v_{21} = v_{12} a^2 v_{12}^* = \nu^2 a^2$.

Now we have to distinguish between the cases $\nu \neq 0$ and $\nu = 0$ corresponding to $\mu \neq 1/(1-\lambda)$ and $\mu = 1/(1-\lambda)$.

Assume first that $\nu \neq 0$. Then one shows that a is strictly positive and from this one deduces $v_{12} = -v_{21}^*$.

Hence with $v := v_{21}$ the unitary U takes the form $U = \begin{pmatrix} \nu a & -v^*\cdot\sqrt{\mathbb{1}-a^2} \\ \sqrt{\mathbb{1}-a^2}\cdot v & a \end{pmatrix}$

By evaluating $U^*\cdot p\cdot U \in \mathfrak{A}_{0,1}$ we conclude that U is contained in $\mathfrak{A}_{0,1}$. If we put $e_n := v^n\cdot v^{*n} - v^{n+1}\cdot v^{*n+1}$ $(n \in \mathbb{N}_0)$ then $(\mathbb{1}\otimes e_n)_{n\in\mathbb{N}_0}$ forms a set of orthogonal projections in $\mathfrak{A}_{0,1} \cap E_0$ summing up to $\mathbb{1}^0$. Moreover, $ve_nv^* = e_{n+1}$ and $a = \sum_{n=0}^\infty \nu^n e_n$.

Clearly the W^*-algebra generated by v is isomorphic to $\mathcal{L}(l^2(\mathbb{N}_0))$. From $\mathbb{1}\otimes v \in E_1$ we infer $\hat{\varphi}(\mathbb{1}\otimes e_n) = \psi(e_n) = (1-\kappa)\kappa^n$ $(n \in \mathbb{N}_0)$ which implies the assertion.

Now assume $\nu = 0$. Then a is a projection denoted by e_0 and $U = \begin{pmatrix} 0 & v_{12} \\ v_{21} & e \end{pmatrix}$.

Put $v_0 := v_{12}^* e_0$, $w := u_{12}^* v_{21}$, and $v_{2n} := w^n v_0 w^{*n}$, $v_{2n+1} := v_{2n+2}^* w$ for $n \in \mathbb{N}_0$. For $n \in \mathbb{N}_0$ the operator v_n is a partial isometry and we put $e_n := v_n^* v_n$. Then $\{e_n : n \in \mathbb{N}_0\}$ forms an orthogonal family of projections summing up to $\mathbb{1}$. Now we can define an isometry $v := -\sum_{n=0}^\infty v_n$. Then $\mathbb{1}\otimes v \in \mathfrak{A}_{0,1} \cap E_1$ and $V := \begin{pmatrix} 0 & -v^* \\ v & e \end{pmatrix}$ is a unitary in $\mathfrak{A}_{0,1} \cap E_0$ with $V^*\cdot i(x)\cdot V = U^*\cdot i(x)\cdot U = \hat{T}\circ i(x)$ for $x \in \mathfrak{A}$

From $ve_n v^* = e_{n+1}$ we proceed as in the case $\nu \neq 0$.

For a motivation of this result compare also 1.5.

3.9 Corollary. Assume that T is extremal in $\mathcal{M}(\mathfrak{A},\varphi,\sigma^{\varphi})$ with $\mu \neq 0$. If $(\tilde{\mathfrak{A}},\tilde{\varphi},\tilde{T};P)$ is a dilation of first order of T then the same assertion as in 3.8 holds.

Proof: Since $\mathfrak{A} = M_2$ a dilation of first order is necessarily a tensor dilation of first order (cf. 1.4). Hence it extends to a Markov dilation by the procedure of ([5],4.2.2) which canonically contains a minimal Markov dilation $(\hat{\mathfrak{A}},\hat{\varphi},\hat{T};P)$. Since $\tilde{\mathfrak{A}}_{0,1}$ and $\mathfrak{A}_{0,1}$ can be canonically identified the assertion follows from 3.8.

In particular, assume that $(\tilde{\mathfrak{A}},\tilde{\varphi},\tilde{T};P)$ is a minimal dilation of first order, i.e., $\tilde{\mathfrak{A}} = \tilde{\mathfrak{A}}_{0,1}$. Then by the above result there exists a unitary U in the centralizer of $\tilde{\mathfrak{A}}$ such that $U^* \cdot i(x) \cdot U = \tilde{T}\, i(x)$ for $x \in \mathfrak{A}$ and $(\tilde{\mathfrak{A}},\tilde{\varphi},AdU;P)$ is isomorphic to the dilation of first order constructed in 3.2. Since $AdU^* \circ \tilde{T}$ is an automorphism of $(\tilde{\mathfrak{A}},\tilde{\varphi})$ leaving $i(\mathfrak{A}) = \mathfrak{A} \otimes \mathbb{1} \subseteq \mathfrak{A} \otimes \mathcal{C}_1$ fixed, there exists a unitary V in the centralizer of $\mathcal{C}_1 = \mathcal{L}(l^2(\mathbb{N}_0))$ with respect to ψ_1 such that $AdU^* \circ \tilde{T} = Ad(\mathbb{1} \otimes V)$. Therefore, a minimal dilation of first order is uniquely determined up to an inner automorphism induced by a unitary in the maximal commutative centralizer of $\mathcal{C}_1$ with respect to ψ_1.

Finally we point out that by the above result a dilation of first order for such a morphism T necessarily leads to an infinite dimensional W^*-algebra. Therefore, it is impossible to use the minimal Stinespring construction for T (which may be considered as a dilation of first order for T in a purely C^*-algebraic framework) in order to obtain such a dilation, since the Stinespring construction in that situation leads only to finite dimensional algebras.

References

[1] Arveson, W.B.: Subalgebras of C^*-Algebras. Acta Math. 123 (1969), 141 - 224.

[2] Choi, M.-D.: Completely Positive Linear Maps on Complex Matrices. Lin. Alg. and Applic. 10 (1975), 285 - 290.

[3] Evans, D.E.: Completely Positive Quasi-Free Maps on the CAR Algebra. Commun. Math. Phys. 70 (1979), 53 - 68.

[4] Kern, M.; Nagel, R.; Palm, G.: Dilations of Positive Operators: Construction and Ergodic Theory. Math. Z. 156 (1977), 265 - 277.

[5] Kümmerer, B.: Markov Dilations on W*-Algebras. Submitted to Journal of Functional Analysis.

Some results of this paper are summarized in Kümmerer, B.: Markov Dilations of Completely Positive Operators on W*-Algebras. In Gr. Arsene (Ed.), "Dilation Theory, Toeplitz Operators, and Other Topics", Operator Theory: Advances and Applications, Vol 11 (Timisoara 1982), Birkhäuser Verlag, Basel, 251 - 259.

[6] Kümmerer, B.: Examples of Markov Dilations over the 2×2 Matrices. In L. Accardi, A. Frigerio, V. Gorini (Ed.), "Quantum Probability and Applications to the Quantum Theory of Irreversible Processes" (Frascati 1982), to appear in Lecture Notes in Mathematics, Springer Verlag, Berlin - Heidelberg - New York.

[7] Kümmerer, B. Representations of Completely Positive Operators. In preparation.

[8] Kümmerer, B.; Schröder, W.: A Markov Dilation of a Non - Quasifree Bloch Evolution. Commun. Math. Phys. 90 (1983), 251 - 262.

[9] Kümmerer, B.; Schröder, W.: A Survey of Markov Dilations for the Spin - 1/2 - Relaxation and Physical Interpretation. Semesterbericht Funktionalanalysis. Wintersemester 1981/82, 187 - 213.

[10] Størmer, E.: Positive Linear Maps of Operator Algebras. Acta Math. 110 (1963), 233 - 278.

[11] Sz.-Nagy, B.; Foias, C.: Harmonic Analysis of Operators on Hilbert Spaces. North Holland, Amsterdam 1970.

[12] Takesaki, M.: Conditional Expectations in von Neumann Algebras. J. Functional Analysis 9 (1971), 306 - 321.

[13] Takesaki, M.: The Structure of a von Neumann Algebra with a Homogeneous Periodic State. Acta Math. 131 (1973) 79 - 121.

[14] Takesaki, M.: Theory of Operator Algebras I. Springer-Verlag, Berlin - Heidelberg - New York 1979.

[15] Varilly, J.C.: Dilations of a Non-Quasifree Dissipative Evolution. Lett. Math. Phys. 5 (1981), 113 - 116.

SOME PROBLEMS AND RESULTS ON REFLEXIVE ALGEBRAS

by E. Christopher Lance

Communication presented at the OATE conference held at Busteni (Romania), 29 August - 9 September 1983.

The aim of this lecture is to make some propaganda for the theory of nonselfadjoint operator algebras. In the fifty years since F.J. Murray and J. von Neumann started their investigations on rings of operators, the theory of selfadjoint operator algebras has developed enormously. As the proceedings of this conference show, it is now a major branch of mathematics, with a substantial body of theory and several important applications. By contrast, the nonselfadjoint theory is still at the "Murray-von Neumann" stage of development. A few intrepid pioneers (notably W.B. Arveson, R.V. Kadison and J.R. Ringrose) have laid down the foundations for the subject. Some deep results have been established and it is clear that there is enormous potential for future development. But many basic questions remain unresolved, and some areas are still unexplored. For example, practically all the work to date has been concerned with weakly closed algebras (though a very promising start has just been made on the study of nonselfadjoint AF algebras and Cuntz algebras [18]).

An excellent survey of nonselfadjoint operator algebras has appeared quite recently [5], so I shall not make any attempt to be comprehensive, but instead will concentrate on a few specific problems in the hope that this will give a sample of recent achievements and an indication of some of the areas where more work is needed.

Let L denote a complete lattice of subspaces of a (separable) Hilbert space H. If $e \in L$ then we shall also use e to denote the projection onto the subspace e, and in general we shall not distinguish notationally between a projection and its range. (By a projection we always mean an orthogonal idempotent.)

Define an algebra $\operatorname{Alg} L$ of operators on H by

$$\operatorname{Alg} L = \{t \in B(H) : te = ete \quad (e \in L)\}.$$

Conversely, given an algebra A of operators on H, define a lattice $\operatorname{Lat} A$ by

$$\operatorname{Lat} A = \{e : te = ete \quad (t \in A)\}.$$

We say that L [resp. A] is reflexive if $L = \operatorname{Lat}\operatorname{Alg} L$ [resp. $A = \operatorname{Alg}\operatorname{Lat} A$].

If the lattice L is complemented then $\operatorname{Alg} L$ is a von Neumann algebra, and every von Neumann algebra A arises in this way (with L the lattice of projections in the commutant A'). At the opposite extreme, if L is totally ordered then we call it a nest, and $\operatorname{Alg} L$ a nest algebra. The two simplest examples of nest algebras occur when L is maximal, and is ordered by the integers and by the reals. We shall denote these algebras by $N(\mathbb{Z})$ and $N(\mathbb{R})$. To construct $N(\mathbb{R})$, take $H = L^2(\mathbb{R})$ and let e_t denote the subspace of H consisting of functions with support in $(-\infty, t)$. If $L = \{e_t : t \in \mathbb{R}\}$ then $\operatorname{Alg} L = N(\mathbb{R})$. The algebra $N(\mathbb{Z})$ can be realised in a similar way on the space $\ell^2(\mathbb{Z})$.

A lattice L consisting of mutually commuting projections is called a commutative subspace lattice (CSL) and $\operatorname{Alg} L$ is then called a CSL algebra. It was proved by Arveson [3] that every CSL is reflexive. A CSL is said to be of finite width if it is generated by a finite number of (commuting) nests.

Given Hilbert spaces H_1, H_2, we can form their Hilbert space tensor product $H_1 \otimes H_2$. If A_1, A_2 are weakly closed algebras of operators on H, K respectively then let $A_1 \bar{\otimes} A_2$ denote their weakly closed tensor product, acting on $H_1 \otimes H_2$. If L_1, L_2 are lattices of subspaces of H_1, H_2 respectively then let $L_1 \otimes L_2$ denote the smallest complete lattice of subspaces of $H_1 \otimes H_2$ which contains $\{e_1 \otimes e_2 : e_1 \in L_1, e_2 \in L_2\}$.

PROBLEM 1. Is the tensor product of two reflexive algebras [resp. lattices] again reflexive?

The answer will be positive if the appropriate one of the following tensor product formulae holds:

(ATPF) $\text{Alg}\, L_1 \bar{\otimes}\, \text{Alg}\, L_2 = \text{Alg}\,(L_1 \otimes L_2)$

(LTPF) $\text{Lat}\, A_1 \otimes \text{Lat}\, A_2 = \text{Lat}\,(A_1 \bar{\otimes} A_2)$

The formula (LTPF) can fail if A_1 or A_2 is not reflexive, so we consider only reflexive algebras here. With this restriction, no counterexample to either of these formulae is known, but the classes of lattices [algebras] for which the formulae have been proved are quite limited. If L_1 and L_2 are both complemented then (ATPF) just becomes the commutation theorem for tensor products of von Neumann algebras : so in this case the formula is true but not at all trivial. The status of (LTPF) is unknown even when A_1 and A_2 are both von Neumann algebras; but it does hold if A_1 and A_2 are both injective von Neumann algebras [8]. Much work on this problem has been done recently and many partial results have been proved ([6],[7],[8],[9],[10],[12],[13],[14]). For (ATPF), the most incisive results at present are that the formula holds for arbitrary L_2 provided that L_1 is a CSL which either has finite width [9] or satisfies a lattice-theoretic condition called complete distributivity [14].

Our second problem is only of interest for nest algebras. We say that two algebras A_1, A_2 acting on Hilbert spaces H_1, H_2 are similar if there is an invertible bounded map $t : H_1 \to H_2$ such that $A_2 = tA_1t^{-1}$.

PROBLEM 2. If two nest algebras are similar, are they unitarily equivalent?

This question was first posed by J. Ringrose about twenty years ago and has become known as the Ringrose problem. It is easy to see that any nest algebra which is similar to $N(Z)$ must be unitarily equivalent to it. Indeed, let $L = \{e_n : n \in Z\}$ be a nest of subspaces of H_1 such that each e_n has codimension one in e_{n+1} and let $A_1 (=N(Z))$ be the corresponding nest algebra.

If $A_2 = tA_1t^{-1}$ then each of the spaces $te_{n+1} \ominus te_n$ has dimension one, so there is a unitary $u : H_1 \to H_2$ such that

$$u(e_{n+1} \ominus e_n) = te_{n+1} \ominus te_n.$$

It follows that $uA_1u^{-1} = \mathrm{Alg}\,(u\,L) = \mathrm{Alg}\,(t\,L) = tA_1t^{-1}$, so that A_2 is unitarily equivalent to A_1.

Now consider the case $A_1 = \mathcal{N}(\mathbb{R})$, $A_2 = M_2 \otimes \mathcal{N}(\mathbb{R})$ (the algebra of 2×2 matrices over $\mathcal{N}(\mathbb{R})$). These algebras are not unitarily equivalent, since the diagonal of A_1 (that is, $A_1 \cap A_1^*$) is commutative, being isomorphic to $L^\infty(\mathbb{R})$, whereas the diagonal of A_2 is not commutative. On the other hand, the argument outlined below shows that A_1 and A_2 are similar. Thus the solution to the Ringrose problem turns out to be negative.

Suppose that A_1 and A_2 are as in the previous paragraph, and denote the corresponding nests by $\{e_t\}$ and $\{f_t\}$. Let K denote the algebra of compact operators on a separable Hilbert space. A deep result of Andersen [1] shows that $A_1 + K$ and $A_2 + K$ are unitarily equivalent, and furthermore for any $\varepsilon > 0$ the unitary u implementing the equivalence can be chosen so that

$$(*) \qquad \|ue_tu^* - f_t\| < \varepsilon \qquad (t \in \mathbb{R}).$$

As explained by Arveson in his lecture at this conference (see also [4]), Andersen's theorem can be regarded as a nonselfadjoint version of Voiculescu's noncommutative Weyl-von Neumann theorem.

The inequality (*) says that the nests $\{ue_tu^*\}$ and $\{f_t\}$ are close together. It follows from [15] that the nest algebras uA_1u^* and A_2 are close together in the sense of perturbation theory. This implies, by well-known results in perturbation theory, that uA_1u^* and A_2 will be similar provided that certain Hochschild cohomology groups vanish. This is indeed the case for nest algebras, as is shown in [15]. Thus uA_1u^* and A_2 (and hence A_1 and A_2) are similar, as required.

The solution to the Ringrose problem was first given by Larson

([16],[17]) using Andersen's theorem together with some intricate analysis. The fact that the result could be deduced directly from Andersen's theorem by the perturbation-theoretic results of [15] was observed by Andersen [2].

Our final problems seem to be still unresolved even in the very simplest cases. They amount to asking for a determination of the K-theory of a nest algebra. To date, however, the methods of K-theory, which have been so spectacularly successful in some parts of C^*-algebra theory, have not led to any new progress on these problems.

PROBLEM 3. Describe the similarity classes of idempotents in a nest algebra.

PROBLEM 4. Is the invertible group of a nest algebra connected in the norm topology?

For the simplest case, when the algebra is $N(\mathbb{Z})$, the answer to Problem 3 is given by the following proposition, which is due to D.R. Larson.

PROPOSITION. In the algebra $N(\mathbb{Z})$, every idempotent is similar to a projection.

Proof (Larson, unpublished). Suppose that p is an idempotent in $A = N(\mathbb{Z})$, and let $q = 1 - p$. Let $r = (p^*p + q^*q)^{\frac{1}{2}} \in B(H)$. It is easy to see that r is invertible and that $r^2p = p^*p$. Thus $rpr^{-1} = r^{-1}p^*pr^{-1}$. This shows that, in $B(H)$, p is similar to a projection.

If $L = \text{Lat}\,A$ then rL is similar to L and therefore (see the remarks on Problem 2) unitarily equivalent to L. So there is a unitary u in $B(H)$ with $uAu^{-1} = rAr^{-1}$. But this means that tpt^{-1} is a projection in A, where $t = u^{-1}r$. This completes the proof.

It is shown by Larson in [17] that the above Proposition is false for the nest algebra $N(\mathbb{R})$.

For Problem 4, even less information is available. This roblem is still unsolved even for $\mathcal{N}(Z)$, and appears to be ifficult. It was observed by Knowles and Saeks [11] that the nvertible group of $\mathcal{N}(Z)$ is connected in the strong operator opology. They also introduced a subalgebra A_0 of $\mathcal{N}(Z)$ which ontains all the elements of $\mathcal{N}(Z)$ which arise in applications to ystems theory and whose invertible group is norm-connected.

If one represents the operators in $\mathcal{N}(Z)$ by infinite matrices hen those matrices which are constant along each diagonal correspond under the Fourier transform) to multiplication by H^∞ functions. his leads one to think of $\mathcal{N}(Z)$ as a "noncommutative version of H^∞"; nd the subalgebra A_0 corresponds exactly to a "noncommutative isc algebra". In this light, the results of Knowles and Saeks are ot surprising, and they lead one to conjecture that the answer to roblem 4 should be negative for $\mathcal{N}(Z)$. More precisely, the natural njection i from H^∞ into $\mathcal{N}(Z)$ should give rise to an injection $_*$ at the level of K-theory. One might hope to show this by using n element of a "nonselfadjoint Kasparov group" $KK(\mathcal{N}(Z),H^\infty)$ to onstruct a left inverse for i_*.

ACKNOWLEDGEMENT

I should like to thank F. Gilfeather, A. Hopenwasser and G. Knowles for some discussions which were very helpful in the preparation of this aterial.

REFERENCES

1. N.T. ANDERSEN, Compact perturbations of reflexive algebras, J. Funct. Anal. 38 (1980) 366-400.

2. N.T. ANDERSEN, Similarity of continuous nests, Bull. London Math. Soc. 15 (1983) 131-132.

3. W.B. ARVESON, Operator algebras and invariant subspaces, Ann. of Math. 100 (1974) 433-532.

4. W.B. ARVESON, Perturbation theory for groups and lattices, J. Funct. Anal. 53 (1983) 22-73.

5. J.A. ERDOS, Nonselfadjoint operator algebras, Proc. Roy. Irish Acad. 81 (1981) 127-145.

6. F.L. GILFEATHER, A.L. HOPENWASSER and D.R. LARSON, Reflexive algebras with finite width lattices : tensor products, cohomology, compact perturbations, J. Funct. Anal. (to appear).

7. K.J. HARRISON, Reflexivity and tensor products for operator algebras and subspace lattices, J. London Math. Soc. (to appear).

8. A.L. HOPENWASSER, Tensor products of reflexive subspace lattices, (to appear).

9. A.L. HOPENWASSER and J. KRAUS, Tensor products of reflexive algebras II, J. London Math. Soc. 28 (1983) 359-362.

10. A.L. HOPENWASSER, C. LAURIE and R.L. MOORE, Reflexive algebras with completely distributive subspace lattices, J. Operator Theory (to appear).

11. G.J. KNOWLES and R. SAEKS, On the structure of invertible operators in a nest-subalgebra of a von Neumann algebra, Integral Equations Operator Theory (to appear).

12. J. KRAUS, W^*-dynamical systems and reflexive operator algebras, J. Operator Theory 8 (1982) 181-194.

13. J. KRAUS, The slice map problem for σ-weakly closed subspaces of von Neumann algebras, Trans. Amer. Math. Soc. 279 (1983) 357-376.

14. J. KRAUS, Tensor products of reflexive algebras, J. London Math. Soc. 28 (1983) 350-358.

15. E.C. LANCE, Cohomology and perturbations of nest algebras, Proc. London Math. Soc. 43 (1981) 334-356.

16. D.R. LARSON, A solution to a problem of J.R. Ringrose, Bull. Amer. Math. Soc. 7 (1982) 243-246.

17. D.R. LARSON, Nest algebras and similarity transformations, Ann. of Math. (to appear).

18. S.C. POWER, On ideals of nest-subalgebras of C^*-algebras, Proc. London Math. Soc. (to appear).

APPROXIMATION FOR ACTIONS OF AMENABLE GROUPS AND TRANSVERSAL AUTOMORPHISMS

A.A.Lodkin and A.M.Vershik

INTRODUCTION. The aim of the paper is an approach to the problem of approximation of amenable group actions by transformations of a certain type of a space realized as a Markov compactum. For any action of $\mathbb{Z}$ by measure-preserving (m.p.) transformations there exists, as it was shown by one of the authors in [14], a so-called adic realization *). The notion of adic transformation makes more precise the idea of a sequence of towers which approximate given automorphism T. We can view a tower as a representation of T as an integral automorphism over automorphism T_A induced by T on some measurable set A. Taking a subset of A, we can construct another tower for T_A. Continuing this process, we get a sequence of towers which are compatible in a sense. If ξ_n is the partition of the Markov compactum X into elements which consist of all sequences $(x_i)_{i\in N}$ in X with the same x_i for $i > n$, then an adic transformation (see §2 for the exact definition) is a limit of transformations leaving ξ_n fixed and acting as a permutation on a finite partition complementary to ξ_n (a Markov property of action). It follows that the orbit partition $\omega(T)$ of an adic transformation T coincides with the tail partition $\bigcap_n \xi_n$ of X.

For the group $\mathbb{Z}$ an adic realization is very useful for studying metric properties of an action. The most important appli-

*) The notion of adic transformation is a development of an idea [10, p.14] to represent an automorphism by a sequence of random permutations (cf. Def.2.1 - 2.4 here). The authors were recently informed by M.Gordin that Sh.Ito [4] has considered a particular case of adic transformation, similar to a certain example of R.Chacon.

cation, however, is a theorem on uniform finite-dimensional approximation of shift and multiplication operators [13], which extends the result of [5] (cf. [6]).

Meanwhile, the similar problem for amenable groops seems to be difficult. As it was shown in an important paper by A.Connes, J.Feldman and B.Weiss [3], for an action of countable amenable group one can construct a finite partition playing the role of a tower for a prescribed finite set of elements of this group. We investigate a possibility of finding a sequence of such towers which are compatible in the same sense as in the case of $\mathbb{Z}$.

It turned out that one cannot demand that every automorphism T in the group G have an adic realization (on the same Markov compactum) simply because $\omega(T)$ does not coincide with the orbit partition of the group, $\omega(G)$, which must be a tail partition. It is the reason of extending the notion of adic transformation. Apparently, there are several possibilities for generalizations, of which we distinguish two (see §2).

The first one, the notion of transversal transformation, is the narrowest possible. In terms of towers, the only difference with the adic transformation is that approximating permutations may induce noncyclic permutations in one element of ξ_n. The definition of transversal transformation given in §2 is based on the introduction of some order (called Markov order here). The definition provides a rich class of interesting transformations.

What we want is to represent each T in G as a transversal transformation with a prescribed tail partition $\psi \prec \omega(T)$, keeping in mind the case $\psi = \omega(G)$. For arbitrary ψ it is a hard problem even for individual transformations.

The other possible adjustments of the notion of adic transformation are a generalized transversal and an approximately cylindrical transformations. These classes of transformations coincide with he class P introduced in [14]. One of our main results, Theorem 2 §4), shows that a m.p. action of an amenable group admits a realiation by transformations of this class. The other, Theorem 1 (§3), ;ives a criterion of a possibility to realize an individual m.p. ransformation as transversal with a prescribed tail partition.

. BASIC NOTIONS CONCERNING PARTITION THEORY.

Let $(X, \mathfrak{A}, \mu)$ be a Lebesgue space *), $P(X)$ ($P_f(X)$) - the family of all (finite) measurable partitions of X, $P_c(X)$ - the family of $\xi \in P(X)$ which are cofinite, i.e. ess sup $\{\text{card } C \mid C \in \xi\} < \infty$. We remind that if ξ is a partition of X and $x \in X$, then $\xi(x)$ is an element of ξ containing x, $x \overset{\xi}{\sim} y$ means that $\xi(x) = \xi(y)$. We write $\xi_1 \prec \xi_2$ if $\xi_1(x) \supset \xi_2(x)$, $x \in X$. We note that for each $\xi \in P_c(x)$ there exists $\eta \in P_f(X)$ such that $\xi \vee \eta = \varepsilon$ (= the partition into points). Call such ξ, η complementary (we do not demand that $\xi \wedge \eta$ be trivial). If $\xi_1 \succ \xi_2 \succ \ldots$ is a decreasing sequence in $P(X)$, its set-theoretic intersection (generally, nonmeasurable, unlike infimum in $P(X)$) $\tau = \bigcap_n \xi_n$ is defined by the formula $\tau(x) = \bigcup_n \xi_n(x)$. When $\xi_1 \prec \xi_2 \prec \ldots$, $\xi_n \in P(X)$, $\bigvee_n \xi_n = \varepsilon$ (supremum in $P(X)$), we write $\xi_n \nearrow \varepsilon$.

A set $A \subset X$ is called a ξ-set if $x \in A$ implies $\xi(x) \subset A$. If $B \subset X$, then $\xi \cap B$ denotes the partition $\{C \cap B \mid C \in \xi\}$ of B, and T_B denotes a transformation induced on B by a transformation T. Finally, if $\xi \prec \zeta$, then ξ/ζ means the partition $\{C/\zeta \mid C \in \xi\}$ of X/ζ.

If a group G or a transformation T act on X, we denote by $\omega(G)$ or $\omega(T)$ the corresponding orbit partitions.

Let us recall (cf. [11], [12]) that a partition ψ of X is called tame, or approximable, if it satisfies the following equivalent conditions:

(i) ψ coincides with $\omega(T)$ for some nonsingular T

(ii) there exist such $\xi_n \in P_c(X)$ $(n \in \mathbb{N})$, $\xi_n \succ \xi_{n+1}$, that $\psi = \bigcap_n \xi_n$.

---

*) Measure theory objects are understood here as classes mod 0.

It is known [3] that for G discrete and amenable, $\omega(G)$ is tame (see[2],[1]for $G = \mathbb{Z}$). It follows from [2],[7] that in ergodic m.p. case all tame orbit partitions are isomorphic.

DEFINITION 1.1. Let X be a measure space, T its automorphism, $B \subset X$ a measurable set. If for some $h \in \mathbb{N}$ the relation $\bigcup_{j=0}^{h-1} T^j B = X$ holds, then there arises a natural partition $\zeta = \zeta(X, B, T) = \{ B_j \mid 0 \leq j < h \}$, where $B_0 = B$, $B_j = TB_{j-1} \setminus B_{j-1}$, $j > 0$, called a <u>tower</u> with the <u>base</u> B. For $x \in B$ let $h(x) = \max \{ j \mid T^{j-1}x \notin B \}$. We denote by $\tau(\zeta)$ a partition $\tau \in P_c(X)$ of X into the sets $\tau(x) = \{ T^j x \mid 0 \leq j < h(x) \}$, $x \in B$. Let π_τ be a projection $X \to B$, $\pi_\tau(x) \in \tau(x) \cap B$.

We warn the reader that in §3 we shall use towers associated with T^{-1} rather than with T, their bases being "on the top".

2. MARKOV COMPACTUM, TRANSVERSAL AND APPROXIMATELY CYLINDRICAL TRANSFORMATIONS

2.1. MARKOV COMPACTUM. Let D_i, $i \in \mathbb{N}$, be arbitrary finite sets with $r_i = \text{card}\, D_i \geqslant 2$. We recall that a <u>Markov compactum</u> is, by definition, a weakly closed subspace $X \subset \prod_i D_i$ defined by a sequence of transition $r_i \times r_{i+1}$ -matrices $M_i = \{ m^i_{de} \mid d \in D_i,$ $e \in D_{i+1}$, $m^i_{de} \in \{0, 1\} \}$ by the formula

$$X = \{ x = (x_i)_{i \in \mathbb{N}} \in \prod_i D_i \mid m^i_{x_i x_{i+1}} = 1 \}.$$

A Markov compactum is called stationary if D_i and M_i do not depend on i.

Let us denote by η_i, $i \in \mathbb{N}$, partitions of X into thin cylinders $C_d = \{ x \in X \mid x_i = d \}$, and write

$$(2.1) \qquad \xi_n = \bigvee_{i=n+1}^{\infty} \eta_i, \quad \varkappa_n = \bigvee_{i=1}^{n} \eta_i .$$

PROPOSITION 2.1 ([14]). If X is a Markov compactum, then partitions η_i, ξ_i, $\varkappa_i$ are Borel measurable and satisfy the following properties:

(i) $\bigvee_{i=1}^{\infty} \eta_i = \varepsilon$

(ii) $\xi_1 \succ \xi_2 \succ \dots$

(iii) for every $n \in \mathbb{N}$ the partitions $\varkappa_n$ and ξ_n are complementary

(iv) (the Markov property): for every $n \in \mathbb{N}$ and $C \in \eta_n$ there is a homeomorphism $C \to C/\varkappa_{n-1} \times C/\xi_n$, given by $x \in C \mapsto ((\eta_1(x),\dots,\eta_{n-1}(x)),(\eta_{n+1}(x),\dots))$

Conversely, if X is a separable totally disconnected compactum, $(\eta_i)_{i\ N}$ is a sequence of its finite partitions consisting of open-closed subsets, ξ_n and $\varkappa_n$ are defined by (2.1), and these partitions satisfy (i) - (iv), then X is a Markov compactum.

PROOF. The necessity is clear from the definition of partitions. Condition (iv) means that if $x,y \in C$, then there exists $z \in C$ with $\eta_i(z) = \eta_i(x)$ for $i<n$ and $\eta_i(z) = \eta_i(y)$ for $i>n$. The second part of the statement is true because one can take $D_i = X/\eta_i$, so that a point $d \in D_i$ can be regarded as an element of η_i. Then $M_i = \{m^i_{de} \mid d \in D_i, e \in D_{i+1}\}$ is given by the formula: m^i_{de} equals to 0 if $d \cap e$ is empty and to 1 otherwise. Q.E.D.

2.2. ORDERING. For the purpose of describing two classes of transformations of a Markov compactum X we introduce two patterns of partial ordering of X, such that X should become a union of chains (future pieces of trajectories of an action). We demand, on the one hand, that each chain lie in one element of $\psi = \bigcap_n \xi_n$. So x and y are comparable only if $x \overset{\xi_n}{\sim} y$ for some n and, evidently, to define an order it suffices to do it in all elements of all ξ_n. On the other hand, we impose a condition that this order depend only on the class of ξ_n/η_{n+1}. Hence we may procede by induction.

DEFINITION 2.1. Let $d \in D_{n+1}$ and $C^n_d = \{(x_1,\dots,x_{n+1}) \mid x_i \in D_i, m^i_{x_i x_{i+1}} = 1$ for $1 \le i \le n$ and $x_{n+1} = d\}$. This set may be identified with an element of ξ_n/η_{n+1}. We define an order in each C^n_d by induction. First define some order in all C_d^1, such that C_d^1 is a disjoint union of maximal linearly ordered subsets (chains).

If for some n we have already defined an order in every C_d^n, such that $C_d^n = \bigcup_j Q_{d,j}^n$, where $Q_{d,j}^n$, $j \in J_d$, are maximal chains, then we define an order in the set C_e^{n+1}, $e \in D_{n+2}$. Denote by $S_{d,j}$ and $F_{d,j}$ the least and the greatest (we should rather say the starting and the final) points of $Q_{d,j}^n$, and let $S_d^n = \{S_{d,j} \mid j \in J_d\}$, $F_d^n = \{F_{d,j} \mid j \in J_d\}$. For $e \in D_{n+2}$ let $\mathcal{F}_e^n$ be some subset in $\bigcup\{F_d^n \mid d \in D_{n+1} \text{ and } m_{de}^{n+1} = 1\}$, $\mathcal{S}_e^n = \bigcup\{S_d^n \mid d \in D_{n+1} \text{ and } m_{de}^{n+1} = 1\}$. We shall call a gluing map any injection $\Phi_e^n: \mathcal{F}_e^n \to \mathcal{S}_e^n$.

We say that a point $\bar{v} = (v_1,\dots,v_{n+1},e) = (v,e)$ follows $\bar{u} = (u_1,\dots,u_{n+1},e) = (u,e)$ $(\bar{u},\bar{v} \in C_e^{n+1})$ if either $u_{n+1} = v_{n+1} = d$ and v follows u in some chain $Q_{d,j}^n$, or $u \in \mathcal{F}_e^n$ and $v = \Phi_e^n(u)$. This rule defines chains in C_e^{n+1}, and thus introduces an order in it.

As we have already noted, it gives us an order in every element of ψ and in X. We call this order a general Markov order.

REMARK. We realize that our rule does not exclude the possibility of gluing the final point to the starting point of the same chain, thus forming a cycle, so we have defined a relation of succeeding rather than an order, but in what follows we shall be interested in the case when there is a continuous Markov Borel measure on X with respect to which the union of cycles is forming a negligible subset.

DEFINITION 2.2. Suppose that for every $n \geq 1$ and $e \in D_{n+1}$ the set $\{d \in D_n \mid m_{de}^n = 1\}$ is ordered and appears to be a union of maximal chains $q_{e,j}^n$, $j \in J_e$. We denote by $s_{e,j}$, $f_{e,j}$ the starting and the final points of $q_{e,j}^n$, correspondingly, and set $s_e^n = \{s_{e,j} \mid j \in J_e\}$, $f_e^n = \{f_{e,j} \mid j \in J_e\}$. Let $\varphi_{e\bar{e}}^n: f_e^n \to s_{\bar{e}}^n$ be some bijections defined for some ordered pairs $(e,\bar{e}) \in D_{n+1} \times D_{n+1}$.

We say that X is provided with Markov order, if it has a general Markov order, the gluing mappings Φ_e^n, $n \in \mathbb{N}$, $e \in D_{n+2}$, having the form

$$\Phi_e^n(u_1,\dots,u_{n+1},e) = (\bar{u}_1,\bar{u}_2,\dots,\bar{u}_{n+1},e),$$

where $\bar{u}_{n+1}$ is a successor for u_{n+1} in its chain $q_{e,j}^n$, $\bar{u}_n = \varphi^n_{u_{n+1}\bar{u}_{n+1}}(u_n)$, ... , $\bar{u}_1 = \varphi^1_{u_2\bar{u}_2}(u_1)$, and gluing bijections $\varphi^k_{u_{k+1}\bar{u}_{k+1}}$ are defined whenever $(u_1,\dots,u_{n+1}) \in \mathcal{F}_e^n$.

2.3. TRANSVERSAL AND GENERALIZED TRANSVERSAL TRANSFORMATIONS.

DEFINITION 2.3. A <u>transversal</u> (<u>generalized transversal</u>) transformation is a transformation of a Markov compactum $\mathbf{X}$, possessing a Markov (general Markov) order, which consists in passing from a point to its successor (if any).

If, for $x \in X$, we set $l(x) = \min\{n \in \mathbb{N} \mid x_n \notin f^n_{x_{n+1}}\}$, then the transversal mapping T is defined by the formula $y = Tx$, where

$$(2.2) \qquad y_i = \begin{cases} x_i, & i > l(x) \\ x_i', & i = l(x) \\ \varphi^{i+1}_{x_{i+1}y_{i+1}}(x_i), & i = l(x)-1,\ l(x)-2,\dots,1 \end{cases}$$

(x_i' is the successor of x_i in $q^i_{x_{i+1}} \ni x_i$).

REMARK. If D_n are linearly ordered and for every $e \in D_{n+1}$ there is only one chain in $\{d \mid m^n_{de} = 1\} \subset D_n$, $n \in \mathbb{N}$, our definition is reduced to that of an <u>adic transformation</u>*) introduced in [14]. If, moreover, X is a stationary Markov compactum with $r_i \equiv p$ and $m^i_{de} \equiv 1$, our transformation is isomorphic to the transformation $x \mapsto x+1$ in the group of p-adic integers (which is the origin of the term "adic").

EXAMPLE 2.1. Let X be a stationary Markov compactum with $r_i \equiv 4$, $D_i \equiv \{1,2,3,4\}$, $m^i_{de} \equiv 1$. Let $q^i_{e,j}$, $j = 0, 1$, be given by the table

j \ e	1	2	3	4
0	(1,2)	(4,2)	(4,1)	(3,4)
1	(3,4)	(1,3)	(2,3)	(1,2)

*) A term "odometer" is also appropriate.

and let $\varphi^{i}_{\bar{e}e}(f^{i}_{e,j}) = s^{i}_{\bar{e},j}$, $j = 0, 1$. Here is a piece of an orbit:
$(1111\ldots) \mapsto (2111\ldots) \mapsto (4211\ldots) \mapsto (2211\ldots) \mapsto (3421\ldots) \mapsto$
$\mapsto (4421\ldots) \mapsto (4221\ldots) \mapsto (2221\ldots) \mapsto \ldots$.

DEFINITION 2.4. By <u>n-cylindrical</u> transformation we mean a transformation T of a Markov compactum X (defined on a subset of X) leaving $\varkappa_n$ invariant and ξ_n fixed, $n \in \mathbb{N}$, i.e. every $C \in \eta_{n+1}$ is an invariant set and $T|_C$ is a product of a permutation and an identity map. We call T <u>approximately cylindrical</u> if there exist a sequence T_n of n-cylindrical transformations and an increasing sequence of measurable sets X_n, such that T is defined on $\bigcup_n X_n$ and $T|_{X_n} = T_m|_{X_n}$, $0 < n \le m$.

We state an evident fact:

PROPOSITION 2.2. The class of generalized transversal transformations and the class of approximately cylindrical transformations coincide.

Let us establish some properties of a transversal transformation T on a Markov compactum X. First, let

$$A_n = \{x \in X \mid l(x) \geq n\}, \quad n \in \mathbb{N} \cup \{\infty\} \tag{2.3}$$

(cf. Def. 2.3). It is clear that $A_1 = X$, $A_n = \{x \mid x_i \in f^{i}_{x_{i+1}},\ 1 \le i < n\}$ and $A_1 \supset A_2 \supset \ldots$. Let

$$\begin{aligned} \zeta_n &= \zeta(X, A_{n+1}, T^{-1}),\ \tau_n = \tau(\zeta_n), \\ \zeta'_n &= \zeta(A_n, A_{n+1}, (T_{A_n})^{-1}),\ \tau_n = \tau(\zeta'_n) \end{aligned} \tag{2.4}$$

(see Def. 1.1).

PROPOSITION 2.3. Given a transversal transformation T of a Markov compactum X, we have

(i) $\tau_1 \succ \tau_2 \succ \ldots$, $\bigcap_n \tau_n = \omega(T)$

(ii) $\tau_n \succ \xi_n$, $n \in \mathbb{N}$

(iii) $\xi_n \cap A_{n+1}$ is invariant under $T_{A_{n+1}}$, $n \in \mathbb{N}$

(iv) A_{n+k} is a $\xi_n \cap A_{n+1}$ -set for $n \in \mathbb{N}$, $k \geq 2$

(v) $\zeta'_{n+k} \prec \xi_n \cap A_{n+1}$ for $n \in \mathbb{N}$, $k \geq 1$

PROOF. (i),(ii) are clear. To prove (iii), observe that $x, y \in A_{n+1}$,

$x \overset{\xi_n}{\sim} y$ means that $x_i \in f^i_{x_{i+1}}$, $y_i \in f^i_{x_{i+1}}$ for $i = 1,\dots,n$ and $x_i = y_i$ for $i \geqslant n+1$. It follows that $l(x) = l(y) = l \geqslant n+1$ and, consequently, $\eta_i(Tx) = \eta_i(Ty) = \eta_i(x) = \eta_i(y)$ for $i > l$, $\eta_i(Tx) = \eta_i(Ty)$ for $n+1 \leqslant i \leqslant l$, from (2.2).

Let $j(x)$ and $k(y)$ be such naturals that $T^{j(x)}x = T_{A_{n+1}}x$, $T^{k(y)}y = T_{A_{n+1}}y$. Then, again by (2.2), $\eta_i(T^j x) = \eta_i(T^k y)$ for $i \geqslant n+1$, $0 \leqslant j \leqslant j(x)$, $0 \leqslant k \leqslant k(y)$. Hence $T_{A_{n+1}}x \overset{\xi_n}{\sim} T_{A_{n+1}}y$.

To check (iv), observe that if $x \in A_{n+k}$, $y \in A_{n+1}$, $x \overset{\xi_n}{\sim} y$, then $y \in A_{n+k}$.

Finally, to verify (v), we take any $C \in \zeta'_{n+k}$, $k \geqslant 1$. By the definition of ζ'_{n+k}, $C = (T_{A_{n+k}})^{-p}(A_{n+k+1})$ for some $p \geqslant 0$. If $x \in C$, $y \in A_{n+k}$, $y \overset{\xi_n}{\sim} x$, then $y \overset{\xi_{n+k-1}}{\sim} x$ and, by (iii), $T_{A_{n+k}}x \overset{\xi_{n+k-1}}{\sim} T_{A_{n+k}}y$. By (iv), $T_{A_{n+k}}x$ is not in A_{n+k+1} unless $T_{A_{n+k}}y$ is. So, after p such steps, we get $T^p_{A_{n+k}}y \in A_{n+k+1}$, that is, $y \in C$. Q.E.D.

REMARK. It follows from Definitions 2.1 - 2.3 that transformations introduced here are defined on some subsets of X. We are interested only in the cases when domains of such transformations are dense subsets of the second category.

3. THE TRANSVERSAL REALIZATION THEOREM.

Here we give some conditions sufficient for a given m.p. transformation to assume a Markov realization with a prescribed tail partition.

THEOREM 1. Let T be an automorphism of a Lebesgue space (X,μ) and ψ a tame partition of X. Suppose there exist a decreasing sequence of partitions $\xi_n \in P_f(X)$, $n \in \mathbb{N}$, with intersection ψ, and a decreasing sequence of measurable sets A_n, $n \in \mathbb{N}$ ($A_1 = X$), $\bigcap_n A_n = \emptyset$ (mod 0), such that

(a) $\psi \prec \omega(T)$

(b) for $n \in \mathbb{N}$ there is such $h_n \in \mathbb{N}$ that

$A_n = \bigcup\{T^{-j}A_{n+1} \mid 0 \leq j < h_n\}$ (mod 0)

(c) $\xi_n \prec \tau_n = \tau(\zeta(X, A_{n+1}, T^{-1}))$, $n \in \mathbb{N}$

(d) $\xi_n \cap A_{n+1}$ is invariant under $T_{A_{n+1}}$, $n \in \mathbb{N}$

(e) A_{n+2} is a $\xi_n \cap A_{n+1}$-set, $n \in \mathbb{N}$

Then there exists a Markov compactum structure on X, compatible with (ξ_n) and making T a transversal transformation, that is, there exist a set $\bar{X} \subset X$, $(X \setminus \bar{X}) = 0$, and partitions $\eta_n \in P_f(\bar{X})$, $n \in \mathbb{N}$, such that

(i) $\xi_n = \bigvee_{i=n+1}^{\infty} \eta_i$, $n \in \mathbb{N}$

(ii) η_n generate a totally disconnected topology on $\bar{X}$

(iii) the sequence (η_n) defines a Markov compactum structure on $\bar{X}$ with the tail partition equal to ψ (mod 0)

(iv) T induces a transversal transformation on $\bar{X}$.

PROOF. At first take some $\sigma_n \in P_f(X)$, $\sigma_n \nearrow \varepsilon$. Partitions η_n will be defined inductively.

<u>First step</u>. Take a tower ζ_1 and $\tau_1 \in P_f(X)$ (see (2.4)). Find some $\eta_1^o \in P_f(A_2)$ which is complementary to $\xi_1 \cap A_2$ and set

$$\eta_1 = \pi_{\tau_1}^{-1}(\eta_1^o) \vee \zeta_1 .$$

<u>General step</u>. Assume that for $n > 1$ we have already partitions η_i, $1 \leq i \leq n-1$, such that $\varkappa_k = \bigvee_{i=1}^{k} \eta_i$, $1 \leq k \leq n-1$, satisfy (iii), (iv) from proposition 2.1. We shall define η_n such that $\eta_n \prec \xi_{n-1}$, η_n/ξ_{n-1} be complementary to ξ_n/ξ_{n-1} and η_n majorize some θ_{n-1}:

Let us say that $C, D \in \xi_{n-1}$ are of the same combinatorial type if there exists a bijection $\varphi: C \to D$ such that $\varphi(E \cap C) = E \cap D$ for any $E \in \sigma_n$, $\varphi(\eta_i \cap C) = \eta_i \cap D$ $(i = 1, \ldots, n-1)$ and φ preserves the order induced on each η_i by the linear order on $\zeta_i \prec \eta_i$. We denote by θ_{n-1} the partition of X into the classes of those $C \in \xi_{n-1}$ which belong to the same combinatorial type.

We continue by taking some partition $\eta_n^{oo} \in P_f(A_{n+1}/(\xi_{n-1} \cap A_{n+1}))$ which is complementary to $(\xi_n/\xi_{n-1}) \cap A_{n+1}$ (note that A_{n+1} is a

$(\xi_{n-1} \cap A_n)$ -set, by (e)). Then put

(3.1) $$\eta_n^o = \pi_{\tau_n'}^{-1}(p_n^{-1}(\eta_n^{oo})) \vee \zeta_n',$$

(3.2) $$\eta_n = \pi_{\tau_n}^{-1}(\eta_n^o) \vee \theta_{n-1},$$

where $p_n : A_{n+1} \longrightarrow A_{n+1}/(\xi_{n-1} \cap A_{n+1})$ is a natural projection.

Verification of the properties of (η_n).

(i) The relation $p_n^{-1}(\eta_n^{oo}) \prec \xi_{k-1} \cap A_{k+1}$, $k \geq 2$, is evident. Substituting n by $k-1$ in (d) and (e), we get $\zeta_k' \prec \xi_{k-1} \cap A_k$. From the evident relation $\tau_k' \succ \xi_k \cap A_k \succ \xi_{k-1} \cap A_k$ and (3.1) we conclude that $\eta_k^o \prec \xi_{k-1} \cap A_k$. Since $\tau_{k-1} \succ \xi_{k-1}$ and $\pi_{\tau_{k-1}}^{-1}(A_k) = X$, we have $\pi_{\tau_{k-1}}^{-1}(\eta_k^o) \prec \xi_{k-1}$ which, together with $\theta_{k-1} \prec \xi_{k-1}$, gives $\eta_k \prec \xi_{k-1}$. By monotonicity of (ξ_n), we have $\eta_k \prec \xi_n$ for $k > n$ and $\xi_n \succ \bigvee_{k=n+1}^{\infty} \eta_k$.

To show the opposite inequality observe that $\eta_k \succ \theta_{k-1}$ and that, by the definition of θ_{k-1}, $\eta_1 \vee \dots \vee \eta_{k-1} \vee \theta_{k-1} \succ \sigma_k$. Hence $\bigvee_{i=1}^{k} \eta_i \succ \sigma_k$, $k \geq 2$, and $\bigvee_{i=1}^{\infty} \eta_i \succ \lim \sigma_k = \varepsilon$. Passing to X/ξ_n, we can, by similar arguments, show that $\bigvee_{i=n+1}^{\infty} \eta_i \succ \xi_n$.

(ii) Consider elements of η_n, $n \in \mathbb{N}$, and a topology generated by them (evidently, totally disconnected). Since $\bigvee_n \eta_n = \varepsilon$, the mapping $\gamma : x \longrightarrow (\eta_n(x))_{n \in \mathbb{N}}$ from X to $Y = \prod_n \eta_n$ is an injection defined on a subset $\bar{X}$ of a total measure. Elements of η_n define cylinders in Y, so γ is a homeomorphism if we consider the weak topology in Y. It is easy to see that $\gamma(\bar{X})$ is closed in Y.

(iii) We know already that conditions (i) - (iii) of Prop. 2.1 are true, and show (iv). It follows from the definition of θ_{n-1} that every its element, say D, has a product structure: $D \cong D/\varkappa_{n-1} \times D/\xi_{n-1}$. Since $\theta_{n-1} \prec \eta_n \prec \xi_{n-1}$, every element of η_n also has a similar product structure, $C \cong C/\varkappa_{n-1} \times C/\xi_n$ (see (3.2)).

(iv) We have to verify that "coordinates" $\eta_i(Tx)$ can be traced from $(\eta_i(x))_{i \in \mathbb{N}}$ according to the rules (2.2).

LEMMA. Let $n,k \in \mathbb{N}$, $E_i \in \eta_{n+i}$, $0 \leq i \leq k+1$ – cylinders with nonvoid intersection and such that $E_{n+i} \cap A_{n+i} \subset A_{n+i+1}$, $0 \leq i < k$, $E_{n+k} \cap A_{n+k} \not\subset A_{n+k+1}$. Then $\eta_j(Tx)$ does not depend on $x \in \bigcap_{i=0}^{k+1} E_{n+i}$, $A_{n-1} = F$, $n \leq j \leq n+k$.

Proof. We see that $\min\{i \mid \eta_i(x) \neq \eta_i(Tx)\} = n+k$, so $x \in A_{n+k} \setminus A_{n+k+1}$ and $\xi_{n+k+1}(x) = \xi_{n+k+1}(Tx)$ $(= C_x)$. Let $y \in F$ and $C_y = \xi_{n+k+1}(y)$. Then, for $1 \leq i \leq n+k+1$, $x \overset{\eta_i}{\sim} y$. Since $\eta_{n+k+1} \succ \theta_{n+k}$, it follows that $C_x \overset{\theta_{n+k}}{\sim} C_y$. By complementarity of $\varkappa_{n+k}$ and ξ_{n+k}, we claim that $\varkappa_{n+k} \cap C_x = \varepsilon$, $\varkappa_{n+k} \cap C_y = \varepsilon$, so x and y are uniquely determined by (η_i) and actually coincide up to θ_{n+k} - equivalence. From the definition of θ_{n+k}, the same can be said about Tx and Ty. Q.E.D.

We shall call $D \in \eta_k$ final if $D \cap A_k \subset A_{k+1}$. To define the gluing mappings $\varphi^n_{E\bar{E}}$ (and Markov order) consider two cases.

a) $E \in \eta_{n+1}$ is not final, $\bar{E} \in \eta_{n+1}$, $\bar{E} \cap A_{n+1} = T_{A_{n+1}}(E \cap A_{n+1})$ $(n \in \mathbb{N})$. It follows from (3.1),(3.2) that for $D \in \eta_n$ which is not final, $T_{A_n}(D \cap A_n) \in \eta_n \cap A_n$. Since each $E \in \eta_{n+1}$ is a ξ_n -set and, consequently, a τ'_n -set, we have $T_{A_n}(D \cap E) \in \eta_n \cap E$. Hence $\eta_n \cap A_n \cap E$ is divided into T_{A_n} -orbits $q^n_{E,j}$ (cf. Def.2.2).

Let $k_E = \max\{\operatorname{card} C \mid C \in \xi_n \cap E \cap A_{n+1}\}$. By the choice of η^{oo}_n, we see that $\eta_n \cap E \cap A_{n+1}$ is complementary to $\xi_n \cap E \cap A_{n+1}$, hence E intersects k_E elements of η_n. By (d) and (3.1), so does $\bar{E}$.

Now take such a final $D \in \eta_n$ that $D \cap E \neq \emptyset$, and some $x \in D \cap E \cap A_n$. Then $T_{A_n}x \in \bar{E}$, and $\eta_n(T_{A_n}x)$ does not, by Lemma, depend on x. So the formula

(3.3) $$\varphi^n_{E\bar{E}}(D) = \eta_n(T_{A_n}x)$$

defines a mapping from $f^n_E = \{D \in \eta_n \mid D \cap E \cap A_{n+1} \neq \emptyset\}$ to $s^n_{\bar{E}} = \{\bar{D} \in \eta_n \mid \bar{D} \cap \bar{E} \neq \emptyset,\ T^{-1}_{A_n}(\bar{D} \cap A_n) \subset A_{n+1}\}$.

b) $E \in \eta_{n+1}$ is final, but there exist $k > 0$ and such sets $E_i, \bar{E}_i \in \eta_{n+i}$ $(E_1 = E,\ \bar{E}_1 = \bar{E})$, $1 \leq i \leq k+1$, that E_i are final except E_{n+k+1}, $\varphi^n_{E_{i+1}\bar{E}_{i+1}}: f^n_{E_{i+1}} \longrightarrow s^n_{\bar{E}_{i+1}}$ are defined for $1 \leq i \leq k$ and $\bar{E}_i = \varphi_{E_{i+1}\bar{E}_{i+1}}(E_i)$.

Applying the Lemma once more, we conclude that the formula (3.3), where now $D \in \eta_n$, $D \cap E \neq \emptyset$, $x \in D \cap E_1 \cap \ldots \cap E_{n+k+1} \cap A_n$, correctly defines the mapping $\varphi^n_{E\bar{E}}: f^n_E \to s^n_{\bar{E}}$. By the arguments similar to those exploited in the case a), it is a bijection. One can check that when we pass from x to Tx, coordinates $\eta_i(x) = x_i$ change according to the rules (2.2). Q.E.D.

REMARK. We do not know whether for every T and tame ψ there exist (ξ_n) and (A_n) satisfying the conditions of our theorem unless $\psi = \omega(T)$, in which case we find ourselves in the conditions of the theorem of [14].

EXAMPLE 3.1. Let X be a stationary dyadic compactum with $D_i \equiv \{0,1\}$, $m^i_{de} \equiv 1$, $\mu = \prod_1 (1/2,\ 1/2)$. Let us introduce two different Markov orderings given by two lists of chains:

(T) $\qquad q^i_0 = q^i_1 = (0,1)$

(S) $\qquad q^i_0 = (0,1),\ q^i_1 = (1,0)$

We omit the index j because it can take only one value; $\varphi^i_{e\bar{e}}$ degenerate.

These orderings define m.p. transformations T and S. T is actually adic, ergodic and has a dyadic spectrum. If we take (ξ_n) and (A_n) defined by (2.1),(2.3) with S instead of T, then the algorithm of construction of a transversal realization for S, given by Theorem 1, leads us to T, whence S and T are isomorphic.

4. THE REALIZATION OF ACTIONS OF AMENABLE GROUPS.

Here we show that the class of generalized transversal transformations is wide enough for our purposes.

THEOREM 2. Let G be a countable group acting on a Lebesgue space X by m.p. transformations with tame orbit partition. Then one can introduce a Markov compactum structure on a subset $\bar{X} \subset X$, $\mu(X \setminus \bar{X}) = 0$, in such a way that elements of G act as generalized transversal transformations of X.

COROLLARY. If G is a countable amenable group acting on X by m.p. transformations, then this action can be represented by generalized transversal transformations of some Markov compactum.

PROOF. By virtue of [3], this action satisfies conditions of Theor.2.

PROOF of the theorem. For an action with a tame orbit partition one can show that the following property is fulfilled:

for any $\varepsilon > 0$ and finite subset $K \subset G$ there exists such a $\xi \in P_c(X)$ that

$$\int_{X/\xi} \mu_C\Big(\bigcap_{g \in K} gC\Big)\, d\mu_\xi(C) > 1 - \varepsilon,$$

where μ_ξ is a projection of μ onto X/ξ, μ_C is a normalized conditional measure on $C \in \xi$.

This property has been demonstrated, for example, in [12]. In the case of amenable group actions it follows directly from [3]. We shall call $\xi = \xi(\varepsilon, K)$ a Følner partition.

It is easily seen that if we have a sequence $\varepsilon_n \searrow 0$ and an increasing sequence K_n, $\bigcup_n K_n = G$, then $\xi_n = \xi(\varepsilon_n, K_n)$ can be chosen to form a decreasing sequence with $\bigcap_n \xi_n = \omega(G)$. For such a sequence of Følner partitions we shall construct a sequence (η_n) which defines a Markov structure on X.

<u>Construction</u>. Take some $\sigma_n \in P_f(X)$ $(n \in \mathbb{N})$, $\sigma_n \nearrow \varepsilon$. Let us enumerate elements in every σ_n and define functions $\nu_n : X \to \mathbb{N}$ assigning to each x a number of the element $\sigma_n(x)$ (cf. a notion of universal projection in [8],[9]). These functions are measurable.

Fix some n and consider any $C \in \xi_n$. Let $C^g = \{x \in C \mid gx \in C\}$, $g \in K_n$. We call $C_1, C_2 \in \xi_n$ (K_n, σ_n)-equivalent if there is an isomorphism $i^n_{C_1C_2} : C_1 \to C_2$, such that $i^n_{C_1C_2}(C_1^g) = C_2^g$, $\nu_n \cdot i^n_{C_1C_2} = \nu_n$, $g i^n_{C_1C_2}(x) = i^n_{C_1C_2}(gx)$ for $g \in K_n$, $x \in C_1^g$. Since $\operatorname{ess\,sup}_{\xi_n} \operatorname{card} C < \infty$, there is only finitely many (K_n, σ_n)-equivalency classes which form a partition θ_n of X. For each element $D \in \theta_n$ there exists a set C_D and a family of 1-1 mappings $i_C : C \in \xi_n \cap D \to C_D$ such that $i^n_{C_1C_2} = (i_{C_2})^{-1} i_{C_1}$, $C_1, C_2 \in \xi_n \cap D$.

Define ζ_D as a partition of D into the sets $\{i_C^{-1}(z) \mid C \in D \cap \xi_n\}$, $z \in C_D$. We can choose i_C so that ζ_D be measurable. Observe that ζ_D is complementary to ξ_n and these two partitions introduce a product structure in D: $D = D/\zeta_D \times D/\xi_n$, all elements of K_n acting in the same way in all $C \in \xi_n \cap D$. Finally, we define ζ_n as a union of partitions ζ_D, $D \in \theta_n$. By the definition of θ_n, we have $\zeta_n \succ \sigma_n$.

At the first step we simply put $\eta_1 = \zeta_1$. If $\eta_1, \ldots, \eta_{n-1}$ are already given and $\varkappa_k = \eta_1 \vee \ldots \vee \eta_k$ is complementary to ξ_k, $1 \leq k < n$, we take ξ_n and construct ζ_n as above, imposing one more restriction on the choice of isomorphisms i_C : when building θ_n, consider $(K_n, \sigma_n \vee \theta_{n-1})$ -equivalence instead of (K_n, σ_n) -equivalence (which is defined in the obvious way) and take care that the isomorphisms $i^n_{C_1C_2}$ should extend the corresponding isomorphisms defined on the preceding step.

Then put $\varkappa_n = \zeta_n$. Our additional restrictions on ζ_n lead to the relation $\varkappa_{n-1} \prec \varkappa_n$. The product character of elements of θ_n makes it possible to define $\eta_n = \varkappa_n \wedge \xi_{n-1}$ and prove that $\varkappa_n = \varkappa_{n-1} \vee \eta_n$.

The <u>verification</u> of the properties of (η_n) is easy. To show relations $\bigvee_{k=1}^{\infty} \eta_k = \varepsilon$, $\bigvee_{k=n+1}^{\infty} \eta_k = \xi_n$ it is sufficient to check the inequality $\varkappa_n \succ \sigma_n$ and then repeat the arguments used in the proof of Theorem 1. We see that, by our construction, each element $g \in G$ belongs to some K_n and acts on each $D \in \theta_n$ as a product of a (partial) permutation and an identity. This property is retained by g when n is increased (stabilization). So g is approximately cylindrical and hence generalized transversal. Q.E.D.

BIBLIOGRAPHY

1. R.M.Belinskaja, Partitionings of a Lebesgue space into trajectories defined by ergodic automorphisms, Funkcional. Anal. i Priložen. 2(1968), N°3, 4-16 (Russian).
2. H.Dye, On groups of measure preserving transformations,I,II, Amer. J. Math. 81(1959),119-159; 85(1963),551-576.
3. A.Connes, J.Feldman, B.Weiss, An amenable equivalence relation is generated by a single transformation, Ergod. Th. & Dynam. Sys. 1(1981), 431-450.
4. Sh.Ito, A construction of transversal flows for maximal Markov automorphisms, Second Japan-USSR Symp. on Probab. Theory, Kyoto, v.3(1972), 12-17.
5. M.Pimsner, D.Voiculescu, Imbedding the irrational rotation C^*-algebra into an AF-algebra, J. Operator Theory 4(1980), 201-210 .
6. M.Pimsner, Embedding some transformation group C^*-algebras into AF-algebras,
7. A.M.Vershik, On lacunary isomorphism of sequences of dyadic partitions, Funkcional. Anal. i Priložen. 2(1968), N°3, 17-21 (Russian).
8. A.M.Vershik, Decreasing sequences of measurable partitions and their applications, DAN SSSR 193(1970)(Russian) = Sov.Math.Dokl. 11(1970), 1007-1011.
9. A.M.Vershik, A continuum of pairwise nonisomorphic dyadic sequences, Funkcional. Anal. i Priložen.5(1971),N°3,16-18 (Russian).
10. A.M.Vershik, Four definitions of a scale of an automorphism, Funkcional. Anal. i Priložen. 7(1973), N°3, 1-17 (Russian).
11. A.M.Vershik, Nonmeasurable partitions, a theory of orbit partitions, and operator algebras, DAN SSSR 199(1971),1004-1007 (Russian) = Sov. Math. Dokl. 12(1971),1218-1222.
12. A.M.Vershik, Approximation in measure theory, Dissertation, Leningrad, 1974.
13. A.M.Vershik, Uniform algebraic approximation of shift and multiplication operators, DAN SSSR 259(1981),526-529 (Russian) = Sov. Math. Dokl. 24(1981), 97-100.
14. A.M.Vershik, A theorem on Markov periodic approximation in ergodic theory, Zap.Nauch.Sem.LOMI 115(1982), 72-82 (Russian) = Ergod.Th. and Related Topics, Math.Research 12(1982), 195-206.

DEPARTMENT OF MATHEMATICS & MECHANICS, LENINGRAD STATE UNIVERSITY

REMARKS ON PSEUDONORMALCY

Roberto Longo (*)

Dipartimento di Matematica, Università di Roma "La Sapienza"

I - 00185 Roma / Italy

A particular aspect of the theory of Standard Inclusions of von Neumann Algebras [6] concerns the relationship between normalcy properties of subalgebras of von Neumann algebras and the analytical structure given by the Tomita-Takesai theory [10]

Recall that a (semi)-standard W*-inclusion is a triple $\Lambda = (A,B,\Omega)$ when $A \subset B$ are von Neumann algebras and Ω is a cyclic separating vector for A,B and $A' \wedge B$ (cyclic and separating for $A' \wedge B$). Λ is said to be pseudo-normal if there exists a unique von Neumann algebra N_Λ between A and B which is globally stable for the action implemented by the modular conjugation $J = J_\Lambda$ associated with $A' \wedge B, \Omega$. Therefore

(1) $$N_\Lambda = A \vee JAJ = B \wedge JBJ$$

namely (1) extends the formulas defining the natural type I factor associated with a standard split W*-inclusion (Λ splits if $A \vee B'$ is naturally isomorphic to $A \otimes B'$) [6].

Note that N_Λ depends on Ω only up to conjugacy by a unitary in $A' \wedge B$, thus pseudo-normalcy is an algebraic property of the pair (A,B). The normalcy problem [6] proposes to characterize the pseudonormalcy of Λ by the normalcy and conormalcy of A in B (A is conormal in B if B' is normal in A') provided $A' \wedge B$ is a factor.

Indeed normalcy plays a crucial role in the theory. For example we have showed in [9] that if Λ is standard and A is normal and conormal in B then the unitary $\Gamma = J_A J_B$ implements a mixing endomorphism of B, provided A,B are factors.

This note exemplifies the applicability of the pseudonormalcy techniques by a computation made in the Free Scalar Massless Field Theory.

Let $A(\mathcal{O})$, $\mathcal{O} \subset \mathbb{R}^4$, be the net of local von Neumann algebras associated with the Free Massless Scalar Field Theory. Denote by $\mathcal{O}^t$ the time-like complement of a region $\mathcal{O} \subset \mathbb{R}^4$, namely $\mathcal{O}^t \equiv \{x \in \mathbb{R}^4,\ (x-y)^2 > 0,\ y \in \mathcal{O}\}$ where $x^2 \equiv x_0^2 - x_1^2 - x_2^2 - x_3^2$. The next proposition shows that, although time-like duality holds for double cones (with the proper definition for $R(\mathcal{O})$) [7] and forward cones [1], the relative time-like duality fails.

(*) Supported by CNR-GNAFA and Ministero della Pubblica Istruzione.

1 PROPOSITION With $A(\mathcal{O})$ as above, let $\mathcal{O}_1$ be the unit double cone and $\mathcal{O}_2 = r\mathcal{O}_1$, $r>0$. . We have

(2) $$A(\mathcal{O}_1)' \wedge A(\mathcal{O}_2) \supsetneqq A(\mathcal{O}_1^t \cap \mathcal{O}_2).$$

PROOF Let $\Lambda = (A(\mathcal{O}_1), A(\mathcal{O}_2), \Omega)$ where Ω is the vacuum vector. By the Reeh-Schlieder theorem Ω is a standard W*-inclusion. By Buchholz theorem [2] (see also [3]) Λ splits. This entail that $\Lambda^c = (A(\mathcal{O}_1)' \wedge A(\mathcal{O}_2), A(\mathcal{O}_2), \Omega)$ is a pseudo-normal standard W*-inclusion [6, prop. 7.3] . Since Λ splits $A(\mathcal{O}_1)$ is the relative commutant of $A(\mathcal{O}_1)' \wedge A(\mathcal{O}_2)$ in $A(\mathcal{O}_2)$, therefore $J = J_{\Lambda^c}$ is the modular conjugation associated with $A(\mathcal{O}_1)$, Ω . By [7] J is the ray inversion operator (multiplied by the time-reversal operator). By the pseudomormalcy condition (1), the equality in (2) would entail

(3) $$\begin{aligned} R_1 &\equiv A(\mathcal{O}_a) \vee A(\mathcal{O}_b) \vee A(-\mathcal{O}_a) \vee A(-\mathcal{O}_b) = \\ &= A(\mathcal{O}_2) \wedge A((\tfrac{1}{r}\mathcal{O}_1)^t) \equiv R_2 \end{aligned}$$

where $\mathcal{O}_a$, $\mathcal{O}_b$ are the double cones with vertices on (1,0), (r,0) and (r^{-1},0), (1,0). With $\bar{\mathcal{O}}$ the double cone spanned by $\mathcal{O}_a \cup \mathcal{O}_b$ i.e. with vertices (r^{-1},0) and (r,0), we have

$$A(\mathcal{O}_a) \vee A(\mathcal{O}_b) \subset A(\bar{\mathcal{O}})$$

$$R_1 \subset A(\bar{\mathcal{O}}) \vee A(-\bar{\mathcal{O}}) \subset R_2;$$

by the split property and time-like commutativity, $A(\bar{\mathcal{O}})$ $A(-\bar{\mathcal{O}})$ is naturally isomorphic to $A(\bar{\mathcal{O}}) \times A(-\bar{\mathcal{O}})$; therefore the equality (3) would imply

$$A(\bar{\mathcal{O}}) = A(\mathcal{O}_a) \vee A(\mathcal{O}_b)$$

which is not possible [7].

The interest of the above proposition relies on the fact that the relative space-like duality holds, namely

(4) $$A(\mathcal{O}_1)' \wedge A(\mathcal{O}_2) = A(\mathcal{O}_1' \cap \mathcal{O}_2).$$

It has long been realized the equality (4) characterizes the theories without superselection sectors, see [4]. This fact can be checked under the split assumtion [6, Cor 9.5] , being related to the construction of the local charges [4,5].

For those who are tempted to attack the normalcy problem, we mention that normalcy and conormalcy are not equivalent properties, as shown the following example constructed by S. Popa following a suggestion of R.V. Kadison.

Let R_i (i = 1, 2, 3) be II_1-factors and put

$B \equiv (R_1 \otimes R_2) * R_3$

$A_1 \equiv (R_1 \otimes 1) * 1$

$A_2 \equiv (1 \otimes R_2) * 1$

where * denoted the free product. Then $A_1' \wedge B = A_2$ and $A_2' \wedge B = A_1$, thus A_1 is normal in B. On the other hand $A_1 \vee A_2 \neq B$, thus A_1 is not conormal in B.

ACKNOWLEDGEMENTS. We would like to thank Professor D. Buchholz and R. Haag for the hospitality extended to us at the University of Hamburg during October-November 1983 where this note has been completed.

REFERENCES

1 BUCHHOLZ, D.: in Proc. of the Int. Conf. on "Operator Algebras Ideals and their Applications in Theoretical Physics", Baumgärtel, Lassner, Pietch, Uhlmann eds., Leipzig: Teubner Verlagsgesellschaft 1978

2 BUCHHOLZ, D.: "Product States for Local Algebras" - Commun. Math. Phys. 85 (1982) 73

3 D'ANTONI, C., LONGO, R.: "Interpolation by Type I Factors and the Flip Automorphism" - J. Funct. Anal. 51 (1983) 361

4 DOPLICHER, S.: "Local Aspects of Superselection Rules" - Commun. Math. Phys. 85 (1982) 73

5 DOPLICHER, S., LONGO, R.: "Local Aspects of Superselection Rules" - Commun. Math. Phys. 88 (1983) 399

6 DOPLICHER, S., LONGO, R.: "Standard and Split Inclusions of von Neumann Algebras" - Invent. Math. (to appear)

7 HISLOP, P.D., LONGO, R.: "Modular Structure of the Local Algebras Associated with the Free Massless Scalar Field Theory" - Commun. Math. Phys. 84 (1982) 71

8 KADISON, R.V., POPA, S.: Private Communications

9 LONGO, R.: "Solution of the Factorial Stone-Weierstrass Conjecture. An Application of the Theory of Standard Split W*-inclusions" - Invent. Math. (to appear)

10 TAKESAKI, M.: "Tomita's Theory of Modular Hilbert Algebras and its Applications". Lecture Notes in Math. 128, Springer-Verlag, 1970

Groupoid Dynamical Systems and Crossed Product

Tetsuya MASUDA

Research Institute for Mathematical Sciences
Kyoto University, Kyoto 606, JAPAN

Abstract

By analogy with W*-dynamical system, we define a W*-groupoid dynamical system (M,Γ,ρ) where M is a W*-algebra, Γ is a locally compact measured groupoid, and $\rho:\Gamma \to \mathrm{Aut}(M)$ is a continuous groupoid homomorphism. The groupoid crossed product $M\times_\rho\Gamma$ is defined by making use of the non-commutative integration theory of A. Connes i.e. integration theory over singular quotient spaces, and is shown to have similar properties as the case of a group action. As a special case of this situation, if ρ is a continuous homomorphism from Γ to a locally compact group G, we obtain groupoid dynamical system $(L^\infty(G),\Gamma,\rho)$. In this case, there exists a co-action $\hat{\rho}$ of G on $\mathrm{End}_\Lambda(\Gamma)$ and the groupoid crossed product $L^\infty(G)\times_\rho\Gamma$ is isomorphic to the co-crossed product $\mathrm{End}_\Lambda(\Gamma)*_{\hat{\rho}}G$ of $\mathrm{End}_\Lambda(\Gamma)$ by G in the sense of Nakagami and Takesaki. Similar results hold for the C*-algebraic framework. This note is a short report on [7], [8].

§1. Direct Integral of Hilbert Spaces over Singular Spaces

In this section, we describe the concept of a direct integral of Hilbert spaces over singular space under a certain condition. Our scheme of construction and notion are the same as those of Connes [3], Kastler [6], and Bellissard-Testard [1]. Throughout this section, we fix a measured groupoid Γ with a Haar system (ν,Λ,δ). We use the notation "Hilb" to denote the category of Hilbert spaces with unitary mappings as morphisms.

Definition 1.1. We call $H=\{H_x\}_{x\in\Gamma^{(0)}}$ a Hilbert Γ-bundle if $H=\{H_x\}_{x\in\Gamma^{(0)}}$ is a measurable family of Hilbert spaces over $\Gamma^{(0)}$ and there exists a measurable covariant functor $U:\Gamma\to \text{Hilb}$ with $U(x)=H_x$. It means that

(i) if $\gamma\in\Gamma_x^y$, then $U(\gamma)$ is a unitary mapping of H_x onto H_y,

(ii) $U(\gamma)U(\gamma_1)$ if γ and γ_1 are composable,

(iii) $\gamma\mapsto(U(\gamma)\xi_{s(\gamma)},\eta_{r(\gamma)})$ is measurable for measurable sections $\xi=\{\xi_x\}_{x\in\Gamma^{(0)}}$, $\eta=\{\eta_x\}_{x\in\Gamma^{(0)}}$ of $H=\{H_x\}_{x\in\Gamma^{(0)}}$.

Example 1.2. Let $H_x=L^2(\Gamma^x,\nu^x)$ with $U(\gamma):L^2(\Gamma^{s(\gamma)},\nu^{s(\gamma)})\to L^2(\Gamma^{r(\gamma)},\nu^{r(\gamma)})$ defined by $[U(\gamma)\xi](\tilde{\gamma})=\xi(\gamma^{-1}\tilde{\gamma})$, $\xi\in L^2(\Gamma^{s(\gamma)},\nu^{s(\gamma)})$. Then we obtain a Hilbert Γ-bundle which we call canonical Γ-bundle.

<u>Definition 1.3.</u> A measurable section $\xi = \{\xi_x\}_{x \in \Gamma^{(0)}}$ of a Hilbert Γ-bundle $H = \{H_x\}_{x \in \Gamma^{(0)}}$ is said to be <u>covariant</u> if $U(\gamma)\xi_y = \delta(\gamma)^{1/2}\xi_x$, $\gamma \in \Gamma_y^x$, for almost all $\gamma \in \Gamma$ with respect to $(\Lambda_\nu \circ \nu)$.

From now on, we are going to deal with Hilbert Γ-bundles of the form $H^c \otimes H = \{H_x^c \otimes H_x\}_{x \in \Gamma^{(0)}}$ with unitaries $\{U^c(\gamma) \otimes U(\gamma)\}_{\gamma \in \Gamma}$ where $H^c = \{H_x^c\}_{x \in \Gamma^{(0)}}$ with unitaries $\{U^c(\gamma)\}_{\gamma \in \Gamma}$ is the canonical Γ-bundle and $H = \{H_x\}_{x \in \Gamma^{(0)}}$ with unitaries $\{U(\gamma)\}_{\gamma \in \Gamma}$ is a Hilbert Γ-bundle. Let $\xi = \{\xi(\gamma)\}_{\gamma \in \Gamma}$ for a measurable section of $\tilde{H} \equiv H^c \otimes H$, $\tilde{H}_x \cong L^2(\Gamma^x, H_x, \nu^x)$, where $\xi(\gamma)$ is a $H_{r(\gamma)}$-valued measurable function satisfying

$$(1.1) \quad \int_{\Gamma^x} \| \xi(\gamma) \|_x^2 \, d\nu^x(\gamma) < \infty$$

for almost all x $\Gamma^{(0)}$ with respect to Λ_ν. We define singular direct integral of the Hilbert Γ-bundle $\tilde{H}$ by

$$(1.2) \quad \int_{\Gamma^{(0)}/\sim}^{\oplus} \tilde{H}_* d\Lambda(*) \equiv \{ \text{ set of all } L^2\text{-covariant section } \xi = \{\xi(\gamma)\}_{\gamma \in \Gamma} \text{ of } \tilde{H} \text{ satisfying } \int_{\Gamma^{(0)}} \{ \int_{\Gamma^x} \| \xi(\gamma) \|_x^2 f(\gamma) \, d\nu^x(\gamma) \} d\Lambda_\nu(x) \text{ is finite } \},$$

where f is a partition of unity associated with ν i.e. $f \in \mathfrak{J}^+(\Gamma)$ such that $\int_{\Gamma^{r(\gamma)}} f(\tilde{\gamma}^{-1}\gamma)d^{\,r(\gamma)}(\tilde{\gamma}) = 1$ for almost all $\gamma \in \Gamma$ with respect to $(\Lambda_\nu \circ \nu)$. The square root of the value of the formula in the right hand side of (1.2) is defined to be the norm $\|\xi\|_\Lambda$ of $\xi = \{\xi(\gamma)\}_{\gamma \in \Gamma}$ which is shown to be independent of the choice of the partition of unity f.

<u>Theorem 1.4.</u> $\displaystyle\int_{\Gamma^{(0)}/\sim}^{\oplus} \tilde{H}_* d\Lambda(*) \cong \int_{\Gamma^{(0)}}^{\oplus} H_x d\Lambda_\nu(x)$.

§2. Crossed Product by Groupoid and Random Operator Field

In a close analogy with the W*-dynamical system defined by a locally compact group, we will define W*-groupoid dynamical system and the corresponding groupoid crossed product.

<u>Definition 2.1.</u> The triplet (M,Γ,ρ) is called W*-groupoid dynamical system (or W*-groupoid system for short) if M is a W*-algebra, Γ is a locally compact measured groupoid with a Haar system (ν,Λ,δ) and $\rho:\Gamma \to \mathrm{Aut}(M)$ is a continuous groupoid homomorphism.

The groupoid crossed product $M \times_\rho \Gamma$ associated with (M,Γ,ρ) is defined by the W*-algebra on $L^2(\Gamma,(\Lambda_\nu \circ \nu)) \otimes H$ generated by

(2.1) $[\pi(a)\xi](\gamma) = \rho_{\gamma^{-1}}(a)\xi(\gamma)$, $a \in M$,

(2.2) $[\lambda(f)\xi](\gamma) = \int_{\Gamma^{(0)}} f(\tilde{\gamma})\xi(\tilde{\gamma}^{-1}\gamma)d\nu^{r(\gamma)}(\tilde{\gamma})$, $f \in \mathfrak{A}(\Gamma)$,

where $\xi \in L^2(\Gamma,(\Lambda_\nu \circ \nu))\otimes H$ and $\mathfrak{A}(\Gamma)$ is a left Hilbert algebra associated with $\mathrm{End}_\Lambda(\Gamma)$ and H is a representation Hilbert space of M.

Now, we shall give a description of $M\times_\rho\Gamma$ using random operator fields. We choose H to be a standard representation Hilbert space and let $U(\gamma)$ be the canonical implementation of ρ_γ on H. Then the constant field $H_x = H$, $x \in \Gamma^{(0)}$ with unitaries $\{U(\gamma)\}_{\gamma\in\Gamma}$ gives Hilbert Γ-bundle. Now, we consider Hilbert Γ-bundle $\tilde{H}_x = H_x^c\otimes H_x^c\otimes H$, $x\in\Gamma^{(0)}$ with unitaries $\tilde{U}(\gamma) = U^c(\gamma)\otimes U^c(\gamma)\otimes U(\gamma)$, $\gamma\in\Gamma$, where $H^c = \{H_x^c\}_{x\in\Gamma^{(0)}}$ with unitaries $\{U^c(\gamma)\}_{\gamma\in\Gamma}$ is the canonical Γ-bundle (see Example 1.2.). Here, remember that $B(H_x^c)$ acts on $H_x^c\otimes H_x^c$ standardly.

Theorem 2.2. $M\times_\rho\Gamma$ is isomorphic to the W*-algebra of random operator fields:

(2.3) $\{T = \{T_x\}_{x\in\Gamma^{(0)}}$: essentially bounded measurable field of operators on $\Gamma^{(0)}$ such that $T_x \in B(H_x^c)\otimes M$ and $(\mathrm{Ad}_{U^s(\gamma)}\otimes\rho_\gamma)(T_x) = T_y$, $\gamma\in\Gamma_x^y$ $\}$,

where $U^s(\gamma) = U^c(\gamma)\otimes U^c(\gamma)$, $\gamma\in\Gamma$. In this expression, $\|T\| = \operatorname{ess.sup}_{x\in\Gamma^{(0)}}\|T_x\|$.

By Theorem 1.4, singular direct integral of Hilbert Γ-bundle $\tilde{H}$ is isomorphic to $L^2(\Gamma,(\Lambda_\nu \circ \nu)) \otimes H$ which is seen to be a standard representation Hilbert space of $M \times_\rho \Gamma$.

Next, we shall give some properties of our groupoid crossed product.

Proposition 2.3.

(1) If $\rho,\sigma:\Gamma \to \mathrm{Aut}(M)$ are mutually cohomologous in the sense that there exists a continuous mapping $\tau:\Gamma^{(0)} \to \mathrm{Aut}(M)$ such that $\rho_\gamma = \tau_{r(\gamma)} \circ \sigma_\gamma \circ \tau^{-1}_{s(\gamma)}$. Then $M \times_\rho \Gamma \cong M \times_\sigma \Gamma$.

(2) If $\rho,\sigma:\Gamma \to \mathrm{Aut}(M)$ are mutually one cocycle equivalent in the sense that there exists a strongly continuous unitary valued mapping $u:\Gamma \to M$ such that

(2.4) $\rho_\gamma(a) = u_\gamma \sigma_\gamma(a) u^*_\gamma \ , \ a \in M \ ,$

(2.5) $u_{\gamma_1\gamma_2} = u_{\gamma_1} \sigma_{\gamma_1}(u_{\gamma_2}) \ , \ s(\gamma_1) = r(\gamma_2)$

then $M \times_\rho \Gamma \cong M \times_\sigma \Gamma$.

(3) If Γ is the graph groupoid $X \times_\alpha G$ of a topological transformation group (X,G,α) with a non-singular Borel measure on X and $r(x,g) = x$, $s(x,g) = \alpha_{g^{-1}}(x)$ for $(x,g) \in \Gamma$, then $M \times_\rho \Gamma$ is isomorphic to a crossed product of $L^\infty(X) \otimes M$ by G with the action $\hat{\beta}$, where

(2.6). $\hat{\beta}_g[f](x) = \rho_{(x,g)}(f(\alpha_{g^{-1}}(x))) \ , \ f \in L^\infty(X) \otimes M \ , \ g \in G \ .$

Remark 2.4. If $\rho:\Gamma \to \mathrm{Aut}(M)$ is of G-split type in the sense that there exists a continuous homomorphism $\beta:G \to \mathrm{Aut}(M)$ such that the diagram

$$\text{(2.7)} \qquad \begin{array}{ccc} \Gamma & \xrightarrow{\rho} & \mathrm{Aut}(M) \\ p\downarrow & \nearrow_{\beta} & \\ G & & \end{array}$$

commutes, where $p:\Gamma \ni (x.g) \mapsto g \in G$ is the canonical projection, then the action of G on $L^{\infty}(X) \otimes M$ is of product type $\alpha \otimes \beta$.

We shall comment on the relation between groupoid crossed product and skew product. In fact, the graph groupoid of a dynamical system constructed from skew product is viewed as a special case of the graph groupoid of a transformation groupoid with groupoid which is a graph of some transformation group. We call the triple (Ω,Γ,ρ) locally compact transformation groupoid if ρ is a continuous action of locally compact groupoid Γ on a locally compact space Ω. Further, (Ω,Γ,ρ) is said to be measured if Γ is measured by a Haar system (ν,Λ,δ) and Ω is equipped with a Γ-quasi invariant Borel measure .

For a given locally compact transformation groupoid (Ω,Γ,ρ), we construct its graph as a locally compact groupoid $\tilde{\Gamma} = \Omega \times_{\rho} \Gamma$. The groupoid $\tilde{\Gamma}$ is defined as the product of Ω and Γ as a topological space with the unit space $\tilde{\Gamma}^{(0)} = \Omega \times \Gamma^{(0)}$ and $\{\tilde{\gamma} = (\omega,\gamma)\} = \tilde{\Gamma}$ is given by the following groupoid structure i.e. $r(\tilde{\gamma}) = (\omega, r(\gamma)) \in \tilde{\Gamma}^{(0)}$, $s(\tilde{\gamma}) = (\rho_{\gamma^{-1}}(\omega), s(\gamma)) \in \tilde{\Gamma}^{(0)}$ and $\tilde{\gamma}^{-1} = (\rho_{\gamma^{-1}}(\omega), \gamma^{-1})$, $\tilde{\gamma}_1\tilde{\gamma}_2 = (\omega_1, \gamma_1\gamma_2)$ for $\tilde{\gamma}_j = (\omega_j, \gamma_j)$, $j = 1,2$

with $\rho_{\gamma_1^{-1}}(\omega_1) = \omega_2$. The transverse function $\tilde{\nu} = \{\tilde{\nu}^{(\omega,x)}\}_{(\omega,x)\in\tilde{\Gamma}^{(0)}}$ of $\tilde{\Gamma}$ is naturally given by $\tilde{\nu}^{(\omega,x)}(\tilde{\gamma}) = \nu^x(\gamma)$ if $\tilde{\gamma} = (\omega,\gamma)$.
If (Ω,Γ,ρ) is measured, then a Haar system $(\tilde{\nu},\tilde{\Lambda},\tilde{\delta})$ of $\tilde{\Gamma}$ is given by the relations $\tilde{\Lambda}_{\tilde{\nu}} = \mu\otimes\Lambda_\nu$ and

$$(2.8)\quad \tilde{\delta}(\omega,\gamma) = \delta(\gamma)\frac{d\mu(\omega)}{d\mu\circ\rho_{\gamma^{-1}}(\omega)} \quad .$$

<u>Proposition 2.5. Let</u> (Ω,Γ,ρ) <u>be a locally compact measured transformation groupoid and</u> $\tilde{\Gamma} = \Omega\times_\rho\Gamma$ <u>be its graph</u>. <u>Then</u>, $\mathrm{End}_{\tilde{\Lambda}}(\tilde{\Gamma}) \cong L^\infty(\Omega,\mu)\times_\rho\Gamma$.

§3. Groupoid Crossed Product and Co-action

Here, we consider a W*-groupoid dynamical system (M,Γ,ρ) with $M = L^\infty(G)$, where G is a locally compact group together with the action ρ determined by the left translation of G through continuous groupoid homomorphism $\rho:\Gamma \to G$ i.e. $[\rho_\gamma(X)](g) = X(\rho(\gamma)^{-1}g)$, $X\in L^\infty(G)$. (This also gives an example of locally compact transformation groupoid.)

<u>Theorem 3.1. There exists a co-action</u> $\hat{\beta}$ <u>of</u> G <u>on</u> $\mathrm{End}_\Lambda(\Gamma)$ <u>i.e.</u> $\hat{\beta}$ <u>is an injective *-homomorphism</u> $\hat{\beta}:\mathrm{End}_\Lambda(\Gamma) \to \mathrm{End}_\Lambda(\Gamma)\otimes W_r^*(G)$ <u>such that the following diagram commutes:</u>

$$(3.1)\qquad \begin{array}{ccc} \mathrm{End}_\Lambda(\Gamma) & \xrightarrow{\hat\rho} & \mathrm{End}_\Lambda(\Gamma)\otimes W_r^*(G) \\ \downarrow{\scriptstyle\hat\rho} & & \downarrow{\scriptstyle\hat\rho\otimes 1} \\ \mathrm{End}_\Lambda(\Gamma)\otimes W_r^*(G) & \xrightarrow{1\otimes\delta_G} & \mathrm{End}_\Lambda(\Gamma)\otimes W_r^*(G)\otimes W_r^*(G) \end{array}$$

where $W_r^*(G)$ is the W*-algebra generated by the left regular representation $\lambda(g)$ of $g\in G$ and $\delta_G(\lambda(g)) = \lambda(g)\otimes\lambda(g)$, $g\in G$. Furthermore, $L^\infty(G)\times_\rho\Gamma \cong \mathrm{End}_\Lambda(\Gamma) *_{\hat\rho} G$, where $*_{\hat\rho}$ denotes the crossed product by co-action $\hat\rho$.

Remark 3.2. If $\rho,\sigma:\Gamma\to G$ are cohomologous, then the two co-actions $\hat\rho,\hat\sigma$ are cohomologous in the sense of Nakagami and Takesaki [9].

Remark 3.3. If we consider the special case that Γ and G are locally compact abelian groups, then $L^\infty(G)\times_\rho\Gamma \cong W_r^*(\Gamma)\times_{\hat\rho}\hat G \cong L^\infty(\hat\Gamma)\times_{\hat\rho}\hat G$. This duality can be viewed as the Plancherel transformation of abelian groupoid, see Bellissard-Testard [1].

§4. Examples

Here, we collect some examples.

Example 4.1. Let Γ be a locally compact measured groupoid with a Haar system (ν,Λ,δ) and assume $\log\delta:\Gamma\to\mathbb{R}$ is continuous. Then we obtain groupoid dynamical system $(L^\infty(\mathbb{R})$,

$\Gamma,\log\delta$) from locally compact transformation groupoid $(\mathbb{R},\Gamma,\log\delta)$ and its W*-groupoid crossed product $L^\infty(\mathbb{R})\times_{\log\delta}\Gamma \cong \mathrm{End}_{\tilde\Lambda}(\tilde\Gamma)$, $\tilde\Gamma = \mathbb{R}\times_{\log\delta}\Gamma$, is isomorphic to the modular crossed product of $\mathrm{End}_\Lambda(\Gamma)$. Actually, the $\mathbb{R}$-action given by $\log\delta$ is the modular action of the weight given by (ν,Λ,δ). The groupoid $\tilde\Gamma$ is known to be the Poincaré suspension of Γ (see [3], [10]).

<u>Example 4.2.</u> Let Γ be as above and further we assume that $\mathrm{End}_\Lambda(\Gamma)$ is hyperfinite III_1-factor (for example, see [2], [4]). Now, $(K,\mathbb{R},\theta)$ is a measure preserving ergodic $\mathbb{R}$-flow on a compact space K. Then $L^\infty(K)\times_{\theta\circ\log\delta}\Gamma \cong \mathrm{End}_{\tilde\Lambda}(\tilde\Gamma)$, $\tilde\Gamma = K\times_{\theta\circ\log\delta}\Gamma$ is a Krieger factor with the smooth flow of weight isomorphic to $(K,\mathbb{R},\theta)$. If we take $K = S^1$ and θ is the translational action of $\mathbb{R}$ on S^1 with period T. Then $\mathrm{End}_{\tilde\Lambda}(\tilde\Gamma) \cong \mathrm{End}_\Lambda(\Gamma)\times_\beta\mathbb{Z}$ and the action β is the restriction of the modular automorphism of $\mathrm{End}_\Lambda(\Gamma)$ to $\mathbb{Z}$. In this case, $\mathrm{End}_{\tilde\Lambda}(\tilde\Gamma)$ is hyperfinite factor of type III_λ, $\lambda = \exp(-2\pi/T)$.

<u>Example 4.3.</u> Let G be a locally compact group with closed subgroups H and K. We assume that the group has a continuous factorization $G \cong (G/K)\times K$ as a topological space. Let (M,K,α) be a W*-dynamical system and $(\tilde M,G,\tilde\alpha) = \underset{K\uparrow G}{\mathrm{ind}}(M,K,\alpha)$ (for the definition of induced action, see [11]). Let $\Gamma = (G/K)\times H$ be a locally compact measured groupoid defined by the topological transformation group $((G/K),H,\text{left multiplication})$. Then there exists a continuous action ρ of Γ on M such that $\tilde M\times_{\tilde\alpha}H \cong M\times_\rho\Gamma$. (The left hand side is a usual crossed product.)

§5. Concluding Remarks

The discussions in the C*-algebraic framework are just the same as those of W*-algebraic framework except for some minor points. The action of groupoid on W*-algebra is also discussed by Jones and Takesaki [5].

We can give some geometric interpretation for our groupoid crossed product in some particular cases. For example, our situation in Section 3 is closely related to the operation of lifting up the foliation on the base spase to the total space of principal fiber bundle. In this situation, the groupoid Γ corresponds to the holonomy groupoid of the foliation on the base space, the group G corresponds to the structure group of the principal fiber bundle, and the groupoid homomorphism $\rho:\Gamma \to G$ corresponds to (partially) flat connection on the bundle.

References

[1] Bellissard, J.-Testard, D., Almost periodic Hamiltonians: an algebraic approach, preprint 1981.

[2] Connes, A., The von Neumann algebra of a foliation, Lecture Notes in Phys., 80 (1978), Springer, 145-151.

[3] Connes, A., Sur la theorie non commutative de l'integration, Lecture Notes in Msth., 725 (1979), Springer, 19-143.

[4] Connes, A., A survey of foliations and operator algebras, Proc. Symp. Pure Math., 38 (1982), part 1, 521-628.

[5] Jones, V.-Takesaki, M., Actions of compact abelian groups on semifinite injective factors, preprint 1982.

[6] Kastler, D., On A. Connes' non-commutative integration theory, Comm. Math. Phys., 85 (1982), 99-120.

[7] Masuda, T., Groupoid dynamical systems and crossed product I -The case of W*-systems-, preprint 1983.

[8] Masuda, T., Groupoid dynamical systems and crossed product II -The case of C*-systems-, preprint 1983.

[9] Nakagami, Y.-Takesaki, M., Duality for crossed products of von Neumann algebras, Lecture Notes in Math., 731 (1979), Springer.

[10] Series, C., The Poincaré flow of a foliation, Amer. J. Math., 102 (1980), 93-128.

[11] Takesaki, M., Duality for crossed products and the structure of von Neumann algebras of typeIII, Acta Matn., 131 (1973), 249-310.

Z_2 - EQUIVARIANT K - THEORY

William L. Paschke

We show how the K - groups for the crossed product of a C^*-algebra by an action of Z_2 are related to those of the fixed point algebra and of the ideal in the fixed point algebra generated by products of elements in the (-1) - eigenspace.

For a fixed locally compact group G, the construction that associates to a C^* - dynamical system (A,G,α) the crossed product $A\times_\alpha G$ [6] is a functor from (G-dynamical systems ; equivariant *-homomorphisms) to (C^*-algebras ; *-homomorphisms). This functor is sufficiently well-behaved that its composition with $K_\#$ for C^*-algebras yields a pair of functors obeying equivariant versions of the rules for ordinary K-theory. (We remark that when G is compact and A is abelian, $K_\#(A\times_\alpha G)$ coincides with the Atiyah-Segal topological G-equivariant K-theory of the spectrum of A [8] [5] [4].) It is thus feasible to compute $K_\#(A\times_\alpha G)$ on an *ad hoc* basis in many instances. For special choices of G (e.g. $\mathbb{F}_n$, $\mathbb{R}$), there are also elegant results relating $K_\#(A\times_\alpha G)$ to $K_\#(A)$ [7] [3], but in general the sort of information in terms of which the K-groups of the crossed product can most conveniently be computed will depend on what sort of group G is. The treatment below of the case $G = Z_2$ illustrates what happens for finite cyclic groups except that here the bookkeeping complications are minimal.

1. Exact sequence for $K_{\#}(A\times_{\alpha}Z_2)$.

Henceforth, A will be a C^*-algebra and α will be an automorphism of A such that $\alpha^2 = id_A$. We let $A_0 = \{a \in A : \alpha(a) = a\}$, $A_1 = \{a \in A : \alpha(a) = -a\}$, and $J = A_1^2$ (the closed linear span of $\{xy : x, y \in A_1\}$). Notice that $A = A_0 + A_1$ via the projections $E_i : A \to A_i$ defined by $E_i(x) = \frac{1}{2}(x + (-1)^i\alpha(x))$ $(i = 0, 1)$. Also, A_1 is a 2-sided A_0-module, J is a closed 2-sided ideal of A_0, and $A_1 + J$ is a closed 2-sided ideal of A. The crossed product $Ax_{\alpha}Z_2$ consists of functions $f : Z_2 \to A$ with multiplication $(fg)(i) = f(0)g(i) + f(1)\alpha(g(i+1))$ and involution $f^*(i) = \alpha^i(f(i))^*$ $(i = 0, 1)$. It is straightforward to check that the map

$$f \longrightarrow \begin{pmatrix} E_0(f(0) + f(1)) & E_1(f(0) - f(1)) \\ E_1(f(0) + f(1)) & E_0(f(0) - f(1)) \end{pmatrix}$$

is an isomorphism of $A\, x_{\alpha}\, Z_2$ with the C^*-subalgebra $\begin{pmatrix} A_0 & A_1 \\ A_1 & A_0 \end{pmatrix}$ of $A \otimes M_2$.

In the theorem below $\mathcal{K}$ is the algebra of compact operators, and $M(\cdot)$ denotes the multiplier algebra. The maps $i_* : K_j(J) \to K_j(A_0)$ and $\partial : K_j(A_0/J) \to K_{1-j}(J)$ come from the cyclic exact sequence of K-groups produced by $0 \to J \to A_0 \to A_0/J \to 0$. (For this an related matters, see [9].) As will be shown in the proof of the theorem, the maps $K_j(A_0) \to K_j(A_0\, x_{\alpha}\, Z_2)$ come from $\begin{pmatrix} 0 & 0 \\ 0 & A_0 \end{pmatrix} \hookrightarrow \begin{pmatrix} A_0 & A_1 \\ A_1 & A_0 \end{pmatrix}$.

<u>Theorem 1</u>: Let A, α, A_0, A_1, and J be as above, and suppose that A_0 and J have strictly positive elements. There is an exact sequence

$$\begin{array}{ccccc} K_0(A_0) & \longrightarrow & K_0(A\times_\alpha Z_2) & \longrightarrow & K_0(A_0/J) \\ \gamma\uparrow & & & & \downarrow\gamma \\ K_1(A_0/J) & \longleftarrow & K_1(A\times_\alpha Z_2) & \longleftarrow & K_1(A_0). \end{array} \tag{1}$$

The vertical maps γ are obtained as follows. There is a unitary u in $\mathcal{M}((A_1 + J) \otimes \mathcal{K})$ such that

(*) u multiplies $A_1 \otimes \mathcal{K}$ to $J \otimes K$ and *vice-versa*.

Let ρ be the automorphism of $J \otimes \mathcal{K}$ obtained by restricting ad(u) to $J \otimes \mathcal{K}$. Then $\gamma = i_*\rho_*\partial$.

Any unitary multiplier of $(A_1 + J) \otimes \mathcal{K}$ satisfying (*) yields the same map ρ_* on $K_\#(J)$, and $\rho_* = \rho_*^{-1}$.

<u>Proof</u>: We begin by tensoring everything with $\mathcal{K}$, replacing A by $A \otimes \mathcal{K}$, α by $\alpha \otimes \mathrm{id}_{\mathcal{K}}$, and so forth. Consider the exact sequence

$$0 \longrightarrow \begin{pmatrix} J & A_1 \\ A_1 & A_0 \end{pmatrix} \longrightarrow \begin{pmatrix} A_0 & A_1 \\ A_1 & A_0 \end{pmatrix} \longrightarrow A_0/J \longrightarrow 0. \tag{2}$$

Using an approximate unit for J, we have $JA_1 = A_1 = A_1J$ and hence

$$\begin{pmatrix} J & A_1 \\ A_1 & A_0 \end{pmatrix} \begin{pmatrix} 0 & 0 \\ 0 & A_0 \end{pmatrix} \begin{pmatrix} J & A_1 \\ A_1 & A_0 \end{pmatrix} = \begin{pmatrix} J & A_1 \\ A_1 & A_0 \end{pmatrix} .$$

(The products here mean closed linear span of products of elements in the indicated sets.) In other words, $\begin{pmatrix} 0 & 0 \\ 0 & A_0 \end{pmatrix}$ is a full corner of $\begin{pmatrix} J & A_1 \\ A_1 & A_0 \end{pmatrix}$. The latter is stable and has a strictly positive element, so by 2.6 of [1], these two algebras are isomorphic. We thus obtain the sequence (1) as the cyclic exact sequence of K-groups arising from (2).

The main problem is that of decoding the vertical arrows in (1). In the terminology of [2], A_1 is a J - J equivalence bimodule with right and left inner products $\langle x,y\rangle = y^*x$ and $[x,y] = xy^*$ (x,y in A_1). Since J is stable and has a strictly positive element, 3.5 of [2] yields an automorphism ρ of J such that A_1 is isomorphic to the J - J equivalence bimodule J_ρ. (The latter is J as a left J-module with left inner product $[a,b] = ab^*$, while the action on the right by c in J is right multiplication by $\rho^{-1}(c)$ and the right inner product is $\langle a,b\rangle_\rho = \rho(b^*a)$.) It follows from 3.3 of [2] and its proof that there is a unitary u in $M(A_1 + J)$ satisfying (*) such that $\rho = ad(u)|_J$. Moreover, if w is another unitary in $M(A_1 + J)$ for which (*) holds, then w^*u multiplies J unitarily to J, so that $ad(w)|_J$ induces the same map on $K_\#(J)$ as does ρ. In particular, taking $w = u^*$ shows that $\rho_*^{-1} = \rho_*$.

We conclude the proof by showing that $\gamma = i_*\rho_*\partial$. The isomorphism of $\begin{pmatrix} J & A_1 \\ A_1 & A_0 \end{pmatrix}$ with its full corner $\begin{pmatrix} 0 & 0 \\ 0 & A_0 \end{pmatrix}$ is implemented by an isometry W in $M\begin{pmatrix} J & A_1 \\ A_1 & A_0 \end{pmatrix}$. The restriction of W to the ideal $\begin{pmatrix} J & A_1 \\ A_1 & J \end{pmatrix}$ is an isometric multiplier (call it $\tilde{W}$) of the latter. Let $\sigma : J \longrightarrow J$ be the *-homomorphism defined by

$$\tilde{W}\begin{pmatrix} b & 0 \\ 0 & 0 \end{pmatrix}\tilde{W}^* = \begin{pmatrix} 0 & 0 \\ 0 & \sigma(b) \end{pmatrix} \text{ (b in J).}$$

It is clear from the definition of the boundary maps in the K-theory exac sequence that γ is obtained by sending $K_j\begin{pmatrix} A_0/J & 0 \\ 0 & 0 \end{pmatrix}$ to $K_{1-j}\begin{pmatrix} J & A_1 \\ A_1 & A_0 \end{pmatrix}$ via ∂ in the (1,1)-slot and then applying the map to $K_{1-j}\begin{pmatrix} 0 & 0 \\ 0 & A_0 \end{pmatrix}$ induced by ad(W). This means $\gamma = i_*\sigma_*\partial$. With u as in the preceding paragraph, let $V = \begin{pmatrix} 0 & 0 \\ u & 0 \end{pmatrix}$, regarded as multiplier of $\begin{pmatrix} J & A_1 \\ A_1 & J \end{pmatrix}$. We have $V^*V = \begin{pmatrix} 1 & 0 \\ 0 & 0 \end{pmatrix}$, $VV^* = \begin{pmatrix} 0 & 0 \\ 0 & 1 \end{pmatrix}$, and $V\begin{pmatrix} b & 0 \\ 0 & 0 \end{pmatrix}V^* = \begin{pmatrix} 0 & 0 \\ 0 & \rho(b) \end{pmatrix}$ for b in J. Since $(V\tilde{W}^*)(\tilde{W}V^*) = \begin{pmatrix} 0 & 0 \\ 0 & 1 \end{pmatrix} \geq (\tilde{W}V^*)(V\tilde{W}^*)$, there is an isometry s in $M(A_1 + J)$, multiplying J into J and A_1 into A_1, such that $\tilde{W}V^* = \begin{pmatrix} 0 & 0 \\ 0 & s \end{pmatrix}$. We have $s\rho(b)s^* = \sigma(b)$ for b in J. Two homomorphisms that differ by an isometric multiplier (in this case $s|_J$) induce the same map on K-theory, so $\rho_* = \sigma_*$ and hence $\gamma = i_*\rho_*\partial$ as required.

We remark that if there is a unitary multiplier v of $(A_1 + J) \otimes M_n$ for some n multiplying $A_1 \otimes M_n$ to $J \otimes M_n$ and *vice-*

versa, then v can be ampliated to u in $M((A_1 + J) \otimes K)$ satisfying (*), so that the automorphism ρ in the theorem can be replaced by the automorphism $ad(v)|_{J \otimes M_n}$ of $J \otimes M_n$.

It frequently happens that ρ_* is the identity map on $K_\#(J)$, in which case $\gamma = 0$ and $K_\#(A \times_\alpha Z_2)$ is an extension of $K_\#(A_0)$ by $K_\#(A_0/J)$. In the next section, we discuss an example in which $\gamma \neq 0$.

2. The commutative case

We turn now to the case $A = C_0(\Omega)$, where Ω is a locally compact T_2 space. The period-2 automorphism α of A comes from an involution, which we will also call α, of Ω. Let Ω_0 be the set of fixed points for α and Ω_1 the complement of Ω_0. It is apparent that for every ω in Ω_1, there is a function f in A_1 such that $f(\omega) \neq 0$. This means that $J = \{f \in A_0 : f(\Omega_0) = \{0\}\}$ and that $A_1 + J = \{f \in A : f(\Omega_0) = \{0\}\}$. Notice also that A_0 is isomorphic to $C_0(X)$, where X is the space obtained from Ω by identifying each ω with $\alpha(\omega)$, and that A_0/J is isomorphic to $C_0(\Omega_0)$. The condition in Theorem 1 concerning strictly positive elements is satisfied if Ω is separable, or if Ω is compact and Ω_1 is separable.

The behavior of the maps γ in (1) has to do with certain unitary-valued continuous functions on Ω_1. We write U(n) for the group of unitary operators (with the norm topology) on a Hilbert space of dimension n ($1 \leq n \leq \infty$).

<u>Proposition 2</u>: (a) If $u : \Omega_1 \longrightarrow U(n)$ is continuous and satisfies $u(\alpha(\omega)) = -u(\omega)$ for all ω in Ω_1, then the map ρ_* of Theorem 1 is induced by the automorphism of $(A_1 + J) \otimes M_n$ (if $n < \infty$, or $(A_1 + J) \otimes K$ if $n = \infty$) obtained from conjugation by u.

(b) If n is finite, then $n(\rho_* - id)$ is the zero map on $K_\#(J)$, and hence $n\gamma = 0$.

(c) If Ω_1 is a subset of $\mathbb{C}^k$ then there exists a function u as in (a), with $n = 2^{k-1}$.

<u>Proof</u>: (a) We may regard $(A_1 + J) \otimes M_n$ (resp. $\otimes$ K) as the algebra of continuous M_n - valued (resp. K-valued) functions on Ω vanishing on Ω_0 and at infinity. The function u acts as a multiplier of this algebra, *viz.*

$$(uf)(\omega) = \begin{cases} u(\omega)\ f(\omega) & \omega \in \Omega_1 \\ 0 & \omega \in \Omega_0 \end{cases}.$$

The condition we have imposed on u causes it to multiply $A_1 \otimes (M_n$ or $K)$ to $J \otimes (M_n$ or $K)$ and *vice-versa.*

(b) As in part (a), we regard u as a unitary multiplier of $(A_1 + J) \otimes M_n$, so ρ_* is induced by $ad(u)|_{J \otimes M_n}$. Define $\mu, \theta: J \otimes M_n \longrightarrow J \otimes M_n \otimes M_n$ by $\mu(f) = f \otimes 1_n$ and $\theta(f) = (u \otimes 1_n)\, \mu(f)(u^* \otimes 1_n)$ for f in $J \otimes M_n$, so $\eta\rho_* = \theta_*$ and $\eta(id) = \mu_*$. Let $H = H^*$ in $M_n \otimes M_n$ be such that $ad(e^{iH})$ is the flip automorphism $S \otimes T \longmapsto T \otimes S$ of $M_n \otimes M_n$. For each t in $[0,1]$, $1_J \otimes e^{itH} = V_t$ is a multiplier of $J \otimes M_n \times M_n$. Define a path $\{\theta_t\}$ of homomorphisms from $J \otimes M_n$ to $J \otimes M_n \otimes M_n$ by $\theta_t(f) = (u \otimes 1_n)\, V_t\, (f \otimes 1_n)\, V_t^*(u^* \otimes 1_n)$. Then $\theta_0 = \theta$ and, since $V_1(f \otimes 1_n)\, V_1^* \in J \otimes 1_n \otimes M_n$ and $u \otimes 1_n \in M(J) \otimes M_n \otimes 1_n$ and J is commutative, we have at the other end that $\theta_1 = \mu$. Thus $\theta_* = \mu_*$.

(c) We construct $\psi : \Omega_1 \longrightarrow S^{2n-1} \subseteq \mathbb{C}^n$ and $\tau: S^{2n-1} \longrightarrow U(2^{n-1})$ such that $\psi \circ \alpha = -\psi$ and $\tau(-s) = -\tau(s)$ (s in S^{2n-1}). Let $\xi_i : \Omega_1 \longrightarrow \mathbb{C}$ be the co-ordinate functions ($i = 1, \ldots, n$). Notice that the functions $\xi_i - \xi_i \circ \alpha$ cannot vanish simultaneously at any point of Ω_1. We let $\psi_0(\omega) = (\xi_1(\omega) - \xi_1(\alpha(\omega)), \ldots, \xi_n(\omega) - \xi_n(\alpha(\omega)))$ and $\psi(\omega) = \|\psi_0(\omega)\|^{-1}\, \psi_0(\omega)$ for ω in Ω_1. Now define $\sigma_j : \mathbb{C}^j \longrightarrow M_{2^{j-1}}$ by $\sigma_1(\eta) = \eta$ for $j = 1$ and then inductively by

$$\sigma_j(\eta_1,\ldots,\eta_j) = \begin{pmatrix} \sigma_{j-1}(\eta_1,\ldots,\eta_{j-1}) & -\bar{\eta}_j 1_{2^{j-2}} \\ \eta_j 1_{2^{j-2}} & \sigma_{j-1}(\eta_1,\ldots,\eta_{j-1})^* \end{pmatrix}.$$

The desired map τ is the restriction of σ_n to S^{2n-1}. Now let $u = \tau \circ \psi$. This completes the proof.

It follows from (b) and (c) above that the maps γ will be zero if Ω_1 is a subset of $\mathbb{C}$ or if Ω_1 is finite-dimensional and $K_\#(J)$ is torsion-free. These constraints provide some guidance in our search for an example in which $\gamma \neq 0$. The one we will present below (example B) is $\Omega = S^3$ with α multiplying by -1 in the first three co-ordinates and fixing the last. For convenience, we precede it with a preliminary example (example A).

Example A: Let $\Omega = S^2$ and let α be the antipodal map $s \longmapsto -s$. There are no fixed points here, so $J = A_0$ and therefore $\gamma = 0$, but we shall see that $\rho_* \neq \mathrm{id}$. We have $J = \{f \in C(\Omega) : f(-s) = f(s) \ \forall\, s \in \Omega\}$, which is isomorphic to the algebra of continuous functions on real projective 2-space $\mathbb{R}\mathbb{P}^2$. Standard computations show that $K_0(J)$ is $Z \oplus Z_2$, where the first direct summand is generated by $[1]$ and the second is generated by $[1] - [p]$, with p the projection in $J \otimes M_2$ defined by

$$p(s_1,s_2,s_3) = \begin{pmatrix} s_1^2 + s_2^2 & s_3(s_1 + i\, s_2) \\ s_3(s_1 - i\, s_2) & s_3^2 \end{pmatrix}.$$

Define $u : S^2 \longrightarrow U(2)$ by

$$u(s_1,s_2,s_3) = \begin{pmatrix} s_1 + i\, s_2 & -s_3 \\ s_3 & s_1 - i\, s_2 \end{pmatrix}.$$

We have $u \circ \alpha = -u$, so ρ_* comes from conjugation by u. Observe that $u\begin{pmatrix}1 & 0\\ 0 & 0\end{pmatrix}u^* = p$. This means that $\rho_*\,[1] = [p]$ and, since $\rho_*^{-1} = \rho_*$, that $\rho_*[p] = [1]$. It follows that ρ_*, viewed as an automorphism of $Z \oplus Z_2$, sends (1,0) to (1,1) and (0,1) to (0,1).

Example B : This is essentially the "two-point suspension" of example A. Here we let $\Omega = S^3 \subseteq \mathbb{R}^4$ and define $\alpha : \Omega \longrightarrow \Omega$ by $\alpha(s_1,s_2,s_3,s_4) = (-s_1, -s_2, -s_3, s_4)$. Topologically, Ω is $(S^2 \times [-1,1])/\sim$, where $\sim$ identifies the subsets $S^2 \times \{\pm 1\}$ to points e_+ and e_-, and α is the antipodal map on each slice $S^2 \times \{t\}$ $(-1 < t < 1)$, with e_+ and e_- fixed. Thus, A_0 is the algebra of continuous functions on the two-point suspension of $\mathbb{R}\mathbb{P}^2$, while J consists of those functions in A_0 vanishing at the two endpoints. This means that $J = SC(\mathbb{R}\mathbb{P}^2)$, where S denotes the C^*-algebraic suspension. Identifying Ω_1 with $S^2 \times (-1,1)$, we define $u_S : \Omega_1 \longrightarrow U(2)$ by $u_S(s_1,s_2,s_3,t) = u(s_1,s_2,s_3)$, with u as in Example A. Since $u_S \circ \alpha = -u_S$, we see that ρ_* in the present example is just the suspension of the ρ_* in Example A. Thus, ρ_* maps $K_1(J)$ $(\approx Z \oplus Z_2)$ by sending (1,0) to (1,1) and (0,1) to itself. Consider the boundary map $\partial : K_0(A_0/J) \longrightarrow K_1(J)$.

This takes $\mathbb{Z} \oplus \mathbb{Z}$ (since A_0/J is $\mathbb{C}^2$) to $\mathbb{Z} \oplus \mathbb{Z}_2$, and one checks easily that $\partial(m,n) = (m+n, 0)$. Since $K_1(A_0/J) = 0$, the map $i_* : K_1(J) \longrightarrow K_1(A_0)$ is the projection of $\mathbb{Z} \oplus \mathbb{Z}_2$ onto $\mathbb{Z}_2$. We conclude that $\gamma = i_* \rho_* \partial$ takes $(1,0)$ in $K_0(A_0/J)$ to the generator of $K_1(A_0)$, so $\gamma \neq 0$. This in turn forces the map in (2) from $K_1(A_0)$ to $K_1(A \times_\alpha \mathbb{Z}_2)$ to vanish, so the latter group is 0. Since $K_0(A_0)$ is $\mathbb{Z}$, we also obtain $K_0(A \times_\alpha \mathbb{Z}_2) \approx \mathbb{Z} \oplus \mathbb{Z} \oplus \mathbb{Z}$.

REFERENCES

1. L.G. Brown, Stable isomorphism of hereditary subalgebras of C^*-algebras, Pacific J. Math. 71 (1977), 335 - 348.

2. L.G. Brown, P. Green, and M. Rieffel, Stable isomorphism and strong Morita equivalence of C^*-algebras, Pacific J. Math. 71 (1977), 349 - 363.

3. A. Connes, An analogue of the Thom isomorphism for crossed products of a C^*-algebra by an action of $\mathbb{R}$, Advances in Math. 39 (1981), 31 - 55.

4. P. Green, Equivariant K-theory and crossed product C^*-algebras, Proc. Symp. Pare Math. Vol. 38 (ed. R.V. Kadison), part 1, 337 - 338.

5. P. Julg, K-théorie équivariante et produits croisés, C.R. Acad. Sc. Paris, Ser. I t. 292 (1981), 629 - 632.

6. G.K. Pedersen, C^*-Algebras and their Automorphism Groups, Academic Press, New York, 1979.

7. M. Pimsner and D. Voiculescu, K-groups of reduced crossed products by free groups, J. Operator Theory 8 (1982), 131 - 156.

8. G. Segal, Equivariant K-theory, Publ. Math. IHES 34 (1968), 129 - 151.

9. J. Taylor, Banach algebras and topology, in "Algebras in Analysis" (ed. J.H. Williamson) Academic Press, New York, 1975.

Research for this paper was supported in part by NSF grant MCS 8002138.

University of Kansas
Lawrence, KS. 66045

RANGES OF TRACES ON K_0 OF REDUCED CROSSED PRODUCTS BY FREE GROUPS

Mihai V. Pimsner

The aim of this paper is to give a formula for computing the range of a trace on K_0 of a reduced crossed product by a free group on m generators. Even the case when the group has one generator, i.e. the case of crossed products by $\mathbb{Z}$, is of interest since it includes the case of the irrational rotation C*-algebras A_θ , which were the initial example for the general problem we consider. In [17] M.A. Rieffel had found examples of projections in A_θ which showed that the range of the trace on $K_0(A_\theta)$ containes $\mathbb{Z}+\theta\mathbb{Z}$ and led lim to conjecture that this range actually coincided with $\mathbb{Z}+\theta\mathbb{Z}$. This was proved later to be the case in [13] by an embedding argument. After the computations of the K-groups of crossed products by $\mathbb{Z}$ in [14], it seemed that the computation of the range of the trace on K_0 would be much easier. Indeed new proofs for computing the range of the trace on $K_0(A_\theta)$ appeared in [7] and [14], both of which however used some particular feature of the irrational rotation algebra. The first natural approach to our problem is due to A. Connes, who combined his formula for crossed products by $\mathbb{R}$ with the "dual trace" [2], to get results for crossed products by $\mathbb{Z}$. Moreover his "differential geometry" approach to the problem [3] and the discovery that the traces are the elements of order zero in a cohomology theory for algebras [4], [5], [6] are crucial for this problem.

In the present paper we combine the results of [14], [15] with those of [5], [6] to get results for the case of reduced crossed products by free groups.

Section 0 recalls very briefly the results concerning the Toeplitz extension of [15].

For the convenience of the reader we have treated the 0-dimensional case separately in Section 1. We show that, the range of a trace τ on $K_0(A \times_{\alpha_r} \mathbb{F}_m)$ is very roughly speaking the subgroup generated by he values of τ on $K_o(A)$ and of a certain 1-trace on $K_1(A)$. This al-eady shows that higher dimensional traces naturally occur in this roblem. To make this section as selfcontained as possible, we have voided any reference to [4], [5], [6], and have used instead the no-ion of determinant associated to τ, introduced by P. de la Harpe and . Skandalis in [10]. As a corollary of the above results we get a slight) generalization of a theorem of N. Riedel [16] and the compu-ation of the range of the trace on $K_o(C(\mathbb{T}) \times_{\alpha_T} \mathbb{T})$ where $T:\mathbb{T} \to \mathbb{T}$ is ny orientation preserving homeomorphism of the unit circle, in terms f the rotation number of T. This section serves as well as an illus-ration of the basic ideas that are behind the proofs of Section 2.

Section 2 treats the case of higher dimensional traces and relies eavily on A. Connes' papers [5] and [6]. To illustrate our results we how that the theorem of G.A. Elliott concerning the range of the trace n K_o of a noncommutative torus [9] can be obtained by a simple induc-ion from our Theorem 15, in the same way their K-groups are obtained y an iteration of the exact sequence of [14].

§0

By F_m we shall denote the free group on m generators $g_1,\dots,g_m$. Consider a C*-algebra A with an action

$$\alpha: F_m \to \mathrm{Aut}(A).$$

The reduced crossed product of A by α, $A \times_{\alpha_r} F_m$ will be identified with the C*-algebra of operators on $\ell^2(F_m,H) \simeq \ell^2(F_m) \otimes H$, generated by the operators

$$1 \otimes \rho(a) \qquad a \in A$$

$$(u_g k)h = v_g k(g^{-1}h) \qquad g,h \in F_m \text{ and } k \in \ell^2(F_m,H)$$

where (ρ, v_g) is any covariant faithful representation of A on the Hil-bert space H [12].

By $\Gamma_k \subset F_m$ we shall denote the subset of F_m consisting of ele-ments $g_{i_1}^{m_1}\dots g_{i_s}^{m_s}$ $(s\geq 0,\ i_1 \neq i_2,\dots,i_{s-1}\neq i_s\ ,\ m_1\neq 0,\dots,m_s\neq 0)$ such that $m_s>0$ if $i_s=k$. Remark that the neutral element e of F_m is in Γ_k and $g_j\Gamma_k=\Gamma_k$ if $j\neq k$, while $g_k\Gamma_k=\Gamma_k\setminus\{e\}$.

On $\bigoplus_{k=1}^{m} \ell^2(\Gamma_k,H) \subset \bigoplus_{k=1}^{m} \ell^2(F_m,H)$, denote by $d(a)$ and s_i the restrictions of $\bigoplus_{k=1}^{m} (1\otimes\rho(a))$ respectively $\bigoplus_{k=1}^{m} u_{g_i}$.

The Toeplitz algebra T is by definition the C*-algebra generated by $d(a)$ and $s_1,\dots,s_m$.

It is shown in [15] that the closed two-sided ideal generated in T by the projections $1-s_i s_i^*$ is isomorphic to $\bigoplus_{k=1}^{m} A\otimes K(\ell^2(\Gamma_k))$ and that the quotient of T by this ideal is isomorphic to $A \times_{\alpha_r} F_m$. Thus there is an exact sequence

$$0 \to (A\otimes K)^m \xrightarrow{i} T \xrightarrow{\pi} A \times_{\alpha_r} F_m \to 0$$

called the Toeplitz extension of $A \times_{\alpha_r} F_m$.

Moreover the map $\pi\circ d: A \to A \times_{\alpha_r} F_m$ coincides with the usual embedding of A into the crossed product. The main result of [15] shows that d_* induces isomorphisms between $K_i(A)$ and $K_i(T)$ and that if $j: A^m \to (A\otimes K)^m \simeq \bigoplus_{k=1}^{m} A\otimes K(\ell^2(\Gamma_k))$ denotes the map

$$j((a_k)_{k=1}^m) = \bigoplus_{k=1}^{m} a_k \otimes e(e,e)$$

($e(g,g')$ is the natural matrix unit of $K(\ell^2(\Gamma_k))$) that induces an isomorphism of $K_*(A)^m$ with $K_*((A\otimes K)^m)$, then the sequence

$$\begin{array}{ccccc}
K_0(A)^m & \xrightarrow{\beta} & K_0(A) & \xrightarrow{(\pi\circ d)_*} & K_0(A \times_{\alpha_r} F_m) \\
\uparrow \partial & & & & \downarrow \partial \\
K_0(A \times_{\alpha_r} F_m) & \xleftarrow{(\pi\circ d)_*} & K_1(A) & \xleftarrow{\beta} & K_1(A)^m
\end{array}$$

where $\beta((x_i)_{i=1}^m) = \sum_{i=1}^{m} (x_i - \alpha_{g_i^{-1}}{}_*(x_i))$ and $\partial = j_*^{-1}\circ\delta$ (δ being the boundary map of the Toeplitz extension) is exact [15, Theorem 3.5].

§ 1

Let A be a C*-algebra and $M_n(A)$ the C*-algebra of nxn matrices ver A. If A has a unit we shall denote by $U_n(A)$ the group of unitary lements in $M_n(A)$. If A has no unit, let $\tilde{A}$ be the C*-algebra obtained y adjoining a unit to A. Then $U_n(A)$ denotes the subgroup of $U_n(\tilde{A})$ conisting of elements of the form $1-x$ with $x \in M_n(A)$. $U_n(A)$ is a topologial group with the topology induced by the norm topology of $M_n(A)$. Its onnected component of the identity will be denoted by $U_n^o(A)$, while the discrete) group of connected components will be denoted by $\pi_o(U_n(A))$. onsidering $U_n(A)$ as a subgroup of $U_{n+1}(A)$ via the map $u \to \begin{pmatrix} u & 0 \\ 0 & 1 \end{pmatrix}$ which ends $U_n^o(A)$ into $U_{n+1}^o(A)$, we get maps

$$\varphi_{n,n+1}:\pi_o(U_n(A)) \to \pi_o(U_{n+1}(A))$$

nd

$$\varphi_n:\pi_o(U_n(A)) \to K_1(A)$$

here $K_1(A)$ is by definition the inductive limit of the system $\pi_o(U_n(A)),\varphi_{n,n+1})$. Moreover we shall denote by $K_o(A)$ the algebraic K roup of A. If $\alpha:A \to B$ is a *-homomorphism we shall denote by α_* the aps induced on either K_o or K_1, keeping the notation α_{*n} for the map

$$\alpha_{*n}:\pi_o(U_n(A)) \to \pi_o(U_n(B))$$

nduced by α. Moreover we shall denote also by α the map

$$\alpha \otimes 1 : A \otimes M_n \to B \otimes M_n .$$

If τ is a trace state on A, we shall denote also by τ its extenion $\tau \circ Tr$ to $M_n(A)$, where $Tr:M_n(A) \to A$ is the usual map $Tr((a_{ij}))=\sum a_{ii}$ nd by $\underline{\tau}$ the induced group homomorphism from $K_o(A)$ to $\mathbb{R}$. Corresponding o τ P. de la Harpe and G. Skandalis defined a determinant with values in $\mathbb{R}/K_o(A)$. Let us briefly recall their construction [10].

If $[\alpha_1,\alpha_2]$ is a compact interval of the real line and $\xi:[\alpha_1,\alpha_2] \to U_n(A)$ a piecewise continuous differentiable path, they defined

$$\tilde{\Delta}_\tau(\xi)=(1/2\pi i)\int_{\alpha_1}^{\alpha_2} \tau(\dot{\xi}(\alpha)\xi(\alpha)^{-1})d\alpha .$$

Since $\tilde{\Delta}_\tau(\xi)$ rests unchanged when ξ is composed with the inclusion $U_n(A) \to U_{n+1}(A)$, $\tilde{\Delta}_\tau(\xi)$ depends only on the homotopy class (with fixed end points) of ξ, and $\tilde{\Delta}_\tau$ restricted to loops around the identity coincides via the Bott isomorphism with

$$\underline{\tau}:K_o(A)\simeq\lim_{\rightarrow}\pi_1(U_n(A)) \to \mathbb{R}$$

one gets a well defined map

$$\Delta_\tau: \bigcup_n U_n^o(A) \to \mathbb{R}/\underline{\tau}(K_o(A))$$

by $\Delta_\tau(u)=q(\tilde{\Delta}_\tau(\xi))$, where $\xi:[\alpha_1,\alpha_2] \to U_n(A)$ is any piecewise continuous differentiable path such that $\xi(\alpha_1)=1, \xi(\alpha_2)=u$ and where

$$q:\mathbb{R} \to \mathbb{R}/\underline{\tau}(K_o(A))$$

is the natural projection.

The determinant is a group homomorphism and if u is of the form $\exp(2\pi i\ a)$ with $a=a^*$ then $\Delta_\tau(u)=q(\tau(a))$.

Let us prove two easy properties of the determinant.

Let $u_i,v_i\in\mathcal{U}(A)$ $i=1,\dots,m$, *be unitaries such that* $\prod_1^m u_i v_i \in U^o(A)$. *Let* s_i , $i=1,\dots,m$ *be isometries and denote by* p_i *the projections* $1-s_is_i^*$. *Then*

$$\left(\prod_1^m u_i(s_iv_is_i^*+p_i)\right)\oplus 1_3 \in U_4^o(A)$$

and

$$\Delta_\tau\left(\prod_1^m u_i(s_iv_is_i^*+p_i)\right)=\Delta_\tau\left(\prod_1^m v_iv_i\right).$$

PROOF. Let $s_i'\in U_2(A)$ be the unitaries

$$s_i'=\begin{bmatrix} s_i & p_i \\ 0 & s_i^* \end{bmatrix}.$$

Since

$$\begin{bmatrix} s_iv_is_i^*+p_i & 0 \\ 0 & 1 \end{bmatrix} = \begin{bmatrix} s_i & p_i \\ 0 & s_i^* \end{bmatrix} \begin{bmatrix} v_i & 0 \\ 0 & 1 \end{bmatrix} \begin{bmatrix} s_i^* & 0 \\ p_i & s_i \end{bmatrix}$$

we may write

$$\left(\prod_1^m u_i(s_iv_is_i^*+p_i)\right)\oplus 1 = \prod(u_i\oplus 1)s_i'(v_i\oplus 1)s_i'^*\ .$$

Passing now to $U_4(A)$ we may replace s_i' with $s_i''=\begin{bmatrix} s_i' & 0 \\ 0 & s_i'^* \end{bmatrix}\in U_4^o(A)$ to get

$$(\prod_1^m v_i(s_i v_i s_i^* + p_i)) \oplus 1_3 = \prod_1^m (u_i \oplus 1_3) s_i''(v_i \oplus 1_3) s_i''^* .$$

hoosing a path joining s_i'' to the identity we get a path ξ connecting $\prod_1^m u_i(s_i v_i s_i^* + p_i) \oplus 1_3$ with $(\prod_1^m u_i v_i) \oplus 1_3$ such that $\tilde{\Delta}_\tau(\xi)=0$. Q.E.D.

If A *is commutative and* $\det : M_n(A) \to A$ *is the usual determinant,* *hen* $\Delta_\tau \circ \det = \Delta_\tau$.

PROOF. If $u \in U_k^o(A)$, then u is a finite product of exponentials [8], o it is enough to suppose $u = \exp(2\pi i\ a)$. Then

$$\Delta_\tau(\det u) = \Delta_\tau(\exp(\mathrm{Tr}(2\pi i\ a))) = q(\tau \circ \mathrm{Tr}(a)) = \Delta_\tau(\exp(2\pi i\ a)) = \Delta_\tau(u).$$

Q.E.D.

LEMMA 1. *Let*

$$0 \longrightarrow J \xrightarrow{i} B \xrightarrow{\pi} A \longrightarrow 0$$

e an exact sequence of C**-algebras and* τ *a trace state on* A. *If* $u \in U_n(J)$ *s a unitary such that* $[i(u)]_1 = 0$ *in* $K_1(B)$, *then* $\Delta_{\tau\circ\pi}(i(u))$ *depends only n the class of* u *in* $K_1(J)$.

PROOF. It is enough to show that if $u \in U_n^o(J)$ then $\Delta_{\tau\circ\pi}(i(u)) = 0$. Let $\xi : [0,1] \to U_n(J)$ be a path of class C^1 such that $\xi(0)=1$, $\xi(1)=u$. Since $\xi(\alpha) - 1 \in M_n(J)$ for $\alpha \in [0,1]$ it follows that $\dot{\xi}(\alpha) \in M_n(J)$, so that $\tau\pi(i(\dot{\xi}(\alpha)) i(\xi(\alpha)^{-1})) = 0$ for every $\alpha \in (0,1)$. Thus $\tilde{\Delta}_{\tau\circ\pi}(i \circ \xi) = 0$. Q.E.D.

The preceding lemma shows that the map

$$\underline{\Delta}_\tau : \ker i_* \to \mathbb{R}/\underline{\tau\circ\pi}(K_0(B))$$

defined for $[u]_1 \in \ker i_* \subset K_1(J)$ by

$$\underline{\Delta}_\tau([u]_1) = \Delta_{\tau\circ\pi}(i(u))$$

is a well defined group homomorphism. (The condition $[u]_1 \in \ker i_*$ is needed for $\Delta_{\tau\circ\pi}(i(u))$ to make sense.)

PROPOSITION 2. *Let*

$$0 \longrightarrow J \xrightarrow{i} B \xrightarrow{\pi} A \longrightarrow 0$$

be an exact sequence of C^**-algebras and* τ *a trace state on* A.

Then

$$0 \longrightarrow \underline{\tau\circ\pi}(K_o(B)) \longrightarrow \underline{\tau}(K_o(A)) \xrightarrow{q} \underline{\Delta}_\tau(\ker i_*) \longrightarrow 0$$

where the first map is the inclusion of the two subgroups of $\mathbb{R}$ *and* q *is the restriction of the map*

$$q:\mathbb{R} \to \mathbb{R}/\underline{\tau\circ\pi}(K_o(B))$$

to $\underline{\tau}(K_o(A))$, *is an exact sequence.*

Moreover if $p\in M_n(A)$ *is a projection then*

$$q(\tau(p))\in\underline{\Delta}_\tau(\ker i_{*n})$$

where

$$i_{*n}:\pi_o(U_n(J)) \to \pi_o(U_n(B)) .$$

PROOF. If $p\in M_n(A)$ is a projection and $a\in M_n(B)$ is a selfadjoint element such that $\pi(a)=p$, then $\exp(2\pi i\ a)\in U_n^o(J)$ and since $\pi(\exp(2\pi i\ a))=$ $=\exp(2\pi i\ p)=1$ there is a unitary $u\in U_n(J)$ such that $i(u)=\exp(2\pi i\ a)$. By definition $\underline{\Delta}_\tau(u)=\Delta_{\tau\circ\pi}(\exp(2\pi i\ a))=q(\tau\circ\pi(a))=q(\tau(p))$. This proves that $q(K_o(A)) \subset \Delta_\tau(\ker i_*)$ and the last part of the proposition. Finally recall that the boundary map $\delta:K_o(A) \to K_1(J)$ is defined exactly by $\delta[p]_o=[u]_1$ where u is related to p in the above manner, so that the equality $q(K_o(A))=\underline{\Delta}_\tau(\ker i_*)$ follows from the fact that $\delta(K_o(A))=\ker i_*$ which is part of the exactness of the six term sequence in K-theory.

Q.E.D.

We shall now apply the preceding proposition to crossed products by free groups.

THEOREM 3. *Let* $\alpha:F_m \to \mathrm{Aut}(A)$ *be an action of the free group on* m *generators* $g_1,\dots,g_m$ *on the* C^**-algebra* A *and* $A \times_{\alpha_r} F_m$ *the correspondind reduced crossed-product. Suppose* τ *is a trace state on* $A \times_{\alpha_r} F_m$, *and denote also by* τ *its restriction to* A.

a) *If* $\beta:K_1(A)^m \to K_1(A)$ *denotes the map*

$$\beta((x_i)_{i=1}^m)=\sum_{i=1}^m (x_i-\alpha_{g_i^{-1}*}(x_i)),$$

hen the map

$$\underline{\Delta}_\tau^\alpha : \ker \beta \to \mathbb{R}/\underline{\tau}(K_o(A))$$

efined by

$$\underline{\Delta}_\tau^\alpha(([u_1],\dots,[u_m])) = \Delta_\tau\Big(\prod_1^m u_i \alpha_{g_i}^{-1}(u_i^{-1})\Big)$$

here u_i *are unitaries in some* $U_n(A)$, *such that* $\prod_1^m u_i\alpha_{g_i}^{-1}(u_i^{-1}) \varepsilon U_n^o(A)$ *is well defined group homomorphism.*

b) *The sequence*

$$0 \to \underline{\tau}(K_o(A)) \to \underline{\tau}(K_o(A \times_{\alpha_r} F_m)) \xrightarrow{q} \underline{\Delta}_\tau^\alpha(\ker \beta) \to 0$$

where the first map is the inclusion of the two subgroups of $\mathbb{R}$) *is xact.*

PROOF. It is an easy exercise to get a) from the properties of he determinant using the fact that τ is a F_m invariant trace on A. lowever both a) and b) follow from the preceding Proposition applied o the Toeplitz extension

$$0 \to (A \otimes \mathcal{R})^m \xrightarrow{i} T \xrightarrow{\pi} A \times_{\alpha_r} F_m \to 0$$

and from the results of [15].

Since $d: A \to T$ induces an isomorphism of $K_o(A)$ with $K_o(T)$ and $\pi o d$ is the usual embedding of A into $A \times_{\alpha_r} F_m$, all we have to prove is that

$$\underline{\Delta}_\tau^\alpha = \underline{\Delta}_\tau \circ j_* .$$

Let $u_i \varepsilon U_n(A)$, $1 \le i \le m$, be unitaries such that

$$\prod_1^m u_i \alpha_{g_i}^{-1}(u_i^{-1}) \varepsilon U_n^o(A) .$$

Then

$$i \circ j((u_i)_{i=1}^m) = \prod_{i=1}^m d(u_i)(s_i d(\alpha_{g_i}^{-1}(u_i^{-1}))s_i^* + 1 - s_i s_i^*)$$

so that the desired equality follows from the previously proved property of the determinant. Q.E.D.

REMARK. Since the ideal appearing in the Toeplitz extension is

$(A\otimes K)^m$ we lost the information of Proposition 2 concerning the possibility of choosing the unitary corresponding to a projection in $M_k(A \times_{\alpha_r} F_m)$ in $U_k(A)^m$. However if the C*-algebra A is commutative one can get the following result.

COROLLARY 4. *In the conditions of the preceding theorem suppose A is commutative. Then*

$$q(\tilde{\underline{\tau}}(K_o(A \times_{\alpha_r} F_m))) = \underline{\Delta}^*_\tau(\ker \beta_1)$$

where $\beta_o : \pi_o(U_1(A))^m \to \pi_o(U_1(A))$ *is defined by*

$$\beta_o((x_i)_{i=1}^m) = \sum_1^n (x_i - (\alpha_{g_i^{-1}})_{*1}(x_i)) .$$

Thus in order to compute the range of the trace on K_o of the crossed product we need only to know the action induced on $\pi_o(U(A)) \simeq \check{H}^1(X,\mathbb{Z})$ where $\check{H}^1(X,\mathbb{Z})$ is the first Cech cohomology group of the maximal ideal space of A and to compute determinants.

The corollary is a direct consequence of the theorem once we show that $\Delta^*_\tau(\ker \beta_1) \supset \Delta^*_\tau(\ker \beta)$, the other inclusion being obvious.

Let $u_1, \ldots, u_n$ be unitaries in some $U_k(A)$ such that $\prod_1^m u_i \alpha_{g_i^{-1}}(u_i)^{-1} \in U_k^o(A)$. Then the usual determinant $\det : M_n(A) \to A$ (acting pointwise when $A \simeq C(X)$) maps u_i in $U(A)$ and $\prod_1^m u_i \alpha_{g_i^{-1}}(u_i)^{-1}$ in $U^o(A)$ so that $(\det(u_i))_{i=1}^m \in \ker \beta_1$, and since $\Delta_\tau \circ \det = \Delta_\tau$ the conclusion follows.

Q.E.D.

The preceding results are especially easy to apply either if $K_1(A)$ (respectively $\check{H}^1(X,\mathbb{Z})$) is trivial or if the action of F_m has discrete spectrum.

For example one gets the following extension of a result of N. Riedel [16].

Let Γ be a discrete abelian group, G its dual (compact) group and μ the normalized Haar measure on G. Denote by $\alpha_\rho \in \mathrm{Aut}(C(G))$ the automorphism determined by translation with a fixed element $\rho \in G$ and by $A_{\Gamma,\rho}$ the crossed product $C(G) \times_{\alpha_g} \mathbb{Z}$. The trace state τ determined by μ extends to a trace state $\tilde{\tau}$ of $A_{\Gamma,\rho}$. For each finite group $F \subset \Gamma$ (including $F=\{0\}$), let $|F|$ denote its cardinality and p_F the period of $\rho|F$, that is the smallest positive integer p such that $(\rho|F)^p$ is the rivial character of F.

Let $\Gamma_{\rho,F}=\{t\in\mathbb{R}|\exp(2\pi i\frac{|F|}{p_F}t)<\Gamma,\rho>\}$ and note that $\Gamma_{\rho,F_1}\subset\Gamma_{\rho,F_2}$ whenever $F_1\subset F_2$, since $\frac{|F_1|}{p_{F_1}}$ divides $\frac{|F_2|}{p_{F_2}}$. Define $\Gamma_\rho=\bigcup_{F\subset\Gamma}\Gamma_{\rho,F}$ which is a roup that depends functorial on (Γ,ρ). Moreover if ρ is faithful then ach finite subgroup $F\subset\Gamma$ is cyclic, $p_F=|F|$, and $\Gamma_\rho=\{t\in\mathbb{R}|\exp(2\pi it)\in <\Gamma|\rho>\}$ so that in this case the next result is Theorem 3.6 of [16].

THEOREM 5. *The range of the trace* $\tilde{\tau}$ *on* $K_o(A_{\Gamma,\rho})$ *is* Γ_ρ.

PROOF. Since both $A_{\Gamma,\rho}$ and Γ_ρ commute with direct limits, it is ufficient to consider the case when Γ is finitely generated, so that e may suppose $\Gamma=\mathbb{Z}^m\times F$, for some $m\in\mathbb{N}\cup\{0\}$ and some finite group F. (Note that in this case $\Gamma_\rho=\Gamma_{\rho,F}$.) It follows that G, the dual of Γ, is somorphic to $\mathbb{T}^m\times\hat{F}$, where $\mathbb{T}^m$ is the m-dimensional torus and $\hat{F}$ is isoorphic to F. This shows on one hand that $\underline{\tau}(K_o(C(G)))=\frac{1}{|F|}\mathbb{Z}$ and on the other hand that $\pi_o(U_1(C(G)))$ is isomorphic to $(\mathbb{Z}^m)^{\hat{F}}$ by regarding an $|\hat{F}|$-tuple $(m_f)_{f\in\hat{F}}$ as the continuous function

$$(m_f)_{f\in\hat{F}}(t,h)=<m_h|t>\in\mathbb{T}. \qquad (\mathbb{Z}^m=\hat{\mathbb{T}}^m).$$

Note that if $\rho=(\rho_1,\rho_2)\in\mathbb{T}^m\times\hat{F}$ is the decomposition of ρ then

$$(\alpha_\rho)^{-1}((m_f)_{f\in\hat{F}})(t,h)=(m_f)_{f\in\hat{F}}(\rho_1t,\rho_2h)=$$

$$=<m_{\rho_2h}|\rho_1t>=<m_{\rho_2h}|\rho_1><m_{\rho_2h}|t>=<m_{\rho_2h}|\rho_1>(m_{\rho_2f})_{f\in\hat{F}}(t,h).$$

We are now in position to apply Corollary 4 for $F_1=\mathbb{Z}$. The above computation of α_ρ shows first that an $|\hat{F}|$-tuple $(m_f)_{f\in\hat{F}}$ is in ker β_o iff $m_{\rho_2f}=m_f$ for every $f\in\hat{F}$, and in this case

$$(m_f)_{f\in\hat{F}}\cdot\alpha_\rho^{-1}((m_f)_{f\in\hat{F}}^{-1})(t,h)=<m_h|t>\cdot\overline{<m_{\rho_2h}|\rho_1>}\,\overline{<m_{\rho_2h}|t>}=$$

$$=\overline{<m_h|(\rho_1)>}=<-m_h|\rho_1>.$$

Since m_f only depends on $\tilde{f}\in\hat{F}/(\rho_2)$, the orbit of f, ($(\rho_2)$ is the subgroup generated by $\rho_2=\rho|F$) it follows that

$$(m_f)_{f\in\hat{F}}\cdot\alpha_\rho^{-1}((m_f)_{f\in\hat{F}}^{-1})=\sum_{\tilde{f}\in\hat{F}/(\rho_2)}<-m_{\tilde{f}}|\rho_1>\chi(\mathbb{T}^m\times\tilde{f}).$$

Thus

$$\Delta_\tau^*([(m_f)_{f\varepsilon\hat{F}}])=q(\sum_{\tilde{f}\varepsilon\tilde{F}/(\rho_2)}\frac{p_F}{|\hat{F}|}t_{\tilde{f}})=q(\sum_{f\varepsilon\tilde{F}/(\rho_2)}\frac{p_F}{|F|}t_{\tilde{f}})$$

where $t_{\tilde{f}}\varepsilon\mathbb{R}$ satisfies $\exp(2\pi\, it_{\tilde{f}})=\langle -m_{\tilde{f}}|\rho_1\rangle$.

Collorary 4 now implies that $\tilde{\tau}(K_o(A_{\Gamma,\rho}))$ is the subgroup of $\mathbb{R}$ generated by $\frac{1}{|F|}$ and $\frac{p_F}{|F|}t$ where $\exp(2\pi\, it)\varepsilon\langle\mathbb{T}^m|\rho_1\rangle=\langle\mathbb{T}^m\times\{0\}|\rho\rangle$.

Since the period of $\rho_2=\rho|F$ is p_F , it follows that there exists $a\varepsilon F$ such that $\langle a|\rho\rangle=\langle a|\rho_2\rangle=\exp(2\pi\, i\frac{1}{p_F})=\exp(2\pi\, i\frac{|F|}{p_F}\cdot\frac{1}{|F|})$. This shows that the above group is contained in $\Gamma_\rho=\Gamma_{\rho,F}$. Conversely if $t\varepsilon\mathbb{R}$ satisfies $\exp(2\pi i\frac{|F|}{p_F}t)=\langle\gamma|\rho\rangle$, for some $\gamma\varepsilon\Gamma$, write $\gamma=(\gamma_1,\gamma_2)$ so that $\langle\gamma|\rho\rangle=$ $=\langle\gamma_1|\rho_1\rangle\langle\gamma_2|\rho_2\rangle=\langle\gamma_1|\rho_1\rangle\exp(2\pi i\frac{a}{p_F})=\langle\gamma_1|\rho_1\rangle\exp(2\pi i\frac{|F|}{p_F}\cdot\frac{a}{|F|})$ for some $a\varepsilon\mathbb{Z}$.

It follows that $t=\frac{a}{|F|}+\frac{p_F}{|F|}t_1$ where $\exp(2\pi it_1)=\langle\gamma_1|\rho_1\rangle$ so that $\Gamma_\rho=\Gamma_{\rho,F}\subset\tilde{\tau}(K_o(A_{\Gamma,\rho}))$ which concludes the proof. Q.E.D.

Another application of Theorem 3 is the following.

Let $\mathbb{T}$ be the one dimensional torus and $T:\mathbb{T}\to\mathbb{T}$ an orientation preserving homeomorphism. The rotation number of T is defined in the following way (see for instance [19]): choose a continuous increasing function $f:\mathbb{R}\to\mathbb{R}$ such that $e^{2\pi if(x)}=T(e^{2\pi ix})$ for every $x\varepsilon\mathbb{R}$ (this function is unique up to an integer constant). One shows that the limit $\lim_{n\to\infty}\frac{f^{(n)}(x)}{n}=\theta$ (where $f^{(n)}$ is the n-times composition of f with itself) exists and is independent of the point x_o. Since $f(x+m)=f(x)+m$ for every $m\varepsilon\mathbb{Z}$, θ is uniquely defined by T up to an integer. The particular $\theta\varepsilon[0,1)$ is called the rotation number of T.

PROPOSITION 6. *Let* $T:\mathbb{T}\to\mathbb{T}$ *be an orientation preserving homeomorphism of the unit circle with rotation number* θ *and* μ *any* T *invariant probability measure on* $\mathbb{T}$.

Let α_T *be the automorphism* $\alpha_T(h)=h\circ T^{-1}$ *of* $C(\mathbb{T})$ *and* $\tau=\tau_\mu$ *be the induced trace on* $C(\mathbb{T})\times_{\alpha_T}\mathbb{Z}$.

Then the range of the trace τ *on* $K_o(C(\mathbb{T})\times_{\alpha_T}\mathbb{Z})$ *is* $\mathbb{Z}+\theta\mathbb{Z}$.

PROOF. We shall apply Theorem 3. Let $z:\mathbb{T} \to \mathbb{T}$ be the identity function which is the generator of $K_1(C(\mathbb{T}))=K^1(\mathbb{T})$. Since T is orientation preserving $[z]_1=[\alpha_{T^{-1}}(z)]_1$, so that ker $\beta=K_1(C(\mathbb{T}))\simeq\mathbb{Z}$. Moreover $(K_o(C(\mathbb{T})))\simeq\mathbb{Z}$ and since $\underline{\Delta}^*_\tau$ is a group homomorphism it is sufficient to compute $\underline{\Delta}^*_\tau([z]_1)$. Let $f:\mathbb{R} \to \mathbb{R}$ be the continuous increasing function such that $e^{2\pi i f(x)}=T(e^{2\pi i x})$ and $\lim\limits_{n\to\infty} \frac{f^{(u)}(x)}{x}=\theta$. Defining $g:\mathbb{T} \to \mathbb{R}$ by $(e^{2\pi i x})=x-f(x)$ it follows that $e^{2\pi i g}=z\cdot\alpha_{T^{-1}}(z)^{-1}$ and so

$$\underline{\Delta}^*_\tau([z]_1)=\Delta_\tau(z\cdot\alpha_{T^{-1}}(z)^{-1})=q(\tau(g))=q(\int_{\mathbb{T}} g d\mu)$$

here $q:\mathbb{R} \to \mathbb{R}/\mathbb{Z}$ is the natural projection. Note that $g\circ T^k(e^{2\pi i x})=$ $g(e^{2\pi i f^{(k)}(x)})=f^{(k)}(x)-f^{(k+1)}(x)$ and since μ is T invariant it follows that

$$\int_{\mathbb{T}} g d\mu=\frac{1}{k}\int_{\mathbb{T}} (g+g\circ T+\ldots+g\circ T^{k-1})d\mu=-\theta$$

since $\frac{1}{k}\sum\limits_{0}^{k-1} g\circ T^i(e^{2\pi i x})=\frac{x-f^{(k)}(x)}{k}$ are uniformly bounded and converge to $-\theta$ for every x.

§2

Recall from [5] that for any algebra A, $\Omega(A)$ denotes the universal graded differential algebra associated to A. Thus if τ is an n+1 linear functional on A, it extends to a linear functional $\hat{\tau}$ on $\Omega^n(A)$, with

$$\hat{\tau}(a^o da^1 da^2\ldots da^n)=\tau(a^o,a^1,\ldots,a^n) \qquad a^i\in A \quad [5].$$

τ is called a cyclic cocycle iff $\hat{\tau}$ is a closed graded trace, i.e. if it satisfies

$$\hat{\tau}(d\omega)=0, \quad \forall\,\omega\in\Omega^{n-1} \qquad \text{and} \qquad \hat{\tau}(\omega\omega')=(-1)^{\deg\omega\,\deg\omega'}\,\hat{\tau}(\omega'\omega) .$$

The set of cyclic cocycles of A is denoted by $Z^n_\lambda(A)$. Let us quote from [6] the definition of an n-trace on a Banach algebra.

DEFINITION. Let B be a Banach algebra. By an *n-trace on* B we mean an n+1 linear functional τ on a dense subalgebra A of B such that

a) τ is a cyclic cocycle on A.

b) For any $a^i \varepsilon A$, $i=1,\dots,n$, there exists $C=C_{a^1,\dots,a^n}<\infty$ such that

$$|\hat{\tau}((x^1da^1)(x^2da^2)\dots(x^nda^n))|\leq C||x^1||\dots||x^n|| \qquad \forall x^i \varepsilon \tilde{A}\ .$$

Note that if τ is the character of an n-dimensional cycle $(\Omega,d,\int)$ [5] that satisfies b) (in Ω), then τ is an n-trace. The main property of n-traces that we will be concerned with, is that they determine maps from $K_*(B)$ to $\mathbb{C}$ (see [6], Theorem 7).

Recall also from [5] the defintion of the cup product $\varphi \# \psi \varepsilon$ $\varepsilon Z_\lambda^{n+m}(A\otimes B)$ of the two cyclic cocycles $\varphi\varepsilon Z_\lambda^n(A)$ and $\psi\varepsilon Z_\lambda^m(B)$, and that the map $S:H_\lambda^n(A) \to H_\lambda^{n+2}(A)$ is induced by the cup product with the generator $\sigma\varepsilon H_\lambda^2(\mathbb{C})$ of $H_\lambda^*(\mathbb{C})$ characterized by

$$\sigma(1,1,1)=2i\pi.$$

Let us note for further use the explicit formula for $S^k:H_\lambda^n(A) \to$ $\to H_\lambda^{n+2k}(A)$.

LEMMA 7. *Let* $\varphi\varepsilon Z_\lambda^n(A)$ *be a cyclic cocycle and* k *a natural number. We shall denote by* $\mathcal{D}_{n,k}$ *the collection of subsets* $D\subset\{1,\dots,n+2k\}$ *consisting of* 2k *elements with the property that the maximal intervals appearing in the decomposition of* D *contain an even number of elements. If* $\omega=x^0dx^1\dots dx^{n+2k}$ *and* $D\varepsilon\mathcal{D}_{n,k}$ *denote by* ω_D *the* n*-form obtained by replacing for each* $i\varepsilon D$ dx^i *with* x^i *in the expression of* ω. *Then*

$$S^k\varphi(\omega)=(2i\pi)^k k!\sum_D\hat{\varphi}(\omega_D)$$

the sum running over all elements of $\mathcal{D}_{n,k}$.

PROOF. The properties of the cup product imply that S^k is given by the cup product with $\sigma^k\varepsilon H^{2k}(\mathbb{C})$. By [5],Corollary 10, σ^k is the 2k--cocycle characterized by

$$\sigma^k(1,1,\dots,1)=(2i\pi)^k k!\ .$$

The above formula is now an easy consequence of the definition of the cup product. Q.E.D.

The n-traces that we will consider on crossed products will arise s characters of some particular cycles, so let us state some of their roperties.

LEMMA 8. *Let* $(\Omega,d,\hat{\varphi}_n)$ *be a cycle of dimension* n. *Suppose that* Ω^o *is unital and that* $d1=0$. *Let* A *be a subalgebra of* Ω^o *containing the nit of* Ω^o *and* $s\in\Omega^o$ *an invertible element that normalizes* A. *Suppose hat* $sdas^{-1}=d(sas^{-1})$ $\forall a\in A$. *Then*

1) 1 *is the unit of* Ω; $ds^{-1}=-s^{-1}dss^{-1}$.

2) *If* ω *is a form in the graded algebra generated by* A *then* $s^{-1}ds\omega=(-1)^{\deg\omega}\omega s^{-1}ds$.

3) *For each* $k\in\mathbb{N}$, $k<n$ *the equality*

$$\varphi_{n-k}(a^o,a^1,\ldots,a^{n-k})=\hat{\varphi}_n(a^o da^1\ldots da^{n-k}\underbrace{s^{-1}ds\ldots s^{-1}ds}_{k\text{-times}})$$

$a^i\in A$, *defines a cyclic cocycle* $\varphi_{n-k}\in Z_\lambda^{n-k}(A)$. *However* $\varphi_{n-k}=0$ *when* k *is even.*

PROOF. 1) Follows from $1dx=dx-d1x=dx$ and similarly $dx1=dx$, and from $0=d1=d(ss^{-1})=dss^{-1}+sds^{-1}$.

2) Since $sdas^{-1}=d(sas^{-1})$ one gets $dsas^{-1}+sads^{-1}=0$. This shows that $s^{-1}dsa=as^{-1}ds$ $\forall a\in A$. Differentiating the same quality one gets $0=dsdas^{-1}-sdads^{-1}=dsdas^{-1}+sdas^{-1}dss^{-1}$ so that we also have $s^{-1}dsda=-das^{-1}ds$, $\forall a\in A$.

3) The equality

$$\hat{\varphi}_n(a^o da^1\ldots da^{n-k}s^{-1}\underbrace{ds\ldots s^{-1}ds}_{k\text{-times}})=$$

$$=(-1)^{n-k}\hat{\varphi}_n(s^{-1}ds\, a^o da^1\ldots da^{n-k}\underbrace{s^{-1}ds\ldots s^{-1}ds}_{k-1\text{-times}})=$$

$$=(-1)^{n-k}(-1)^{n-1}\hat{\varphi}_n(a^o da^1\ldots da^{n-k}\underbrace{s^{-1}ds\ldots s^{-1}ds}_{k\text{-times}})$$

where the first follows from 2) and the second from the fact that $\hat{\varphi}_n$ is a graded trace, shows that $\varphi_{n-k}=0$ when k is even. Note next that $d((s^{-1}ds)^m)=-m(s^{-1}ds)^{m+1}$ so that $\hat{\varphi}_n(d\omega(s^{-1}ds)^k)=(-1)^k k\hat{\varphi}_n(\omega(s^{-1}ds)^{k+1})$ for any form ω in the graded algebra generated by A. Thus if k is odd then $\hat{\varphi}_{n-k}$ is closed. To show that $\hat{\varphi}_{n-k}$ is graded let ω and ω' be forms in the graded algebra generated by A. Then

$$\hat{\varphi}_n(\omega\omega'(s^{-1}ds)^k)=(-1)^{\deg\omega'\cdot k}\hat{\varphi}_n(\omega(s^{-1}ds)^k\omega')=$$

$=(-1)^{\deg \omega'\cdot k}(-1)^{\deg \omega'(\deg \omega+k)}\hat{\varphi}_n(\omega'\omega(s^{-1}ds)^k)=$

$=(-1)^{\deg \omega' \deg \omega}\hat{\varphi}_n(\omega'\omega(s^{-1}ds)^k)$

which gives the desired result. Q.E.D.

Let now $\mathbb{T}$ denote the one dimensional torus and ε the 1-trace on on $C(\mathbb{T})$ (the C*-algebra of continuous functions on $\mathbb{T}$) given by

$$\varepsilon(f^o,f^1)=\int f^o df^1$$

defined on the dense subalgebra of smooth functions. Given any Banach algebra A and any n-trace φ on A, the cup product $\varphi \# \varepsilon$ defines an n+1-trace on $A\otimes C(\mathbb{T})$. The explicit formula of this cup product is as follows. If $a^i(t)$ are smooth functions on $\mathbb{T}$ with values in the domain of φ, then

$$(\varphi \# \varepsilon)(a^o,a^1,\dots,a^{n+1})=$$

$$=\sum_{i=1}^{n+1}(-1)^{n+1-i}\int\hat{\varphi}(a^o(t)da^1(t)\dots da^{i-1}(t)\dot{a}^i(t)da^{i+1}\dots da^{n+1})dt\ .$$

The next lemma is the main computational result of this section.

LEMMA 9. *Let φ_n be an n-trace on the Banach algebra B with domain $\mathcal{B}$ and $A \subset B$ a Banach subalgebra such that $\mathcal{A}=A \cap \mathcal{B}$ is dense in A. Suppose that B is unital, that the unit lies in $\mathcal{A}$ and that there exists an invertible $s\in\mathcal{B}$ that normalizes $\mathcal{A}$. Suppose moreover that φ_n arises as the character of an n-cycle (Ω,d,φ_n) with the properties*

a) $\Omega^o=\mathcal{B}$

b) $d1=0$; $d(sas^{-1})=sdas^{-1}$ *for every* $a\in\mathcal{A}$.

Let $v_t\in M_2(\mathcal{B})$ be the invertible element

$$v_t=\begin{bmatrix} s & 0\\ 0 & 1\end{bmatrix}\begin{bmatrix}\cos t & \sin t\\ -\sin t & \cos t\end{bmatrix}\begin{bmatrix}s^{-1} & 0\\ 0 & 1\end{bmatrix}$$

and for any $x\in\mathcal{A}$ let $x_t=v_t\begin{bmatrix}0 & 0\\ 0 & x\end{bmatrix}v_t^{-1}\in M_2(\mathcal{B})$. Then for every $x^o,x^1,\dots,x^{n+1}$ belonging to $\mathcal{A}$ one gets

$$\sum_{i=1}^{n+1}(-1)^{n-i+1}\int_o^{\pi/2}\varphi_n \# \mathrm{Tr}(x_t^o dx_t^1\dots dx_t^{i-1}\dot{x}_t^i dx_t^{i+1}\dots dx_t^{n+1})dt=$$

$$= \sum_{k=1}^{[\frac{n+1}{2}]} c_k S^k \varphi_{n-2k+1}(x^o, x^1, \ldots, x^{n+1})$$

here $c_k = (-1)^k \frac{1}{(2i\pi)^k} \frac{1}{2^{k-1}} \frac{1}{(2k-1)(2k-3)\ldots 3\cdot 1}$ *and* φ_{n-2k+1} *is defined in he preceding lemma. In particular the left hand side of the above equaity defines an* n+1-*trace.*

PROOF. Let us denote by p_t the projection $\begin{bmatrix} s^2 & sc \\ sc & c^2 \end{bmatrix}$ by $\dot{p}_t$ its erivative which is the unitary $\begin{bmatrix} 2sc & c^2-s^2 \\ c^2-s^2 & 2sc \end{bmatrix}$ and by $e=\begin{bmatrix} 0 & 1 \\ -1 & 0 \end{bmatrix}$, here s=sint t and c=cos t. Note that

$$p_t^2 = p_t \ , \ p_t\dot{p}_t = \dot{p}_t(1-p_t) \ , \quad (1-p_t)\dot{p}_t = \dot{p}_t p_t$$

$$ep_t = (1-p_t)e \ , \quad e\dot{p}_t = -\dot{p}_t e$$

and that $\mathrm{Tr}(p_t\dot{p}_t e)=1$ for every t. A direct computation shows that

$$x_t = \begin{bmatrix} s & 0 \\ 0 & 1 \end{bmatrix} p_t \otimes x \begin{bmatrix} s^{-1} & 0 \\ 0 & 1 \end{bmatrix}$$

$$\dot{x}_t = \begin{bmatrix} s & 0 \\ 0 & 1 \end{bmatrix} \dot{p}_t \otimes x \begin{bmatrix} s^{-1} & 0 \\ 0 & 1 \end{bmatrix}$$

$$dx_t = \begin{bmatrix} s & 0 \\ 0 & 1 \end{bmatrix} (p_t \otimes dx + sce \otimes (s^{-1}ds\ x)) \begin{bmatrix} s^{-1} & 0 \\ 0 & 1 \end{bmatrix}$$

so that we have to compute

$$\sum_{i=1}^{n+1} (-1)^{n+1-i} \varphi_n \# \mathrm{Tr}(p_t \otimes x^o (p_t \otimes dx^1 + sce \otimes (s^{-1}dsx^1)) \ldots$$

$$\ldots (\dot{p}_t \otimes x^i) \ldots (p_t \otimes dx^{n+1} + sce \otimes (s^{-1}dsx^{n+1}))) \ .$$

In order to expand the above sum recall from Lemma 1 the definition of the sets belonging to $\mathcal{D}_{n,k}$ and note that the formulae connecting p_t, $\dot{p}_t$ and e ensure that only terms for which $\dot{p}_t$ and all the e's belong to some $\mathcal{D}_{n-2k+1,k}$ appear. To be more precise, denote by $\mathcal{D}_{n-2k+1,k,i}$ the collection of pairs (D,i) with $D\varepsilon\mathcal{D}_{n-2k+1,k}$, $i\varepsilon\{1,\ldots,n+1\}$, such that $i\varepsilon D$. For each such pair denoty by $r_{t,D,i}$ the matrix obtained from $r^o r^1\ldots r^{n+1}$ by replacing r^i with $\dot{p}_t$, r^j for $j\varepsilon D\setminus\{i\}$ with e and r^j for $j\notin D$ with p_t. Similarly denote by $\omega_{D,i}$ the n-form obtained from $x^o dx^1\ldots dx^{n+1}$ by replacing dx^j with

$$\begin{cases} x^i & \text{for } j=i \\ s^{-1}dsx^j & \text{for } j\varepsilon D\setminus\{i\} \\ dx^j & \text{for } j\notin D\,. \end{cases}$$

The definition of the cup product with Tr and the above discussion shows that the desired sum equals

$$\sum_{(D,i)} (-1)^{n+1-i}(sc)^{2k-1}\mathrm{Tr}(r_{t,D,i})\varphi_n(\omega_{D,i}) \;,$$

the sum running over $\bigcup_{k,i}\mathcal{D}_{n-2k+1,k,i}$.

Note next that property 2) of Lemma 2 implies that

$$\varphi_n(\omega_{D,i})=(-1)^{n-i+\varepsilon}\varphi_{n-2k+1}(\omega_D)$$

where

$$\varepsilon=\begin{cases} 0 & \text{if there is an odd number of elements in D greater than i} \\ 1 & \text{otherwise} \end{cases}$$

(ω_D is defined in Lemma 1).

Note further that since $e^2=-1$ and $\mathrm{Tr}(p_t\dot{p}_t e)=1=-\mathrm{Tr}(p_t e\dot{p}_t)$ one gets

$$\mathrm{Tr}(r_{t,D,i})=(-1)^{k-1+\varepsilon}$$

so that out sum becomes

$$\sum_{(D,i)}(-1)^k(sc)^{2k-1}\varphi_{n-2k+1}(\omega_D)=$$

$$=\sum_k(-1)^k(sc)^{2k-1}2k\Big(\sum_{D\varepsilon\mathcal{D}_{n-2k+1,k}}\varphi_{n-2k+1}(\omega_D)\Big)=$$

$$=\sum_k(-1)^k(sc)^{2k-1}(2i\pi)^{-k}2k(k!)^{-1}S^k\varphi_{n-2k+1}(x^o,x^1,\ldots,x^{n+1})$$

by Lemma 1.

The proof is complete once we notice that

$$\int_0^{\pi/2} (\sin t \cos t)^{2k-1} dt = \frac{1}{2^k} \cdot \frac{(k-1)!}{(2k-1)\cdot(2k-3)\ldots 3\cdot 1} . \qquad \text{Q.E.D.}$$

The next lemma expresses the fact that the cup product with ε is compatible with Bott periodicity.

LEMMA 10. *Let* φ *be an* n-*trace on the Banach algebra* B, *with domain* $\mathcal{B}$.

a) n=2m. *Let* $p \in \mathrm{Proj}\, M_k(\mathcal{B})$ *be a projection and* $u_p \in Gl_k(\mathcal{B} \otimes C^\infty(\mathbb{T}))$ *the invertible element* $u_p(t) = \exp(2i\pi t)p + (1-p)$. *Then*

$$<[p],[\varphi]> = <[u_p],[\varphi \# \varepsilon]> .$$

b) n=2m-1. *Let* $u \in Gl_k(\mathcal{B})$ *be an invertible element and* $v_t \in Gl_{2k}(\mathcal{B} \otimes C^\infty([0,1]))$ *any differentiable path such that*

$$v_o = \begin{bmatrix} 1 & 0 \\ 0 & 1 \end{bmatrix} \qquad v_1 = \begin{bmatrix} u^{-1} & 0 \\ 0 & u \end{bmatrix} .$$

If e *denotes the constant projection* $e = \begin{bmatrix} 0 & 0 \\ 0 & 1 \end{bmatrix}$ *and* p_u *denotes the projection* $p_u(t) = v_t e v_t^{-1} \in \mathrm{Proj}\, M_{2k}(\mathcal{B} \otimes C^\infty(\mathbb{T}))$, *then*

$$<[u],[\varphi]> = <[e]-[p_u],[\varphi]>$$

(<,> *denotes the pairing of* $H^*(A)$ *with* $K_*(A)$ *defined in* [5] *and* [6].)

PROOF. a) Since $\mathrm{Tr} \# (\varphi \# \varepsilon) = (\mathrm{Tr} \# \varphi) \# \varepsilon$ it is sufficient to consider k=1. Note that $u_p - 1 = (\exp(2i\pi t)-1)p$, $u_p^{-1} - 1 = (\exp(-2i\pi t)-1)p$ so that

$$(\varphi \# \varepsilon)(u_p^{-1}-1, u_p-1, \ldots, u_p^{-1}-1, u_p-1) =$$

$$= \sum_{i=1}^{n+1} (-1)^{n+i-1} \hat{\varphi}(p\, dp..dp\; p\; dp..dp) \int_0^1 (\exp(2i\pi t)-1)^m (\exp(-2i\pi t)-1)^m \cdot$$

$$\cdot (-1)^{i-1} 2i\pi \exp((-1)^{i-1} 2i\pi t)(\exp((-1)^i 2i\pi t)-1)\, dt .$$

Since

$$\hat{\varphi}(p\, dp..dp\; p\; dp..dp) = \begin{cases} 0 & \text{when } i \text{ is even} \\ \varphi(p,\dots,p) & \text{when } i \text{ is odd} \end{cases}$$

[5] the above sum becomes

$$\sum_{\substack{i=1\\ i=\text{odd}}}^{n+1} \varphi(p,..,p)\int_0^1 (\exp(-2i\pi t)-1)^{2m+1}(-1)^m \exp(2i\pi t(m+1))2i\pi dt =$$

$$=2i\pi(m+1)\frac{(2m+1)!}{(m+1)!m!}\varphi(p,\dots,p) = 2i\pi\frac{(2m+1)!}{m!\;m!}\varphi(p,\dots,p)\quad .$$

Thus

$$<[u_p],[\varphi \# \varepsilon]> = \frac{1}{(2i\pi)^{m+1}}\cdot\frac{1}{2^{2m+1}}\frac{1}{(m+\frac{1}{2})(m-\frac{1}{2})\dots\frac{1}{2}}\cdot$$

$$\cdot 2i\pi\frac{(2m+1)!}{m!\;m!}\varphi(p,..,p) = \frac{1}{(2i\pi)^m m!}\varphi(p,..,p) = <[p],[\varphi]>\quad .$$

b) Again it is sufficient to consider the case k=1. Since the result does not depend on the path of invertible elements, we will choose the following particular one

$$v_t = \begin{cases} \begin{bmatrix} u & 0 \\ 0 & 1 \end{bmatrix}\begin{bmatrix} \cos t & \sin t \\ -\sin t & \cos t \end{bmatrix}\begin{bmatrix} u^{-1} & 0 \\ 0 & 1 \end{bmatrix} & \text{for } t\in[0,\frac{\pi}{2}] \\ \begin{bmatrix} \sin t & \cos t \\ -\cos t & \sin t \end{bmatrix}\begin{bmatrix} 0 & u \\ -u^{-1} & 0 \end{bmatrix} & \text{for } t\in[\frac{\pi}{2},\pi]. \end{cases}$$

Note next that we may use the same trick as in [5] to suppose that $1\in\mathcal{B}$ and that $d1=0$. That is, we replace (even if $\mathcal{B}$ is already unital) $\mathcal{B}$ by $\tilde{\mathcal{B}}$ and extend φ to $\tilde{\mathcal{B}}$ by

$$\tilde{\varphi}(x^o+\lambda^o 1, x^1+\lambda^o 1,..,x^n+\lambda^n 1) = \varphi(x^o,..,x^n)\;.$$

In this case $<[e],[\varphi]>=0$.

To compute $<[p_u],[\varphi]>$ we shall apply Lemma 9 with $\mathcal{A}$ equal to the scalar multiples of the identity. The integral up to $\frac{\pi}{2}$ appearing in the formula of $\varphi \# \varepsilon$ thus equals

$$\sum_{k=1}^{m} c_k S^k \varphi_{n-2k+1}(1,1,\ldots,1)$$

and again since $d1=0$, only

$$c_m S^m \varphi_o(1,\ldots,1)$$

is different from 0. The integral from $\frac{\pi}{2}$ to π is again of the type considered in Lemma 9 but this time with s equal to 1. Using once again $d1=0$ it follows that $\varphi_{n-k}=0$ for every k. Thus

$$(\varphi \# \varepsilon)(p_u,\ldots,p_u)=c_m S^m \varphi_o(1,\ldots,1) \ .$$

Since the operator S does not affect the value of $<[p],[\varphi]>$ ([5], Proposition 14)

$$<[p_u],[\varphi \# \varepsilon]>=c_m<[1],[\varphi_o]>=c_m\hat{\varphi}(u^{-1}du\ldots u^{-1}du) \ .$$

Using the fact that $du^{-1}=-u^{-1}du\, u^{-1}$ the last term equals

$$(-1)^{m-1}c_m\varphi(u^{-1},u,\ldots,u^{-1},u)=-<[u],[\varphi]> \ ,$$

which concludes the proof. Q.E.D.

Let us restate the preceding lemma as

PROPOSITION 11. *If* φ *is an* n-*trace on the Banach algebra* B, *and* $b:K_n(B) \to K_{n+1}(B\otimes C(\mathbb{T}))$ *is the Bott map, then*

$$<b(x),[\varphi \# \varepsilon]>=<x,[\varphi]>$$

for every $x\varepsilon K_n(B)$.

We have now the technical tools in order to extend the results of the preceding paragraph to higher dimensional traces.

Let us first consider the case of short exact sequences

$$0 \longrightarrow I \xrightarrow{i} B \xrightarrow{\pi} A \longrightarrow 0$$

of Banach algebras.

Denote by B_I the algebra of continuous functions $f:[0,1] \to B$ such that $f(0)=0$, $f(1)\varepsilon I$, and denoty by $\rho:B_I \to I$ the map given by evaluation at 1. The kernel of ρ is $C_o(\mathbb{T},B)$, the algebra of continuous functions on $\mathbb{T}$ with values in B that vanish at 1. Note that the exact sequence

$$0 \longrightarrow C_o(T,B) \longrightarrow B_I \xrightarrow{\rho} I \longrightarrow 0$$

gives rise to the exact sequence

$$(*) \qquad K_m(C_o(T,B)) \longrightarrow K_m(B_I) \xrightarrow{\rho_*} \ker i_* \longrightarrow 0$$

for every m.

Suppose that φ is an n-trace on A. Since there is an obvious map

$$\tilde{\pi}: B_I \to A \otimes C(\mathbb{T})$$

we get an n+1-trace on B_I by $\tilde{\pi}^*(\varphi \# \varepsilon)$. Let us denote this n+1-trace by φ_π.

PROPOSITION 12. a) *The range of* φ_π *restricted to* $K_{n+1}(C_o(\mathbb{T},B))$ *coincides with the range of* $\pi^*\varphi$ *on* $K_n(B)$.

b) *The* n+1-*trace* φ_π *determines a well defined group homomorphism*

$$K_{n+1}(I) \supset \ker i_* \xrightarrow{\underline{\varphi}_\pi} \mathbb{C}/\langle K_n(B),[\pi^*\varphi]\rangle$$

by the formula

$$\underline{\varphi}_\pi(\rho_*(x)) = q(\langle x,[\varphi_\pi]\rangle)$$

where $q:\mathbb{C} \to \mathbb{C}/\langle K_n(B),[\pi^*\varphi]\rangle$ *is the natural projection.*

c) *If* $\delta: K_n(A) \to K_{n+1}(I)$ *denotes the boundary map determined by the considered exact sequence, then*

$$q(\langle x,[\varphi]\rangle) = \underline{\varphi}_\pi(\delta(x)).$$

In particular, the sequence

$$0 \longrightarrow \langle K_n(B),[\pi^*\varphi]\rangle \longrightarrow \langle K_n(A),[\varphi]\rangle \xrightarrow{q} \underline{\varphi}_\pi(\ker i_*) \to 0$$

is exact.

PROOF. a) Follows from the preceding proposition once we notice that the restriction of φ_π to $K_{n+1}(C_o(\mathbb{T},B))$ coincides with $(\pi^*\varphi) \# \varepsilon$.

b) Is a direct consequence of a) and of the exact sequence (*).

c) Recall the definition of δ to see that with the above nota-

ions $\delta(x)=\rho_*(z)$, where $z\in K_{n+1}(B_I)$ is any element such that $\tilde{\pi}_*(z)=b(x)$.
hus

$$\underline{\varphi}_\pi(\delta(x))=q(<z,[\varphi_\pi]) = q(<\tilde{\pi}_*(z),[\varphi \# \varepsilon]>) = q(<b(x),[\varphi \# \varepsilon]>)$$

o that by the preceding proposition

$$\underline{\varphi}_\pi(\delta(x))=q(<x,[\varphi]>) \ .$$

he exact sequence now follows from the six term exact sequence asso-
iated to the considered short sequence. Q.E.D.

As in Section 1 we shall apply the above proposition to the
'oeplitz extension associated to reduced crossed products by free
roups.

Let $\alpha: F_m \to \mathrm{Aut}(A)$ be an action of the free group on m genera-
ors $g_1,\dots,g_m$ on the unital C*-algebra A and $A \times_{\alpha_r} F_m$ the correspond-
ng reduced crossed product.

Denote by $A_\alpha \subset M_m(C([0,1],A))$ the C*-algebra of those $M_m(A)$ va-
ued functions f such that f(0) and f(1) are diagonal matrices whose
entries satisfy the conditions

$$f(1)_{ii}=\alpha_{g_i}(f(0)_{ii}) \ .$$

Evaluation at 1 determines a map $\eta: A_\alpha \to A^m$, whose kernel is
$M_m(C_o(\mathbb{T},A))$, the C*-algebra of continuous functions on $\mathbb{T}$ with values
in $M_m(A)$ that vanish at 1. Note that the sequence

$$0 \longrightarrow M_m(C_o(\mathbb{T},A)) \longrightarrow A_\alpha \xrightarrow{\eta} A^m \longrightarrow 0$$

gives rise to the exact sequence

$$(**) \quad K_n(C_o(\mathbb{T},A)) \longrightarrow K_n(A_\alpha) \xrightarrow{\eta_*} \ker\beta \longrightarrow 0$$

where $\beta: K_n(A)^m \to K_n(A)$ is the map

$$\beta((x)_{i=1}^m) = \sum_{i=1}^m (x_i - \alpha_{g_i *}^{-1}(x_i)) .$$

Suppose that φ is an n-trace on $A \times_{\alpha_r} F_m$, arising as the character of an n-cycle $(\Omega,d,\hat{\varphi})$ with the following properties:

i) $\Omega^o \subset A \times_{\alpha_r} F_m$ contains the immage of the free group (in particular it contains the unit of $A \times_{\alpha_r} F_m$).

ii) the algebra $A = A \cap \Omega^o$ is dense in A and is normalised by u_g, $\forall g \in F_m$.

iii) $d1=0$; $d(u_g a u_g^{-1}) = u_g da u_g^{-1}$ $\forall a \in A$ and $\forall g \in F_m$.

This implies in particular that the restriction of φ to A (still denoted by φ) is an invariant n-trace, i.e.

$$\varphi(u_g a^o u_g^{-1}, \dots, u_g a^n u_g^{-1}) = \varphi(a^o, a^1, \dots, a^n)$$

for every $a^i \in A$ and $g \in F_m$.

An n-trace with the above properties on a crossed product will be called natural.

LEMMA 13. *If* φ *is a natural* n-*trace on* $A \times_{\alpha_r} F_m$, *then the formula*

$$\varphi_\alpha(f^o, f^1, \dots, f^{n+1}) =$$

$$= \sum_{i=1}^{n+1} (-1)^{n-i+1} \int \hat{\varphi} \# \mathrm{Tr}(f^o(t) df^1(t) \dots df^{i-1}(t) \dot{f}^i(t) df^{i+1}(t) \dots df^{n+1}(t)) dt$$

$\forall f^i \in M_m(C^\infty[0,1], A)$, *defines an* n+1-*trace on* A_α, *whose restriction to* $M_m(C_o(T,A))$ *coincides with* $(\varphi \# \mathrm{Tr}) \# \varepsilon$.

PROOF. It is an easy computation to show that φ_α is a Hochschild cocycle. (In fact it is the restriction to A_α of $(\varphi \# \mathrm{Tr}) \# \tilde{\varepsilon}$, where $\tilde{\varepsilon}$ is the Hochschild cocycle

$$\tilde{\varepsilon}(f^o, f^1) = \int f^o df^1 \quad \text{on} \quad C^\infty[0,1].)$$

Note next that since 1 is the unit of Ω, and $d1=0$

$$\varphi_\alpha(1, f^1, \dots, f^{n+1}) = \sum_1^{n+1} (-1)^{n-i+1} \int \varphi \# \mathrm{Tr}(df^1 \dots df^{i-1} \dot{f}^i df^{i+1} \dots df^{n+1}) =$$

$$= \sum_1^{n+1} (-1)^n \int \varphi \# \mathrm{Tr}(f^1, \dots, f^{i-1}, \dot{f}^i, f^{i+1}, \dots f^{n+1}) =$$

$$= (-1)^n (\varphi \# \mathrm{Tr}(f^1(1), \dots, f^{n+1}(1)) - \varphi \# \mathrm{Tr}(f^1(0), \dots, f^{n+1}(0))) = 0$$

(since φ is invariant). This now implies that φ_α is cyclic:

$$\varphi_\alpha(f^o,f^1,..,f^{n+1})=\varphi_\alpha(1,f^of^1,..,f^{n+1})-\varphi_\alpha(1,f^o,f^1f^2,..,f^{n+1})+...$$

$$...+(-1)^{n+1}\varphi_\alpha(f^{n+1},f^o,..,f^n) .$$

he rest is now easy. Q.E.D.

Recall that $\mathcal{T}_{(A\otimes K)^m}$ is the algebra of continuous functions $:[0,1] \to \mathcal{T}$ such that $f(0)=0$ and $f(1)\in(A\otimes K)^m$. Let us denote it from ow on with $\mathcal{T}_*$. Consider also $\mathcal{T}_{**}$ the algebra of continuous functions $:[0,1] \to \mathcal{T}$ such that $f(0)-f(1)\in(A\otimes K)^m$. One thus gets an exact sequence

$$0 \to \mathcal{T}_* \to \mathcal{T}_{**} \to \mathcal{T} \to 0 .$$

The obvious map $\tilde{\tilde{\pi}}$ from $\mathcal{T}_{**}$ to $(A \times_{\alpha_r} F_m)\otimes C(\mathbb{T})$ induces the n+1-trace $^*(\varphi \# \varepsilon)$ which will be denoted by φ_π (note that its restriction to $\mathcal{T}_*$ coincides with the previous φ_π).

In order to get from Proposition 12 the result for the reduced crossed product by the free group, we shall construct an embedding of A_α into $M_{2m+1}\otimes\mathcal{T}_{**}$. Let (e_{ij}), $i,j=0,1,...,2m$, be the matrix units in M_{2m+1} and $f\in A_\alpha$. The embedding will be done in four steps.

1) $t\in[-2,-\frac{\pi}{2}]$. Consider the diagonal matrix $f(1)\in M_m(A)$ and define

$$\sigma_o(f)=\sum_{i=1}^m d(f(1)_{ii})\otimes e_{ii}$$

where $d:A \to \mathcal{T}$ is the natural inclusion. It is then easily seen that there exists a smooth path of unitaries $w_t\in M_{2m+1}\otimes\mathcal{T}$, $t\in[-2, -\frac{\pi}{2}]$ such that $w_{-2}=1$, $(1\otimes\pi)(w_t)=1$ for $t\in[-2, -\frac{\pi}{2}]$ and

$$w_{-\frac{\pi}{2}}\sigma_o(f(1))w^*_{-\frac{\pi}{2}}=i\circ j(f(1))\otimes e_{oo}+\sum_{i=1}^m d(f(1)_{ii})s_is_i^*\otimes e_{ii}$$

where $i:(A\otimes K)^m \to \mathcal{T}$ is the inclusion map, $j:A^m \to (A\otimes K)^m$ the natural embedding (so that $i\circ j((a_i)_i)=\sum_{i=1}^m(1-s_is_i^*)d(a_i))$, and s_i the isometries that satisfiy $\pi(s_i)=u_i$, $s_id(a)=d(\alpha_{g_i}(a))s_i$.

Let

$$\sigma_t'(f)=w_t\sigma_o(f(1))w_t^* .$$

2) $t\varepsilon[-\frac{\pi}{2},0]$. Let w_t be the isometries

$$w_t=1\otimes e_{oo}+\sum_{i=1}^{m}(s_i\otimes e_{ii}+1\otimes e_{2i2i})(\cos(-t)\otimes e_{ii}+$$

$$+\sin(-t)\otimes e_{i,2i}-\sin(-t)\otimes e_{2i,i}+\cos(-t)\otimes e_{2i,2i})(s_i^*\otimes e_{ii}+1\otimes e_{2i,2i}) .$$

Note that the isometries w_t satisfy the following relations

$$w_o(i\circ j(f(1))\otimes e_{oo}+\sum_{i=1}^{m}d(f(0)_{ii})\otimes e_{2i,2i})w_o^*=$$

$$=i\circ j(f(1))\otimes e_{oo}+\sum_{i=1}^{m}d(f(0)_{ii})\otimes e_{2i,2i}$$

and

$$w_{-\frac{\pi}{2}}(i\circ j(f(1))\otimes e_{oo}+\sum_{i=1}^{m}d(f(0)_{ii})\otimes e_{2i,2i})w^*_{-\frac{\pi}{2}}=$$

$$=i\circ j(f(1))\otimes e_{oo}+\sum_{i=1}^{m}d(\alpha_{g_i}(f(0)_{ii}))s_is_i^*\otimes e_{ii}=$$

$$=i\circ j(f(1))\otimes e_{oo}+\sum_{1}^{m}d(f(1)_{ii})s_is_i^*\otimes e_{ii} .$$

Define

$$\sigma_t'(f)=w_t(i\circ j(f(1))\otimes e_{oo}+\sum_{1}^{m}d(f(0))_{ii})\otimes e_{2i,2i})w_t^* .$$

3) $t\varepsilon[0,1]$. Define

$$\sigma_t'(f)=i_\circ j(f(1))+\sum_{i,j=1}^{m}d(f(t)_{ij})\otimes e_{2i,2j} .$$

4) $t\varepsilon[1,2]$. Let w_t be the unitaries

$$w_t=1\otimes e_{oo}+\sum_{i=1}^{m}(\sin\tfrac{\pi}{2}t\otimes e_{ii}+\cos\tfrac{\pi}{2}t\otimes e_{i,2i}-\cos\tfrac{\pi}{2}t\otimes e_{2i,i}+\sin\tfrac{\pi}{2}t\otimes e_{2i,2i})$$

and define

$$\sigma_t'(f)=w_t(i\circ j(f(1)+\sum_{i=1}^{m}d(f(1)_{ii})\otimes e_{2i,2i}))w_t^* .$$

The desired embedding is given by the formula

$$\sigma_t(f)=\sigma_{3t-2}'(f) .$$

Note that

$\sigma_1(f)-\sigma_o(f)=i\circ j(f(1))\otimes e_{oo}$.

LEMMA 14. *Let* φ *be a natural* n-*trace on* $A \times_{\alpha_r} F_m$ *and* φ_π , φ_α , $\eta:A_\alpha \to A^m$ *and* $\sigma:A_\alpha \to M_{2m+1}\otimes T_{**}$ *as above. For each* $k\le n$ *and* $i=1,\dots,m$ *consider the* n-k-*traces on* A

$$\varphi^i_{n-k}(a^o,a^1,\dots,a^{n-k})=\varphi(a^o da^1\dots da^{n-k}(u_{g_i}du^{-1}_{g_i})^k)$$

(defined in Lemma 8), *and* φ_{n-k} *the* n-k *trace on* A^m

$$\varphi_{n-k}=\sum_{i=1}^{m}\varphi^i_{n-k} .$$

Then

$$\sigma^*(Tr \# \varphi_\pi)=\varphi_\alpha+\sum_{k=1}^{[\frac{n+1}{2}]} c_k\eta^*(S^k\varphi_{n-2k+1})$$

where $c_k=(-1)^k\frac{1}{(2i\pi)^k}\frac{1}{2^{k-1}}\frac{1}{(2k-1)(2k-3)\dots3\cdot1}$.

PROOF. Let $f^o,\dots,f^{n+1}\in A_\alpha$ be smooth functions with values in A. Since $\sigma(f^i)$ are piecewise smooth functions we have to compute

$$\sum_{i=1}^{n+1}(-1)^{n-i+1}\int_0^1(Tr \# \hat\varphi)(\pi(\sigma_t(f^o))d(\pi(\sigma_t(f^1)))\dots\widehat{\pi(\sigma_t(\dot f^i))}\dots d(\pi(\sigma_t(f^{n+1}))))dt=$$

$$=\sum_{i=1}^{n+1}(-1)^{n-i+1}\int_{-2}^{2}(Tr \# \hat\varphi)(\pi(\sigma'_t(f^o))d(\pi(\sigma'_t(f^1)))\dots\widehat{\pi(\sigma'_t(\dot f^i))}\dots d(\pi(\sigma'_t(f^{n+1}))))dt.$$

Note that the integral from -2 to $-\frac{\pi}{2}$ is 0 since in this case $\pi(\sigma'_t(f))$ is a constant function.

The integral from $-\frac{\pi}{2}$ to 0 is of the type considered in Lemma 9. More precisely if $u\in M_m\otimes(A \times_{\alpha_r} F_m)$ denotes the unitary

$$u=\sum_{i=1}^{m}u_i\otimes e_{ii}$$

and we identify M_{2m} with $M_2\otimes M_m$, then we have to compute

$$\sum_{i=1}^{n+1}(-1)^{n-i+1}\int_{-\pi/2}^{0}Tr \# (Tr \# \hat\varphi)(v_{-t}x^o v^{-1}_{-t}d(v_{-t}x^1v^{-1}_{-t})\dots)dt=$$

$$=\sum_{i=1}^{n+1}(-1)^{n-1}\int_0^{\pi/2}Tr \# (Tr \# \hat\varphi)(v_t x^o v^{-1}_t d(v_t x^1 v^{-1}_t)\dots)dt$$

where

$$v_t = \begin{bmatrix} u & 0 \\ 0 & 1 \end{bmatrix} \begin{bmatrix} \cos t & \sin t \\ -\sin t & \cos t \end{bmatrix} \begin{bmatrix} u^{-1} & 0 \\ 0 & 1 \end{bmatrix}$$

and $x^i \varepsilon M_m \otimes (A \times_{\alpha_r} F_m)$ is

$$x^i = \sum_{j=1}^{m} f^i(0)_{jj} \otimes e_{jj} .$$

Lemma 9 now applies to yield

$$\sum_{k=1}^{[\frac{n+1}{2}]} -c_k S^k (\mathrm{Tr} \# \varphi)'_{n-2k+1}(x^o, x^1, \ldots, x^{n+1}) =$$

$$\sum_{k=1}^{[\frac{n+1}{2}]} -c_k S^k \Big(\sum_{i=1}^{m} \varphi'^{\,i}_{n-2k+1} \Big) (f(0)_{ii}, \ldots, f^{n+1}(0)_{ii}) ,$$

where

$$\varphi'^{\,i}_{n-2k+1}(a^o, \ldots, a^{n-k}) = \varphi(a^o da^1 \ldots da^{n-k} (u_{g_i}^{-1} du_{g_i})^k) .$$

But since

$$\varphi'^{\,i}_{n-2k+1}(u_{g_i}^{-1} a^o u_{g_i}, u_{g_i}^{-1} a^1 u_{g_i}, \ldots, u_{g_i}^{-1} a^{n-2k+1} u_{g_i}) =$$

$$= \varphi(a^o da^1 \ldots da^{n-2k+1} u_{g_i} (u_{g_i}^{-1} du_{g_i})^{2k+1} u_{g_i}^{-1}) =$$

$$= -\varphi^i_{n-2k+1}(a^o, a^1, \ldots, a^{n-2k+1})$$

we get that the integral from $-\frac{\pi}{2}$ to 0 is equal to

$$\sum_{k=1}^{[\frac{n+1}{2}]} c_k S^k \Big(\sum_{i=1}^{m} \varphi^i_{n-2k+1} \Big) (f^o(1)_{ii}, \ldots, f^{n+1}(1)_{ii}) .$$

The integral from 0 to 1 obviously coincides with

$$\varphi_\alpha(f^o, \ldots, f^{n+1})$$

while the integral from 1 to 2 is again of the type considered in Lemma 9, but this time with the invertible element equal to 1 and so equals 0. Q.E.D.

THEOREM 15. *Let* φ *be a natural* n-*trace on* $A \times_{\alpha_r} F_m$. *Let* $A_\alpha = \{f \in M_m(C([0,1],A));\ f(1)_{ii} = \alpha_{g_i}(f(1)_{ii}),\ f(0)_{ij} = f(1)_{ij} = 0$ *for every* $i,j = 1,\dots,m,\ i \neq j\}$ *and* φ_α *the* n+1 *trace on* A_α *defined in* Lemma 13. *Then:*

a) *The* n+1 *trace* φ_α *induces a well defined group homomorphism*

$$\underline{\varphi}_\alpha : \ker \beta \to \mathbb{C}/\langle K_n(A),[\varphi]\rangle$$

by the formula

$$\underline{\varphi}_\alpha(\eta_*(x)) = q(\langle x,[\varphi_\alpha]\rangle)$$

where $\beta : K_{n+1}(A)^m \to K_{n+1}(A)$ *is the map* $\beta((x_i)_i) = \sum_{i=1}^{m} (x_i - \alpha_{g_{i*}^{-1}}(x_i))$, $\eta : A_\alpha \to A^m$ *is the map given by evaluation at* 1 *and* $q : \mathbb{C} \to \mathbb{C}/\langle K_n(A),[\varphi]\rangle$ *is the natural projection.*

b) *For each generator* $g_i \in F_m$ *consider the* n-k *traces on* A, $\varphi^i_{n-k}(a^0,\dots,a^{n-k}) = \varphi(a^0 da^1 \dots da^{n-k}(u_g du_g^{-1})^k)$ (*see* Lemma 8) *and let* φ_{n-k} *be the* n-k *trace on* A^m, $\varphi_{n-k} = \sum_{i=1}^{m} \varphi^i_{n-k}$. *Let also* $\partial : K_n(A \times_{\alpha_r} F_m) \to K_{n+1}(A)^m$ *be the map which composed with* $j_* : K_{n+1}(A)^m \to K_{n+1}((A \otimes K)^m)$ *gives the boundary map* δ *associated to the Toeplitz extension. Then*

$$q(\langle x,[\varphi]\rangle) = \underline{\psi}_\alpha(\partial(x))$$

where

$$\underline{\psi}_\alpha = \underline{\varphi}_\alpha + q\Big(\sum_{k=1}^{[\frac{n+1}{2}]} c_k \varphi_{n-2k+1}\Big)$$

and

$$c_k = (-1)^k \frac{1}{(2i\pi)^k} \frac{1}{2^{k-1}} \frac{1}{(2k-1)(2k-3)\dots 3\cdot 1}.$$

In particular the sequence

$$0 \to \langle K_n(A),[\varphi]\rangle \to \langle K_n(A \times_{\alpha_r} F_m),[\varphi]\rangle \to \underline{\psi}_\alpha(\ker \beta) \to 0$$

is exact.

PROOF. a) Follows imediately from the exact sequence (**), from Lemma 13 and from Proposition 12.

b) Since the inclusion of A in T induces an isomorphism of $K_n(A)$

with $K_n(\mathcal{T})$, [15, Lemma 3.4], we know already by Proposition 12 that

$$q(\langle x,[\varphi]\rangle)=\underline{\varphi}_\pi(\delta(x)) \ .$$

So all we have to do is to compute $\underline{\varphi}_\pi(j_*\partial(x))$.

Consider the diagram

$$\begin{array}{ccccccc} 0 \longrightarrow & K_*(\mathcal{T}_*) & \longrightarrow & K_*(\mathcal{T}_{**}) & \longrightarrow & K_*(\mathcal{T}) & \longrightarrow 0 \\ & \rho_* \big\downarrow & & \big\uparrow \sigma_* & & & \\ & K_*((A\otimes\mathcal{K})^m) & & K_*(A_\alpha) & & & \end{array}$$

and note that the horizontal sequence is exact and split. This leads to a map $r:K_*(\mathcal{T}_{**}) \to K_*(\mathcal{T}_*)$. Recall the definition of σ and η to see that

$$\sigma_1(f)=i\circ j\circ\eta(f)\otimes e_{oo}\oplus\sigma_o(f) \ .$$

This easily implies that $\rho_*\circ r\circ\sigma_*=j_*\circ\eta_*$ so that $\underline{\varphi}_\pi(j_*\circ\eta_*(x))=$ $=q(\langle r\circ\sigma_*(x),[\varphi_\pi]\rangle)$. Note next that since $y-r(y)$ may be represented by a constant function, for every $y\in K_*(\mathcal{T}_{**})$,

$$\langle y,[\varphi_\pi]\rangle=\langle r(y),[\varphi_\pi]\rangle$$

so that

$$\underline{\varphi}_\pi(j_*\ \eta_*(x))=q(\langle\sigma_*(x),[\varphi_\pi]\rangle)=q(\langle x,[\sigma^*\varphi_\pi]\rangle),$$

so that the conclusion follows by the preceding lemma.

Again the exact sequence is now a direct consequence of the six term exact sequence in K-theory. Q.E.D.

REMARKS.- Since the proof depends only on the results concerning the Toeplitz extension [15] and on Proposition 12 it works without the naturality condition on the n-trace φ. However, $\sigma^*\varphi_\pi$ is no longer a sum of traces obtained in an easy way from φ.

- The only term in the expression of $\underline{\psi}_\alpha$ that does not stem from a trace on A is $\underline{\varphi}_\alpha$. However if there is a cross-section to the map $\eta:A_\alpha \to A^m$ (for example if there is for each generator $g_i\in F_m$ a continuous path of automorphisms connecting α_{g_i} to the identity) this term also appears as an n+1 trace on A (the pull back of φ_α).

- The analogue of Corollary 4 is a consequence of the interpretation given in [5], [6] to the cohomology $H^*_\lambda(A)$, in the case A is the C*-algebra of continuous functions on a smooth manifold, in terms of the homology (with complex coefficients) of the manifold.

As an application of the preceding theorem we shall give a new proof of the result of G.A. Elliott [9] concerning the range of the trace on noncommutative tori. Given a discrete abelian group G and a character ρ of the second exterior power $G\wedge G$ one defines the C*-algebra A_ρ as follows (see [9] and the references given there). Choose a 2-cocycle α on G with values in $\mathbb{T}$ such that

$$\alpha(g,h)\alpha(h,g)^{-1}=\rho(g\wedge h)$$

and consider the envelopping C*-algebra of unitaries $(u_g)_{g\varepsilon G}$ with the relations

$$u_g u_h=\alpha(g,h)u_{g+h}$$

(see for example [12]). The resulting C*-algebra is independent of the cocycle α and is denoted A_ρ. It is the universal C*-algebra generated by a projective representation of G such that

$$u_g u_h u_g^{-1} u_h^{-1}=\rho(g\wedge h) \ .$$

When G is isomorphic to $\mathbb{Z}^n$ it is easy to see that A_ρ is obtained by an iteration of n ordinary crossed products by actions of $\mathbb{Z}$, the first action on $\mathbb{C}$. Thus we may apply Theorem 15 recursively.

Let $\{e_i\}$ be the canonical basis in $\mathbb{Z}^n$, and $(\theta_{ij})_{i,j}\varepsilon M_n(\mathbb{R})$ the antisymetric matrix given by

$$\rho(e_i\wedge e_j)=e^{2i\pi\theta_{ij}} \ .$$

In this case A_ρ is generated by n unitaries $u_1,\ldots,u_n$ satisfying the relations

$$u_i u_j=e^{2i\pi\theta_{ij}}u_j u_i \ .$$

We shall denote by $A_{\rho,k}$, $0\leq k\leq n$ the C*-algebra generated by $\{u_{k+1},\ldots,u_n\}$ $(A_{\rho,n}=\mathbb{C})$. This is a decomposition series of A_ρ in the sense that

$$A_\rho=A_{\rho,o}\supset A_{\rho,1}\supset\ldots\supset A_{\rho,n}=\mathbb{C}$$

and

$$A_{\rho,k-1}=A_{\rho,k}\times_{\alpha_k}\mathbb{Z}$$

where $\alpha_k\varepsilon \mathrm{Aut}(A_{\rho,k})$ is the automorphism induced by the unitary u_k , which normalizes $A_{\rho,k}$.

Moreover the universal property of A_ρ shows that there exists an action

$$\alpha: \mathbb{R}^n \to \mathrm{Aut}\ (A_\rho)$$

defined on each generator by

$$\alpha_{t_1,\ldots,t_n}(u_l) = (\exp(2\pi i \sum_j \theta_{jl} t_j)) u_l \ .$$

Note that α_k is the restriction of α_{f_k} on $A_{\rho,k}$, where $f_1,\ldots,f_n$ is the canonical basis in $\mathbb{R}^n$. The algebra of smooth elements under this action, i.e. of those elements a such that

$$\mathbb{R}^n \ni t \to \alpha_t(a) \varepsilon A_\rho$$

is smooth, will be denoted by A_ρ^∞.

As in [2] and [6] we shall consider the action of the Lie algebra of $\mathbb{R}^n$ on A_ρ^∞, i.e. the derivations

$$\delta_j : A_\rho^\infty \to A_\rho^\infty \qquad j=1,\ldots,n \ ,$$

corresponding for fixed j to the one parameter group of automorphisms

$$\mathbb{R} \ni t \to \alpha_{f_j t} \varepsilon \mathrm{Aut}\ (A_\rho)$$

and the corresponding graded diferential algebra

$$\Omega = A_\rho^\infty \otimes \Lambda \mathbb{R}^n$$

where the differential $d: A_\rho^\infty \otimes \Lambda^p \mathbb{R}^n \to A_\rho^\infty \otimes \Lambda^{p+1} \mathbb{R}^n$ is defined in the following way:

$$d(a \otimes \omega) = \sum_{i=1}^{n} \delta_i(a) \otimes (f_i \wedge \omega) \ .$$

Note that the generators $u_1,\ldots,u_n \varepsilon A_\rho^\infty$ and that

$$\delta_j(u_i) = 2i\pi\theta_{ji} u_i \ .$$

Suppose now that τ is any trace state on A_ρ.

Again as in [6] we shall consider for each $\omega^* \varepsilon \Lambda^p(\mathbb{R}^n)^*$ the closed graded p-trace τ_{ω^*} on Ω defined by

$$\tau_{\omega^*}(a\otimes\omega)=\tau(a)\cdot\omega^*(\omega) .$$

Let us denote the dual basis in $(\mathbb{R}^n)^*$ by $f_1^*,\dots,f_n^*$.

The p-traces corresponding to the elements

$$f_{i_1}^*\wedge f_{i_2}^*\wedge\dots\wedge f_{i_p}^* \qquad p>0$$

will be denoted by $\tau_{i_1,\dots,i_p}$. Note that the explicit formula of $\tau_{i_1,\dots,i_p}$ is:

$$\tau_{i_1,\dots,i_p}(a^0,a^1,\dots,a^p)=\sum_{\sigma\in\mathfrak{S}_p}\varepsilon(\sigma)\tau(a^0\delta_{i_{\sigma(1)}}(a^1)\dots\delta_{i_{\sigma(p)}}(a^p))$$

$a^0,a^1,\dots,a^p\in A_\rho^\infty$.

For $p=0$ we get $\tau_{1^*}=\tau$.

The inclusion $A_{\rho,k-1}^\infty \to A_\rho^\infty$ thus defines p-dimensional cycles $(\Omega,d,\tau_{i_1,\dots,i_p})$ satisfying

$$d1=0$$

$$d(u_k a u_k^{-1})=u_k da u_k^{-1} \qquad a\in A_{\rho,k}^\infty ,$$

so that the restriction of $\tau_{i_1,\dots,i_p}$ to each $A_{\rho,k-1}$ is a natural p-trace with respect to the decomposition

$$A_{\rho,k-1}=A_{\rho,k}\times_{\alpha_k}\mathbb{Z} .$$

Moreover since $u_k du_k^{-1}=2i\pi\sum_{i=1}^{n}-\theta_{ik}1\otimes f_i$, from all the p-m traces

$$\hat{\tau}_{i_1,\dots,i_p}(a^0da^1\dots da^{p-m}(u_kdu_k^{-1})^m)$$

only $\hat{\tau}_{i_1,\dots,i_p}(a^0da^1..da^{p-1}(u_kdu_k^{-1}))$ is different from 0 and equals

$$2i\pi\sum_{j=1}^{p}(-1)^{p-j+1}\theta_{i_j,k}\tau_{i_1,\dots,\hat{i}_j,\dots,i_p}(a^0da^1\dots da^{p-1}).$$

Let us also compute the map $\underline{\varphi}_{\alpha_k}$ of Theorem 15. Note that there exists a map

$$\gamma:A_{\rho,k}\to(A_{\rho,k})_{\alpha_k}$$

(where $(A_{\rho,k})_{\alpha_k}$ is by definition the C*-algebra of continuous functions

$f:[0,1] \to A_{\rho,k}$ such that $f(1)=\alpha_k(f(0))$ given by the formula

$$\gamma(a)(t)=\alpha_{k,t}(a)$$

where with our previous notations $\alpha_{k,t}=\alpha_{f_k t}$.

Recall the definition of $\underline{\varphi}_{\alpha_k}$ to see that we have to compute

$$\tau_{i_1,\dots,i_p;\alpha_k}(x^o,x^1,\dots,x^{p+1})=$$

$$=\sum_{i=1}^{p+1}(-1)^{p-i+1}\int_o^1 \hat{\tau}_{i_1,\dots,i_p}(\alpha_{k,t}(x^o)d(\alpha_{k,t}(x^1))\dots d(\alpha_{k,t}(x^{i-1}))\cdot$$

$$\cdot\dot{\widehat{\alpha_{k,t}(x^i)}}d(\alpha_{k,t}(x^{i+1}))\dots d(\alpha_{k,t}(x^{p+1}))dt=\sum_{i=1}^{p+1}(-1)^{p-i+1}\int_o^1\sum_{\sigma\in\mathfrak{S}_p}\varepsilon(\sigma)\cdot$$

$$\cdot\tau(\alpha_{k,t}(x^o)\delta_{i_{\sigma(1)}}(\alpha_{k,t}(x^1))\dots\delta_{i_{\sigma(i-1)}}(\alpha_{k,t}(x^{i-1}))\delta_k(\alpha_{k,t}(x^i))\cdot$$

$$\cdot\delta_{i_{\sigma(i)}}(\alpha_{k,t}(x^{i+1}))\dots\delta_{i_{\sigma(p)}}(x^{p+1}))dt=\int_o^1\sum_{\sigma\in\mathfrak{S}_{p+1}}\varepsilon(\sigma)\cdot$$

$$\cdot\tau(\alpha_{k,t}^{(x^o)}\delta_{i_{\sigma(1)}}(\alpha_{k,t}(x^1))\dots\delta_{i_{\sigma(p+1)}}(\alpha_{k,t}(x^{p+1}))dt=$$

$$=\tau_{i_1,\dots,i_p,k}(x^o,x^1,\dots,x^{p+1}),$$

since the derivations commute with the automorphisms and τ is α invariant.

Thus $\underline{\tau}_{i_1,\dots,i_p;\alpha_k}=q\circ\tau_{i_1,\dots,i_p,k}$.

Theorem 15 thus says that the range of $\tau_{i_1,\dots,i_p}$ on $K_p(A_{\rho,k-1})$ is equal to the subgroup generated by the range of $\tau_{i_1,\dots,i_p}$ on $K_p(A_{\rho,k})$ and the range of

$$\tau_{i_1,\dots,i_p,k}+\sum_{j=1}^{p}(-1)^{p-j}\theta_{i_j,k}\tau_{i_1,\dots,\hat{i}_j,\dots,i_p}$$

on $\ker\beta=K_{p+1}(A_{\rho,k})$.

THEOREM 16. (G.A. Elliott). *Let* G *be a torsion free abelian group and* $\theta:G\wedge G\to\mathbb{R}$ *be a homomorphism. Let* ρ *be the character* $\rho=\exp\circ\theta$ *of* $G\wedge G$, *and* A_ρ *the corresponding* C*-*algebra.*

Given any trace τ *on* A_ρ, *the range of* τ *on* $K_o(A_\rho)$ *is the range* of the map

$$\exp_\Lambda \theta = 1+\theta+\frac{1}{2}(\theta\wedge\theta)+\ldots : \Lambda G \to \mathbb{R}\ .$$

PROOF. It is sufficient to consider the case when $G=\mathbb{Z}^n$ for some n. Consider then the decomposition series of A_ρ

$$A_\rho = A_{\rho,o} \supset A_{\rho,1} \supset \cdots \supset A_{\rho,n} = \mathbb{C}\ ,$$

$$A_{\rho,k-1} = A_{\rho,k} \times_{\alpha_k} \mathbb{Z}\ .$$

An easy induction argument, using the above form of Theorem 15, shows that the range of τ on A_ρ is the subgroup generated by the following 2^n traces on K_o of $A_{\rho,n}=\mathbb{C}$. For each $i_1<i_2<\ldots<i_p$

$$\varphi_{i_1,\ldots,i_p} = \tau_{i_1,\ldots,i_p} + \frac{1}{(p-2)!2!}\sum_{\sigma\in\mathfrak{S}_p}\varepsilon(\sigma)\tau_{i_{\sigma(1)}\ldots i_{\sigma(p-2)}}\cdot$$

$$\cdot\theta_{i_{\sigma(p-1)}i_{\sigma(p)}} + \frac{1}{2!}\frac{1}{(p-4)!2!2!}\sum_{\sigma\in\mathfrak{S}_p}\varepsilon(\sigma)\tau_{i_{\sigma(1)}\ldots i_{\sigma(p-4)}}\theta_{i_{\sigma(p-1)}i_{\sigma(p-2)}}\cdot$$

$$\cdot\theta_{i_{\sigma(p-1)}i_{\sigma(p)}} + \ldots\frac{1}{m!}\frac{1}{(p-2m)(2!)^m}\sum_{\sigma\in\mathfrak{S}_p}\varepsilon(\sigma)\tau_{i_{\sigma(1)}\ldots i_{\sigma(p-2m)}}\cdot$$

$$\cdot\theta_{i_{\sigma(p-2m+1)}i_{\sigma(p-2m+2)}}\cdots\theta_{i_{\sigma(p-1)}i_{\sigma(p)}} + \ldots$$

where $\mathfrak{S}_p$ is the symmetric group and $\varepsilon(\sigma)$ the signature of the permutation σ. Since $K_o(\mathbb{C})=\mathbb{Z}$ with generator 1 all traces $\tau_{i_1\ldots i_p}$ for $p>0$ vanish on $K_o(\mathbb{C})$ so that only the traces $\varphi_{i_1,\ldots,i_p}$ with p even, $p=2k$, are different from 0 on $K_o(\mathbb{C})$ and their value equals

$$\frac{1}{k!(2!)^k}\sum_{\sigma\in\mathfrak{S}_p}\varepsilon(\sigma)\theta_{i_{\sigma(1)}i_{\sigma(2)}}\cdots\theta_{i_{\sigma(p-1)}i_{\sigma(p)}} =$$

$$= \frac{1}{k!}\underbrace{\theta\wedge\theta\wedge\ldots\wedge\theta}_{k\text{-times}}(e_{i_1}\wedge\ldots\wedge e_{i_p})\ .$$

Q.E.D.

REFERENCES

1. Atiyah, M.F.: *K-Theory*, Benjamin, 1967.

2. Connes, A.: An analogue of the Thom isomorphism for crossed products of a C*-algebra by an action of $\mathbb{R}$, *Adv. in Math.*, 39 (1981), 31-55.

3. Connes, A.: C*-algèbras et géometrie differentielle, *C.R. Acad. Sci. Paris, Ser. A*, 290 (1980), 599-604.

4. Connes, A.: Non commutative differential geometry. I: The Chern character in K-homology, *Publ. IHES*, to appear.

5. Connes, A.: Non commutative differential geometry. II: De Rham homology and noncommutative algebra, Preprint IHES, 1983.

6. Connes, A.: Cyclic cohomology and the transverse fundamental class of a foliation, Preprint IHES, 1984.

7. Cuntz, J.: K-theory for certain C*-algebras. II, *J. Operator Theory*, 5 (1981), 101-108.

8. Douglas, R.G.: *Banach algebra techniques in operator theory*, Academic Press, 1972.

9. Elliott, G.A.: On the K-theory of the C*-algebra generated by a projective representation of a torsion free discrete abelian group, in *Operator Algebras and Group Representations*, vol.I, Pitman, 1983, pp.157-184.

10. de la Harpe, P.; Skandalis, G.: Determinant associé a une trace sur une algèbre de Banach, Preprint,1982.

11. Karoubi, M.: *K-theory. An introduction*, Grundlehren der Math. Wiss. no.226, Springer Verlag, 1976.

12. Pedersen, G.K.: *C*-algebras and their automorphism groups*, Academic Press, 1979.

13. Pimsner, M.; Voiculescu, D.: Imbedding the irrational rotation C*-algebra into an AF algebra, *J. Operator Theory*, 4 (1980), 201-210.

14. Pimsner, M.; Voiculescu, D.: Exact sequences for K-groups and Ext-groups of certain cross-product C*-algebras, *J. Operator Theory*, 4 (1980), 93-118.

15. Pimsner, M.; Voiculescu, D.: K-groups of reduced crossed products by free groups, *J. Operator Theory*, 8 (1982), 131-156.

16. Riedel, N.: Classification of the C*-algebras associated with minimal rotations, *Pacific J. Math.*, 101 (1982), 153-161.

17. Rieffel, M.A.: *Irrational rotation C*-algebras*, short communication at the International Congress of Mathematicians, 1978.

18. Rieffel, M.A.: C*-algebras associated with irrational rotations, *Pacific J. Math.*, 93 (1981), 415-429.

19. Sinai, Ia.G.; Kornfeld, I.P.; Fomin, S.V.: *Ergodic theory* (Russian), Moscow, 1980.

20. Taylor, J.L.: Banach algebras and topology, in *Algebras in Analysis*, Academic Press, 1975, pp. 118-186.

M. Pimsner

Department of Mathematics, INCREST

Bdul Păcii 220, 79622 Bucharest

Romania.

K-theory of the reduced C^*-algebra of $SL_2(Q_p)$

R.J. Plymen

Introduction

We consider SL_2 over the p-adic field Q_p with $p \neq 2$, and compute the K-groups of the reduced C^*-algebra of SL_2. The K-groups K_o and K_1 are both free abelian of infinite rank. We describe how the special representation, the supercuspidal representations, and the principal series representations parametrize generators of K_o or K_1 (see the Theorem at the end of section 2).

Julg and Valette associate, to the action of SL_2 on its tree, a Fredholm SL_2-module, and prove that SL_2 is K-amenable [10, 11]. This implies that the quotient map $C^*(SL_2) \longrightarrow C^*_r(SL_2)$ induces an isomorphism of K-groups:

$$K(C^*(SL_2)) \cong K(C^*_r(SL_2)).$$

The K-groups K_o and K_1 of the full C^*-algebra $C^*(SL_2)$ are therefore both free abelian of infinite rank.

I would like to thank the referee for valuable comments on an earlier version of this article.

1. The reduced dual of SL_2.

Let F denote the p-adic field Q_p where $p \neq 2$. We have $F^\times \cong Z \times U$, where U is the group of p-adic units, and the dual of $F^\times$ is therefore

$$(F^\times)^\wedge \cong T \times \hat{U}$$

where T is the circle group. The group $\mathbb{Z}$ has a unique character of order 2, sending n to $(-1)^n$. The group U has a unique character of order 2 : the Legendre character λ which sends u to the Legendre symbol of the mod p reduction of u. The structure of $F^\times$ is described in Serre [16, p.17].

Thus $F^\times$ has 3 unitary characters of order 2, given by

$$\phi_1(p^n u) = (-1)^n$$

$$\phi_2(p^n u) = \lambda(u)$$

$$\phi_3(p^n u) = (-1)^n \lambda(u) \qquad u \in U.$$

It is elementary to check that ϕ_1 is the character of order 2 on $F^\times$ whose kernel is $\mathrm{Norm}(V^\times)$ where V is the unramified quadratic extension of F. The characters ϕ_2, ϕ_3 correspond similarly to the 2 ramified quadratic extensions of F.

The principal series of unitary representations of SL_2 is given by

$$\pi_\phi(g).f(x) = f\left(\frac{ax+c}{bx+d}\right).\phi(bx+d).|bx+d|^{-1} \qquad g = \begin{pmatrix} a & b \\ c & d \end{pmatrix}$$

where $f \in L^2(F)$ and ϕ is a unitary character of $F^\times$.

The representation π_ϕ is irreducible unless ϕ is a character of order 2. Let ϕ be a character of order 2. Let H_ϕ^- be the subspace of all f in $L^2(F)$ which vanish on the kernel of ϕ; and let H_ϕ^+ be the orthogonal complement of H_ϕ^- in $L^2(F)$. The subspaces H_ϕ^+, H_ϕ^- determine irreducible subrepresentation of SL_2; see [7, p.164].

The representation π_ϕ is unitarily equivalent to π_ψ if and only if $\phi = \psi$ or $\phi = \psi^{-1}$. The Weyl group of SL_2 is $W = Z/2Z$. The Weyl group acts on the diagonal subgroup of SL_2, hence on $F^\times$ and its unitary dual. Let w be the generator of W. Now

$$(F^\times)^\wedge \cong T \times \hat{U} .$$

If $\phi = (z, \mu) \in T \times \hat{U}$ then $w\phi = (z^{-1}, \mu^{-1})$. There are two cases:

(i) The discrete parameter μ is fixed by W: in which case $\mu = 1$ or λ, where λ is the Legendre character. Then $\pi_\phi \cong \pi_\psi$ where $\phi = (z, \mu)$ and $\psi = (z^{-1}, \mu)$.

(ii) The discrete parameter μ is not fixed by W. Then $\pi_\phi \cong \pi_\psi$ with $\phi = (z, \mu)$, $\psi = (z^{-1}, \mu^{-1})$ and $\mu \neq \mu^{-1}$.

The parameter space of the principal series is $(T \times \hat{U})/W$. This is a Hausdorff space in which each connected component is either a circle or a copy of the closed unit interval: there are countably many circles and two intervals.

The special representation of SL_2 may be viewed as the natural representation of SL_2 on the space H of square-integrable harmonic 1-forms on the tree of SL_2. A square-integrable 1-form is harmonic if it is annihilated by the simplicial Laplace operator. The character of the special representation is the Steinberg character $\theta - 1$, where θ is the induced character from the trivial representation of a minimal parabolic subgroup. These statements are a special case of results of Borel [1].

By a result of Casselman [3, p.416], the special representation σ is $L^{1+\varepsilon}$ for each $\varepsilon > 0$. According to Gulizia [9], the Kunze-Stein phenomenon holds good for SL_2, and there is a continuous embedding

$$L^p(G) \longrightarrow C_r^*(G) \qquad 1 < p < 2$$

where $G = SL_2$. Let d be the formal degree of σ, and let ξ be a unit vector in the Hilbert space H. Let

$$e = d(\sigma(.)\xi, \xi).$$

Then e is a minimal projection in $C_r^*(G)$. As in Valette [15], the singleton $\{\sigma\}$ is therefore open in $\hat{G}_r$, the dual of $C_r^*(G)$. Since $\hat{G}_r$ is a T_1-space, σ is an isolated point in $\hat{G}_r$.

Gelbart describes in [8, §7] the construction of the supercuspidal representations of SL_2 in terms of the Weil representation of the symplectic group over a quadratic extension V of F. The representations so constructed are denoted $\Pi(\pm 1, \psi, V)$ where ψ is a unitary character of the norm one group V^1 of V. If $\psi \neq 1$, then $\Pi(\pm 1, \psi, V)$ (or its irreducible components when ψ is the character of order 2) is a supercuspidal representation [13].

The matrix coefficient function of a supercuspidal representation Π has compact support. Hence Π is integrable. Therefore Π is an isolated point in the reduced dual, by Dixmier [5, 18.4.2].

There are 3 quadratic extensions V of F. If ψ is the trivial character, then the representations $\Pi(\pm 1, \psi, V)$ are equivalent to the irreducible components of the reducible representations in the principal series for SL_2; see [13]. The 3 pairs of representations $\Pi(\pm 1, 1, V)$, corresponding to

$\psi = 1$, we shall call the limit supercuspidal representations, by analogy with the case of $SL(2, \mathbb{R})$.

The special representation, the supercuspidal representations and the principal series representations comprise the reduced dual of SL_2 (the support of the Plancherel measure). This can be inferred from the Plancherel Theorem for SL_2; see [7, p.212] and [14].

2. The reduced C^*-algebra of SL_2

Any square-integrable representation π of G gives an element $[\pi]$ in $\mathrm{Ext}_o\, C_r^*(G)$, and a minimal projection e in $C_r^*(G)$. If $[e]$ is the class of e in $K_0(C_r^*(G))$, the pairing between K-homology and K-theory sends $[e] \otimes [\pi]$ to $\dim \pi(e) = 1$. This shows that $[e]$ has infinite order in $K_0(C_r^*(G))$. Thus any square-integrable representation of $G = SL_2$ contributes a generator to $K_0(C_r^*(G))$.

We now consider the contribution to $K\,(C_r^*(G))$ made by the principal series.

Let Q be the parameter space of the principal series. Thus Q is the quotient by the Weyl group $\mathbb{Z}/2\mathbb{Z}$ of $(F^\times)\hat{}$. Here $(F^\times)\hat{}$ is given the Pontryagin dual topology, so that $(F^\times)\hat{}$ has countably many connected components: and each connected component is a copy of the circle T in its Euclidean topology. As in section 1, the $\mathbb{Z}/2\mathbb{Z}$-identifications are such that each connected component in Q is a circle or a copy of the closed unit interval: there are countably many circles and two intervals. Let $\hat{G}_P$ be the union of the irreducible principal series and the irreducible components of the reducible principal series. If $\phi_n \longrightarrow \phi$ in Q and ϕ is not of order 2 then $\pi_{\phi_n} \longrightarrow \pi_\phi$ in the natural topology on $\hat{G}_P$; if $\phi_n \longrightarrow \phi$

in Q and ϕ is of order 2 then $\pi_{\phi_n} \longrightarrow \pi_\phi^+$ and $\pi_{\phi_n} \longrightarrow \pi_\phi^-$ in the <u>natural topology</u> on $\hat{G}_p$. The natural topology, so defined, is non-Hausdorff with 3 double-points.

<u>Lemma 1</u>. The natural topology and the hull-kernel topology on $\hat{G}_p$ coincide.

Let $H(G)$ be the Hecke algebra of locally constant compactly supported functions on G, and let D be the set of minimal central projections in $L^1(SL_2(Z_p))$. Since $H(G)$ is dense in $C_r^*(G)$, the set B of all finite linear combinations of functions in $D * H(G) * D$ is dense in $D * C_r^*(G) * D$. Hence, B is a dense * subalgebra of $C_r^*(G)$ all of whose elements are boundedly represented in $\hat{G}_r$, i.e.

$$\text{rank}\ \pi(f) \leq N \quad \text{all} \quad \pi \epsilon \hat{G}_r$$

where $f \epsilon B$: cf. [12, p.401].

We now apply a result of Fell [6, p.391] to give

<u>Lemma 2</u> (Fell). Let $\{\pi_n : n = 1,2,3,\ldots\}$ and $V_1, \ldots, V_k$ be elements of $\hat{G}_r$ (not necessarily distinct) such that

$$\lim_{n\to\infty} \text{tr}\ \pi_n(f) = \sum_{i=1}^{k} \text{tr}\ V_i(f) \quad \text{for all } f \text{ in } B.$$

Then for any π in $\hat{G}_r$, $\pi_n \longrightarrow \pi$ in the hull-kernel topology if and only if $\pi = V_i$ for some $i \epsilon \{1,\ldots,k\}$.

<u>Lemma 3</u>. Suppose $q_n \longrightarrow q$ in Q (with the natural topology). Then for each f in $H(G)$

$$\int_G f(g)\theta_n(g)dg \longrightarrow \int_G f(g)\theta(g)dg$$

where $\theta_n (n \in N)$ and θ are the Harish-Chandra characters corresponding to q_n and q.

<u>Proof</u>. Clearly if $q_n \longrightarrow q$ in the natural topology, we may assume that q_n and q lie on the same circle $T \subset Q$. The character formula [7, p.200] is given by

$$(*) \qquad \theta_n(g) = \begin{cases} \dfrac{\phi_n(\lambda_g)+\phi_n(\lambda_g^{-1})}{|\lambda_g - \lambda_g^{-1}|} & \lambda_g \in F \\ 0 & \text{otherwise} \end{cases}$$

where λ_g and λ_g^{-1} are the eigenvalues of $g \in SL_2$. Here

$$\phi_n(x) = \exp(2\pi i m\alpha_n)\chi(u)$$

$$\phi\ (x) = \exp(2\pi i m\alpha)\ \chi(u)$$

$$x = (m, u) \in \mathbb{Z} \times U \quad .$$

and $\alpha_n \longrightarrow \alpha$ in T.

By $(*)$ we see that $\theta_n(g) \longrightarrow \theta(g)$ for any fixed g in G, so the lemma will be proved if we can apply the dominated convergence theorem. However, by $(*)$

$$|\theta_n(g)| \leqslant \frac{2}{|\lambda_g - \lambda_g^{-1}|}, \qquad (\forall n \in N)$$

The right hand side is locally integrable, so the result follows, as f has compact support.

<u>Proof of Lemma 1</u>. Let $\{\pi_n\}$ be a sequence in $\hat{G}_P$ and let $\pi \in \hat{G}_P$. Let $q_n = p(\pi_n)$ and $q = p(\pi)$. We need to prove that $\pi_n \longrightarrow \pi$ in the hull-kernel topology if and only if $q_n \longrightarrow q$ in the natural topology. Here, p is the projection of $\hat{G}_P$ onto Q.

(1) Suppose $q_n \longrightarrow q$. Then for all f in B, $\lim_{n\to\infty} \operatorname{tr} q_n(f) = \operatorname{tr} q(f)$ by Lemma 3. Thus by Lemma 2, $\pi_n \longrightarrow \pi$ in the hull-kernel topology.

(2) Suppose $\pi_n \longrightarrow \pi$ in the hull-kernel topology.

If $\{q_n\}$ has no cluster points, then $q_n \longrightarrow \infty$ and so

$$\operatorname{tr} q_n(f) \longrightarrow 0$$

by the Riemann-Lebesgue Lemma. Thus, by Lemma 2, π_n has no limit, a contradiction. So we assume $\{q_n\}$ has a cluster point, say $\bar{q}$. By choosing a subsequence if necessary, we may assume that $q_n \longrightarrow \bar{q}$. But then by part (1) above, $\bar{q} = p(\pi)$. Hence $\bar{q} = q$ as required.

Note that SL_2 is a liminal group [7, p.227].

Let

$$\mathbb{I} = \{x \in \mathbb{R} : -1 \leqslant x \leqslant 1\}$$

$$k = C^*\text{-algebra of compact operators on } L^2(F)$$

$$A = \{f \in C(\mathbb{I},k) : f(-1) \text{ leaves } H_1^+ \text{ and } H_1^- \text{ invariant}\}$$

$$B = \{f \in C(\mathbb{I},k) : f(1) \text{ leaves } H_2^+ \text{ and } H_2^- \text{ invariant}, \quad f(-1) \text{ leaves } H_3^+ \text{ and } H_3^- \text{ invariant}\}$$

where $H_j^{\pm} = H_{\phi_j}^{\pm}$ $\qquad j = 1,2,3$

$$X_1 = \{\pi_1^{\pm}\} \cup (-1,1]$$

$$X_2 = \{\pi_2^{\pm}\} \cup (-1,1) \cup \{\pi_3^{\pm}\}$$

$$X_n = T \qquad n = 3,4,5,\ldots.$$

Let I_n denote the ideal in $C_r^*(G)$ with $\hat{I}_n = X_n$. We now imitate the proof in Boyer and Martin [2, p.373]. Each I_n is a direct summand of $C_r^*(G)$. Let J_1 be the ideal of I_1 with $\hat{J}_1 = (-1,1]$. Then J_1 is a C^*-algebra with continuous trace. Since $H^3((-1,1]; \mathbb{Z}) = 0$ it follows that J_1 is isomorphic to $C_0((-1,1], k)$. Let J_2 be the ideal of I_2 with $\hat{J}_2 = (-1,1)$. Then J_2 is isomorphic to $C_0((-1,1), k)$. By the extension theory of Delaroche [4] we conclude that I_1 is isomorphic to A, and I_2 is isomorphic to B.

Thus I_1 is an extension of the cone Ck by k^2 : the short exact sequence is

$$0 \longrightarrow Ck \longrightarrow I_1 \xrightarrow{\alpha} k^2 \longrightarrow 0$$

where $\alpha(f) = f(-1)$. Since $K(Ck) = 0$, it is clear from the 6-term exact sequence that $K_o(I_1) = K_o(k^2) = \mathbb{Z}^2$, $K_1(I_1) = K_1(k^2) = 0$.

Let $\{v_t : t \in \Pi\}$ be a continuous field of unitary operators on $L^2(F)$ such that $v_{-1} = I$, $v_1(H_2^+) = H_3^+$. This field induces an isomorphism of B onto

$$B' = \{f \in C(\Pi, k) : f(1), f(-1) \text{ leave } H_2^+, H_2^- \text{ invariant}\}.$$

Now B' is an extension of the suspension Sk by k^4 : the short exact sequence is

$$0 \longrightarrow Sk \longrightarrow B' \xrightarrow{\beta} k^4 \longrightarrow 0$$

where $\beta(f) = f(1) \oplus f(-1)$. Now $K_o(Sk) = 0$, $K_1(Sk) = \mathbb{Z}$, and the connecting map $K_o(\mathbb{Z}^4) \longrightarrow K_1(Sk)$ is the alternating sum from $\mathbb{Z}^4$ to $\mathbb{Z}$. Hence $K_o(B') = \mathbb{Z}^3$, $K_1(B') = 0$.

When $n \geq 3$, $\hat{I}_n$ is homeomorphic to the circle T. Since $H^3(T; \mathbb{Z}) = 0$, we have $I_n \cong C(T, k)$. Thus $K_o(I_n) = \mathbb{Z} = K_1(I_n)$. Each square-integrable representation determines an isolated point in the reduced dual, and therefore contributes a direct summand k to the reduced C^*-algebra.

This establishes the following result.

<u>Theorem</u>.

(i) The reduced dual of SL_2 is the non-Hausdorff space

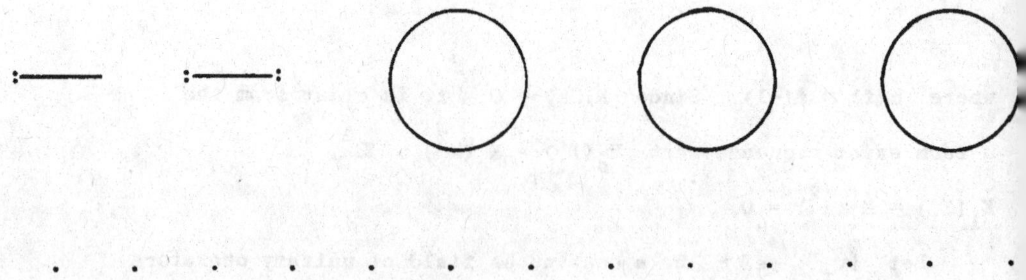

(This diagram is infinitely extended to the right).

The 3 double-points in the reduced dual represent the 3 pairs $\{\pi_\phi^+, \pi_\phi^-\}$ where ϕ is a character of order 2. The interval with 1 double-point has discrete parameter $1 \in \hat{U}$; the interval with 2 double-points has discrete parameter $\lambda \in \hat{U}$; each circle has, as discrete parameter, a W-orbit in $\hat{U}$ on which W acts freely.

(ii) The reduced C^*-algebra of SL_2 is isomorphic to

$$I_1 \oplus I_2 \oplus I_3 \oplus \ldots \oplus k \oplus k \oplus k \oplus \ldots$$

(iii) The table of K-groups is

	I_1	I_2	I_n	k
K_0	$\mathbb{Z}^2$	$\mathbb{Z}^3$	$\mathbb{Z}$	$\mathbb{Z}$
K_1	0	0	$\mathbb{Z}$	0

$n \geqslant 3$

REFERENCES

1. A. Borel. Admissible representations of a semisimple group over a local field with vectors fixed under an Iwahori subgroup. Inventiones Math. 35 (1976) 233-259.

2. R. Boyer and R. Martin. The regular group C^*-algebra for real-rank one groups. Proc. Amer. Math. Soc. 59 (1976) 371-376.

3. W. Casselman. The Steinberg character as a true character. Proc. Symp. Pure Math. 26 (1973) 413-418.

4. C. Delaroche. Extensions des C^*-algèbres. Bull. Soc. Math. France, Mémoire 29 (1972).

5. J. Dixmier. C^*-algebras. North Holland, Amsterdam 1977.

6. J.M.G. Fell. The dual spaces of C^*-algebras. Trans. Amer. Math. Soc. 94 (1960) 365-403.

7. I.M. Gelfand, M.I. Graev, I.I. Pyatetskii-Shapiro. Representation theory and automorphic functions. Saunders, Philadelphia, 1969.

8. S.S. Gelbart. Automorphic forms on adèle groups. Annals of Math. Studies 83, Princeton, 1975.

9. C.L. Gulizia. Harmonic analysis of SL(2) over a locally compact field. J. Functional Analysis 12 (1973) 384 - 400.

10. P. Julg and A. Valette. K-moyennabilité pour les groupes opérant sur les arbres. C.R. Acad. Sci. Paris 296 (1983) 977-980.

11. —————— . Groups acting on trees and K-amenability. These proceedings 1983.

12. R.L. Lipsman. The dual topology for the principal and discrete series on semisimple groups. Trans. Amer. Math. Soc. 152 (1970) 399-417.

13. P.J. Sally. Character formulas for SL_2. Proc. Symp. Pure Math. 26 (1973) 395-400.

14. P.J. Sally and J.A. Shalika. The Plancherel formula for SL_2 over a local field. Proc. Nat. Acad. Sci. U.S.A. 63 (1969) 661-667.

15. A. Valette. Minimal projections, integrable representations and property (T). Preprint 1983.

16. J-P. Serre. A course in arithmetic. Springer-Verlag, Berlin, 1973.

Mathematics Department

The University

Manchester M13 9PL

England.

HYPERFINITE SUBALGEBRAS NORMALIZED BY A GIVEN AUTOMORPHISM AND RELATED PROBLEMS

Sorin Popa

Department of Mathematics, INCREST
Bdul Pacii 220, 79622 Bucharest
Romania.

INTRODUCTION

When Murray and von Neumann proved the uniqueness of the hyperfinite II_1 factor R they also showed that any type II_1 factor contains copies of it as a subfactor. Since the properly infinite factors split the I_∞ factor in which R acts, it follows that R is actually contained in any infinite dimensional factor. However if one requires in addition that R be the range of a normal conditional expectation of the ambient factor (abreviated NCE subalgebra) then the problem becomes much more complicated and is solved affirmatively only as a consequence of A.Connes' discrete decomposition (for III_λ, $0 \leq \lambda < 1$, factors) and of the Connes-Stormer transitivity theorem (for III_1 factors) in [2] and [6]. Thus any (weakly) separable factor contains R as an NCE subalgebra or, equivalently, has a normal faithful state whose centralizer contains R.

The existence of a normal conditional expectation of an algebra M onto a subalgebra $N \subset M$ is a natural condition to ask and, for properly infinite factors, the least that can be said about N to relate it with M. We consider in this paper only such subalgebras.

An important additional condition to require for a hyperfinite II_1 subfactor in a factor M is to have trivial relative commutant. But then we have to exclude the cases when M is II_∞ or III_0: such algebras don't have NCE subfactors of type II_1 with trivial relative comutant (cf [2], and 3.6 below). For III_1 factors nothing is known on this problem. For II_1 and III_λ, $0 < \lambda < 1$, factors the answer is yes: by [12] and the discrete decomposition any such factor M contains a copy of R which is the range of a normal conditional expectation of M and has trivial relative commutant in M.

One can then ask a more general problem: what hyperfinite NCE subfactors R_0 (not necessary of type II_1) with trivial relative commutant are contained in a given type III factor M? When M is of type III_0 a particularly nice answer can be given: if (N,θ) is a discrete decomposition of M such that θ acts as an odometer on the center of N (cf. 5.3.6 in [2]) then N has a hyperfinite II_∞ NCE subalgebra $R_0 \subset N$ with $R_0' \cap N = Z(N) = Z(R_0)$ such that R_0 is normalized by an inner perturbation of θ. Thus in particular, if $R_1 = R_0 \times \theta$ then R_1 is a hyperfinite III_0 NCE subfactor of M with trivial relative commutant (see the remark in the introductory part of [15]).

This fact brings into attention the following type of problem: given a type II_1 (or II_∞) factor M and θ an automorphism of it does there always exists an NCE hyperfinite subfactor $R_0 \subset M$ such that $R_0' \cap M = \mathbf{C}$ and $\theta(R_0) = R_0$?

This is not true in general, since in [14] we construct an example of a type II_1 factor M with an automorphism θ such that if $B \subset M$ is an invariant hyperfinite subalgebra (not necessary a factor), then $B = \mathbf{C}$. So we study the problem up to perturbations of θ by inner automorphisms.

The main result we prove in this paper is that for any aperiodic automorphism θ of a separable II_1 factor M there exists a unitary element $u \in M$ such that $(\mathrm{Ad}\,u)\circ\theta$ normalizes a hyperfinite II_1 subfactor $R \subset M$ with $R' \cap M = \mathbb{C}$ (Theorem 2.1). As a consequence, when θ has modulus one it follows a similar result for II_∞ factors. When θ has modulus less than one, we show that there exists a hyperfinite II_∞ NCE subfactor normalized by θ (Theorem 3.1).

Aside these results the rest of the paper is expository. In Section 1 we discuss the II_1 case of the iterative technique of constructing maximal abelian $*$-subalgebras and hyperfinite subalgebras with trivial relative commutant that we introduced in [12]. In the last part of Section 3 we detail the above mentioned construction of the hyperfinite III_0 NCE subfactor with trivial relative commutant in an arbitrary type III_0 factor.

In the last section we comment on some related open problems.

§1. CONSTRUCTION OF HYPERFINITE SUBALGEBRAS

Let M be a separable finite von Neumann algebra with a fixed normal faithful trace τ, $\tau(1) = 1$. We denote $\|x\|_2 = \tau(x^*x)^{\frac{1}{2}}$, $x \in M$. If $B \subset M$ is a von Neumann subalgebra then E_B is the unique normal faithful τ-preserving conditional expectation of M onto B (cf. [17]).

The next lemma gives a criterion for an abelian von Neumann subalgebra in M to be maximal abelian.

1.1. LEMMA. *Let $A \subset M$ be an abelian von Neumann subalgebra.*

(i). *There exists an increasing sequence of finite dimensional von Neumann subalgebras $A_n \subset A$, $n \geq 1$, such that $A = \overline{\bigcup_n A_n}^w$. If in addition A is diffuse then A_n, $n \geq 1$, can be chosen such that for each n the minimal projections of A_n have equal traces.*

(ii). *If A_n is an increasing sequence of finite dimensional von Neumann subalgebras in A with $A = \overline{\bigcup_n A_n}^w$, then A is maximal abelian in M if and only if*

$$\|E_{A_n' \cap M}(x) - E_{A_n}(x)\|_2 \to 0$$

for all $x \in M_0$, where $M_0 \subset M$ is a total subset.

PROOF. (i) is an easy consequence of the separability of M. (ii) follows if we first note that

$$\|E_{A_n}(x) - E_A(x)\|_2 \to 0 \quad , \quad \|E_{A_n' \cap M}(x) - E_{A' \cap M}(x)\|_2 \to 0$$

for all $x \in M$. Thus $\|E_{A_n' \cap M}(x) - E_{A_n}(x)\|_2 \to 0$ is equivalent to $E_{A' \cap M}(x) = E_A(x)$, $x \in M_0$. Since M_0 is total in M this last condition is equivalent to $E_{A' \cap M} = E_A$ and thus to $A' \cap M = A$. But this means that A is maximal abelian in M. **Q.E.D.**

Why is it important to have such criteria for our problems ? The reason is quite simple. To construct in a factor M a hyperfinite subfactor (or subalgebra) $R \subset M$ with trivial relative commutant (or more generally $R' \cap M \subset R$) we construct R as an increasing sequence of finite dimensional subalgebras such that their diagonals increase to a maximal abelian subalgebra A in M. Then this implies that $R' \cap M \subset A' \cap M = A \subset R$. In particular if we can show (by other methods) that R is a factor then $R' \cap M = \mathbb{C}$.

To get an idea of how 1.1 can be used to make this work let's prove here a particular case of Theorem 3.2 in [12].

1.2. THEOREM. *Let* M *be a separable type* II_1 *factor, then* M *contains a hyperfinite* II_1 *subfactor* R *having a Cartan subalgebra* A *which is maximal abelian in* M. *In particular the subfactor* R *has trivial relative commutant in* M.

PROOF. Let $(x_n)_n \subset M$ be a sequence of elements dense in M in the norm $\|\cdot\|_2$. We construct inductively finite dimensional subfactors $M_n \subset M$ with matrix units $\{e_{ij}^n\}_{i,j}$ such that if $A_n = \operatorname{span}\{e_{ii}^n\}_i$ then:

(a) $M_n \supset M_{n-1}$, $A_n \supset A_{n-1}$;

(b) each e_{ij}^{n-1} is the sum of some e_{rs}^n;

(c) $\| E_{A_n' \cap M}(x_j) - E_{A_n}(x_j) \|_2 < 2^{-n}$, $1 \leq j \leq n$.

Assume we have constructed $M_1, \ldots, M_{n-1}$ with the above properties. Let B^o be a maximal abelian subalgebra in $e_{11}^{n-1} M e_{11}^{n-1}$ and let $B_m^o \subset B^o$ an increasing sequence of finite dimensional von Neumann subalgebras in B^o such that $\overline{\bigcup_m B_m^o}^w = B^o$ and such that each B_m^o has mutually equivalent minimal projections (cf. 1.1 (i)). It follows that $B = \sum_i e_{i1}^{n-1} B^o e_{1i}^{n-1}$ is maximal abelian in M and $B_m = \sum_i e_{i1}^{n-1} B_m^o e_{1i}^{n-1}$ increase (with m) to B. By 1.1 (ii) we have $\| E_{B_m' \cap M}(x) - E_{B_m}(x) \|_2 \underset{m}{\to} 0$ for all $x \in M$. Thus for sufficiently large m we have $\| E_{B_m' \cap M}(x_j) - E_{B_m}(x_j) \|_2 < 2^{-n}$, $1 \leq j \leq n$. Let $A_n = B_m$ and $\{f_{rs}\}_{r,s}$ be a matrix unit in $e_{11}^{n-1} M e_{11}^{n-1}$ such that $\operatorname{span}\{f_{ss}\}_s = B_m^o$ (this is possible because the minimal projections of B_m^o are mutually equivalent in M). It follows that if we denote $\{e_{ij}^n\}_{i,j} = \{e_{pq}^{n-1} f_{rs}\}_{p,q,r,s}$ then $\{e_{ij}^n\}_{i,j}$ and the algebra M_n generated by them satisfy (a), (b) and (c).

Let $A = \overline{\bigcup_n A_n}^w$ and $R = \overline{\bigcup_n M_n}^w$. By (c) and 1.1 A is maximal abelian in M. By (b) it follows that the normalizer of A in R generates R, so A is a Cartan subalgebra in R. The algebra R has a unique trace, since it is a limit of factors each one with a unique trace. Thus R is a factor and since it contains a maximal abelian subalgebra of M, $R' \cap M = \mathbb{C}$. Q.E.D.

1.3. COROLLARY. *If* M *is a separable type* III_λ *factor* $0 < \lambda < 1$, *then* M *contains a hyperfinite type* II_1 *subfactor* R, *range of a normal conditional expectation and having a Cartan subalgebra which is maximal abelian in* M. *In particular* R *has trivial relative commutant in* M.

PROOF. By [2] M has a type II_1 NCE subfactor $N_o \subset M$ such that any maximal abelian subalgebra of N_o is maximal abelian in M. By 1.2 if we take $A \subset R \subset N_o$ such that A is a Cartan subalgebra of the hyperfinite II_1 factor R and is maximal abelian in N, then it is also maximal abelian in M. Since there exist normal conditional expectations of M onto N_o and of N_o onto R, R is an NCE subalgebra of M. Q.E.D.

We mention that in fact by Theorem 3.2 in [12] the condition $R'\cap M = \mathbf{C}$ for a hyperfinite $\mathbf{II}_1$ NCE subfactor $R\subset M$ is equivalent to the existence of a Cartan subalgebra in R, maximal abelian in M.

§2. INVARIANT HYPERFINITE SUBALGEBRAS IN $\mathbf{II}_1$ FACTORS

In this section we prove the following equivariant form of 1.2:

2.1. THEOREM. *Let* M *be a separable type* $\mathbf{II}_1$ *factor and* θ *an aperiodic automorphism of* M. *For any* $\varepsilon>0$ *there exists a unitary element* u *in* M *such that:*

(i). $\|u-1\|_2<\varepsilon$;

(ii). $(\mathrm{Ad}\,u)\circ\theta$ *normalizes a hyperfinite subfactor* $R\subset M$ *and a Cartan subalgebra* A *of* R *which is maximal abelian in* M *(so that in particular* $R'\cap M=\mathbf{C}$*).*

The construction of A and R will be similar to that of 1.2. But now we have the additional difficulty of making them invariant to θ. To do this we use A.Connes' Rohlin type theorem ([4]). The unitary u and the factor R will be constructed so that the restriction θ' of $(\mathrm{Ad}\,u)\circ\theta$ to R will be an aperiodic product type action. With slightly more care in the proof we may actually get θ' to be a certain model action (see 2.4 below).

Before we proceed with the proof of the theorem let's describe an example of a type $\mathbf{II}_1$ factor with an automorphism that normalizes no hyperfinite subalgebras other than $\mathbf{C}$. So, let $\mathbf{F_Z}$ be the free group on countable many generators indexed by $\mathbf{Z}$ and let $L(\mathbf{F_Z})$ be the type $\mathbf{II}_1$ factor associated with it. Let θ be the unique automorphism of $L(\mathbf{F_Z})$ that shifts the generators by one. Then it is shown in ([14], 3.5, 2^0) that if $B\subset L(\mathbf{F_Z})$ is a hyperfinite subalgebra and $\theta(B)=B$ then $B=\mathbf{C}$.

The proof of 2.1 splits in two parts: first we construct matrix algebras whose diagonals are "close to be maximal abelian in M" (in the sense of 1.1) and which are normalized by small perturbations of θ. Then we proceed by induction as in the proof of 1.2.

2.2. LEMMA. *Let* N *be a type* $\mathbf{II}_1$ *factor,* σ *an aperiodic automorphism of* N, $y_1,\dots,y_m\in N$ *and* $\delta>0$. *There exist a unitary element* $u\in N$, *a finite dimensional subfactor* $P\subset N$ *with diagonal subalgebra* $B\subset P$ *such that:*

(1) $\|u-1\|_2<\delta$;

(2) $(\mathrm{Ad}\,u)\circ\sigma$ *normalizes* B *and* P;

(3) $\|E_{B'\cap N}(y_i)-E_B(y_i)\|_2<\delta$, $1\le i\le m$.

PROOF. By Connes - Rohlin type theorem for any $s\in\mathbf{N}$ there exist a unitary element $v\in N$ and mutually orthogonal equivalent projections $f_1,\dots,f_s\in N$ such that $\|v-1\|_2<\delta/2$, $\sum_j f_j=1$ and $(\mathrm{Ad}\,v)\circ\sigma(f_i)=f_{i+1}$, $1\le i\le s$, $f_{s+1}=f_1$. In particular we can choose s so that $4s^{-1}<\delta/2$. Denote $\sigma_0=(\mathrm{Ad}\,v)\circ\sigma$. By 1.1 there exists a finite dimensional abelian von Neumann subalgebra B_1 in f_1Nf_1, with mutually equivalent minimal projections, such that in the reduced algebra $f_1Nf_1=N_1$ the following inequalities are satisfied for all $1\le r\le m$, $0\le j\le s-1$:

$$\| E_{B_1'\cap N_1}(\sigma_o^{-j}(f_{j+1}y_rf_{j+1})) - E_{B_1}(\sigma_o^{-j}(f_{j+1}y_rf_{j+1}))\|_2 < \delta.$$

Let $B = \sum_{j=0}^{s-1}\sigma_o^j(B_1)$. If we denote by $B_{j+1} = \sigma_o^j(B_1) = Bf_{j+1}$ and $N_j = f_jNf_j$, $0 \le j \le s-1$, then we have:

$$\| E_{B'\cap N}(y_r) - E_B(y_r)\|_2 = \sum_{j=1}^{s} \| f_j(E_{B'\cap N}(y_r) - E_B(y_r))f_j\|_2^2 =$$

$$= \sum_j \| E_{B'\cap N}(f_jy_rf_j) - E_B(f_jy_rf_j)\|_2^2 = \sum_j \tau(f_j)\| E_{B_j'\cap N_j}(f_jy_rf_j) - E_{B_j}(f_jy_rf_j)\|_2^2 =$$

$$= \sum_j \tau(f_j)\| E_{B_1'\cap N_1}(\sigma_o^{-j+1}(f_jy_rf_j)) - E_{B_1}(\sigma_o^{-j+1}(f_jy_rf_j))\|_2^2 < \delta\sum_j\tau(f_j) = \delta.$$

Thus B satisfies (3). Let k be the number of minimal projections in $B_1 = Bf_1$. We let $\{e_i\}_{1\le i\le ks}$ be an enumeration of the minimal projections of B such that $\sum_{j=0}^{k-1} e_{js+1} = f_1$ and $\sigma_o^i(e_{js+1}) = e_{js+i+1}$ for $s-1 \ge i \ge 0$. This is possible because for all $i \le s-1$ we have $\sigma_o^i(Bf_1) = Bf_{i+1} = \sigma_o(Bf_i)$. Since $\sigma_o(f_s) = f_1$ it follows that $\sigma_o(B(1-f_s)) = B(1-f_1)$ and $\sigma_o(Bf_s)$ is an abelian $*$-subalgebra supported by f_1 whose minimal projections have traces equal to those of Bf_1, namely $(ks)^{-1}$. Hence there exists a unitary element w in f_1Nf_1 such that $w\sigma_o(Bf_s)w^* = Bf_1$ and such that $w\sigma_o(e_{js})w^* = e_{js+1}$ for $j=1,2,\ldots,k-1$ and $w\sigma_o(e_{ks})w^* = e_1$. Thus, if we let $v_1 = w + (1-f_1)$ then v_1 is a unitary element in N, $\|v_1 - 1\|_2 \le 2s^{-1}$ and $(\mathrm{Ad}\, v_1)\circ\sigma_o(e_i) = e_{i+1}$, $1 \le i \le n$, where $n = ks$ and $e_{n+1} = e_1$.

Finally denote by $\sigma_1 = (\mathrm{Ad}\, v_1)\circ\sigma_o = \mathrm{Ad}(v_1v)\circ\sigma$ and by w_1 a partial isometry in N such that $w_1w_1^* = e_1$, $w_1^*w_1 = e_2$. Let P be the matrix algebra generated by B and $\sigma_1^k(w_1)$, $0 \le k \le n-2$. This algebra is an $n\times n$ matrix factor and has a unique matrix unit $\{e_{ij}\}_{i,j}$ such that $e_{ii} = e_i$, $e_{i,i+1} = \sigma_1^{i-1}(w_1)$, $1 \le i \le n-1$. Let $w_2 = e_{1n}\sigma_1(e_{n-1,n}) = e_{1n}\sigma_1(\sigma_1^{n-2}(w_1)) = e_{1n}\sigma_1^{n-1}(w_1)$ and $v_2 = w_2 + (1-e_1)$. Since $\sigma_1(e_{n-1}) = e_n$ and $\sigma_1(e_n) = e_1$, v_2 is a unitary element in N and for $0 \le k \le n-3$ we have

$$(\mathrm{Ad}\, v_2)\circ\sigma_1(e_{k+1,k+2}) = (\mathrm{Ad}\, v_2)\circ\sigma_1(\sigma_1^k(w_1)) =$$
$$= (\mathrm{Ad}\, v_2)(\sigma_1^{k+1}(w_1)) = \sigma_1^{k+1}(w_1) = e_{k+2,k+3} \in P.$$

Moreover we have

$$(\mathrm{Ad}\, v_2)\circ\sigma_1(e_{n-1,n}) = (\mathrm{Ad}\, v_2)\circ\sigma_1(\sigma_1^{n-2}(w_1)) = v_2\sigma_1^{n-1}(w_1)v_2^* =$$
$$= \sigma_1^{n-1}(w_1)v_2^* = \sigma_1^{n-1}(w_1)(\sigma_1^{n-1}(w_1))^*e_n = e_n.$$

Thus if $u = v_2v_1v$ then u satisfies (2) for B and P defined as above. Since

$$\|u-1\|_2 \le \|v_2v_1 - 1\|_2 + \|v-1\|_2 \le \|v_2 - 1\| + \|v_1 - 1\|_2 + \|v-1\|_2 \le$$
$$\le 2n^{-1} + 2s^{-1} + \delta/2 \le 4s^{-1} + \delta/2 < \delta$$

condition (1) is also satisfied. **Q.E.D.**

2.3. REMARK. Let $k \geq 1$ and denote by p_k the k'th prime number. It is easily seen that in the proof of 2.2 we may choose the $n\times n$ matrix algebra P so that n be a product of consecutive prime numbers starting with p_k, i.e. $n = p_k p_{k+1} \cdots$. Indeed, since $n = s\times n_1$, where $4s^{-1} < \delta/2$ and n_1 is the linear dimension of B_1, we can first take $s = p_k p_{k+1} \cdots p_r$ a product of consecutive prime numbers. Next, since the sequence $p_{r+1}, p_{r+1}p_{r+2}, \ldots$ is infinite we can slightly modify B_1 (by increasing its dimension if necessary) such that $n_1 = p_{r+1} \cdots p_m$. Thus $n = sn_1 = p_k p_{k+1} \cdots p_m$. Let's also note that by the construction in the proof of 2.2 $(\mathrm{Ad}\, u)\circ\sigma$ acts transitively on the abelian finite dimensional algebra B. Thus $(\mathrm{Ad}\, u)\circ\sigma|P$ is of the form $\mathrm{Ad}\, v$ for some unitary element $v \in P$ of spectrum $\{\exp(2\pi ik/n\}_{0\leq k\leq n-1}$. Further on, if $n = p_k p_{k+1} \cdots p_m$ as above then there exists a decomposition $P = P_k \times \ldots \times P_m$, where P_i are $p_i \times p_i$ factors, and unitary elements $v_i \in P_i$ of spectrum $\{\exp(2\pi ik/p_i\}_k$, such that v is the product of $v_k, \ldots, v_m$.

PROOF OF 2.1. Let $\{x_n\}_n$ be a sequence of elements in M dense in the norm $\|\cdot\|_2$. We construct recursively unitary elements u_n, matrix algebras M_n with matrix units $\{e^n_{ij}\}_{i,j}$ and diagonal algebras $A_n = \mathrm{span}\{e^n_{ii}\}_i$ such that:

(a) $\|1 - u_n\| < \varepsilon 2^{-n}$ and u_n commutes with M_{n-1};

(b) $M_n \supset M_{n-1}$, $A_n \supset A_{n-1}$ and each e^{n-1}_{rs} is the sum of some e^n_{ij};

(c) $\mathrm{Ad}(u_n u_{n-1} \ldots u_1)\circ\theta$ normalizes M_n and A_n;

(d) $\|E_{A'_n\cap M}(x_i) - E_{A_n}(x_i)\|_2 < 2^{-n}$, $1 \leq i \leq n$.

Suppose we have constructed these objects up to n-1. Then we apply the lemma for $N = M'_{n-1}\cap M$, $\sigma = \mathrm{Ad}(u_{n-1} \ldots u_1)\circ\theta$, $\{y_k\}_k = \{\sum_i e^{n-1}_{ij} x_k e^{n-1}_{ji} \mid j,k\}$, $\delta = \varepsilon 2^{-n}$, to get a unitary $u_n \in N$, a finite dimensional subfactor P with diagonal algebra $B\subset P$ such that $\|u_n - 1\|_2 < \varepsilon 2^{-n}$, $(\mathrm{Ad}\, u_n)\circ\sigma$ normalizes B and P and $\|E_{B'\cap N}(y_k) - E_B(y_k)\|_2 < 2^{-n}$. Let $\{f_{ij}\}_{ij}$ be a matrix unit for P such that $B = \mathrm{span}\{f_{ii}\}_i$. Denote by $\{e^n_{ij}\}_{ij} = \{f_{kr} e^{n-1}_{st}\}_{k,r,s,t}$ and $M_n = \mathrm{span}\{e^n_{ij}\}_{ij}$, $A_n = \mathrm{span}\{e^n_{ii}\}_i$. Then clearly $M_n = (P\cup M_{n-1})''$, $A_n = (B\cup A_{n-1})''$. Since u_n acts identically on M_{n-1} (because $u_n \in N = M'_{n-1}\cap M$) and $\sigma = \mathrm{Ad}(u_{n-1} \ldots u_1)\circ\theta$ normalizes M_{n-1} and A_{n-1} it follows that $(\mathrm{Ad}\, u_n)\circ\sigma$ normalizes M_{n-1}, A_{n-1}, P and B and thus it normalizes M_n and A_n.

To show that A_n satisfies (d) fix k and let $z_j = \sum_i e^{n-1}_{ij} x_k e^{n-1}_{ji} \in N = M'_{n-1}\cap M$. Then we have:

$$E_{A'_n\cap M}(x_k) - E_{A_n}(x_k) = \sum_j (E_{A'_n\cap M}(z_j) - E_{A_n}(z_j))e^{n-1}_{jj}.$$

But $E_{A'_n\cap M}(z_j) - E_{A_n}(z_j) = E_{B'\cap N}(z_j) - E_B(z_j)$ so that

$$\|E_{A'_n\cap M}(x_k) - E_{A_n}(x_k)\|_2^2 = \sum_j \|E_{B'\cap N}(z_j) - E_B(z_j)\|_2^2 \tau(e^{n-1}_{jj}) < 2^{-2n}\sum_j \tau(e^{n-1}_{jj}) = 2^{-2n}$$

which proves (d).

Let now $R = \overline{\bigcup_n M_n}^w$, $A = \overline{\bigcup_n A_n}^w$. By (d) and 1.1 A is maximal abelian in M and by (b) he normalizer of A in R generates R, so A is a Cartan subalgebra of R. By (a) it follows that he sequence $(u_n u_{n-1} \dots u_1)_n$ is Cauchy in $\|\cdot\|_2$ so it is convergent to a unitary element $u \in M$ nd $\|u-1\|_2 < \varepsilon$. Moreover, since $u_n \in M_k'$ for all $n > k$, by (c) it follows that $\mathrm{Ad}(u_n \dots u_1) \circ \theta$ ormalize M_k and A_k for all $n \geq k$, thus $(\mathrm{Ad}\,u) \circ \theta$ normalizes M_k and A_k for all k. Therefore it lso normalizes M and A. **Q.E.D.**

2.4. REMARK. In the preceding proof the restriction θ' of $(\mathrm{Ad}\,u) \circ \theta$ to R is a product ype action. Indeed if $N_k = M_{k-1}' \cap M_k$, with $M_0 = \mathbf{C}$, then each N_k is normalized by θ' so there xist unitary elements $V_k \in N_k$ such that $\theta' | N_k = \mathrm{Ad}\, V_k$. Thus $R = \otimes_k N_k$ and $\theta' = \otimes_k \mathrm{Ad}\, V_k$. Since in he construction we use 2.2 it follows (cf. 2.3) that $\mathrm{Ad}\, V_k$ acts transitively on $A \cap N_k = A_k \cap N_k \subset N_k$. Moreover, by Remark 2.3, for given m and k we may construct M_k such hat if n_k is the dimension of N_k (i.e. N_k is an $n_k \times n_k$ matrix algebra) then n_k is the product of consecutive prime numbers starting with p_m. We can therefore include in the proof of 2.1 an additional induction argument to construct $M_1, \dots, M_k$ (and thus $N_1, \dots, N_k$) such that $n_1 n_2 \dots n_k = p_1 p_2 p_3 \dots$. By 2.3 it follows that there exist mutually commuting subfactors P_m of dimension p_m, $m \geq 1$, such that $P_1, \dots, P_k$ generate M_k, and unitary elements $v_k \in P_k$ with spectrum $\{\exp(2\pi i s/p_k)\}$, such that $R = \bigotimes_k P_k$ and $\theta' = \bigotimes_k \mathrm{Ad}\, v_k$.

§3. INVARIANT HYPERFINITE SUBALGEBRAS IN II_∞ VON NEUMANN ALGEBRAS

In this section we discuss some tipe II_∞ versions of Theorem 2.1. Via the discrete decomposition this has some consequences for embedding problems in III_λ factors, $0 \leq \lambda < 1$.

We are interested only in subalgebras B of type II_∞ von Neumann algebras that are ranges of normal conditional expectations (i.e. NCE subalgebras). For $B \subset N$ to be an NCE subalgebra of the von Neumann algebra N it is sufficient to be generated by finite projections of N (in other words that any normal semifinite trace on N restricts to a semifinite trace on B). We show later (3.6) that if $B' \cap N \subset B$ these two conditions are actually equivalent.

3.1. THEOREM. *Let* N *be a separable type* II_∞ *factor and* θ *an aperiodic automorphism of* N.

(i) *If* θ *has modulus* 1 *then there exists a unitary element* $u \in N$ *such that* $(\mathrm{Ad}\,u) \circ \theta$ *normalizes a type* II_∞ *hyperfinite subfactor* $R_0 \subset N$, *generated by finite projections of* N, *and a Cartan subalgebra* A *of* R_0 *which is maximal abelian in* N.

(ii) *If* θ *has modulus* $\lambda < 1$ *then* θ *normalizes:*

1°. *A hyperfinite von Neumann subalgebra* $R_1 \subset N$ *isomorphic to* $R \bar{\otimes} l^\infty(\mathbf{Z})$ (R *is as usual the hyperfinite* II_1 *factor) such that* $R_1' \cap N = Z(R_1)$ *and such that the minimal projections of* $Z(R_1)$ *are finite projections in* N.

2°. *A hyperfinite* II_∞ *subfactor* $R_0 \subset N$, *generated by finite projections of* N *(but not necessary* $R_0' \cap N = \mathbf{C}$).

PROOF. (i) If we write N as $N_o \otimes F_\infty$ where N_o is a type II_1 factor and F_∞ is the type I_∞ factor then there exists a unitary element $u_1 \in N$ such that $(\mathrm{Ad}\, u_1)\circ\theta = \theta_o \otimes I$ for some aperiodic automorphism θ_o of N_o. By 2.1 there exists $u_o \in N_o$ such that $(\mathrm{Ad}\, u_o)\circ \theta_o$ normalizes a hyperfinite II_1 factor $R \subset N_o$ and a Cartan subalgebra of it, $A_o \subset R$, such that A_o is maximal abelian in N. Thus if A_1 is a diagonal subalgebra of F_∞ then $u = (u_o \otimes I)u_1$, $R_o = R \otimes F_\infty$ and $A = A_1 \otimes A_2$ satisfy the conditions.

(ii). 1°. By [2] there exists a partition of the unity $\{e_n\}_{n\in\mathbf{Z}}$ of finite projections in N such that $\theta(e_n) = e_{n+1}$. Let $R \subset e_o N e_o$ be a hyperfinite II_1 factor such that $R' \cap e_o N e_o = \mathbf{C} e_o$ and denote by $R_1 = \bigoplus_{n\in\mathbf{Z}} \theta^n(R)$. Then R_1 satisfies the conditions.

(ii). 2°. We consider only the case $\lambda = 1/2$, when the formalism is considerably simplified. Let τ be a normal semifinite trace on N. Then $\tau(e_1) = \tau(\theta(e_o)) = \tau(e_o)/2$ (with e_o, e_1 as in (ii) 1°). It is sufficient to show that there exist a type II_1 hyperfinite subfactor $R \subset e_o N e_o$ and a partial isometry $v \in N$ such that $v^*v = e_1$, $vv^* \leq e_o$, $vv^* \in R$ and $v\theta(R)v^* = vv^*Rvv^*$. To do this let first $\{e^1_{ij}\}_{1\leq i,j\leq 2}$ a matrix unit in $e_o N e_o$ and v a partial isometry in N with $v^*v = e_1$, $vv^* = e^1_{11}$. Denote by M_1 the 2×2 matrix algebra generated by e^1_{ij}. Let $e^2_{ij} = v\theta(e^1_{ij})v^* + e^1_{21} v\theta(e^1_{ij})v^* e^1_{12}$, $1\leq i,j\leq 2$. Then $\{e^2_{ij}\}_{1\leq i,j\leq 2}$ is a matrix unit in $e_o N e_o$ and the 2×2 matrix algebra M_2 generated by it commutes with M_1. Moreover $\theta(M_2)$ commutes with $\theta(M_1)$ so that if we denote by $e^3_{ij} = v\theta(e^2_{ij})v^* + e^1_{21} v\theta(e^2_{ij})v^* e^1_{12}$ then $\{e^3_{ij}\}$ is the matrix unit of a 2×2 matrix algebra $M_3 \subset (M_1 \cup M_2)' \cap e_o N e_o$. The induction procedure is now clear. We finally let $R = (\bigcup_k M_k)'' \subset e_o N e_o$ and v as above and R_o is the von Neumann algebra generated by $\theta^k(R)$, $\theta^k(v)$, $k \in \mathbf{Z}$.

When λ is rational the pattern is quite clear. If λ is irrational we may write it in diadic expansion $\lambda = 0, a_1 a_2 \ldots$ and use an analogue argument (as in the rational case) but with the matrix units e^n_{ij} supported on some projections $f_n \leq e_o$ with $f_n \to e_o$. We leave the details as an exercise.

Q.E.D.

3.2. COROLLARY. *Let* M *be a separable type* III_λ *factor,* $0<\lambda<1$. *Then* M *contains a hyperfinite type* III_λ NCE *subfactor.*

PROOF. Let (N,θ) be the discrete decomposition of M and R_o N the hyperfinite II_∞ subfactor of (ii) 2° in 3.1. Since $\theta(R_o) = R_o$ and the trace of N restricts to a semifinite trace on R_o, it follows that $\theta|R_o$ has modulus λ. Thus $R_1 = R_o \rtimes \theta \subset N \rtimes \theta = M$ is a hyperfinite III_λ subfactor. Since there exists a normal conditional expectation of N onto R_o it follows that there exists one from $N \rtimes \theta$ onto $R_o \rtimes \theta$, i.e. of M onto R_1.

Q.E.D.

3.3. THEOREM. *Let* N *be a separable type* II_∞ *von Neumann algebra with diffuse center* Z *and* θ *an automorphism of* N *acting as an odometer on* Z *(see [8]). Then there exist a unitary element* $u \in N$, *a type* II_∞ *hyperfinite von Neumann subalgebra* $R_o \subset N$ *and a Cartan subalgebra* $A \subset R_o$ *such that:*

1°. $(\mathrm{Ad}\,u)\circ\theta$ *normalizes* R_o *and* A;

2°. R_o *and* A *are generated by finite projections of* N;

3°. A *is maximal abelian in* N *and* $R_o'\cap N = R_o'\cap R_o = Z$.

The proof of this theorem uses the same iterative construction of [12] which was explained here in Section 1, more precisely the adaption of this technique described in [1], where it is proved the existence of $A\subset R_o$ in N satisfying only 2° and 3° of the above conditions. Then we only need to remark that because of the nice form of the automorphism θ (an odometer on Z) one can easily make the construction such that at each step of the induction the corresponding finite dimensional algebras are normalized by some small inner perturbtions of θ. We shall however present here the arguments in full detail, for the reader's convenience.

The main interest of the theorem stems in the next:

3.4. COROLLARY. *Let* M *be a separable type* $\mathbf{III}_o$ *factor. Then there exists a discrete decomposition* (N,θ) *of* M *such that* θ *normalizes a hyperfinite* $\mathbf{II}_\infty$ *von Neumann subalgebra* $R_o\subset N$ *and a Cartan subalgebra* A *of* R_o *with the properties:*

1°. A *is maximal abelian in* N;

2°. $R_o'\cap N = R_o'\cap R_o = Z(N)$;

3°. A *and* R_o *are generated by finite projections of* N.

In particular $R_1 = R_o\rtimes\theta$ *is a hyperfinite* $\mathbf{III}_o$ *subfactor of* M, $R_1'\cap M = \mathbb{C}$ *and there exists a normal conditional expectation of* M *onto* R_1.

PROOF OF 3.4. By 5.3.6 in [2], M has a discrete decomposition (N,θ_o) such that θ_o acts as an odometer on N. Then Theorem 3.3 applies to get a unitary $u\in N$, a hyperfinite $\mathbf{II}_\infty$ subalgebra R_o N and a Cartan subalgebra $A\subset R_o$ such that $\theta = (\mathrm{Ad}\,u)\circ\theta_o$, R_o and A satisfy 1°, 2°, 3°. Since $R_o'\cap N = Z(N) = Z(R_o)$ it follows that $\theta|R_o$ acts ergodically on the center of R_o. If τ is a normal semifinite faithful trace on N such that $\tau\circ\theta \le \lambda\tau$ for some $\lambda < 1$ then $\tau|R_o$ also satisfies this condition, thus $(R_o, \theta|R_o)$ is a discrete decomposition of a hyperfinite $\mathbf{III}_o$ factor $R_1 = R_o\rtimes\theta\subset N\rtimes\theta = M$ (cf. [2]). Since R_o is the range of a normal conditional expectation of N, it follows that R_1 is the range of a normal conditional expectation of M. Moreover if $x\in R_1'\cap M$ then $x\in R_o'\cap M$, but $R_o'\cap M = Z(N)$ (cf. [2]) and since $\theta(x) = x$, by the ergodicity of θ on Z, it follows that x is a scalar multiple of the identity, so $R_1'\cap M = \mathbb{C}$. Q.E.D.

To prove 3.3 we need to introduce some notations. We fix a normal faithful trace τ on N which takes finite values on the finite projections of N. We denote $N_\tau = \{x\in N \mid \tau(x^*x) < \infty\}$ and $\|x\|_2 = \tau(x^*x)^{\frac{1}{2}}$. If $B\subset N$ is a weakly closed $*$-subalgebra, the support of B is the unit of B regarded as a projection in N and if this support is a finite projection in N, say s, then we denote by E_B the unique normal conditional expectation of N onto B which is τ-preserving on sNs and has support s. We denote by $\mathcal{G}(B)$ the normalizing groupoid of Bs in sNs, i.e. the set $\{v\in sNs \mid$ there exists a unitary element $u\in sNs$ and a projection $e\in B$ such that $v = ue$ and $uBu^* = B\}$. If $B\subset B_1\subset N$ then $\mathcal{G}_{B_1}(B) = \mathcal{G}(B)\cap B_1$.

For the inductive constructions in the proof of 3.3 we need the following local argument:

3.5. LEMMA. *Let* $\varepsilon>0$, $x_1,\dots x_n \in N_\tau$ *and* $N_o \subset N$ *a finite dimensional* $*$*-subalgebra with diagonal subalgebra* $B_o \subset N_o$ *and finite support* s_o. *There exists a finite dimensional* $*$*-subalgebra* $N_1 \subset N$ *with finite support* s_1 *and diagonal subalgebra* B_1 *such that:*

(1). $B_o \subset B_1$, $N_o \subset N_1$, $G_{N_o}(B_o) \subset G_{N_1}(B_1)$;

(2). $\|s_1 x_i s_1 - x_i\|_2 < \varepsilon$ *and* $\|E_{B_1' \cap s_1 N s_1}(x_i) - E_{B_1}(x_i)\|_2 < \varepsilon$, $1 \leq i \leq n$;

(3). *There exists a projection* p *in the center of* N_1 *such that* $\tau(s_1 - p) < \varepsilon$ *and any two minimal projections in* $N_1 p$ *are either equivalent in* N_1 *or have mutually orthogonal central supports in* N.

PROOF. Choose first $s_1 \in N_\tau$, $s_1 \geq s_o$ with $\|s_1 x_i s_1 - x_i\|_2 < \varepsilon$, $1 \leq i \leq n$, and let $M_o = N_o + \mathbf{C}(s_1 - s_o)$, $A_o = B_o + \mathbf{C}(s_1 - s_o)$. For each central projection c in M_o choose a minimal projection f in $A_o c$ and let e be the sum of these f's. Using 1.1 we can find, as in the first part of the proof of 1.2, a finite dimensional abelian von Neumann subalgebra $A_o^1 \subset eNe$ such that if A_1 is the algebra generated by all $vA_o^1 v^*$ with $v \in G_{M_o}(A_o)$ then (2) is satisfied (for A_1 instead of B_1). Note that A_1 is finite dimensional abelian and with support s_1. Obviously $G_{M_1}(A_1) \supset G_{N_o}(B_o)$ where $M_1 = (A_1 \cup M_o)''$. Since $s_1 N s_1$ is a type II_1 von Neumann algebra there exists a projection $p \in M_1' \cap s_1 N s_1$ such that $\tau(s_1 - p) < \varepsilon$ and any two minimal projections in $M_1 p$ are either comensurable or have mutually orthogonal central supports in pNp (and thus in N). This means in particular that we can refine $M_1 p$ to a finite dimensional $*$-subalgebra $M_1 p \subset N_1^o \subset pNp$ such that any two minimal projections in N_1^o are either equivalent in N_1^o or have mutually orthogonal central supports in pNp (and thus in N). Moreover we may clearly do this so that for some apropriate maximal abelian subalgebra $B_1^o \subset N_1^o$ we have $A_1 p \subset B_1^o$ and $G_{M_1 p}(A_1 p) \subset G_{N_1^o}(B_1^o)$. Finally we let $N_1 = N_1^o + M_1(s_1 - p)$, $B_1 = B_1^o + A_1(s_1 - p)$ which satisfy all the conditions. **Q.E.D.**

PROOF OF 3.3. Let $M = N \times \theta$ and $U_\theta \in M$ the unitary element canonically implementing θ on N. Since θ acts as an odometer on Z there exists matrix units $\{E_{ij}^n\}_{1 \leq i,j \leq 2^n}$, $n \geq 0$, in M such that $E_1^o = 1$, $E_{ii}^n \in Z$, $\{E_{ii}^n \mid i,n\}'' = Z$, $E_{ij}^n = E_{2i-1,2j-1}^{n+1} + E_{2i,2j}^{n+1}$, $E_{ij}^n N E_{ji}^n = N E_{ii}^n$ and $U_\theta = E_{12}^1 + \sum_{n \geq 2} E_{2^n-1,2}^n$. Let also $\{x_n\}_n$ be a dense sequence in N_τ in the norm $\|\cdot\|_2$.

We construct by induction the finite dimensional $*$-subalgebras M_n, with support $s_n \in N_\tau$ and diagonal subalgebra A_n, and the unitaries $u_n \in NE_{22}^n$, such that for all $n \geq 1$ we have:

(a) $\|s_n x_i s_n - x_i\|_2 < 2^{-n}$, $1 \leq i \leq n$;

(b) $A_n \supset A_{n-1}$, $M_n \supset M_{n-1}$, $G_{M_n}(A_n) \supset G_{M_{n-1}}(A_{n-1})$;

(c) There exists a projection p_n in the center of M_n such that $\tau(s_n - p_n) < 2^{-n}$ and any two minimal projections in $M_n p_n$ are either equivalent in M_n or have mutually orthogonal central supports in N;

(d) $\|E_{A_n' \cap s_n N s_n}(x_i) - E_{A_n}(x_i)\| < 2^{-n}$, $1 \leq i \leq n$;

(f) $M_n E^n_{ii} \subset M_n$ and $u_k E^k_{21} M_n E^k_{12} u_k^* = M_n E^k_{22}$, $1 \leq k \leq n$.

Suppose we reached the n-1'th step of the induction. Let u_n be a unitary element in NE^n_{22} such that $u_n E^n_{21} s_{n-1} E^n_{12} u_n^*$ is orthogonal to $s_{n-1} E^n_{22}$ (this is possible because $E^n_{21} s_{n-1} E^n_{12}$ and $s_{n-1} E^n_{22}$ are finite projections in the type II_∞ von Neumann algebra NE^n_{22} !) Denote by $\{F^n_{ij}\}_{1 \leq i,j \leq 2^n}$ the matrix unit generated by $F^n_{kk} = E^n_{kk}$ and $F^n_{21} = u_k E^k_{21}$, $1 \leq k \leq n$. It is easily seen that $E^n_{ij} F^n_{ji} \in N$ (it can be explicitely computed in terms of u_k, $1 \leq k \leq n$). We can now apply the preceding lemma for $N_o = M_{n-1} F^n_{11} + F^n_{12} M_{n-1} F^n_{21}$ (by the choice of u_n this is an orthogonal sum) with diagonal $B_o = A_{n-1} F^n_{11} + F^n_{12} A_{n-1} F^n_{21}$ to get a finite dimensional algebra $N_1 \subset NF^n_{11}$ with diagonal B_1 and a projection $p \in N_1' \cap N_1$ satisfying (1), (2), (3) of the lemma for an ε sufficiently small that if we denote $M_n = \sum_i F^n_{i1} N_1 F^n_{1i}$, $A_n = \sum_i F^n_{i1} B_1 F^n_{1i}$, $s_n = \sum_i F^n_{i1} s F^n_{1i}$, $p_n = \sum_i F^n_{i1} p F^n_{1i}$ then (a), (c) and (d) are fulfilled. By the definition of F^n_{ij} and of M_n (f) is also satisfied. Since $M_{n-1} F^{n-1}_{11} = M_{n-1}(F^n_{11} + F^n_{22}) \subset M_{n-1} F^n_{11} + M_{n-1} F^n_{22} = M_{n-1} F^n_{11} + F^n_{21} \times$ $\times (F^n_{12} M_{n-1} F^n_{21}) F_{12} \subset M_n F^n_{11} + F^n_{21} M_n F^n_{12} = M_n F^n_{11} + M_n F^n_{22} = M_n F^{n-1}_{11}$, we obtain that $M_n \supset$ $\supset M_{n-1}$. Similary $A_n \supset A_{n-1}$ and $G_{M_n}(A_n) \supset G_{M_{n-1}}(A_{n-1})$ follows now easily by condition (1) of Lemma 3.5.

Finally we put $R_o = \overline{\bigcup_n M_n}^w$, $A = \overline{\bigcup_n A_n}^w$. By (a), (d) and Lemma 1.1 (see also [13], 1.3), A is maximal abelian in N. By (b) A is a Cartan subalgebra in R_o and both R_o and A are generated by finite projections of N. By (c) we have $R_o' \cap R_o = Z(N)$. These prove 2° and 3° of 3.3. Since each F^n_{ij} (as defined above) is a perturbation of E^n_{ij} by a unitary element in N, it follows that $U_\theta' = F^1_{12} + \sum_{n>1} F^n_{2^n-1,2}$ is a perturbation of U_θ by a unitary element $u \in N$. To show that U_θ' (and thus $(\mathrm{Ad}\, u) \circ \theta$) normalizes R_o and A it is sufficient to obseve that F^1_{12} and each $F^n_{2^n-1,2}$ (as well as their adjoints) normalize M_k and A_k for all $k \geq n \geq 1$. But this follows by (f) and the definition of F^n_{ij}. **Q.E.D.**

In contrast with the type III_λ case, $0 < \lambda < 1$, the type III_o factors contain no NCE II_1 or II_∞ subfactors with trivial relative commutant, so that Corollary 3.4 is the best result in this direction. Indeed, by Connes' results in [2] we have:

3.5. PROPOSITION. *Let* M *be a separable type* III_o *factor and* $N \subset M$ *a semifinite* NCE *subfactor. Then* $N' \cap M \neq \mathbf{C}$.

PROOF. It is sufficient to consider the case when N is of type II_1. Let E be a normal conditional expectation of M onto N, τ the trace on N and $\phi = \tau \circ E$. By [2] (see also [16], 30.7) the centralizer M^ϕ of ϕ satisfies $Z(M^\phi)' \cap M = M^\phi$. Since $N \subset M^\phi$ it follows that $(N' \cap M)' \cap M \subset ((M^\phi)' \cap M)' \cap M \subset ((M^\phi)' \cap M^\phi)' \cap M = Z(M^\phi)' \cap M = M^\phi$. Thus $N' \cap M \neq \mathbf{C}$. **Q.E.D.**

We now prove a similar result for II_∞ factors. It is possibly known, but since we couldn't find an appropriate reference, we include a proof for convenience.

3.6. PROPOSITION. *Let* N *be a semifinite von Neumann algebra and* $B \subset N$ *an* NCE *subalgebra such that* $B' \cap N \subset B$. *Then* B *is generated by finite projections of* N. *In particular if* N *and* B *are factors then they are either both* II_1 *or both* II_∞.

PROOF. We first note that by results of Sakai and Tomiyama (see [16], 10.21) the subalgebra B is necessary semifinite. Now if we assume that $B' \cap N$ is not generated by finite projections then there exists a projection e in the center of B such that Be is finite and Be contains no finite projections of N. So we may assume that B is finite and contains no finite projections of N and moreover that it is countably decomposable. Let E be a normal conditional expectation of N onto B and τ a normal semifinite faithful trace on N. Then arguing as in the proof of 2.4 in [12] (see the proof of 2.3 in [11]) it follows that given any $x \in N_+$ with $\tau(x) < \infty$ and $n \geq 1$ there exist projections $e_1^n, e_2^n, \ldots, e_{k_n}^n \in B$ such that $\sum_i e_i^n = 1$ and $\| \sum_i e_i^n x e_i^n \|_2 < 2^{-n}$. Thus $\{\sum_i e_i^n x e_i^n\}_n$ tends to zero in the strong operator topology. It follows that $\sum_i e_i^n E(x) e_i^n = E(\sum_i e_i^n x e_i^n) \underset{n}{\to} 0$. But since B is finite it has a normal faithful finite trace τ_0 and since $e_i^n \in B$ we get $\tau_0(\sum_i e_i^n E(x) e_i^n) = \tau_0(E(x))$ so that $\tau_0(E(x)) = 0$. Since $E(x) \geq 0$ this implies $E(x) = 0$. But the set $\{x \in N_+ \mid \tau(x) < \infty\}$ is weakly dense in N_+ and since E is normal we obtain that $E = 0$, a contradiction. **Q.E.D.**

§4. SOME OPEN PROBLEMS

In this section we discuss several problems that arise naturally from the preceding results.

PROBLEM 1. Is 2.1 true for (minimal) periodic actions ?

As a comment on this problem let's note that if θ is a minimal periodic automorphism of a type II_1 factor M, with $\theta^n = 1$ and θ^p outer for any $1 \leq p < n$, then the fixed point algebra M^θ has trivial relative commutant in M so that if M is separable then by ([12], 3.2) there exists a hyperfinite II_1 subfactor $R \subset M^\theta$ with a Cartan subalgebra $A \subset R$ which is maximal abelian in M. Since $A \subset R \subset M^\theta$, A and R are pointwise fixed by θ, so they are invariant to θ. This shows that in the preceding problem we need to prove only the case when the Connes outer invariant $\lambda(\theta)$ is nontrivial ([3]).

PROBLEM 2. Let N be a separable type II_∞ factor with θ an automorphism of N that scales the trace. Does there exist a type II_∞ hyperfinite NCE subfactor $R_0 \subset N$ with $R_0' \cap N = \mathbf{C}$, $\theta(R_0) = R_0$? (i.e. 3.1, (ii) 2° with the additional trivial relative commutant condition).

This seems to be an interesting and quite difficult problem. Anyway the construction of such an R_0 seems to be beyond the techniques we used in this paper. Note that by 3.6 R_0 is necessary of type II_∞ and the semifinite trace of N restricts to a semifinite trace on R_0. An affirmative answer to this problem would in particular imply that any separable III_λ factor, $0 < \lambda < 1$, has a III_λ hyperfinite NCE subfactor with trivial relative commutant.

PROBLEM 3. Let M be a separable type III_λ factor, $0 \le \lambda \le 1$ fixed. What NCE subfactor with trivial relative commutant may M contain ? (e.g. can M be of type III_λ, $\lambda \ne 1$, and $N \subset M$, as above, of type III_1 ?).

In conection with this problem we note that one can easily construct a type III_0 factor having type III_λ. NCE subfactors with trivial relative commutant for any $\lambda < 1$. As we have seen 3.5 above) there are no type II_1, II_∞ such subfactors.

PROBLEM 4. Has any separable III_1 factor M an NCE hyperfinite subfactor with trivial relative commutant ? Has it semifinite NCE subalgebras B with $B' \cap M \subset B$?

This is an outstanding problem. Actually, due to results of A.Connes an affirmative answer to the second question would imply the uniqueness of the hyperfinite type III_1 factor.

Let us finally mention that a partial result to the preceding problem has been recently obtained in [9] and [15] where it is shown that any separable type III_1 factor contains hyperfinite subfactors with trivial relative commutant. Unfortunately to get the hyperfinite subfactors to be the range of some normal conditional expectations seems to be beyond the techniques of the above mentioned papers. Moreover, it is not clear what is the type of the hyperfinite factors that can be constructed by [15] and [9].

REFERENCES

1. **C.Akemann ; J.Anderson :** The Stone-Weierstrass problem for C^*-algebras, in *Invariant subspaces and other topics*, **OT 6**, Birkhauser, 1982, pp.15-32.
2. **A.Connes :** Une classification des facteurs de type III, *Ann. Ec. Norm. Sup.*, **6** (1973), 133-252.
3. **A.Connes :** Periodic automorphisms of the hyperfinite factor of type II_1, *Acta Sci. Math. (Szeged)*, **39** (1977), 39-66.
4. **A.Connes :** Outer conjugacy classes of automorphisms of factors, *Ann. Ec. Norm. Sup.*, **8** (1975), 383-419.
5. **A.Connes :** Classification of injective factors, *Ann. of Math.*, **104** (1976), 73-115.
6. **A.Connes ; E.Stormer :** Homogeneity of the state space of factors of type III_1, *J. Functional Analysis*, **28** (1978), 187-196.
7. **J.Dixmier :** *Les algebres d'operateurs dans l'espace hilbertien (Algebres de von Neumann)*, Gauthier-Villars, Paris, 1969.
8. **Y.Katznelson ; B.Weiss :** Notes on orbit equivalence, unpublished.
9. **R.Longo :** Solution of the factorial Stone-Weierstrass conjecture. An application of the theory of standard and split inclusions, *Invent. Math.* 76(1984).
10. **F.Murray ; J.von Neumann :** Rings of operators. IV, *Ann. of Math.* 44(1943), 716-808.
11. **M.Pimsner ; S.Popa :** Entropy and index for subfactors, preprint, 1983.
12. **S.Popa :** On a problem of R.V.Kadison on maximal abelian $*$-subalgebras in factors, *Invent. Math.* **65** (1981), 269-281.
13. **S.Popa :** Singular maximal abelian $*$-subalgebras in continuous von Neumann algebras, *J.Functional Analysis*, **50** (1983), 151-166.
14. **S.Popa :** Maximal abelian injective subalgebras in factors associated with free groups, *Adv. in Math.* **50** (1983), 27-48.
15. **S.Popa :** Semiregular maximal abelian $*$-subalgebras and the solution to the factor states Stone-Weierstrass problem, *Invent. Math.*, **76** (1984).
16. **Ş.Stratila :** *Modular theory in operator algebras*, Abacus Press/Editura Academiei, Turnbridge Wells/Bucuresti, 1981.

Two applications of the dual groupoid of a C^*-algebra

Jean Renault

Introduction.

The notion of dual groupoid of a C^*-algebra was introduced by A. Connes a few years ago and developped in Alami Idrissi's thesis [1] to obtain a Riesz theorem for traces on a C^*-algebra. Indeed the dual groupoid provides a natural desingularisation of the spectrum of the C^*-algebra. Although this notion deserves a greater attention, I shall limit myself to two applications. First it will be shown that diagonal subalgebras, as defined by A. Kumjian in [5], give "good" transversals for the dual groupoid. It will then be easy to recover Kumjian's characterisation of principal discrete groupoid C^*-algebras. Second we shall see how the dual groupoid naturally defines a generalized Dixmier-Douady invariant for an arbitrary C^*-algebra.

The dual groupoid of a C^*-algebra.

I am reproducing here, with his kind permission, some parts and results of the second chapter of Alami's thesis [1]. Let me first establish some notation. Let A be a C^*-algebra. The set of states is denoted by $S(A)$ and the subset of pure states by $P(A)$. For x in $S(A)$, (π_x, H_x, ξ_x) will denote the GNS triple associated with it so that x can be represented as

$$x(a) = (\xi_x \mid \pi_x(a)\xi_x) = \omega_{\xi_x} \circ \pi_x (a).$$

The elements of A form a fundamental family of continuous sections for the field of Hilbert spaces $x \to H_x$. The corresponding topological Hilbert bundle, or rather its restriction to the pure state space $P(A)$, will be denoted by $H(A)$. Every element ϕ of the dual A^* admits, as an element of the predual of the von Neumann algebra A^{**}, a polar decomposition $\phi = xu = uy$ where x and y are positive elements of A^* and u is a partial isometry in A^{**}, which is unique if we insist on u having suppy as its initial support and suppx as its final support. One can see that an element ϕ of the unit ball of A^* is extremal iff x or y is a pure state iff u is a minimal partial isometry in A^{**}.

Alami introduces three equivalent definitions of the dual groupoid.
Call G_1 the set of extremal elements of the unit ball of $A^\star$. Such an element will be written $\phi = xu = uy = (x, u, y)$ according to the above notation.
Call G_2 the set of triples (x, U, y) where x and y are pure states and U is an isometry from H_y to H_x intertwining Π_y and Π_x. We shall soon make use of the fact that U is unique up to a scalar multiple.
Call G_3 the sphere bundle of $H(A)$, that is, the set of (ξ,y)'s where y is a pure state and ξ is a unit vector in H_y.
The correspondences between these sets are given by

$$\phi = (x, u, y) = (x, U, y) = (\xi, y)$$
$$\xi = U^\star \xi_x = \Pi_y(u)\, \xi_y$$
$$x = \omega_\xi \circ \Pi_y$$
$$\phi(a) = (\xi_y \mid \Pi_y(a)\, \xi)\ .$$

Now G_1 (or G_2) has a natural groupoid structure with $P(A)$ as its space of units :

$$(x, u, y)\,(y, v, z) = (x, uv, z)$$
$$(x, u, y)^\star = (y, u^\star, x)$$
$$r(x, u, y) = x \quad \text{and} \quad s(x, u, y) = y\ .$$

This groupoid is called the dual groupoid of the $C^\star$-algebra A and is denoted by $G(A)$. It provides a desingularisation for the singular quotient space $\hat{A} = P(A)/G(A)$.
The $\sigma(A^\star,A)$ - topology on $G(A)$ is compatible with the groupoid structure. When A is separable, $G(A)$ is in fact a polish groupoid.
The circle group S^1 acts freely on $G(A)$ by multiplication. It makes it into a locally trivial principal bundle over $R(A) = G(A)/S^1$, which is the graph of the unitary equivalence relation on $P(A)$. Thus we have the groupoid extension

$$S^1 \times P(A) \to G(A) \to R(A)\ .$$

There is a natural representation of the dual groupoid $G(A)$ on the Hilbert bundle $H(A)$ given by

$$G(A) \star H(A) \to H(A)$$
$$(x, U, y)\,(y, \xi) \to (x, U\xi)\ .$$

Restricted to the sphere bundle of $H(A)$, this is just the groupoid operation $(\phi, \psi) \to \psi\, \phi^\star$. In particular this action is continuous when $H(A)$ is provided with the weak topology.

Every element a of A defines an intertwining operator $(\Pi_x(a))$ of this $G(A)$-Hilbert bundle ; moreover $\|a\| = \sup_{x \in P(A)} \|\Pi_x(a)\|$.

Let me conclude this brief introduction of the dual groupoid by a question. How would one characterize the image of A in the commutant algebra $\mathfrak{A} = \mathrm{End}_{G(A)}\, H(A)$?

According to F. Shultz' [8], G(A) endowed with the uniform structure induced by the $\sigma(A^\star,A)$-topology of the unit ball of $A^\star$ and with function $\phi \to \phi(1)$ is a complete invariant for A . The answer should give a direct proof.

Diagonals in $C^\star$-algebras.

J. Feldman and C. Moore have given in [4] an axiomatization of the von Neumann algebra arising from their generalized group-measure construction. They say that an abelian subalgebra $\mathfrak{B}$ of a von Neumann algebra $\mathfrak{A}$ is a Cartan subalgebra if it is maximal abelian, there is a faithful normal conditional expectation of $\mathfrak{A}$ onto $\mathfrak{B}$ and $\mathfrak{A}$ is generated by the normalizer of $\mathfrak{B}$. They establish a one-to-one correspondence between pairs $(\mathfrak{A},\mathfrak{B})$, where $\mathfrak{B}$ is a Cartan subalgebra of the von Neumann algebra $\mathfrak{A}$, and twisted equivalence relations (R,σ), where R is a discrete measured equivalence relation and σ a cohomology class in $H^2(R;S^1)$.
A. Kumjian has obtained in [5] a $C^\star$-algebra version of this result. He makes a definition equivalent to the following

Definition. An abelian subalgebra B of a C^*-algebra A is called a diagonal if
(i) it has the unique extension property,
(ii) the conditional expectation of A onto B is faithful, and
(iii) A is generated by the normalizer $N(B) = \{ a \in A : aB = Ba \}$.

The unique extension property says that a pure state of B extends uniquely to a pure state of A ; see [2] . It implies that B is maximal abelian and there is a unique conditional expectation onto B . The conditions (i) (ii) and (iii) are independant. For instance, the Cuntz algebra O_n has an abelian subalgebra which satisfies (ii), (iii) but not (i) (see [2]). The full $C^\star$-algebra of a non amenable principal discrete groupoid has an abelian subalgebra satisfying (i), (iii) but not (ii). The reduced $C^\star$-algebra of a free group has an abelian subalgebra which satisfies (i), (ii) but not (iii).

The basic example of a diagonal pair (A,B) is provided by a twisted relation. Let us review this construction. Consider a groupoid extension

$$S^1 \times X \to E \to G$$

where $X = E^{(o)} = G^{(o)}$. The sections of the associated line bundle will be viewed as complex functions on E which satisfy $f(\lambda\sigma) = f(\sigma)\overline{\lambda}$ for λ in S^1 and σ in E . Let $C_c(G,E)$ be the space of continuous sections with compact support of this line bundle.
Suppose next that we are given a right Haar system (λ_x) on G . This will allow

is to construct a Hilbert bundle H and a representation of $C_c(G,E)$.
For each x in X, we let $H_x = L^2(G, E, \lambda_x)$ be the Hilbert space of λ_x-square integrable sections of the hermitian line bundle associated with the extension. The scalar product is given by

$$(\xi \mid \eta) = \int \overline{\xi(\gamma)}\, \eta(\gamma)\, d\lambda_x(\dot{\gamma})$$

where the dot means the quotient map $E \to G$. Choosing $C_c(G,E)$ as a fundamental family of continuous sections, we get a Hilbert bundle H. The groupoid E acts on H: for $\gamma \in E$, $L(\gamma)$ is the isometry from $H_{s(\gamma)}$ to $H_{r(\gamma)}$ given by $L(\gamma)\,\xi\,(\gamma') = \xi(\gamma'\gamma)$. Every f in $C_c(G,E)$ defines an element $\Pi(f)$ of the commutant algebra $\mathrm{End}_E H$, where

$$\Pi_x(f)\,\xi\,(\gamma) = \int f(\gamma\gamma')\,\xi\,(\gamma'^{-1})\, d\lambda^x(\dot{\gamma}') \quad , \text{ with } \lambda^x = (\lambda_x)^{-1}$$

Thus if one defines

$$f \star g\,(\gamma) = \int f(\gamma\gamma')\, g(\gamma'^{-1})\, d\lambda^{s(\gamma)}(\dot{\gamma}')$$

$$f^\star(\gamma) = \overline{f(\gamma^{-1})}\ ,$$

then Π is a $\star$-representation of the $\star$-algebra $C_c(G,E)$. The reduced $C^\star$-algebra $C^\star_{red}(G,E)$ is the completion of $C_c(G,E)$ with respect to the norm $\|f\| = \sup_x \|\Pi_x(f)\|$.

We are now ready to give the basic example of a diagonal pair.

Proposition. Let E be an extension of a principal discrete groupoid G (discrete means that the counting measures on each equivalence class form a Haar system). Then $C_o(G^{(o)})$ is a diagonal in $C^\star_{red}(G,E)$.

The proof appears in [5] (see also [6] and [2]). The elements of $C^\star_{red}(G,E)$ are continuous sections of the line bundle and the elements of $C_o(G^{(o)})$ are those sections which vanish off $G^{(o)}$ (recall that the bundle is trivial over $G^{(o)}$).
It is convenient to check the extension property under the form

$$A = B \oplus \overline{\mathrm{span}}\,[B,A]\ .$$

Suppose now that we are given a $C^\star$-algebra A with a diagonal subalgebra B. Because of the unique extension property, the pure state space $P(B)$ of B will be viewed as a subspace of $P(A)$. Hence we may reduce to $P(B)$ the dual groupoid $G(A)$ and its principal quotient $R(A)$. We denote by $G(A,B)$ and $R(A,B)$ these reductions. The first step to the converse of the previous proposition is the following

Proposition. Let (A,B) be a diagonal pair. Then $R(A,B)$ is a discrete locally compact groupoid.

We shall need two easy lemmas.

Lemma 1. Let (Π,H) be a representation of A. If a is in $N(B)$ (i.e. $aB=Ba$) and ξ is an eigenvector for $\Pi(B)$, then $\Pi(a)\xi$ is again an eigenvector for $\Pi(B)$.

Indeed for every b in B, there is b' in B such that $ba=ab'$. Therefore $\Pi(b)\Pi(a)\xi = \Pi(a)\Pi(b')\xi = \Pi(a)\lambda\xi = \lambda\Pi(a)\xi$, where λ is a scalar □

Lemma 2. Let (Π,H) be an irreducible representation of A. The vector state $\omega_\xi \circ \Pi$ is B-central (see [2]) iff ξ is an eigenvector for $\Pi(B)$. Moreover all eigenspaces of $\Pi(B)$, if they exist, are one-dimensional. If there exists an eigenvector for $\Pi(B)$, there exists a basis of eigenvectors.

Recall from [2] that a state x of A is called B-central if $x(u^\star.u)$ is equal to x for every unitary u in B. Since $\omega_\xi \circ \Pi(u^\star.u) = \omega_{\Pi(u)\xi} \circ \Pi$ and Π is irreducible, $\omega_\xi \circ \Pi$ is B-central iff $\Pi(u)\xi$ is a multiple of ξ for each unitary u in B.
Let ξ and η be two eigenvectors for $\Pi(B)$ with eigencharacters x and y in $P(B)$ respectively. Suppose that they are linearly independant. Then the states $\omega_\xi \circ \Pi$ and $\omega_\eta \circ \Pi$ are distinct. Because of the extension property, x and y are distinct.
Suppose that ξ is an eigenvector for $\Pi(B)$. From lemma 1, the set $\Pi(N(B))\xi$ consists of eigenvectors. Since $N(B)$ generates A, this set is total □

Here comes the proof of the proposition. According to [6,1.2.8] one has to show that the source map $s : R(A,B) \to P(B)$ is a local homeomorphism.
From lemma 2, $G(A,B)$ consists of pairs (ξ,x) where x is a pure state of B and ξ is an eigenvector for $\Pi_x(B)$.
Let $\phi = (\xi,x)$ be fixed in $G(A,B)$. By Kadison's transitivity theorem, there exists a in A such that $\xi = \Pi_x(a)\xi_x$. In fact a can be chosen in $N(B)$. Indeed since the linear span of $N(B)$ is dense in A, there exist $a_1,\dots,a_n$ in $N(B)$ such that $\|a - \sum_1^n a_i\| < 1$. Since then $\|\xi - \sum_1^n \Pi_x(a_i)\xi_x\| < 1$, at least one of the $\Pi_x(a_i)\xi_x$ is not orthogonal to ξ. Since both vectors are eigenvectors for $\Pi(B)$, $\Pi_x(a_i)\xi_x$ is a multiple of ξ. It suffices to replace a by a multiple of a_i.
Let $V = \{ \psi \in G(A) : |\psi(a^\star)| > \frac{1}{2} \}$ and $W = \{ y \in P(A) : y(a^\star a) > \frac{1}{4} \}$.
Then V [resp. W] is an open neighborhood of ϕ [resp. x]. We will show that the restriction of s to $V \cap R(A,B)$ is a homeomorphism onto W.
If $\psi = (\eta,y)$ is in $V \cap G(A,B)$, η is a multiple of $\Pi_y(a)\xi_y$ because

$\Pi_y(a)\,\xi_y \mid \eta) = \psi\,(a^\star) \neq 0$. It results that $|\psi\,(a^\star)| = y(a^\star a)^{1/2}$, hence $y(a^\star a) > \frac{1}{4}$.
hus s maps $V \cap G(A,B)$ into W .
he map is onto because for y in W , $(y(a^\star a))^{-1/2}\ \Pi_y(a)\xi_y, y)$ is in $V \cap G(A,B)$.
he maps is one-to-one in R(A,B). Indeed if $\psi = (\eta, y)$ and $\psi' = (\eta', y)$ both belong o $V \cap G(A,B)$ then η and η' are both multiple of $\Pi_y(a)\xi_y$.
ince s is continuous and open, this restriction of s is a homeomorphism.

□

e are now ready to prove Kumjian's result.

heorem. Let B be a diagonal in A . Then there exists an isomorphism from A onto $^\star_{red}$ (R(A,B) , G(A,B)) which carries B onto $C_o(P(B))$.

roof. Let (A,B) be a diagonal pair. We set G = G(A,B) and R = R(A,B).
le first observe that the Hilbert bundle $\ell^2(R,G)$ of the regular representation entering into the definition of $C^\star_{red}(R,G)$ is isomorphic as a G-Hilbert bundle to the reduction H(A,B) of the GNS Hilbert bundle to P(B). The isomorphism is defined by

$$\ell^2(R_x, G_x) \to H_x \quad , \quad x \text{ in } P(B)$$

$$\varepsilon_{(\xi,x)} \quad \to \xi$$

where $\varepsilon_{(\xi,x)}$ is the function

$$\varepsilon_{(\xi,x)}(\eta,x) = (\eta|\xi) \quad , \text{ with } \xi , \eta \text{ eigenvectors for } \Pi_x(B) .$$

Using this isomorphism we view $C_c(R,G)$ as an algebra of intertwining operators on H(A,B). For f in $C_c(R,G)$ the corresponding intertwining operator $\Pi(f)$ is defined by its matrix coefficients

$$(\eta \mid \Pi_x\,(f)\,\xi) = (\varepsilon_{(\eta,x)} \mid \Pi_x\,(f)\,\varepsilon_{(\xi,x)})$$
$$= f\,(\,(\eta,x)\,(\xi,x)^\star)\ .$$

In fact for every intertwining operator $F = (F_x)$ there exists a function f on G such that

$$(\eta \mid F_x\,\xi) = f\,(\,(\eta,x)\,(\xi,x)^\star)\ .$$

It suffices to set $f(\eta,x) = (\eta \mid F_x\,\xi_x)$. This function f uniquely determines the operator F . We will freely identify the intertwining operator F and the function f. As we have seen, an element a of A defines an intertwining operator $\Pi(a) = (\Pi_x(a))$ on H(A), hence on H(A,B). Since the expectation onto B is faithful, the family of states P(B) is separating and this representation of A as an algebra of intertwining operators on H(A,B) is faithful. The matrix coefficients of the intertwining operator $\Pi(a)$ are given by

$$(\eta \mid \Pi_x(a)\xi) = (\xi,x)\,(\eta,x)^\star\,(a)\ .$$

Therefore the corresponding function on G is

$\hat{a}(\gamma) = \gamma^{\star}(a)$, or equivalently

$\hat{a}(\xi,x) = (\xi \mid \pi_x(a)\xi_x)$.

This isomorphism carries B onto $C_o(P(B))$. (Recall that $C(P(B))$ consists of functions on G supported in the open set $S^1P(B)$.) It is an extension of the usual Gelfand transform.

<u>Claim 1</u>. Every f in $C_c(R,G)$ is of the form $\hat{a}$ for some a in A .

Indeed, we can cover the support of f by finitely many open sets $U_1,\dots,U_n$ of the form $U_i = \{$ multiple of $(\pi_x(a_i)\xi_x, x) : x(a_i^{\star}a_i) > 0 \}$ with a_i in $N(B)$. Let $h_1,\dots,h_n$ be a partition of unity subordinate to $s(U_1),\dots,s(U_n)$. Since $f.h_i os$ has a compact support contained in U_i, there exists k_i in $C_c(P(B))$ such that $f.h_i os = \hat{a}_i.k_i os$. But we know from the Stone-Weierstrass theorem that k_i is of the form $\hat{b}_i$ for some b_i in B . Then

$$f = \sum_1^n f.h_i os = \sum_1^n \hat{a}_i.\hat{b}_i os = \sum_1^n \widehat{a_i b_i} = \hat{a}$$

because for a in A and b in B

$$\begin{aligned} \widehat{ab}(\xi,x) &= (\xi \mid \pi_x(ab)\xi_x) \\ &= (\xi \mid \pi_x(a)\pi_x(b)\xi_x) \\ &= (\xi \mid \pi_x(a)\xi_x)x(b) \\ &= \hat{a}(\xi,x)\,\hat{b}\,(x) . \end{aligned}$$

<u>Claim 2.</u> If a is in $N(B)$, $\hat{a}$ is in the norm closure of $C_c(R,G)$.

We first observe that for a in $N(B)$, $\hat{a}(\xi,x) = (\xi \mid \pi_x(a)\xi_x) = 0$ unless ξ is a multiple of $\pi_x(a)\xi_x$. This is so because $N(B)$ permutes the eigenvectors of $\pi_x(B)$. Applying this to $a^{\star}a$ which is also in $N(B)$, one sees that $\pi_x(a^{\star}a)\xi_x$ is a multiple of ξ_x and that $\widehat{a^{\star}a}(\xi,x) = 0$ unless ξ is a multiple of ξ_x. In other words, $\widehat{a^{\star}a}$ vanishes off $P(B)$. It is also clear that elements of A induce continuous functions on $P(B)$ which vanish at infinity. Therefore $\widehat{a^{\star}a}$ is in $C_o(P(B))$. It can be uniformly approximated by $\widehat{a^{\star}a}.h_i$, where h_i is a continuous function with values in $[0,1]$ and compact support contained in $\{x \in P(B) : \widehat{a^{\star}a}(x) > 0 \}$. There exists b_i in B such that $h_i = \hat{b}_i$. Then $\widehat{ab_i} = \hat{a}.h_i os$ is in $C_c(R,G)$ and

$$\begin{aligned} \|\widehat{ab_i} - \hat{a}\|^2 &= \|b_i a^{\star}ab_i - a^{\star}ab_i - b_i a^{\star}a + a^{\star}a\| \\ &= \|h_i(\widehat{a^{\star}a})h_i - (\widehat{a^{\star}a})h_i - h_i(\widehat{a^{\star}a}) + \widehat{a^{\star}a}\| \end{aligned}$$

tends to zero.

ince $N(B)$ generates A, for every a in A, $\hat{a}$ is in $C^{\star}_{red}(R,G)$.
he map $A \to C^{\star}_{red}(R,G)$ which sends a into $\hat{a}$ is an isomorphism sending B to
$_{o}(P(B))$.

□

n interpretation of the Dixmier-Douady invariant.

We have seen that the dual groupoid $G(A)$ of a $C^{\star}$-algebra A satisfies the roupoid extension

$$S^1 \times P(A) \to G(A) \to R(A) .$$

s usual, the equivalence classes of extensions of $R(A)$ by S^1 form a group, denoted y $Ext(R(A) ; S^1)$. The class of $G(A)$ in $Ext(R(A) ; S^1)$ is our generalized ixmier-Douady invariant.
n the case when A is a continuous trace $C^{\star}$-algebra, a suitable version of xt$(R(A) ; S^1)$ will be identified with the Čech cohomology group $\check{H}^2(\hat{A};\hat{A}xS^1)$. The mage of the class of $G(A)$ in $H^3(\hat{A};\mathbb{Z})$ is the usual Dixmier-Douady invariant of A. ne obtains as a by-product the Dixmier-Douady invariant of the $C^{\star}$-algebra $C^{\star}(R,E)$, here R is a principal and proper groupoid.

One of the difficulties in continuous cohomology is the absence of continuous ections. It will be by-passed through the use of equivalent groupoids, as defined n [7].
et G be a groupoid and M be a G-bundle of abelian groups. An extension of G by M consists of a groupoid E and an exact sequence of groupoids $0 \to \underline{M} \to E \to \underline{G} \to 0$. ere $\underline{G}$ is a groupoid equivalent to G via an equivalence X and $\underline{M}$ is the $\underline{G}$-bundle induced by M via X.
Two extensions E_1 and E_2 are called equivalent if there exists an equivalence Y of the groupoids E_1 and E_2 compatible with the action of M.
The set of equivalence classes of extensions is denoted by $Ext(G;M)$. The addition is defined as follows. Two arbitrary extensions may be replaced by equivalent extensions with the same quotient. Then one can construct their Baer sum as usual. The identity element is the class of the semi-direct product $G \ltimes M$.

We will identify this group in a particular case. Let R be the graph of an equivalence relation on a space X. We assume that
(i) X is paracompact,
(ii) the quotient map $X \to X/R$ is open and admits continuous local sections,
(iii) R is closed in $X \times X$ and endowed with the induced topology.

(iv) X/R is paracompact.

Then X is an equivalence between R and X/R.

Let M_R be an R-bundle of abelian groups. It is induced by a bundle M over X/R. In the definition of $\mathrm{Ext}(R;M_R)$, we add the following hypotheses. First we only consider extensions $0 \to \underline{M} \to E \overset{\pi}{\to} \underline{R} \to 0$ where π admits local sections and $\underline{R}$ satisfies (i)(ii) and (iii). Second we only consider equivalences Y such that the quotient map $Y \to Y/M$ admits local sections. The corresponding group will be denoted by $\mathrm{Ext}'(R;M_R)$.

Proposition. With the above notation, $\mathrm{Ext}'(R;M_R) = \check{H}^2(X/R;M)$.

We just sketch the construction of the isomorphism between these groups.

Let $0 \to \underline{M} \to E \to \underline{R} \to 0$ be an extension in $\mathrm{Ext}'(R;M_R)$. It can be replaced by an equivalent extension which admits a global section. Indeed choose a locally finite open cover $(U_i)_{i\in I}$ of $\underline{X}$ such that for each (i,j) in $I \times I$, there exists a section s_{ij} over $\underline{R} \cap U_i \times U_j$. Then replace

$\underline{X}$ by $Y = \{ (i,x) : i \in I \text{ and } x \in U_i \}$

$\underline{R}$ by $S = \{ (i,x,j,y) \in Y \times Y \text{ with } (x,y) \in \underline{R} \}$

E by $F = \{ (i,e,j) : i,j \in I \text{ , } e \in E \text{ , } r(e) \in U_i \text{ and } s(e) \in U_j \}$

Then F is equivalent to E via

$Z = \{ (i,e) : i \in I \text{ , } e \in E \text{ and } r(e) \in U_i \}$

and the map from Fonto S sending (i,e,j) into $(i,r(e),j,s(e)$ admits the global section sending (i,x,j,y) into $(i,s_{ij}(x,y),j)$. Therefore we assume that $E \to \underline{R}$ has a global section s.

Since the quotient map $\underline{X} \to \underline{X}/\underline{R} = X/R$ has local sections, there exist a locally finite open cover $(V_i)_{i \in J}$ of X/R and sections σ_i over each V_i.

For ω in V_{ijk}, $f_{ijk}(\omega) = s(\sigma_i(\omega),\sigma_j(\omega))s(\sigma_j(\omega),\sigma_k(\omega))s(\sigma_i(\omega),\sigma_k(\omega))^{-1}$ is in M_ω.

It is a routine matter to chek that (f_{ijk}) is a 2-Čech cocycle relative to the the cover (V_i) and that its class in $\check{H}^2(X/R;M)$ depends only on the class of the extension. Conversely, suppose that (f_{ijk}) is a cocycle relative to a locally finite open cover $(V_i)_{i \in J}$. Construct the space

$\underline{X} = \{ (i,\omega) : i \in J \text{ , } \omega \in V_i \}$,

the equivalence relation $\underline{R}$ on $\underline{X}$ given by the quotient map from $\underline{X}$ onto X/R sending (i,ω) into ω and the groupoid

$E = \{ (i,\omega,j,a) : i,j \in J \text{ , } \omega \in V_{ij} \text{ and } a \in M_\omega \}$ with multiplication

$(i,\omega,j,a)(j,\omega,k,b) = (i,\omega,k,af_{ijk}(\omega)b)$

This defines an extension which has a global section

$s(i,\omega,j) = (i,\omega,j,1_\omega)$ where 1_ω is the identity element of M_ω.

he above construction gives back the cocycle (f_{ijk}).

e are going to apply this proposition to the dual groupoid of a $C^\star$-algebra A with ontinuous trace. The following results are well known.

emma. Let A be a separable $C^\star$-algebra with continuous trace. Then the unitary quivalence relation $R(A)$ on the pure state space $P(A)$ satisfies the above onditions (i) to (iv).

or (i) , one observes that $P(A)$ is metrizable.

or (ii) , it is a general property that the quotient map $P(A) \to \hat{A}$ is open. Moreover, ince A has continuous trace, there exists for each π_0 in $\hat{A}$ an element e in A^+ uch that $\pi(e)$ is a rank one projection for each π in some neighborhood of π_0. hen $\pi \to \mathrm{Tr} \circ \pi(e.)$ is a continuous section.

t is also known that $\hat{A}$ is Hausdorff. Since it is locally compact and second countable, iv) is satisfied.

t remains to check that the topology of $R(A)$, defined as a quotient of $G(A)$ oincides with the topology induced from $P(A) \times P(A)$. Since it is sufficient to check his property on each element of an open cover of $\hat{A}$, we may assume that A is efined by a Hilbert bundle H over $\hat{A}$. Then $P(A)$ can be identified with the rojective Hilbert bundle $P(H) = S(H)/S^1$, where $S(H)$ is the sphere bundle of H rovided with the weak topology. On the other hand $G(A) = S(H)\star S(H)/S^1$, where $\star$ enotes the fibered product over $\hat{A}$. Since the quotient map from $S(H)$ to $P(H)$ as continuous local sections one can see that the inclusion map from $R(A)$ into $P(H) \times P(H)$ is a homeomorphism onto its image.

□

Proposition. The Dixmier-Douady invariant of a separable $C^\star$-algebra with continuous trace A is the opposite of the image in $H^3(\hat{A},\mathbb{Z})$ of the extension defined by the dual groupoid $G(A)$.

The crucial fact in the proof is that when A is given by a Hilbert bundle H , the extension $G(A)$ is equivalent to the trivial extension $\hat{A} \times S^1$ via the sphere bundle $S(H)$. Here $G(A) = S(H)\star S(H)/S^1$ acts on $S(H)$ according to

$$\omega_{\xi,\eta}.\zeta = \xi\,(\eta \mid \zeta) \qquad \text{where } \eta \text{ and } \zeta \text{ are collinear.}$$

In the general case, quoting [3 ,10.7.11] , there exist an open cover $(T_i)_{i \in I}$ of $\hat{A}$ and for each i in I a Hilbert bundle H_i and an isomorphism h_i from $K(H_i)$, the algebra of compact operators on H_i onto $A|_{T_i}$, the reduction of A to T_i .

Moreover for each i,j in I, the isomorphism $h_i^{-1} \circ h_j$ from $K(H_j | T_{ij})$ onto $K(H_i | T_{ij})$ is implemented by an isomorphism g_{ij} from $H_j | T_{ij}$ onto $H_i | T_{ij}$. The Dixmier-Douady invariant is defined by the cocycle (u_{ijk}) such that $g_{ij}\, g_{jk} = u_{ijk}\, g_{ik}$.
Construct as above the extension associated with the cocycle $(\bar{u}_{ijk})$.
I claim that it is equivalent to the extension $G(A)$. Indeed the equivalence is given by

$$Z = \{ (i,\xi) : i \in I , \xi \in S(H_i) \}$$

where $G(A)$ acts on Z according to

$$\phi\cdot(i,\xi) = h_i^{\star}(\phi)\cdot\xi .$$

As above this is only defined when ϕ and ξ live on the same point of $\hat{A}$ and the source of $h_i^{\star}(\phi)$ is the state defined by ξ.
The extension

$$E = \{ (i,t,j,\lambda) : i,j \in I , t \in T_{ij} \text{ and } \lambda \in S^1 \}$$

associated with the cocycle $(\bar{u}_{ijk})$ acts on Z according to

$$(i,\xi)(i,t,j,\lambda) = (j, g_{ij}^{-1}(\xi)\lambda) .$$

□

Remark. In fact this proof exhibits an equivalence bimodule between A and the $C^{\star}$-algebra of the extension associated with the cocycle $(\bar{u}_{ijk})$. This gives the well known fact that continuous trace $C^{\star}$-algebras are classified up to Morita equivalence by their Dixmier-Douady invariant.

Corollary. Let R be an equivalence relation on a locally compact space X satisfying the above conditions (i) to (iv) and admitting a Haar system. For every extension E in $Ext'(R;X\times S^1)$, the $C^{\star}$-algebra $C^{\star}(R,E)$ of the twisted relation has continuous trace and its Dixmier-Douady invariant is the opposite of the image of E in $H^3(X/R;\mathbb{Z})$.

Indeed the elements $f^{\star} \star f$, where f is in $C_c(R,E)$, have continuous trace. Moreover the extensions $G(A)$ and E are equivalent via the sphere bundle of $L^2(R,E)$.

□

Aknowledgments.
I would like to thank A. Alami for letting me include results of his thesis [1] and A. Kumjian for many fruitful discussions.

References

1. Alami Idrissi, Sur le théorème de Riesz dans les algèbres stellaires, Thèse de 3ème cycle, Paris 6 (1979).
2. Archbold, Bunce, Gregson, Extensions of states of C^*-algebras, II, Proc. Royal Soc. Edinburgh 92 A (1982) 113-122.
3. Dixmier, Les C^*-algèbres et leurs représentations, Gauthier-Villars.
4. Feldman, Moore, Ergodic equivalence relations, cohomology and von Neumann algebras, I, II, Trans. A.M.S. 234 (1977) 289-359.
5. Kumjian, On C^*-diagonals and twisted relations, Tübingen Semesterbericht, W83.
6. Renault, A groupoid approach to C^*-algebras, Springer Lecture Notes, 793 (1980).
7. Renault, C^*-algebras of groupoids and foliations, Proc. of Symposia in Pure Math., 38 (1982), part 1, 339-350.
8. Shultz, Pure states as a dual object for C^*-algebras, Comm. Math. Phys. 82 (1982) 497-509.

INVARIANTS FOR TOPOLOGICAL MARKOV CHAINS

Norbert Riedel*

Introduction. In the theory of topological Markov chains the main outstanding question is to decide whether shift equivalence implies strong shift equivalence or not (cf. [4],[5]). In the sequel we shall define a new invariant for strong shift equivalent matrices which relies on the splitting process for positive integral matrices. Though we are not able to decide whether our invariant separates shift equivalence and strong shift equivalence or not, we shall show how it reflects the presence of shift equivalence.

For the general theory of topological Markov chains (subshifts of finite type) we refer to [1].

1. We start with the review of some notions and notations. We shall denote by $\mathbb{Z}_+^{(p,q)}$ the set of all $p \times q$ matrices with non-negative integral entries such that each column and each row has at least one component which is non-zero. In case $p = q$ we simply write $\mathbb{Z}_+^{(p)}$ instead of $\mathbb{Z}_+^{(p,q)}$.

1.1 Definition (see [6]). A $p \times q$ matrix A is a subdivision matrix if each row has only one non-zero component and this component is equal to 1. A is called an amalgamation matrix if its transpose is a subdivision matrix.

In the following definition we describe the process of iterated splitting of integral matrices with non-negative entries which is crucial for the definition of our invariant.

1.2 Definition. Let $A = (a_{ij}) \in \mathbb{Z}_+^{(p)}$ be any matrix. Let $q_i = \sum_{k=1}^{i} \sum_{\ell=1}^{p} a_{k\ell}$, for $1 \leq i \leq p$, and let $q_0 = 0$, $q = q_p$. We define an amalgamation matrix $R_A = (r_{ij}) \in \mathbb{Z}_+^{(p,q)}$ as follows,

$$r_{ij} = \begin{cases} 1 & \text{if } q_{i-1} < j \leq q_i \\ 0 & \text{elsewhere} \end{cases}$$

and a subdivision matrix $S_A = (s_{ij}) \in \mathbb{Z}_+^{(q,p)}$ as follows

$$s_{ij} = \begin{cases} 1 & \text{if } q_{k-1} + \sum_{\ell<j} a_{k\ell} < i \leq q_{k-1} + \sum_{\ell \leq j} a_{k\ell}\ , \qquad k=1,\dots,p \\ 0 & \text{elsewhere} \end{cases}$$

(observe that $A = R_A S_A$ holds). We set $A_{[1]} = S_A R_A$. Moreover, we define for each $r \in \mathbb{N}$, $r > 1$,

$$A_{[r]} = \underbrace{((A_{[1]})_{[1]} \cdots)_{[1]}}_{r\text{-times}} \ ,$$

and $A_{[0]} = A$.

Remarks. (1) The entries of $A_{[r]}$ are either 0 or 1 for each $r \in \mathbb{N}$.

(2) For any matrix $A \in \mathbb{Z}_+^{(p)}$ we have $A = A_{[1]}$ if and only if A is a permutation matrix.

(3) For each $A \in \mathbb{Z}_+^{(p)}$ the matrix $A_{[1]}$ satisfies the following uniqueness condition (see [6], 5.3): If R is an amalgamation matrix and S is a subdivision matrix with $RS = A$ then there is a permutation matrix P with $A_{[1]} = PSRP^{-1}$.

Before going on to define our invariant we shall try to give some motivation. Recall that two matrices $A \in \mathbb{Z}_+^{(p)}$, $B \in \mathbb{Z}_+^{(q)}$ are strong shift equivalent if there exist matrices $T_i \in \mathbb{Z}_+^{(p_i)}$ $(0 \leq i \leq n)$ and $R_i \in \mathbb{Z}_+^{(p_i, p_{i+1})}$, $S_i \in \mathbb{Z}_+^{(p_{i+1}, p_i)}$ $(0 \leq i \leq n-1)$, such that $T_0 = A$, $T_n = B$ and $R_i S_i = T_i$, $S_i R_i = T_{i+1}$ $(0 \leq i \leq n-1)$: cf. [6] and [4]. For the matrix A we consider the set of all sequences of non-negative integers, $\Gamma^{(0)}(A)$ say, which is associated with A in the following natural manner: $\{x_n\}_{n\in\mathbb{N}}$ is in $\Gamma^{(0)}(A)$ if and only if there is a pair of indices i,j such that x_n is equal to the (i,j)-component of the matrix A^n for each $n \in \mathbb{N}$. Similarly we define $\Gamma^{(0)}(B)$. Now we are looking for a procedure to enlarge $\Gamma^{(0)}(A)$ $(\Gamma^{(0)}(B))$ to a set $\Gamma(A)$ $(\Gamma(B))$ of sequences of non-negative integers in such a manner that $\Gamma(A) = \Gamma(B)$ holds if A and B are strong shift equivalent and $\Gamma(A)$ $(\Gamma(B))$ is as small as possible. As will become more obvious later, the following definition puts this idea in a precise form.

1.3 Definition. Let $A \in \mathbb{Z}_+^{(p)}$. For each integer r, $r \geq 0$, we define $\Gamma_r(A)$ to be the set of sequences $\{x_n\}_{n \in \mathbb{N}}$ of non-negative integers satisfying the following condition: There is a subdivision matrix S; there are diagonal matrices D_1, D_2 with positive integral entries in the diagonal, satisfying the following identity

$$(S^t D_1^{-1} A^n_{[r]} D_2^{-1} S) = (S^t D_1^{-1} A_{[r]} D_2^{-1} S)^n \quad \text{for all } n \geq 0 ,$$

and there is a pair of indices i,j such that x_n is equal to the (i,j)-component of the matrix $S^t D_1^{-1} A^n_{[r]} D_2^{-1} S$ for each $n \in \mathbb{N}$. Finally, we define $\Gamma(A) = \bigcup_{r=1}^{\infty} \Gamma_r(A)$.

Remark. For each $r \geq 0$ we have $\Gamma_r(A) \subseteq \Gamma_{r+1}(A)$ (cf. 2.3).

Our main purpose is to prove the following theorem. Let us recall that a matrix $A \in \mathbb{Z}_+^{(p)}$ is called irreducible if for any pair of indices i,j ($1 \leq i,j \leq p$) the (i,j)-component of A^n is strictly positive for some $n \in \mathbb{N}$.

1.4 Theorem. If $A \in \mathbb{Z}_+^{(p)}$ and $B \in \mathbb{Z}_+^{(q)}$ are irreducible matrices which are strong shift equivalent, then $\Gamma(A) = \Gamma(B)$ holds.

Remark. Let us recall that two matrices $A \in \mathbb{Z}_+^{(p)}$, $B \in \mathbb{Z}_+^{(q)}$ are shift equivalent if there are matrices $R \in \mathbb{Z}_+^{(p,q)}$, $S \in \mathbb{Z}_+^{(p,q)}$ such that $RS = A^n$, $SR = B^n$ for some $n \in \mathbb{N}$ and $SA = BS$, $AR = RB$ (cf. [6] and [3]). Observe that we always have $\Gamma(A) = \Gamma(A^t)$. This shows that our invariant is not equivalent to shift equivalence (in a trivial manner at least). This shows also that $\Gamma(A)$ is not a complete invariant for strong shift equivalence.

2. We assume that $A = (a_{ij}) \in \mathbb{Z}_+^{(p)}$ is an irreducible 0-1-matrix. With A there is associated a topological Markov chain (X_A, T_A) as follows. X_A is the subset of $\{1,\dots,p\}^{\mathbb{Z}}$ consisting of all sequences $\{x_n\}_{n \in \mathbb{N}}$ with $a_{x_n x_{n+1}} = 1$

for all $n \in \mathbb{N}$. X_A is a compact subspace of $\{1,\dots,p\}^{\mathbb{Z}}$ which is invariant under the two-sided shift. T_A is the restriction of the two-sided shift to X_A. We need another notation.

Notation. Let $A \in \mathbb{Z}_+^{(p)}$, $B \in \mathbb{Z}_+^{(p)}$ be any matrices. We shall write $A \dot{\prec} B$ if there exists a subdivision matrix $S \in \mathbb{Z}_+^{(q,p)}$, and a matrix $R \in \mathbb{Z}_+^{(p,q)}$ such that $A = RS$, $B = SR$; $A \underset{\cdot}{\prec} B$ if $A^t \dot{\prec} B^t$ holds.

In the following we fix an arbitrary irreducible 0-1-matrix $A \in \mathbb{Z}_+^{(p)}$. We shall write (X,T) instead of (X_A,T_A). By a partition of X we mean a finite set $\{Y_1,\dots,Y_n\}$ of closed open subsets of X which are pairwise disjoint, and the union of $Y_1,\dots,Y_n$ is equal to X. For each pair of partitions ξ,η of X we shall write $\xi \leqslant \eta$ if η is a refinement of ξ. Moreover $\xi \vee \eta$ denotes the smallest partition refining ξ and η.

Notation. Let ξ,η be any partitions of X. We shall write

$$\xi \dot{\prec} \eta \quad \text{if} \quad T^{-1}\xi \leqslant \eta \leqslant \xi \vee T^{-1}\xi \ ,$$
$$\xi \underset{\cdot}{\prec} \eta \quad \text{if} \quad \xi \leqslant \eta \leqslant \xi \vee T^{-1}\xi \ .$$

For each partition $\xi = \{Y_1,\dots,Y_n\}$ of X we define a 0-1-matrix $A_\xi = (a_{ij}^{(\xi)})$ as follows:

$$a_{ij}^{(\xi)} = \begin{cases} 1 & \text{if } Y_i \cap T^{-1}Y_j \neq \emptyset \\ 0 & \text{elsewhere} \ . \end{cases}$$

From [5], Proposition 2.3, we obtain the following.

2.1 Lemma. If ξ and η are partitions of X with $\xi \dot{\prec} \eta$ ($\xi \underset{\cdot}{\prec} \eta$) then $A_\xi \dot{\prec} A_\eta$ ($A_\xi \underset{\cdot}{\prec} A_\eta$) holds.

For each partition ξ of X we define

$$\xi_{[1]} = \xi \vee T^{-1}\xi$$

and for each $r \in \mathbb{N}$, $r > 1$

$$\xi_{[r]} = \underbrace{((\xi_{[1]})_{[1]} \cdots)_{[1]}}_{r\text{-times}} .$$

Again from [5], Proposition 2.3, we obtain the following.

2.2 Lemma. For each partition ξ of X the matrices $A_{\xi_{[1]}}$ and $(A_\xi)_{[1]}$ are conjugate via a permutation matrix.

The following lemma will be crucial for the proof of Theorem 1.4.

2.3 Lemma. Let $T_1 \in \mathbb{Z}_+^{(p)}$, $T_2 \in \mathbb{Z}_+^{(q)}$ be any matrices with $T_1 \dot{\prec} T_2$ or $T_1 \prec\!\!\cdot\, T_2$. Then $\Gamma_0(T_1) \subseteq \Gamma_0(T_2)$.

Proof. Suppose that $T_1 \dot{\prec} T_2$. Let $S \in \mathbb{Z}_+^{(q,p)}$ be a subdivision matrix and let $R \in \mathbb{Z}_+^{(p,q)}$ be a matrix such that $T_1 = RS$, $T_2 = SR$. Moreover, we choose any subdivision matrix S_0 and diagonal matrices D_1, D_2 with positive integral entries in the diagonal such that

$$(S_0^t D_1^{-1} T_1 D_2^{-1} S_0)^n = S_0^t D_1^{-1} T_1^n D_2^{-1} S_0$$

for each $n \in \mathbb{N}$. We get

$$T_1^n = (S^t S)^{-1} S^t T_2^n S \quad , \quad n \in \mathbb{N} ,$$

where $S^t S$ is a diagonal matrix. Therefore,

$$(S_0^t D_1^{-1} (S^t S)^{-1} S^t T_2 S D_2^{-1} S_0)^n = S_0^t D_1^{-1} (S^t S)^{-1} S^t T_1^n S D_2^{-1} S_0 .$$

Now, since for any subdivision matrix, say L, and for any diagonal matrix, say D, (D having as many columns as L), there is a diagonal matrix $\bar{D}$ ($\bar{D}$ having as many rows as L) such that $LD = \bar{D}L$, and since the product of two subdivision matrices is again a subdivision matrix, we conclude that $\Gamma_0(T_1) \subseteq \Gamma_0(T_2)$ holds by our last identity. Similarly one gets the conclusion if $T_1 \prec\!\!\cdot\, T_2$.

We are now in a position to give a proof of our main result.

Proof of Theorem 1.4. Let $A \in \mathbb{Z}_+^{(p)}$, $B \in \mathbb{Z}_+^{(q)}$ be irreducible matrices which are strong shift equivalent. As $\Gamma_0(A) \subseteq \Gamma_0(A_{[1]})$ and $\Gamma_{r+1}(A) = \Gamma_r(A_{[1]})$ for each $r \in \mathbb{N}$, we have $\Gamma(A) = \Gamma(A_{[1]})$. Similarly we get $\Gamma(B) = \Gamma(B_{[1]})$. Therefore we may assume that A and B are 0-1-matrices. Again we shall write (X,T) instead of (X_A,T_A). As A and B are strong shift equivalent, it follows from [6] that (X,T) and (X_B,T_B) are topologically conjugate. Thus there exists a partition η of X such that $A_\eta = B$. We also choose a partition ξ with $A_\xi = A$. By [5], Lemma 2.2, there is a finite sequence $\xi_0,\xi_1,\dots,\xi_n$ of partitions such that $\xi_0 = \xi$, $\xi_n = \eta$ and for each i, $0 \leqslant i \leqslant n-1$, at least one of the following relations holds: $\xi_i \mathrel{\cdot\!\!\prec} \xi_{i+1}$, $\xi_i \mathrel{\prec\!\!\cdot} \xi_{i+1}$, $\xi_{i+1} \mathrel{\cdot\!\!\prec} \xi_i$, $\xi_{i+1} \mathrel{\prec\!\!\cdot} \xi_i$. We shall show that $\Gamma(A_{\xi_i}) = \Gamma(A_{\xi_{i+1}})$ for each i, $0 \leqslant i \leqslant n-1$. For convenience we assume that $n = 1$. Let us first consider the case $\xi \mathrel{\prec\!\!\cdot} \eta$, i.e. $\xi \leqslant \eta \leqslant \xi_{[1]}$. From these inequalities we obtain

$$\xi \leqslant \eta \leqslant \xi_{[1]} \leqslant \eta_{[1]} \leqslant \xi_{[2]} \leqslant \eta_{[2]} \leqslant \dots .$$

Thus

$$\xi \mathrel{\prec\!\!\cdot} \eta \mathrel{\prec\!\!\cdot} \xi_{[1]} \mathrel{\prec\!\!\cdot} \eta_{[1]} \mathrel{\prec\!\!\cdot} \xi_{[2]} \mathrel{\prec\!\!\cdot} \eta_{[2]} \mathrel{\prec\!\!\cdot} \cdots .$$

Now Lemma 2.1 and Lemma 2.2 yield

$$A \mathrel{\prec\!\!\cdot} B \mathrel{\prec\!\!\cdot} A_{[1]} \mathrel{\prec\!\!\cdot} B_{[1]} \mathrel{\prec\!\!\cdot} A_{[2]} \mathrel{\prec\!\!\cdot} B_{[2]} \mathrel{\prec\!\!\cdot} \cdots .$$

Finally we can apply Lemma 2.3 in order to obtain

$$\Gamma_0(A) \subseteq \Gamma_0(B) \subseteq \Gamma_1(A) \subseteq \Gamma_1(B) \subseteq \Gamma_2(A) \subseteq \Gamma_2(B) \subseteq \dots .$$

Thus $\Gamma(A) = \Gamma(B)$ holds. In case $\xi \mathrel{\cdot\!\!\prec} \eta$ we consider the inverse T^{-1} in place of T. Then we have $\xi \mathrel{\prec\!\!\cdot} \eta$ and the matrices associated with ξ and η are A^t and B^t respectively. So, by our first case we have $\Gamma(A) = \Gamma(A^t) = \Gamma(B^t) = \Gamma(B)$. The other cases are just the cases we have already considered, the roles of ξ and η being interchanged.

Remarks. (1) By [6], two topological Markov chains (X_A, T_A) and (X_B, T_B) are topologically conjugate if and only if the matrices A and B are strong shift equivalent. Thus Theorem 1.4 shows that the invariant Γ is an invariant for the topological conjugacy of topological Markov chains.

(2) The proof of Theorem 1.4 shows that the following statement holds if A and B are strong shift equivalent matrices: There is a non-negative integer r and a subdivision matrix S, and there are diagonal matrices D_1, D_2 with positive integral entries in the diagonal, such that $B^n = S^t D_1^{-1} A^n D_2^{-1} S$ for each $n \in \mathbb{N} \cup \{0\}$. A different invariant for strong shift equivalence, and probably a much stronger one, could be based on this relation. However, we don't know whether this invariant would be transitive or not.

3. We shall discuss now how the presence of shift equivalence is reflected by the invariant Γ. First we introduce another notation. For each $k \in \mathbb{N}$ we define a mapping Φ_k from $(\mathbb{N} \cup \{0\})^{\mathbb{N}}$ into $(\mathbb{N} \cup \{0\})^{\mathbb{N}}$ as follows: for any sequence $\{x_n\}_{n \in \mathbb{N}}$ the image of $\{x_n\}_{n \in \mathbb{N}}$ via Φ_k is the sequence $\{x_{kn}\}_{n \in \mathbb{N}}$. We shall prove the following proposition.

3.1 Proposition. Let $A \in \mathbb{Z}_+^{(p)}$, $B \in \mathbb{Z}_+^{(q)}$ be irreducible matrices which are shift equivalent. Then there is a positive integer k such that $\Phi_n(\Gamma(A)) = \Phi_n(\Gamma(B))$ for each $n \geq k$.

If two matrices $A, B \in \mathbb{Z}_+^{(p)}$ are conjugate via a permutation matrix then we shall write $A \approx B$. We need the following lemma.

3.2 Lemma. Let $A \in \mathbb{Z}_+^{(p)}$ be any matrix. Then for each $k \in \mathbb{N}$ we have $(A^k)_{[r]} \approx (A_{[kr]})^k$ for each $r \in \mathbb{N}$.

Proof. Assume that k and r are arbitrarily fixed. First we show the following,

(i) $$(A^n)_{[1]} \approx (A_{[n]})^n \qquad \text{for each } n \in \mathbb{N} .$$

Let $n \in \mathbb{N}$ be given. We choose a sequence of amalgamation matrices $R_1,\dots,R_n$ and a sequence of subdivision matrices $S_1,\dots,S_n$ such that

$$R_iS_i = A_{[i-1]}, \qquad S_iR_i = A_{[i]} \quad \text{for } 1 \leqslant i \leqslant n .$$

We define

$$R = R_1 \cdot \ldots \cdot R_n , \qquad S = S_n \cdot \ldots \cdot S_1 .$$

R is an amalgamation matrix and S is a subdivision matrix. Moreover, we have

$$RS = A^n , \qquad SR = (A_{[n]})^n .$$

Hence (see Remark (3) following Definition 1.2), $(A^n)_{[1]} \approx (A_{[n]})^n$. Applying formula (i) r-times we obtain

$$(A^k)_{[r]} \approx ((A^k)_{[1]})_{[r-1]} \approx ((A_{[k]})^k)_{[r-1]} \approx \dots \approx (A_{[kr]})^k ,$$

concluding our argument.

Proof of Proposition 3.1. As A and B are shift equivalent there are matrices $R \in \mathbb{Z}^{(p,q)}$, $S \in \mathbb{Z}^{(q,p)}$ such that $RS = A^k$, $SR = B^k$ for some $k > 0$ and $AR = RB$, $SA = BS$. Hence for each $n \geqslant k$ we have

$$(A^{n-k}R)S = A^n , \qquad S(A^{n-k}R) = SRB^{n-k} = B^n ,$$

i.e. A^n and B^n are strong shift equivalent for each $n \geqslant k$. By Theorem 1.4 this implies $\Gamma(A^n) = \Gamma(B^n)$ for each $n \geqslant k$. By Lemma 3.2 we have for $n \geqslant k$,

$$\Phi_n(\Gamma_{nr}(A)) = \Phi_n(\Gamma_{nr}(B)) , \qquad r \in \mathbb{N} .$$

As $\Gamma_r(A) \subseteq \Gamma_{r+1}(A)$ and $\Gamma_r(B) \subseteq \Gamma_{r+1}(B)$ for each $r \in \mathbb{N}$, we obtain from these equalities

$$\Phi_n(\Gamma(A)) = \Phi_n(\Gamma(B)) \qquad \text{for each } n \geqslant k .$$

4. In this final section we investigate the connection between the topological entropy and the invariant Γ of topological Markov chains.

4.1 Proposition. Let $A \in \mathbb{Z}_+^{(p)}$ be a matrix which is aperiodic and irreducible, i.e. A^n has strictly positive entries for some $n \in \mathbb{N}$. Then for the maximum eigenvalue λ of A and for each sequence $\{x_n\}_{n \in \mathbb{N}} \in \Gamma(A)$ we have

$$(+) \qquad \lambda = \lim_{n \to \infty} \frac{x_{n+1}}{x_n} .$$

In particular (aperiodic and irreducible) topological Markov chains with the same Γ-invariant have the same topological entropy.

Proof. For each $r \geq 0$ the matrix $A_{[r]}$ is also aperiodic and irreducible. Hence by the Perron-Frobenius theory (cf. [2]) the maximum eigenvalue λ of $A_{[r]}$ is strictly larger than the absolute value of the other eigenvalues of $A_{[r]}$, and $A_{[r]}$ has an eigenvector $x^{(r)}$ associated with λ such that all components of $x^{(r)}$ are strictly positive. We assume without loss of generality that the sum of the components of $x^{(r)}$ is equal to 1. It is seen that the sequence $\{\lambda^{-n}A^n\}_{n \in \mathbb{N}}$ converges to a rank 1 idempotent A_0 satisfying $A_0 x^{(r)} = x^{(r)}$. It follows that the column vectors of A_0 are all equal to $x^{(r)}$. Hence the sequence $\{\lambda^{-n}A^{n+1}\}_{n \in \mathbb{N}}$ converges to $AA_0 = \lambda A_0$. We obtain that for any suitable pair of indices i,j the sequence of the (i,j)-components of the powers of $A_{[r]}$ satisfies the condition (+), for each $r \geq 0$. As sums and multiples of sequences satisfying (+) satisfy (+) also, our first claim follows from this.

In order to prove our second claim it suffices to note that the entropy of T_A is equal to $\log \lambda$ (cf [1]).

Remark. Let $A \in \mathbb{Z}_+^{(p)}$ be any matrix and let Q be the minimal polynomial of A. Then the following can be shown easily: For each sequence $\{x_n\}_{n \in \mathbb{N}} \in \Gamma(A)$ there is a positive integer k such that the sequence $\{x_n\}_{n \geq k}$ satisfies a homogeneous linear difference equation whose (constant) coefficients are exactly the coefficients of Q.

eferences

1] M.Denker, C.Grillenberger, K.Sigmund: Ergodic Theory on Compact Spaces, Lecture Notes in Mathematics, No. 527, Springer-Verlag, 1976.

2] F.R.Gantmacher: Matrizenrechnung II, VEB Deutscher Verlag der Wissenschaften, Berlin, 1959.

3] W.Krieger: On dimension functions and topological Markov chains, Invent. Math. **56** (1980), 239-250.

4] W.Parry: The classification of topological Markov chains. Adapted shift equivalence, Isr. J. Math. **38** (1981), 335-344.

5] W.Parry, R.F.Williams: Block coding and a zeta function for finite Markov chains, Proc. London Math. Soc. **35** (1977), 483-495.

6] R.F.Williams: Classification of subshifts of finite type, Ann. of Math. **98** (1973), 120-153; Errata, Ann. of Math. **99** (1974), 380-381.

Department of Mathematics
University of California
Berkeley, California 94720

*Supported by the "Heisenbergprogramm der Deutschen Forschungsgemeinschaft".

"VECTOR BUNDLES" OVER HIGHER DIMENSIONAL "NON-COMMUTATIVE TORI"

by Marc A. Rieffel

We will at first view our subject from a somewhat broader perspective than indicated by the title. Let D be a discrete group, and let σ be a 2-cocycle on D with values in the group T of complex numbers of modulus one. Then we can construct the "twisted" group C^*-algebra $C^*(D, \sigma)$, as discussed, for example, in [12]. The main question we consider in this announcement is: How does one explicitly construct finitely generated projective modules ("vector bundles") over $C^*(D, \sigma)$?

Among the reasons that answers to this question are interesting, are that they give information about the nature of the positive cone of $K_0(C^*(D, \sigma))$ and about cancellation properties of projective modules [10], and also that such modules may be convenient places for studying aspects of non-commutative differential geometry along the lines developed by Connes in [1,2].

We will not review the construction of $C^*(D, \sigma)$ here, since we will find it convenient to use an alternative formulation which we will describe shortly. Also, we will be sloppy about the distinction between full and reduced group C^*-algebras, since we will not carry the discussion far enough for the distinction to make much difference. But let us indicate now some of the examples we have in mind.

EXAMPLE 1. Let $D = Z^n$. If $\sigma \equiv 1$, then $C^*(D, \sigma)$ is isomorphic to $C(T^n)$, an ordinary torus. Then for σ non-trivial, the algebras $C^*(D, \sigma)$ are the ones referred to in the title as "non-commutative tori", as first suggested, we believe, by Elliott [3], who was the first to study their K-theory (for $n \geq 3$). The case $n = 2$ gives the irrational (or rational) rotation C^*-algebras studied in [8, 10]. For use somewhat later, we recall from [3] that, given σ, one can find a real skew-symmetric matrix $\theta = \{\theta_{jk}\}$ such that, if we set $\lambda_{jk} = \exp(2\pi i\theta_{jk})$, and if $U_1, \ldots, U_n$ denote the unitaries in $C^*(D, \sigma)$ corresponding to the standard generators of Z^n, then

$$U_j U_k = \lambda_{jk} U_k U_j ,$$

and $C^*(D, \sigma)$ is the universal C^*-algebra for these relations. Thus it will be convenient to label these algebras by θ, so we will set $A_\theta = C^*(D, \sigma)$. (Many different σ's can correspond to the same θ.)

EXAMPLE 2. Let D be the discrete Heisenberg group, that is, the group with generators U, V and Z such that Z is central and $VU = ZUV$. Then, much as above, if we view U, V and Z as unitary generators for $C^*(D, \sigma)$ for a given σ, then matters can be arranged so that $ZU = \lambda UZ$, $ZV = \mu VZ$, and $VU = ZUV$, for appropriate λ and μ in T. Again λ and μ do not determine σ, but do determine $C^*(D, \sigma)$ up to isomorphism. (Relations between the K-theory for this and similar algebras, and the K-theory for corresponding induced flows, have very recently been studied by Judith Packer, using the foliation techniques of Connes.)

EXAMPLE 3. If we take the point of view that for $D = Z^2$ in Example 1 we are just considering the fundamental group of the ordinary 2-torus, then it is natural to consider also the case in which D is the fundamental group of a 2-holed torus, or, more generally, of a closed surface of higher genus. For the case $\sigma \equiv 1$ Kasparov has very recently indicated [5] that he has a proof that $K_0(C^*(D)) = Z^2$.

We will now give a brief description of a general method for constructing finitely generated projective $C^*(D, \sigma)$-modules. (Details will appear later.) This method involves embedding D as a cocompact discrete subgroup of a larger (perhaps Lie) group, to which the cocycle σ extends. This is more conveniently formulated by using the fact [6] that σ determines a central extension, E, of D by T. If e denotes the function on T defined by $e(z) = z$ for $z \in T$, and if e is viewed as an element of $C^*(E)$, using the fact that T is open in E, then e is a central projection, and we can define $C^*(D, \sigma)$ to be $eC^*(E)$. In this way we may forget cocycles and consider only extensions. Then our aim becomes to embed D as a cocompact subgroup of a group G which has a central extension, H, by T, into which E embeds such that the diagram

$$\begin{array}{ccccccccc}
 & & & \nearrow & E & \dashrightarrow & D & \searrow & \\
0 & \longrightarrow & T & & \updownarrow & & \updownarrow & & 0 \\
 & & & \searrow & H & \longrightarrow & G & \nearrow &
\end{array}$$

commutes. Note that e can be viewed as an element of the center of the double centralizer algebra of $C^*(H)$, so that we can form $eC^*(H)$.

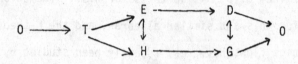

Now the space $C_c(H)$ of continuous functions of compact support, suitably completed, forms [7] an imprimitivity (i.e. equivalence) bimodule, $\overline{C}_c(H)$, between $C^*(E)$ and the transformation group C^*-algebra $C^*(H, H/E)$. It is easily verified that e is central for everything, and that as a consequence, if we set $X = e\overline{C}_c(H)$, then X will be an imprimitivity bimodule between $eC^*(H, H/E)$ and $eC^*(E)$ $(= C^*(D, \sigma))$. The basis for our method is the following easily verified result:

PROPOSITION. Let A and B be C^*-algebras, with B having an identity element, and let X be an A-B-imprimitivity bimodule. Then for any projection p in A, pX is a finitely generated projective B-module.

Thus to construct finitely generated projective $eC^*(E)$-modules, it suffices to arrange matters so that we can see how to find projections in $eC^*(H, H/E)$.

We now use the assumption that E is cocompact in H. This assumption implies that there is a natural homomorphism of $C^*(H)$ into $C^*(H, H/E)$, and so of $eC^*(H)$ into $eC^*(H, H/E)$. Thus one way to find projections in $eC^*(H, H/E)$ is to find projections in $eC^*(H)$. But projections in $eC^*(H)$ correspond, more or less, to square-integrable σ-representations of G. We say "more or less" because, on the one hand we do not insist on the representations being irreducible, and on the other hand not all irreducible square integrable representations give projections [4, 11]. In particular, this is one place where the distinction between full and reduced algebras will begin to be felt.

Anyway, if p is a projection in $eC^*(H)$, then pX will be a finitely generated projective $C^*(D, \sigma)$-module. Note that pX $(= p\overline{C}_c(H))$ is closely related to the restriction to D of the subrepresentation corresponding to p of the right regular σ-representation of G.

Let us now consider Examples 2 and 3 briefly - we will consider Example 1 in greater detail shortly.

If D is the discrete Heisenberg group, then we can embed D as a cocompact subgroup of the group G whose space is $R \times R \times Z$ with multiplication

$$(c, b, a)(c', b', a') = (c + c' + ab' , b + b', a + a'),$$

and it is not difficult to see that cocycles on D will extend to G, and that for most of these cocycles there will be square-integrable σ-representations (not irreducible) which give finitely generated projective $C^*(D, \sigma)$-module.

If, instead, D is the fundamental group of a closed surface of genus 2 or greater, then it is a classical fact that D can be embedded as a cocompact subgroup of $G = SL(2, R)$. Now this group has ordinary irreducible square-integrable representations which give projections in $C_r^*(G)$ (the reduced algebra), and so give finitely generated projective modules over $C_r^*(D)$. If the canonical trace on $C_r^*(D)$ is viewed as defining a homomorphism from $K_0(C_r^*(D))$ to R, then it is not difficult to show that the trace on the class in K_0 of such a module will be the product of the formal degree of the square-integrable representation with the volume of G/D (which product does not depend on the choice of the Haar measure on G). I conjecture that if D is instead taken to be $SL(2,Z)$, which has finite covolume but is not cocompact in $SL(2,R)$, then the corresponding $C^*(D)$-modules are not finitely generated.

We now consider a variation of the general method described above. This comes about from observing that if D is embedded in G as a cocompact subgroup, and if F is any finite group with any homomorphism of D into F, then D embeds into $G \times F$, and cocycles on F will contribute to the corresponding cocycles on D. It can be quite confusing to try to keep track of all the possibilities which can arise in this way. But, in fact, they can be considered to be special cases of the following easily verified construction:

PROPOSITION. If V is a finitely generated projective $C^*(D,\sigma)$-module, if ρ is a 2-cocycle on D, and if M is the Hilbert space of a finite dimensional unitary σ-representation of D, then $V \otimes M$ is a finitely generated projective $C^*(D, \sigma\rho)$-module, where the action is the diagonal action.

Thus our method for constructing finitely generated projective $C^*(D, \sigma)$-modules can be summarized as follows:

STEP 1. Find all 2-cocycles ρ of D for which there are finite dimensional unitary ρ-representations of D, say on Hilbert spaces M.

STEP 2. For each ρ of Step 1, find the various embeddings of D as cocompact subgroups of groups G to which the cocycle $\sigma\rho^{-1}$

extends and for which there are square-integrable $\sigma\rho^{-1}$-representations of a kind which give projections in $C^*(G, \sigma\rho^{-1})$. Form the corresponding finitely generated projective $C^*(D, \sigma\rho^{-1})$-modules, V. But it is probably not necessary to consider groups G which have a finite subgroup as a direct summand, at least if the cocycle also splits.

STEP 3. Form $V \otimes M$. Then take direct sums of various modules constructed in this way for different ρ and different embeddings.

In carrying out the above method for a specific group D, one can obtain a bewilderingly large collection of modules, and so one needs some means for determining which modules are isomorphic, which are submodules of others, etc. The natural way to start is by trying to calculate the trace of (the class in K_0 of) a module, as indicated above. This can be quite successful if the trace is faithful on K_0, but in general it need not be. If it is not, then calculation of the Chern character, along the lines developed by Connes [1,2], can be quite effective, as we will see below.

Let us turn now to Example 1, in which $D = Z^n$, with cocycle σ. To construct appropriate groups G, consider any decomposition $n = 2k + m$ for non-negative integers k and m, and let $A = R^k \times Z^m$, so that the dual group, $\hat{A}$, is $R^k \times T^m$. Let $G = A \times \hat{A}$, and note that G has the standard square-integrable irreducible cocycle representation (the Heisenberg representation) on $L^2(A)$, in which A acts by translation and $\hat{A}$ acts by pointwise multiplication. It is not difficult to see that D can be embedded as a cocompact subgroup of G (because $n = 2k + m$), in such a way that the standard cocycle on G restricts to σ on D (because of the flexibility available in choosing the component of the embedding which goes into T^k). Of course, all this should actually be done for the various $\sigma\rho^{-1}$ as indicated above, but for notational convenience we will continue just with σ. One needs to classify the modules obtained by the above method.

Now Elliott [3] has calculated the range of the Chern character on K_0, pointing out that the Chern character is injective on K_0 even if the trace is not. This suggests that we try to calculate the Chern character for the modules V constructed above. Following Connes' definition [1,2], we let L be the finite-dimensional commutative Lie algebra of (unbounded) derivations of A_θ $(= C^*(Z^n, \sigma))$ spanned by the derivations δ_j defined on the generators by $\delta_j(U_j) = 2\pi i U_j$, and $\delta_j(U_i) = 0$ for $j \neq i$. We need to define a corresponding connection on V. This is easily done by taking suitable linear combinations, depending on θ and the embedding, of the operators defined on the Schwartz space $S(A)$ by

$$(K_s f)(r, q) = -\sum s_i \frac{\partial f}{\partial r_i}(r, q)$$

$$(M_t f)(r, q) = 2\pi i\langle r, t\rangle f(r, q)$$

$$(N_u f)(r, q) = 2\pi i\langle q, u\rangle f(r, q)$$

for $f \in S(A)$, $r, s, t \in R^k$, $q \in Z^m$, $u \in R^m$. When one calculates the corresponding Chern character following [1,2], one finds that it is 0 for dimensions strictly greater than k, and in dimension k it is a decomposible integral $2k$-form, μ_k, on L. Conversely, one can specify μ_k and try to construct a corresponding module, and the only constraint to this turns out to be that the trace on the K_0-class of the module must be positive. But this trace is determined by μ_k and θ, where θ is now conveniently viewed as an element of $\Lambda^2 L$. For any j we will let θ^j denote the jth exterior product, $\theta \wedge \ldots \wedge \theta$, in $\Lambda^{2j}L$. We obtain:

PROPOSITION. Let $n = 2k + m$, and let μ_k be any decomposable element of $\Lambda^{2k}Z^n \subset \Lambda^{2k}L^*$ such that $\langle \theta^k, \mu_k \rangle > 0$ (trace condition). Then there is an action of A_θ on a completion, V, of $S(R^k \times Z^m)$ such that V is a finitely generated projective A_θ-module, and there is a connection on V whose curvature, Ω, is given by

$$\Omega(X,Y) = 2\pi i \langle X \wedge Y \wedge \theta^{k-1}, \mu_k \rangle ((k-1)!)^{-1} (tr(I_V))^{-1} I_V$$

for $X, Y \in L$. The jth Chern class, ch_j, is then given by

$$ch_j(X_1 \wedge \ldots \wedge X_{2j}) = \langle X_1 \wedge \ldots \wedge X_{2j} \wedge \theta^{k-j}, \mu_k \rangle / (k-j)!$$

for $X_1, \ldots, X_{2j} \in L$.

Thus to determine which modules arise from our method of construction, one must play off the ways of expressing forms as sums of decomposable forms, of tensoring with finite dimensional ρ-representations, and of taking direct sums of modules, against the positive trace requirement (which Elliott ellucidated in [3]). The possibilities must then be compared against Elliott's calculation of the range of the Chern character [3] to see what information about the positive cone of $K_0(A_\theta)$ one obtains. So far I have carried this out only for $n = 3$ and $n = 4$, and the results, combined with the techniques developed in [9, 10] (where cancellation for $n = 2$ was settled), yield:

THEOREM. If $n = 2, 3,$ or 4, and if θ is not rational, then the positive cone of $K_0(A_\theta)$ consists of exactly the elements with positive trace, and every such element is represented by a module constructed by the above procedures. Furthermore, finitely generated projective modules which are stably isomorphic are, in fact, isomorphic (the cancellation property). Thus every finitely generated A_θ-module is isomorphic to one constructed by the above procedures.

We remark that we can prove cancellation using the techniques of [9, 10] because it is not difficult to calculate the endomorphism algebras of the modules constructed above. For $n = 3$ Elliott already pointed out [3] that the positive cone is as described above, but cancellation is not evident from his work.

I am presently investigating the situation for $n \geq 5$.

REFERENCES

1. A. Connes, C*-algèbres et géometrie différentielle, C. R. Acad. Sc. Paris 290 (1980) 599-604.
2. ________, A survey of foliations and operator algebras, Operator Algebras and Applications (R. V. Kadison, ed.) Proc. Symp. Pure Math., 38, Amer. Math. Soc., Providence 1982.
3. G. A. Elliott, On the K-theory of the C*-algebra generated by a projective representation of a torsion-free discrete Abelian group, Operator Algebras and Group Representations, Pitman, London, to appear.
4. P. Green, Square-integrable representations and the dual topology, J. Funct. Anal. 35 (1980), 279-294.
5. G. G. Kasparov, Lorentz groups: K-theory of unitary representations and crossed products, preprint.
6. G. W. Mackey, Unitary representations of group extensions, Acta Math. 99 (1958), 265-311.
7. M. A. Rieffel, Induced representations of C*-algebras, Adv. Math. 13 (1974), 176-257.
8. ____________, C*-algebras associated with irrational rotations, Pacific J. Math. 95 (1981) 415-429.
9. ____________, Dimension and stable rank in the K-theory of C*-algebras, Proc. London Math. Soc. 46 (1983), 301-333.
10. ____________, The cancellation theorem for projective modules over irrational rotation C*-algebras, Proc. London Math. Soc. 47 (1983), to appear.

11. J. Rosenberg, Group C^*-algebras and topological invariants, Operator Algebras and Group Representations, Pitman, London, to appear.

12. G. Zeller-Meier, Produits croisés d'une C^*-algèbres par un groupe d'automorphismes, J. Math. Pures Appl., 47 (1968), 101-239.

Department of Mathematics
University of California
Berkeley
California 94720, U.S.A.

PRODUITS TENSORIELS DE $\mathfrak{Z}$- MODULES ET APPLICATIONS

Jean-Luc Sauvageot

Introduction.

Cet article est la version définitive, remaniée et complétée, d'une prépublication sur le produit tensoriel de $\mathfrak{Z}$-modules ([11]). Nous en conservons l'essentiel dans notre premier paragraphe : la définition et les propriétés fonctorielles du produit tensoriel relatif d'espaces de Hilbert au dessus d'une algèbre de von Neumann abélienne $\mathfrak{Z}$, ainsi que les constructions particulières au cas commutatif : $\mathfrak{Z}$-produit d'algèbres de von Neumann comme $\mathfrak{Z}$-modules, $\mathfrak{Z}$-produit d'implémentations d'un automorphisme de $\mathfrak{Z}$.

[Rappelons que le produit tensoriel relatif a, depuis, été généralisé au cas non commutatif (cf. [3] et [12]), de sorte qu'une partie du contenu de [11] doit être considérée comme périmée et ne fera pas l'objet d'une publication définitive].

Le second paragraphe est consacré au produit tensoriel des représentations covariantes d'un système dynamique (qui est un produit interne, généralisant donc le produit de Kronecker des représentations des groupes), pour fournir le cadre commun où s'insèrent un certain nombre d'applications développées au troisième paragraphe :

- somme de Baer des "twisted crossed products" (problème soulevé par C. Sutherland [13]) : si $(\mathfrak{Z},G,\alpha)$ est un système dynamique, nous montrons (3.2) que le produit des classes de cohomologie dans $H^2(G,U(\mathfrak{Z}))$ s'interprète comme un $\mathfrak{Z}$-produit tensoriel des produits croisés tordus.

- interprétation du produit tensoriel des représentations d'un groupoïde analytique (problème suggéré par A. Connes [2]) : au niveau intégral, on obtient une traduction en termes de $\mathfrak{Z}$-produit algébrique. On développe l'étude dans deux directions :

a) la quasi-équivalence $\lambda \sim \lambda \otimes_{\mathfrak{Z}} \lambda$ (λ = représentation régulière gauche, celle qui, pour les systèmes dynamiques, correspond au produit croisé) munit un produit croisé ou une algèbre de groupoïde d'une véritable structure "de Hopf-von Neumann involutive modulo $\mathfrak{Z}$" (3.3) dont nous espérons qu'elle permettra une théorie de la dualité (cf. N. Tatsuuma [15]) et une analyse harmonique (cf. P. Eymard [4]) pour les systèmes dynamiques et les feuilletages.

b) comme une propriété spécifique au cas où il n'y a pas de stabilisateurs (actions libres, feuilletages sans holonomie, relations d'équivalence mesurées), nous montrons que la "comultiplication" de cette structure de Hopf-von Neumann est alors un isomorphisme, ce qui s'écrit plus intrinsèquement, pour deux représentations de carré intégrable dans des espaces aléatoires H et K et une mesure transverse Λ :

$$\mathrm{End}_\Lambda(H \otimes K) = \mathrm{End}_\Lambda(H) \otimes_{\mathfrak{Z}} \mathrm{End}_\Lambda(K).$$

. Produit tensoriel relatif.

1.1. $\mathfrak{Z}$-produit d'espaces de Hilbert.

1.1.1. Définition ([3],[11],[12])

Dans tout l'article $\mathfrak{Z}$ désigne une algèbre de von Neumann abélienne ; τ est une trace normale semi-finie fidèle sur $\mathfrak{Z}$ fixée une fois pour toutes.

Si un espace de Hilbert H est muni d'une structure de $\mathfrak{Z}$-module c'est à dire d'une représentation normale non dégénérée de $\mathfrak{Z}$ dans $L(H)$), on note par $\omega_{\xi,\xi}(\xi\in H)$ la forme linéaire ultra-faiblement continue $z \to \langle z\xi,\xi\rangle$ sur $\mathfrak{Z}$; on désigne par $D(H,\tau)$ l'espace vectoriel (dense) des vecteurs τ-bornés de H, c'est à dire des ξ tels que la dérivée de Radon-Nikodym $\frac{d\omega_{\xi,\xi}}{d\tau}$ soit un opérateur borné (et donc un élément de $\mathfrak{Z}$).

Soient H_1 et H_2 deux espaces de Hilbert munis d'une structure de $\mathfrak{Z}$-module : le produit tensoriel relatif $H_1 \otimes_\tau H_2$ est l'espace de Hilbert séparé complété du produit tensoriel algébrique $D(H_1,\tau)\otimes H_2$ pour le produit scalaire caractérisé par :

$$\langle \xi_1 \otimes \xi_2, \eta_1 \otimes \eta_2 \rangle = \langle \frac{d\omega_{\xi_1,\eta_1}}{d\tau} \xi_2, \eta_2 \rangle \ , \ \xi_1,\eta_1 \in D(H_1,\tau),\ \xi_2,\eta_2 \in H_2.$$

On désigne par $\xi_1 \otimes_\tau \xi_2$ l'image dans $H_1 \otimes_\tau H_2$ du tenseur élémentaire $\xi_1 \otimes \xi_2$ de $D(H_1,\tau)\otimes H_2$. Par continuité, le tenseur $\xi_1 \otimes_\tau \xi_2$ existe dans $H_1 \otimes_\tau H_2$ dès que l'on a

$$\tau\left(\frac{d\omega_{\xi_1,\eta_1}}{d\tau} \frac{d\omega_{\xi_2,\eta_2}}{d\tau}\right) < \infty \ , \ \xi_1,\eta_1 \in H_1,\ \xi_2,\eta_2 \in H_2 \ ;$$

en particulier pour $\xi_2 \in D(H_2,\tau)$, $\forall \xi_1 \in H_1$; de sorte que la construction est symétrique : $H_1 \otimes_\tau H_2$ est canoniquement isomorphe à $H_2 \otimes_\tau H_1$.

1.1.2. Fonctorialité.

Si K_1 et K_2 sont deux autres espaces de Hilbert munis d'une structure de $\mathfrak{Z}$-module, on sait faire le produit tensoriel relatif $T_1 \otimes_{\mathfrak{Z}} T_2$ de deux opérateurs d'entrelacement $T_1 \in L_{\mathfrak{Z}}(H_1,K_1)$ et $T_2 \in L_{\mathfrak{Z}}(H_2,K_2)$, simplement en prolongeant l'action produit sur les tenseurs élémentaires :

$$(T_1 \otimes_{\mathfrak{Z}} T_2)(\xi_1 \otimes_\tau \xi_2) = T_1\xi_1 \otimes_\tau T_2\xi_2 \ , \ \xi_1 \in D(H_1,\tau),\ \xi_2 \in H_2.$$

En particulier, on sait construire dans $L(H_1 \otimes_\tau H_2)$ le produit tensoriel relatif $N_1 \otimes_{\mathfrak{Z}} N_2$ de deux algèbres de von Neumann dans $L(H_1)$ et $L(H_2)$ respectivement, commutant toutes deux à l'image de $\mathfrak{Z}$: c'est l'algèbre de von Neumann engendrée par les tenseurs élémentaires $T_1 \otimes_{\mathfrak{Z}} T_2$, $T_1 \in N_1$, $T_2 \in N_2$; on peut montrer par exemple que les algèbres $L_{\mathfrak{Z}}(H_1) \otimes_{\mathfrak{Z}} \mathbb{C}_{H_2}$ et $\mathbb{C}_{H_1} \otimes_{\mathfrak{Z}} L_{\mathfrak{Z}}(H_2)$ sont exactement le commutant l'une de l'autre dans $L(H_1 \otimes_\tau H_2)$.

De toutes ces affirmations nous ne donnons pas ici la démonstration, celle-ci devant apparaître dans un cadre plus général ([12]). Cet article traite des phénomènes spécifiques au cas commutatif, et en premier lieu celui-ci :

1.1.3. Proposition :

Sur $H_1 \otimes_\tau H_2$, les structures de $\mathfrak{Z}$-module induites par $\mathfrak{Z}_1 \otimes_{\mathfrak{Z}} \mathbb{C}_{H_2}$ et $\mathbb{C}_{H_1} \otimes_{\mathfrak{Z}} \mathfrak{Z}_2$ coïncident (où $\mathfrak{Z}_1$ et $\mathfrak{Z}_2$ sont les images respectives de $\mathfrak{Z}$ dans $L(H_1)$ et $L(H_2)$).

Ce qui s'écrit heuristiquement :

$$z \otimes_{\mathfrak{Z}} 1_{H_2} = 1_{H_1} \otimes_{\mathfrak{Z}} z,\ \forall z \in \mathfrak{Z},$$

de sorte que $H_1 \otimes_\tau H_2$ est muni sans ambiguité d'une structure de $\mathfrak{Z}$-module.

Démonstration :

Comme une conséquence immédiate de la définition, on a

$$<z\xi_1 \otimes_\tau \xi_2, \eta_1 \otimes_\tau \eta_2> = <\xi_1 \otimes_\tau z\xi_2, \eta_1 \otimes_\tau \eta_2> \ , \ \forall \xi_1,\eta_1 \in D(H_1,\tau),\ \forall \xi_2,\eta_2 \in H_2,\ \forall z \in \mathfrak{Z}$$

(les deux membres de l'équation sont égaux à $<\frac{d\omega_{\xi_1,\eta_1}}{d\tau} z\xi_2, \eta_2>$).

1.1.4. Invariance par changement de trace :

Si on remplace τ par une autre trace normale semi-finie fidèle τ' sur $\mathfrak{Z}$, les espaces de Hilbert $H_1 \otimes_\tau H_2$ et $H_1 \otimes_{\tau'} H_2$ sont canoniquement isomorphes, mais l'isomorphisme ne respecte pas les tenseurs élémentaires :

l échange $\xi_1 \otimes_\tau \xi_2$ et $(\frac{d\tau'}{d\tau})^{1/2}\ \xi_1 \otimes_{\tau'} \xi_2$ dès que ces deux expressions ont un sens.

En revanche, et c'est l'essentiel, cet isomorphisme canonique préserve les produits tensoriels relatifs d'opérateurs d'entrelacement $T_1 \otimes_{\mathfrak{Z}} T_2$: ce qui justifie a posteriori la différence que nous avons introduite dans les notations ; il préserve donc les structures de $N_1 \otimes_{\mathfrak{Z}} N_2$-module, de $\mathfrak{Z}(H_1) \otimes_{\mathfrak{Z}} L_{\mathfrak{Z}}(H_2)$-module, et bien sûr de $\mathfrak{Z}$-module.

1.1.5. <u>Cas séparable.</u>

On a $\mathfrak{Z} = L^\infty(X,\mu)$, avec X borélien standard : μ est la mesure σ-finie correspondant à τ. Un espace de Hilbert muni d'une structure de $\mathfrak{Z}$-module est une intégrable hilbertienne, et on a simplement : pour les espaces de Hilbert

$$\{\int_X^\oplus H_1(x)d\mu(x)\} \otimes_\tau \{\int_X^\oplus H_2(x)d\mu(x)\} = \int_X^\oplus \{H_1(x)\otimes H_2(x)\}d\mu(x) \ ;$$

pour les vecteurs

$$\{\int_X^\oplus \xi_1(x)d\mu(x)\} \otimes_\tau \{\int_X^\oplus \xi_2(x)d\mu(x)\} = \int_X^\oplus \{\xi_1(x)\otimes\xi_2(x)\}d\mu(x) \ ;$$

pour les opérateurs décomposables

$$\{\int_X^\oplus T_1(x)d\mu(x)\} \otimes_{\mathfrak{Z}} \{\int_X^\oplus T_2(x)d\mu(x)\} = \int_X^\oplus \{T_1(x)\otimes T_2(x)\}d\mu(x).$$

1.2. <u>$\mathfrak{Z}$-produit d'algèbres de von Neumann.</u>

Soient H_1 et H_2 deux espaces de Hilbert, M_1 et M_2 deux algèbres de von Neumann dans $L(H_1)$ et $L(H_2)$ respectivement, et soient ρ_1 et ρ_2 deux représentations (normales, respectant l'unité) de $\mathfrak{Z}$ dans M_1 et M_2 respectivement : de sorte que H_1 et H_2 sont munis d'une structure de $\mathfrak{Z}$-module.

Le $\mathfrak{Z}$-produit de M_1 et M_2 se définit par double commutation :

1.2.1. <u>Définition</u> :

On appelle <u>$\mathfrak{Z}$-produit</u> de M_1 et de M_2, et on note $M_1 *_{\mathfrak{Z}} M_2$, le commutant dans $L(H_1 \otimes_\tau H_2)$ du produit tensoriel relatif des commutants respectifs de M_1 dans $L(H_1)$ et M_2 dans $L(H_2)$.

Ce qui s'écrit formellement :

$$M_1 *_{\mathcal{Z}} M_2 = L_{L_{M_1}(H_1) \otimes_{\mathcal{Z}} L_{M_2}(H_2)}(H_1 \otimes_\tau H_2).$$

On remarquera que si on change de trace, l'isomorphisme canonique de $H_1 \otimes_\tau H_2$ sur $H_1 \otimes_{\tau'} H_2$ respecte la structure de $L_{M_1}(H_1) \otimes_{\mathcal{Z}} L_{M_2}(H_2)$-module, de sorte que la définition de $M_1 *_{\mathcal{Z}} M_2$ est, à isomorphisme canonique près, indépendante du choix de la trace τ ; on verra qu'elle est également indépendante de la réalisation spatiale de M_1 et M_2.

1.2.2. Cas particuliers.

1°). Si les images $\rho_1(\mathcal{Z})$ et $\rho_2(\mathcal{Z})$ sont dans le centre de M_1 et M_2 respectivement, on retrouve la notion précédente :

$$M_1 *_{\mathcal{Z}} M_2 = M_1 \otimes_{\mathcal{Z}} M_2.$$

2°). Si $\rho_1(\mathcal{Z})$ seulement est dans le centre de M_1, on a alors :

$$M_1 *_{\mathcal{Z}} M_2 = M_1 \otimes_{\mathcal{Z}} (M_2 \cap \rho_2(\mathcal{Z})').$$

En particulier, $\mathcal{Z} *_{\mathcal{Z}} M_2$ est isomorphe à $M_2 \cap \rho_2(\mathcal{Z})'$; et $\mathcal{Z} *_{\mathcal{Z}} \mathcal{Z}$ est canoniquement isomorphe à $\mathcal{Z}$.

1.2.3. Invariance par isomorphisme.

Soient K_1 et K_2 deux espaces de Hilbert, π_1 et π_2 deux représentations de M_1 et M_2 respectivement dans $L(K_1)$ et $L(K_2)$: alors K_1 et K_2 sont munis (via $\pi_1 \circ \rho_1$ et $\pi_2 \circ \rho_2$) d'une structure de $\mathcal{Z}$-module.

1.2.4. Proposition.

1°). Il existe une représentation $\pi_1 *_{\mathcal{Z}} \pi_2$ et une seule de $M_1 *_{\mathcal{Z}} M_2$ dans $L(K_1 \otimes_\tau K_2)$ vérifiant :

$$(T_1 \otimes_{\mathcal{Z}} T_2).X.\xi = (\pi_1 *_{\mathcal{Z}} \pi_2)(X)(T_1 \otimes_{\mathcal{Z}} T_2)\xi \ , \ \forall T_1 \in L_{M_1}(H_1, K_1),$$
$$\forall T_2 \in L_{M_2}(H_2, K_2), \ \forall X \in M_1 *_{\mathcal{Z}} M_2, \ \forall \xi \in H_1 \otimes_\tau H_2.$$

2°). On a $\pi_1 *_{\mathcal{Z}} \pi_2(M_1 *_{\mathcal{Z}} M_2) = \pi_1(M_1) *_{\mathcal{Z}} \pi_2(M_2)$.

3°). Si π_1 et π_2 sont fidèles, $\pi_1 *_{\mathcal{Z}} \pi_2$ est fidèle.

Démonstration.

1°). Les vecteurs $(T_1 \otimes_{\mathfrak{Z}} T_2)\xi$ $(T_1 \in L_{M_1}(H_1,K_1),\ T_2 \in L_{M_2}(H_2,K_2),\ \xi \in H_1 \otimes_\tau H_2)$ sont totaux dans $K_1 \otimes_\tau K_2$: d'où l'unicité. Pour l'existence, il suffit par composition de considérer le cas où l'une des représentations π_i est l'identité, et ou l'autre est, soit une ampliation, soit une réduction.

2°). $\pi_1 *_{\mathfrak{Z}} \pi_2(M_1 *_{\mathfrak{Z}} M_2)$ commute à $\pi_1(M_1)' \otimes_{\mathfrak{Z}} \pi_2(M_2)'$, d'où l'inclusion :

$$\pi_1 *_{\mathfrak{Z}} \pi_2(M_1 *_{\mathfrak{Z}} M_2) \subset \pi_1(M_1) *_{\mathfrak{Z}} \pi_2(M_2).$$

Réciproquement, d'après [1], un opérateur $t_1 \otimes_{\mathfrak{Z}} t_2$ $(t_i \in \pi_i(M_i)' = i=1,2)$ est limite faible de combinaisons linéaires d'opérateurs $t_1' t_1''^* \otimes_{\mathfrak{Z}} t_2' t_2''^*$, avec t_1', t_1'' dans $L_{M_1}(H_1,K_1)$, t_2', t_2'' dans $L_{M_2}(H_2,K_2)$; d'où l'inclusion inverse.

3°). Si π_1 et π_2 sont fidèles, 1°) et 2°) permettent de construire le morphisme inverse $\pi_1^{-1} *_{\mathfrak{Z}} \pi_2^{-1}$ de $\pi_1(M_1) *_{\mathfrak{Z}} \pi_2(M_2)$ dans $M_1 *_{\mathfrak{Z}} M_2$.

1.2.5. Corollaire.

Soient N_1 et N_2 deux algèbres de von Neumann, $\tilde{\rho}_1$ et $\tilde{\rho}_2$ deux représentations de $\mathfrak{Z}$ dans N_1 et N_2 respectivement, π_1 et π_2 deux représentations de M_1 et M_2 respectivement dans N_1 et N_2, vérifiant $\pi_1 \circ \rho_1 = \tilde{\rho}_1$, $\pi_2 \circ \rho_2 = \tilde{\rho}_2$.

Alors il existe une et une seule représentation $\pi_1 *_{\mathfrak{Z}} \pi_2$ de $M_1 *_{\mathfrak{Z}} M_2$ dans $N_1 *_{\mathfrak{Z}} N_2$ ayant la propriété suivante :

pour toute réalisation spatiale de N_1 dans $L(H_1)$ et de N_2 dans $L(H_2)$ (où H_1 et H_2 sont des espaces de Hilbert), les deux structures de $M_1 *_{\mathfrak{Z}} M_2$-module sur $H_1 \otimes_\tau H_2$ induites, d'une part (via π_1 et π_2) par la structure de M_1-module de H_1 et la structure de M_2-module de H_2, d'autre part (via $\pi_1 *_{\mathfrak{Z}} \pi_2$) par la structure de $N_1 *_{\mathfrak{Z}} N_2$-module, coïncident.

1.2.6. Remarque.

Par définition de $M_1 *_{\mathfrak{Z}} M_2$, si M_1 et M_2 sont réalisées dans $L(H_1)$ et $L(H_2)$ respectivement, la représentation naturelle de $\mathfrak{Z}$ dans $L(H_1 \otimes_\tau H_2)$ a son image dans $M_1 *_{\mathfrak{Z}} M_2$: $\mathfrak{Z}$ est donc naturellement représentée dans

$M_1 *_{\mathfrak{Z}} M_2$, et cette représentation n'est autre que $\rho_1 *_{\mathfrak{Z}} \rho_2$.

1.3. $\mathfrak{Z}$-produit d'implémentations.

On ne peut pas espérer faire en toute généralité le $\mathfrak{Z}$-produit dans $M_1 *_{\mathfrak{Z}} M_2$ d'opérateurs de M_1 et M_2 qui ne commutent pas aux images de $\mathfrak{Z}$ (cf. 1.2.2.2°). Toutefois, on va montrer que ce produit existe entre unitaires qui normalisent l'image de $\mathfrak{Z}$ et implémentent le même automorphisme.

1.3.1. Lemme.

1°). Soient H_1 et H_2 deux espaces de Hilbert munis d'une structure de $\mathfrak{Z}$-module, α un automorphisme de $\mathfrak{Z}$, u_1 et u_2 deux unitaires dans $L(H_1)$ et $L(H_2)$ respectivement, implémentant α (i.e. tels que l'on ait $u_i z\xi = \alpha(z)u_i\xi$, $\forall z \in \mathfrak{Z}, \forall \xi \in H_i$, $i=1,2$).

Il existe un unitaire dans $L(H_1 \boxtimes_\tau H_2)$ et un seul, noté $u_1 \boxtimes_{\mathfrak{Z}} u_2$, vérifiant :

$$(u_1 \boxtimes_{\mathfrak{Z}} u_2)(\xi_1 \boxtimes_\tau \xi_2) = \frac{d\tau}{d\alpha\tau}^{1/2} u_1\xi_1 \boxtimes_\tau u_2\xi_2, \quad \xi_1 \in D(H_1,\tau) \cap D(H_1,\alpha^{-1}\tau), \xi_2 \in H_2.$$

2°). $u_1 \boxtimes_{\mathfrak{Z}} u_2$ implémente α.

3°). Soit M_1 une algèbre de von Neumann dans $L(H_1)$ contenant u_1 et l'image de $\mathfrak{Z}$, M_2 une algèbre de von Neumann dans $L(H_2)$ contenant u_2 et l'image de $\mathfrak{Z}$: alors $u_1 \boxtimes_{\mathfrak{Z}} u_2$ appartient à $M_1 *_{\mathfrak{Z}} M_2$.

Démonstration.

1°). Par continuité séparée du produit tensoriel $\xi_1 \boxtimes_\tau \xi_2$, $D(H_1,\tau) \cap D(H_1,\alpha^{-1}\tau)$ est dense dans $H_1 \boxtimes_\tau H_2$; d'où l'unicité.

L'existence résulte immédiatement du calcul suivant : si le vecteur ξ_1 de H_1 est τ-borné, alors $u_1\xi_1$ est $\alpha\tau$-borné, et on a

$$\frac{d\omega_{u_1\xi_1, u_1\xi_1}}{d\alpha\tau} = \alpha\left(\frac{d\omega_{\xi_1,\xi_1}}{d\tau}\right) ;$$

si ξ_1 est en outre $\alpha^{-1}\tau$-borné, alors $u_1\xi_1$ est τ-borné et $u_1\xi_1 \boxtimes_\tau u_2\xi_2$ existe pour tout ξ_2 de H_2 ; on a alors l'équation dans $[0,+\infty]$:

$$\left\| \frac{d\tau}{d\alpha\tau}^{1/2} (u_1\xi_1 \boxtimes_\tau u_2\xi_2) \right\|^2 = \langle \frac{d\tau}{d\alpha\tau} \frac{d\omega_{u_1\xi_1,u_1\xi_1}}{d\tau} u_2\xi_2, u_2\xi_2 \rangle$$

$$= \langle \alpha \left(\frac{d\omega_{\xi_1,\xi_1}}{d\tau}\right) u_2\xi_2, u_2\xi_2 \rangle$$

$$= \| \xi_1 \boxtimes_\tau \xi_2 \|^2 ,$$

ui prouve à la fois que $u_1\xi_1 \boxtimes u_2\xi_2$ est dans le domaine de $\frac{d\tau}{d\alpha\tau}^{1/2}$, et (par olarisation, linéarité et continuité) l'existence de $u_1 \boxtimes_{\mathfrak{Z}} u_2$.

2°) et 3°) se vérifient sans difficulté.

1.3.2. Remarque.

La notation $u_1 \boxtimes_{\mathfrak{Z}} u_2$ présuppose que la construction de cet uni-aire est invariante par changement de trace, ce qui se vérifie aisément.

1.4. Application aux C^*-algèbres abéliennes.

A aucun moment nous n'avons supposé que les $\mathfrak{Z}$-modules envisagés dussent être fidèles. De sorte que tout ce qui précède s'adapte au cadre C^*-algèbre : si A est une C^*-algèbre abélienne, si ρ_1 et ρ_2 sont deux représentations (non dégénérées) de A dans $L(H_1)$ et $L(H_2)$ respectivement, on définira le produit tensoriel relatif $H_1 \boxtimes_\tau H_2$ en fixant une trace (n.s.f.f.) sur $(\rho_1 \oplus \rho_2)(A)''$, et on pourra définir intrinsèquement le A-produit tensoriel d'opérateurs d'entre-lacements, et d'implémentations. Si bien que le paragraphe suivant, qui traite des systèmes dynamiques et des représentations covariantes, aura une traduction en termes de C^*-algèbres.

2. Produit tensoriel des représentations covariantes.

2.1. Construction de produit de Kronecker.

Soit G un groupe localement compact et $\alpha(G \ni s \to \alpha_s)$ un morphisme continu de G dans le groupe des automorphismes de $\mathfrak{Z}$: une représentation (tordue par un 2-cocycle) du système dynamique $(\mathfrak{Z}, G, \alpha)$ dans l'espace de Hilbert H est un triplet $V = (\rho, v, \sigma)$ où

- ρ est une représentation (normale non dégénérée) de $\mathfrak{Z}$ dans $L(H)$;

- σ est un 2-cocycle mesurable, normalisé, de $G\times G$ dans le groupe unitaire $U(\mathfrak{Z})$ de $\mathfrak{Z}$;

- v est un morphisme mesurable $(s \to v(s))$ de G dans le groupe unitaire de $L(H)$ vérifiant :

(i) $v(s)\rho(z)v(s)^* = \rho(\alpha_s(z))$, $s\in G$, $z\in\mathfrak{Z}$;

(ii) $v(s)v(t)v(st)^* = \rho(\sigma(s,t))$, $s,t\in G$.

On dira aussi, pour préciser, que V est une <u>σ-représentation</u> du système dynamique ; on notera par $M(V)$ l'algèbre de von Neumann de $L(H)$ engendrée par $\rho(\mathfrak{Z})$ et $v(G)$.

En particulier, à chaque 2-cocycle σ correspond le σ-produit croisé que nous écrirons comme la représentation régulière $L_\sigma = (\rho_\sigma,\lambda_\sigma,\sigma)$ dans l'espace de Hilbert $L^2(G,K)$ (où K est n'importe quel espace de Hilbert muni d'une structure de $\mathfrak{Z}$-module fidèle), de la manière suivante

- ρ_σ est la représentation de $\mathfrak{Z}$ définie par

$$(\rho_\sigma(z)\xi)(s) = \alpha_{s^{-1}}(z)\xi(s),\ z\in\mathfrak{Z},\ \xi\in L^2(G,K),\ s\in G\ ;$$

- λ_σ est la σ-représentation de G définie par

$$(\lambda_\sigma(s)\xi)(t) = \sigma(t^{-1},s)\xi(s^{-1}t),\ \xi\in L^2(G,K),\ s,t\in G.$$

(cf. [6], [13] : L_σ est indépendante, à quasi-équivalence près, du choix du $\mathfrak{Z}$-module fidèle K ; l'algèbre de von Neumann $M(L_\sigma)$ n'est autre que le "produit croisé tordu", noté par exemple $R(\mathfrak{Z},G,\alpha,\sigma)$ dans [13]).

2.1.1. <u>Proposition et définition</u>.

Soient $V_1 = (\rho_1,v_1,\sigma_1)$ et $V_2 = (\rho_2,v_2,\sigma_2)$ deux représentations du système dynamique $(\mathfrak{Z},G,\alpha)$ dans les espaces de Hilbert H_1 et H_2 respectivement. Alors $(\rho_1 *_{\mathfrak{Z}} \rho_2,\ v_1 \boxtimes_{\mathfrak{Z}} v_2, \sigma_1\sigma_2)$ est une $\sigma_1\sigma_2$-représentation de $(\mathfrak{Z},G,\alpha)$ dans l'espace de Hilbert $H_1 \boxtimes_\tau H_2$ (où $v_1\boxtimes_{\mathfrak{Z}} v_2$ est l'application $s \to v_1(s)\boxtimes_{\mathfrak{Z}} v_2(s)$ définie en 1.3.1), qui sera notée $V_1\boxtimes_{\mathfrak{Z}} V_2$ et appelée <u>$\mathfrak{Z}$-produit tensoriel</u>, ou <u>produit de Kronecker</u> des représentations covariantes V_1 et V_2.

<u>Démonstration</u>.

Par 1.3.1.2°), on sait que, pour tout s de G, pour tout z de $\mathfrak{Z}$, on a

$$(v_1(s)\boxtimes_{\mathfrak{Z}} v_2(s))\rho_1 *_{\mathfrak{Z}} \rho_2(z) = \rho_1 *_{\mathfrak{Z}} \rho_2(\alpha_s(z))(v_1(s)\boxtimes_{\mathfrak{Z}} v_2(s)),$$

d'où la propriété (i) de la définition.

Par composition, on aura pour tous s et t dans G :

$$(v_1(s)\otimes_{\mathcal{Z}} v_2(s))(v_1(t)\otimes_{\mathcal{Z}} v_2(t))(v_1(st)\otimes_{\mathcal{Z}} v_2(st))^* =$$

$$= \rho_1(\sigma_1(s,t))\otimes_{\mathcal{Z}} \rho_2(\sigma_2(s,t))$$

$$= \rho_1 *_{\mathcal{Z}} \rho_2(\sigma_1\sigma_2(s,t)) \quad \text{d'après 1.1.3 et 1.2.6.}$$

Le seul point délicat est le caractère mesurable de $v_1\otimes_{\mathcal{Z}} v_2$. Lorsque G, H_1 et H_2 sont séparables, les projecteurs spectraux de $\frac{d\tau}{d\alpha_s\tau}$ correspondant à un intervalle donné sont des fonctions fortement mesurables de s ; soit $e_n(s)$ cette famille de projecteurs spectraux pour l'intervalle $\left[\frac{1}{n},n\right]$. Pour tout s, pour tout η_1 τ-borné de H_1, l'application de G dans $\mathcal{Z}$

$$s \to e_n(s)\,\frac{d\omega_{v_1(s)\xi_1,\eta_1}}{d\tau}$$

est mesurable ; d'où la mesurabilité de $(v_1(s)\otimes_{\mathcal{Z}} v_2(s))(\xi_1\otimes_\tau \xi_2)$ comme limite des applications mesurables

$$s \to e_n(s)\,\frac{d\tau}{d\alpha_s\tau}^{1/2}\, v_1(s)\xi_1\otimes_\tau v_2(s)\xi_2.$$

Dans le cas général, on écrit G comme limite, pour la topologie de Fell, du filtre de ses-groupes dénombrablement engendrés (nous laissons au lecteur qui en aurait réellement besoin le soin de rédiger une démonstration détaillée).

2.1.2. Remarque.

En vertu de 1.3.1.3, on a l'inclusion :

$$M(V_1\otimes_{\mathcal{Z}} V_2) \subset M(V_1) *_{\mathcal{Z}} M(V_2).$$

2.2. Propriété fondamentale.

Comme dans le cas des représentations de groupe, les "représentations régulières" absorbent toute autre représentation :

2.2.1. Proposition.

Soient σ et σ' deux 2-cocycles de G dans le groupe unitaire de $\mathcal{Z}$, et soit $V = (\rho, v', \sigma')$ une σ'-représentation du système dynamique $(\mathcal{Z}, G, \alpha)$ dans l'espace de Hilbert H.

La $\sigma\sigma'$-représentation de $L_\sigma \otimes_{\mathfrak{Z}} V$ est quasi-contenue dans $L_{\sigma\sigma'}$; si ρ est fidèle, $L_\sigma \otimes_{\mathfrak{Z}} V$ et $L_{\sigma\sigma'}$ sont quasi-équivalentes.

Démonstration.

On choisit ξ et ξ' dans $L^2(G,K)$, η et η' dans $D(H,\tau)$, et on calcule :

$$\langle \xi \otimes_\tau \eta, \xi' \otimes_\tau \eta' \rangle = \langle \rho_\sigma\left(\frac{d\omega_{\eta,\eta'}}{d\tau}\right)\xi, \xi' \rangle$$

$$= \int_G \langle \alpha_t^{-1}\left(\frac{d\omega_{\eta,\eta'}}{\tau}\right) \xi(t), \xi'(t) \rangle \, dt$$

$$= \int_G \langle \frac{d\omega_{v(t^{-1})\eta, v(t^{-1})\eta'}}{d\alpha_{t^{-1}}\tau} \xi(t), \xi'(t) \rangle \, dt$$

soit

$$\langle \xi \otimes_\tau \eta, \xi' \otimes_\tau \eta' \rangle = \int_G \langle \xi(t) \otimes_{\alpha_{t^{-1}}\tau} v(t^{-1})\eta, \xi'(t) \otimes_{\alpha_{t^{-1}}} v(t^{-1})\eta' \rangle \, dt.$$

Pour chaque t de G, on note w_t l'isomorphisme canonique de $K \otimes_{\alpha_{t^{-1}}\tau} H$ $K \otimes_\tau H$ (défini par $w_t(\xi \otimes_{\alpha_{t^{-1}}\tau} \eta) = \frac{d\alpha_{t^{-1}}\tau}{d\tau}^{1/2} (\xi \otimes_\tau \eta)$, dès que la formule a un sens).

Le calcul ci-dessus prouve l'existence d'un isomorphisme W de $L^2(G,K) \otimes_\tau H$ sur $L^2(G, K \otimes_\tau H)$ qui, au tenseur élémentaire $\xi \otimes_\tau \eta$ (η τ-borné) associe la fonction de carré intégrable $t \to w_t(\xi(t) \otimes_{\alpha_{t^{-1}}\tau} v(t^{-1})\eta)$.

On vérifie sans difficulté que W entrelace les représentations $\rho_\sigma *_{\mathfrak{Z}} \rho$ et $\rho_{\sigma\sigma'}$ de $\mathfrak{Z}$, ainsi que les $\sigma\sigma'$-représentations $\lambda_\sigma \otimes_{\mathfrak{Z}} V$ et $\lambda_{\sigma\sigma'}$ de G.

3. Applications.

3.1. Exemple de l'induction.

Si G est un groupe localement compact et S un sous groupe fermé de G, le processus d'induction établit une équivalence entre la catégorie des représentations de S et celle des représentations covariantes (non tordues) du système dynamique $(G, L^\infty(G/X))$ (cf. [9]).

Il est facile de vérifier que l'induction transforme le produit tensoriel ordinaire des représentations de S en produit tensoriel relatif (en $L^\infty(G/S)$-produit tensoriel) des représentations de G : si π_1 et π_2 sont deux représentations de S, on a canoniquement l'équivalence

$$\operatorname{ind}_{S\uparrow G}(\pi_1 \otimes \pi_2) = \operatorname{ind}_{S\uparrow G} \pi_1 \otimes_{L^\infty(G/S)} \operatorname{ind}_{S\uparrow G} \pi_2 .$$

3.2. Somme de Baer des extensions de groupe.

Considérons le cas particulier où $\mathfrak{Z} = L(\hat{\Gamma})$, $\hat{\Gamma}$ étant le dual d'un groupe abélien Γ.

Les extensions de Γ par G sont classifiées, d'une part par les actions de G sur Γ par automorphisme, puis, une telle action étant donnée (qui se transporte en une action α de G sur $\mathfrak{Z}$), par les classes de 2-cohomologie de G dans Γ pour cette action (cf. [7]) :

L'action α étant fixée, à tout 2-cocycle σ de G dans Γ est associée une extension E_σ de Γ par G

$$0 \to \Gamma \to E_\sigma \to G \to 0,$$

et en interprétant σ comme à valeurs dans $U(\mathfrak{Z})$, l'algèbre $M(L_\sigma)$ est l'algèbre de la représentation régulière gauche du groupe E_σ ; en d'autres termes, la représentation covariante L_σ s'identifie à la représentation régulière gauche de E_σ.

Le concept de $\mathfrak{Z}$-produit tensoriel fournit donc une interprétation au niveau de la représentation régulière gauche de la "somme de Baer" de deux extensions de groupe ([7] , p. 113) : la représentation régulière gauche de la somme de Baer $E_\sigma + E_{\sigma'}$ des extensions E_σ et $E_{\sigma'}$ (qui est isomorphe à $E_{\sigma\sigma'}$) s'identifie au $L^\infty(\hat{\Gamma})$-produit tensoriel des représentations régulières gauches de E_σ et $E_{\sigma'}$.

3.2 bis. Extensions d'une algèbre de von Neumann abélienne.

C. Sutherland ([13]) appelle extension de $\mathfrak{Z}$ par G une σ-représentation du système $(\mathfrak{Z}, G, \alpha)$; il pose la question d'une interprétation algébrique du produit des deux cocycles : comment définir en termes de représentations la "somme de Baer" de deux extensions ? comment déduire $L_{\sigma\sigma'}$ de L_σ et $L_{\sigma'}$?

Nous avons exactement la réponse : le produit des (classes de) 2-cocycles dans $H^2(G, U(\mathfrak{Z}))$ s'interprète comme le produit de Kronecker des (classes d') extensions de $\mathfrak{Z}$ par G ; telle est la signification de l'équation :

$$L_{\sigma\sigma'} \sim L_\sigma \boxtimes_{\mathfrak{Z}} L_{\sigma'} \text{ (corollaire de 2.2.1).}$$

3.3. Structure de Hopf-von Neumann sur le produit croisé.

Considérons le cas où σ est le 2-cocycle trivial noté $\circ$. $L_\circ$ est la représentation covariante correspondant au produit croisé, soit $M(L_\circ) = W^*(\mathfrak{Z}, G, \alpha)$.

La remarque 2.1.2 et la proposition 2.2.1 fournissent un morphisme canonique de $W^*(\mathfrak{Z}, G, \alpha)$ dans $W^*(\mathfrak{Z}, G, \alpha) *_{\mathfrak{Z}} W^*(\mathfrak{Z}, G, \alpha)$ qui a toutes les propriétés d'une "comultiplication" :

3.3.1. Proposition

a) Il existe un (unique) morphisme Γ, injectif, de $M(L_\circ) = W^*(\mathfrak{Z}, G, \alpha)$ dans le $\mathfrak{Z}$-produit $W^*(\mathfrak{Z}, G, \alpha) *_{\mathfrak{Z}} W(\mathfrak{Z}, G, \alpha)$ vérifiant :

i) $\Gamma \circ \rho_\circ = \rho_\circ *_{\mathfrak{Z}} \rho_\circ$;

ii) $\Gamma \circ \lambda_\circ = \lambda_\circ \boxtimes_{\mathfrak{Z}} \lambda_\circ$

b) Le diagramme ci-dessous est commutatif (propriété de Hopf-von Neumann)

$$\begin{array}{ccc}
W^*(\mathfrak{Z}, G, \alpha) & \xrightarrow{\quad \Gamma \quad} & W^*(\mathfrak{Z}, G, \alpha) *_{\mathfrak{Z}} W(\mathfrak{Z}, G, \alpha) \\
\Gamma \downarrow & & \downarrow i *_{\mathfrak{Z}} \Gamma \\
W^*(\mathfrak{Z}, G, \alpha) *_{\mathfrak{Z}} W^*(\mathfrak{Z}, G, \alpha) & \xrightarrow[\Gamma *_{\mathfrak{Z}} i]{} & W^*(\mathfrak{Z}, G, \alpha) *_{\mathfrak{Z}} W^*(\mathfrak{Z}, G, \alpha) *_{\mathfrak{Z}} W^*(\mathfrak{Z}, G, \alpha)
\end{array}$$

(où i est l'identité de $W(\mathfrak{Z}, G, \alpha)$).

Démonstration.

2.1.2. montre que l'on a $M(L_\circ \otimes_{\mathcal{Z}} L_\circ) \subset M(L_\circ) *_{\mathcal{Z}} M(L_\circ)$, et 2.2.1 assure que $L_\circ$ est quasi-équivalente à $L_\circ \otimes_{\mathcal{Z}} L_\circ$, ce qui est exactement a). b) se vérifie sans difficulté à partir de a).

On peut pousser plus loin l'analogie avec les algèbres de groupe ([14]) et munir $W^*(\mathcal{Z}, G, \alpha)$ de l'anti-automorphisme κ caractérisé par

$$. \ \kappa \circ \rho_\circ = \rho_\circ$$

$$. \ \kappa(\lambda_\circ)(s)) = \lambda_\circ(s^{-1}), \ s \in G,$$

et qui vérifie $(\kappa *_{\mathcal{Z}} \kappa) \circ \Gamma = \Gamma \circ \kappa$.

Avec Γ et κ, le produit croisé $W(\mathcal{Z}, G, \alpha)$ est muni d'une véritable structure "de Hopf-von Neumann involutive modulo $\mathcal{Z}$" (cf. [14] pour les applications de cette structure à la dualité dans le cas des groupes localement compacts).

On verra un peu plus loin que Γ est un isomorphisme lorsque G agit librement.

3.4. Application aux relations d'équivalence.

3.4.1. Construction de Krieger.

On trouvera une étude détaillée de la construction de Krieger dans ([6]) : le groupe G est dénombrable discret et $\mathcal{Z} = L^\infty(X,\mu)$, où X est un G-espace borélien standard et μ la mesure quasi-invariante correspondant à la trace τ.

L'algèbre de Krieger $M(L_r)$ est celle engendrée par la représentation (sans cocycle) $L_r = (\rho_r, l_r)$ du système dynamique $(\mathcal{Z}, G, \alpha)$ dans l'espace de Hilbert $H = \int_X^\oplus \ell^2(Gx)d\mu(x)$, définie de la manière suivante : ρ_r est la représentation naturelle de $\mathcal{Z}$ dans l'algèbre des diagonalisables ; si $\xi = \int_X^\oplus \xi_x d\mu(x)$ est l'élément générique de H, chaque ξ_x est une fonction de carré intégrable $y \to \xi_x(y)$ sur l'orbite Gx de x, et on pose, pour s dans G :

$$(l_r(s)\xi)_x(y) = \frac{ds\mu}{d\mu}(x)^{1/2} \ \xi_{s^{-1}x}(y).$$

On obtient le même résultat que pour le produit croisé $M(L_\circ)$; L_r est quasi-équivalente à $L_r \otimes_{\mathcal{Z}} L_r$, avec cette propriété supplémentaire que la représentation ainsi obtenue de $M(L_r)$ dans $M(L_r) *_{\mathcal{Z}} M(L_r)$ est un isomorphisme (ce qui fournit une description complète du $\mathfrak{Z}$-produit $M(L_r)*_{\mathfrak{Z}}M(L_r))$.

3.4.1.1. Proposition.

Avec les notations ci-dessus, il existe un (unique) isomorphisme Γ_r de $M(L_r)$ sur $M(L_r)*_{\mathfrak{Z}}M(L_r)$ vérifiant

$$\text{i)}\quad \Gamma_r \circ \rho_r = \rho_r *_{\mathfrak{Z}} \rho_r ,$$

$$\text{ii)}\quad \Gamma_r \circ 1_r = 1_r \boxtimes_{\mathfrak{Z}} 1_r .$$

3.4.1.2. Remarque.

Comme en 3.3.1.b, on a $(i*_{\mathfrak{Z}}\Gamma_r)\Gamma_r = (\Gamma_r *_{\mathfrak{Z}} i)\Gamma_r$, et $(\kappa *_{\mathfrak{Z}} \kappa)\Gamma_r = \Gamma_r \circ \kappa$ (où κ est l'anti-automorphisme naturel de $M(L_r)$, analogue à celui de $M(L_\circ)$).

Démonstration de 3.4.1.1 :

On note W l'isométrie de H dans $H \boxtimes_\tau H = \int_X^\oplus \ell^2(Gx \times Gx)d\mu(x)$ définie par $(W\xi)_x(y,y') = \delta_{y,y'}\xi_x(y)$, où $\delta_{y,y'}$ est le symbole de Kronecker, égal à 1 si $y=y'$, à 0 sinon. W est décomposable : c'est donc un morphisme de $\mathfrak{Z}$-module ; et elle entrelace manifestement 1_r et $1_r \boxtimes_{\mathfrak{Z}} 1_r$. L_r est donc équivalente à une sous-représentation de $L_r \boxtimes_{\mathfrak{Z}} L_r$.

Notons π la représentation de $\mathfrak{Z}$ dans $L(H)$ définie par

$$(\pi(s)\xi)_x(y) = z(y)\xi_x(y), \xi \in H,\ z \in \mathfrak{Z},\ x \in X,\ y \in Gx ;$$

Notons $1'_r$ la représentation de G dans H définie par :

$$(1'_r(s)\xi_x(y) = \xi_x(s^{-1}y),\ s \in G,\ \xi \in H,\ x \in X,\ y \in Gx.$$

$\pi(\mathfrak{Z})$ et $1'_r(G)$ engendrent le commutant de $M(L_r)$ ([6]).

Le projecteur $P=WW^*$ est dans $\pi(\mathfrak{Z}) \boxtimes_{\mathfrak{Z}} \pi(\mathfrak{Z})$ et il est cyclique pour $1'_r(G)'' \boxtimes_{\mathfrak{Z}} 1'_r(G)''$: P est donc un projecteur de support central 1 dans le commutant de $M(L_r)*_{\mathfrak{Z}}M(L_r)$. On aura démontré la proposition si ou prouve que l'on a $W\,M(L_r)W^* = (M(L_r)*_{\mathfrak{Z}}M(L_r))_P$.

L'inclusion dans un sens est évidente ; l'inclusion dans l'autre sens est une conséquence immédiate des équations :

$$W\,\pi(z)\,W^* = P(\pi(z)\otimes_{\mathcal{Z}} 1_H)P,\quad \forall z\in \mathcal{Z},$$
$$W\,1'_r(s)\,W^* = P(1'_r(s)\otimes_{\mathcal{Z}} 1'_r(s))P,\quad \forall s\in G.$$

3.4.1.3. Remarque.

On peut également faire la même démonstration en rajoutant des 2-cocycles, pour retrouver le cadre général des relations d'équivalences mesurées à orbites dénombrables ([5]), et on obtient, pour deux éléments de cohomologie seconde σ_1 et σ_2, l'isomorphisme de $M(L_r(\sigma_1))*_{\mathcal{Z}} M(L_r(\sigma_2))$ avec $M(L_r(\sigma_1\sigma_2))$.

3.4.2. Relations d'équivalence analytiques.

Les concepts et les notations sont ceux de [2], sinon qu'on note R un groupoïde analytique qui est une relation d'équivalence : l'élément générique γ de R est caractérisé par sa source x et son but y dans l'espace des unités $R^{(o)} = X$.

On fixe sur R une mesure transverse Λ.

On se donne deux représentations de carré intégrable U^1 et U^2 de R dans les espaces aléatoires $K^1 = (K^1_x, x\in X)$ et $K^2=(K^2_x, x\in X)$ respectivement.

A chaque fonction transverse fidèle ν sur R est associée une mesure quasi-invariant $\mu = \Lambda_\nu$ sur X, donc une trace τ sur l'algèbre $\mathcal{Z} = L^\infty(X,\mu)$ (qui est indépendante du choix de ν). Pour $i=1,2$, on notera $K^i(\mu)$ l'espace intégral $\int_X^\oplus K^i_x d\mu(x)$: $\mathcal{Z}$ agit dans $K^i(\mu)$ par opérateurs diagonaux, et l'algèbre $End_\Lambda(K^i)$ est représentée fidèlement dans le commutant de $\mathcal{Z}$.

3.4.2.1. Proposition.

On a l'égalité dans $K^1(\mu)\otimes_\tau K^2(\mu) = (K^1\otimes K^2)(\mu)$:

$$End_\Lambda(K^1)\otimes_{\mathcal{Z}} End_\Lambda(K^2) = End_\Lambda(K^1\otimes K^2).$$

Démonstration.

Par fonctorialité, on peut se ramener au cas où K^1 et K^2 sont tous deux l'espace aléatoire $L^2(\nu) = (L^2(\nu^x), x\in X)$, et où U^1 et U^2 sont toutes deux égales à la représentation "régulière gauche" de R dans $L^2(\nu)$.

On sait d'autre part que R s'identifie (modulo Λ) au produit direct d'une relation d'équivalence grossière par une relation d'équivalence à orbites dénombrables (cf. [8]). Or, pour la relation grossière, la proposition est immédiate (les algèbres End_Λ sont de dimension 1), et, pour une relation d'équivalence à orbites dénombrables, la proposition est déjà démontrée (3.4.1.1).

3.4.2.2. Remarques.

A). La proposition reste vraie si on rajoute des 2-cocycles boréliens de R dans le cercle unité.

B). Le commutant de $\mathrm{End}_\Lambda(K^1)$ dans $L(K^1(\mu))$ est l'algèbre $U^1(R)$ engendrée par les opérateurs $U^1_\mu(f\nu)$, où f est une fonction borélienne complexe sur R suffisamment régulière pour que la formule

$$(U^1_\mu(f\nu)\xi^1)_y = \int_{R^y} f(\gamma)U^1(\gamma)\xi^1_x \, d\mu^y(\gamma)$$

définisse un opérateur borné de $K^1(\mu)$.

Avec des notations analogues pour U^2, 3.4.2.1. s'interprète comme un isomorphisme canonique de $U^1_\mu(R) *_{\mathfrak{Z}} U^2_\mu(R)$ avec $(U^1 \boxtimes U^2)_\mu(R)$.

C). Si on ne suppose plus que le groupoïde est une relation d'équivalence, on a simplement au niveau intégral une interprétation du produit tensoriel des représentations, comme un $\mathfrak{Z}$-produit, et

. une inclusion $\mathrm{End}_\Lambda(K^1) \boxtimes_{\mathfrak{Z}} \mathrm{End}_\Lambda(K_2) \subset \mathrm{End}_\Lambda(K^1 \boxtimes K^2)$;

. un morphisme canonique injectif de $(U^1_\mu \boxtimes U^2_\mu)(R)$ dans $U^1_\mu(R) \boxtimes_{\mathfrak{Z}} U^2_\mu(R)$.

3.4.3. Systèmes dynamiques avec action libre.

Comme un corollaire de la proposition 3.4.2.1 et de la remarque 3.4.2.2.B, on obtient le résultat annoncé plus haut :

Proposition.

Soit $(\mathfrak{Z}, G, \alpha)$ un système dynamique. Si G agit librement dans $\mathfrak{Z}$, le morphisme canonique Γ de $W^*(\mathfrak{Z}, G, \alpha) *_{\mathfrak{Z}} W^*(\mathfrak{Z}, G, \alpha)$ défini en 3.3.1.a, est un isomorphisme.

BIBLIOGRAPHIE

[1] A. Connes, On the spatial theory of von Neumann algebras, J.F.A. 35/2 (1980), p. 153-164.

[2] A. Connes, Sur la théorie non commutative de l'intégration, Springer Lecture notes in Maths. n° 725 (1979).

[3] A. Connes, notes manuscrites, 1981.

[4] P. Eymard, L'algèbre de Fourier d'un groupe localement compact, Bull. Soc. Math. Fr. 92 (1964), p. 181-236.

[5] J. FELDMAN et C.C. Moore, Ergodic equivalence relations, cohomology and von Neumann algebras, T.A.M.S. 234 (1977), n° 2, p. 289-359.

[6] A. Guichardet, Systèmes dynamiques non commutatifs, Astérisque n° 13-14, (1974).

[7] S. Mac Lane, Homology, Academic Press (1963).

[8] A. Ramsay, Topologies on measurae groupoïds (preprint).

[9] M.A. Rieffel, Induced representations of C^*-algebras, Advances in Math. 13 (1974), p. 176-257.

[10] J.L. Sauvageot, Image d'un homomorphisme et flot des poids d'une relation d'équivalence mesurée, Math. Scand. 42 (1978), p. 71-100.

[11] J.L. Sauvageot, Produits tensoriels de $\mathbb{Z}$-modules, Publ. Univ. P. & M. Curie n° 23 (1980).

[12] J.L. Sauvageot, Sur le produit tensoriel relatif d'espaces de Hilbert, J.O.T. 9 (1983), p. 237-252.

[13] C. Sutherland, Cohomology and extensions of von Neumann algebras I, II, Publ. Res. Inst. Math. Science 16 (1980) n° 1, p. 105-133, et 135-174.

[14] M. Takesaki, Duality and von Neumann algebras, Lectures on operator algebras, Springer lecture notes in Maths. n° 247 (1972), p. 665-786.

[15] N. Tatsuuma, A duality theorem for locally compact groups, J. Math. Kyoto Univ. 6 (1967), p. 187-293.

Laboratoire de Probabilite
Tour 56 - 3eme etage
Universite P. et M.Curie
4, Place Jussieu
75230 Paris Cedex 05

Cohomology and the absence of strong ergodicity for ergodic group actions

Klaus Schmidt

§1. The main results

Let G be a countable group, $(X,\mathcal{S},\mu)$ a nonatomic Lebesgue probability space, and $(g,x) \to gx$ a nonsingular, ergodic action of G on $(X,\mathcal{S},\mu)$. A sequence $(B_n, n\geq 1)$ in $\mathcal{S}$ is called *asymptotically invariant* (a.i.) if $\lim_n \mu(B_n \Delta gB_n) = 0$ for every $g \in G$. An a.i. sequence $(B_n) \subset \mathcal{S}$ is said to be *trivial* if $\lim_n \mu(B_n)(1-\mu(B_n)) = 0$, and the action of G on $(X,\mathcal{S},\mu)$ is called *strongly ergodic* if every a.i. sequence of G in $\mathcal{S}$ is trivial. For background and details we refer to [2,3,8,10,11]. We denote by $[G]$ the *full group* of G, i.e. the group of all nonsingular automorphisms V of $(X,\mathcal{S},\mu)$ with $Vx \in Gx$ for μ-a.e. $x \in X$. It is not difficult to verify that every a.i. sequence $(B_n) \subset \mathcal{S}$ satisfies

$$\lim_n \mu(B_n \Delta VB_n) = 0$$

for every $V \in [G]$. Strong ergodicity is thus an invariant of orbit equivalence (cf. [10]). If, in particular, the action of G on $(X,\mathcal{S},\mu)$ is approximately finite we may assume that $G = \mathbb{Z}$ and use Rokhlin's theorem to show that the action of G is not strongly ergodic. This implies in particular that amenable groups have no strongly ergodic actions, since every nonsingular action of an amenable group is approximately finite (cf. [1]). The converse is also true: if a countable group G has no strongly ergodic actions on $(X,\mathcal{S},\mu)$, then G is amenable [8,11]. The connection between the absence of strong ergodicity on the one hand and amenability and Rokhlin's lemma on the other hand raises the question whether every ergodic, but not strongly ergodic actions of a countable group satisfies some (possibly very weak) form of amenability or Rokhlin's lemma. This question turns out to have a very natural and simple answer and to be closely connected with certain properties of the first cohomology of the action of G on $(X,\mathcal{S},\mu)$. In order to explain this we need some terminology. Let A be a locally compact, second countable, abelian group. A measureable map $c:G\times X \to A$ is called a *cocycle* if

$$c(g_1,g_2x) + c(g_2,x) = c(g_1g_2,x) \quad \mu\text{-a.e.},$$

for every $g_1,g_2 \in G$. A cocycle $c:G\times X \to A$ is called a *coboundary* if there exists a measurable map $f:X \to A$ with

$$c(g,x) = f(gx) - f(x) \ \mu\text{-a.e.},$$

for every $g \in G$. We denote by $Z^1(G,U(X,A))$ the additive group of all such cocycles and by $B^1(G,U(X,A))$ the subgroup of coboundaries. The group $Z^1(G,U(X,A))$ carries a natural polish topology in which a sequence (c_n) of cocycles converges to a cocycle c_0 if and only if $\lim_n c_n(g,\cdot) = c_0(g,\cdot)$ in μ-measure, for every $g \in G$. As was shown in [10], the subgroup $B^1(G,U(X,A))$ is closed in $Z^1(G,U(X,A))$ if and only if the action of G on $(X,\mathcal{S},\mu)$ is strongly ergodic. In particular, the first cohomology group $H^1(G,U(X,A)) = Z^1(G,U(X,A))/B^1(G,U(X,A))$ is nontrivial whenever the action of G is not strongly ergodic.

The *essential* (or *asymptotic*) *range* $\bar{E}(c)$ of a cocycle $c:G\times X \to A$ is defined as follows: an element $\alpha \in A$ is said to belong to $\bar{E}(c)$ if, for every $B \in \mathcal{S}$ with $\mu(B) > 0$, and for every neighbourhood $N(\alpha)$ of α in A,

$$\mu\big(\bigcup_{g\in G} (B\cap g^{-1}B\cap\{x:c(g,x) \in N(\alpha)\})\big) > 0 ,$$

and, if A is non-compact, we say that $\infty \in \bar{E}(c)$ if, for every $B \in \mathcal{S}$ with $\mu(B) > 0$ and every compact set $K \subset A$,

$$\mu\big(\bigcup_{g\in G} (B\cap g^{-1}B\cap\{x:c(g,x) \notin K\})\big) > 0 .$$

The set $E(c) = \bar{E}(c) \cap A$ is a closed subgroup of A, and c is a coboundary if and only if $\bar{E}(c) = \{0\}$. If c_1 and c_2 differ by a coboundary, then

$\bar{E}(c_1) = \bar{E}(c_2)$ (cf. [4,9]).

Two cases will be of particular interest to us: $\bar{E}(c) = \{0,\infty\}$ and $E(c) = A$. Leaving the second case aside for the moment we recall a result in [10] which states that every cocycle $c:G\times X \to A$ with $\bar{E}(c) = \{0,\infty\}$ must lie in the closure $\overline{B^1(G,U(X,A))}$ of $B^1(G,U(X,A))$ in $Z^1(G,U(X,A))$. The existence of such a cocycle implies therefore that the action of G is not strongly ergodic. Somewhat surprisingly, the converse also holds. We start with some notation.

For every cocycle $c:G\times X \to A$ we can define the skew product action

$$(g,(x,\alpha)) \to g.(x,\alpha) = (gx,\alpha+c(g,x)) \tag{1.1}$$

of G on $(X\times A, \mu\times\lambda)$, where λ denotes the Haar measure of A. The action (1.1) is ergodic if and only if $E(c) = A$. For details and proofs we refer to ([4,9]).

1.1. Theorem

Let G be a countable group, and let $(g,x) \to gx$ be a nonsingular, ergodic action of G on a nonatomic Lebesgue probability space $(X,\mathcal{S},\mu)$. There exists a cocycle $c:G\times X \to \mathbb{Z}$ with $\bar{E}(c) = \{0,\infty\}$ if and only if the action of G is not strongly ergodic.

Proof

As we have seen already, the existence of a cocycle $c:G\times X \to \mathbb{Z}$ with $\bar{E}(c) = \{0,\infty\}$ implies that G is not strongly ergodic. To prove the converse assume that G is not strongly ergodic, i.e. that there exists a nontrivial a.i. sequence $(B_n, n\leq 1) \subset \mathcal{S}$. An easy argument shows that $\lim_n (\mu(A\cap B_n) - \mu(A).\mu(B_n)) = 0$ for every set $A \in \mathcal{S}$, and this fact may be used to prove the existence of an a.i. sequence $(C_n, n\geq 1) \subset \mathcal{S}$ with $\lim_n \mu(C_n) = \frac{1}{2}$ (cf. [2]). We choose and fix an enumeration $(g_1, g_2, \dots)$ of G. By replacing $(C_n, n\geq 1)$ by a subsequence, if necessary, one can assume that the following conditions are satisfied for every $n\geq 1$ and every $k = 1,..,n$.

$$|\mu(C_n) - \tfrac{1}{2}| < 2^{-n}\ , \tag{1.2}$$

$$|\mu(C_k \cap C_{n+1}) - \tfrac{1}{4}| < 2^{-n}\ , \tag{1.3}$$

and

$$\mu(C_n \mathbin{\Delta} g_k C_n) < 2^{-n}\ . \tag{1.4}$$

Define, for every $g \in G$,

$$c(g,\cdot) = \sum_{n=1}^{\infty} 2^n (\chi_{C_n} \cdot g - \chi_{C_n})\ . \tag{1.5}$$

Condition (1.4) implies that $c(g,\cdot)$ converges in measure for every $g \in G$, and the resulting function $c: G \times X \to \mathbb{Z}$ is easily seen to be a cocycle. We define the skew product action (1.1) of G on $(X \times \mathbb{Z}, \mu \times \lambda)$, where λ is the counting measure on $\mathbb{Z}$. If $Y \subset X \times \mathbb{Z}$ is an ergodic component of this action, the set $Y_0 = Y \cap (X \times \{0\})$ must either satisfy $Y_0 \subset C_n$ or $Y_0 \cap C_n = \phi$ for every $n \geq 1$, and (1.3) and (1.4) now imply that $\mu(Y_0) = 0$. All the measures in the ergodic decomposition of $\mu \times \lambda$ for the skew product action (1.1) of G are thus concentrated on null-sets, which implies that $\bar{E}(c) = \{0,\infty\}$ (cf. [9], Proposition 3.12). The theorem is proved. □

Now we turn to the problem of finding a notion of amenability appropriate to actions of G which are not strongly ergodic. Consider the following example: let $(g,x) \to gx$ be a nonsingular ergodic action of G on $(X,\mathcal{S},\mu)$, and put $A = \mathbb{Z}$. Assume that there exists a cocycle $c: G \times X \to \mathbb{Z}$ with $E(c) = \mathbb{Z}$. The ergodicity of the action (1.1) implies the ergodicity of the skew product action

$$(g,(x,t)) \to (gx, t+\beta c(g,x) \pmod 1) \tag{1.6}$$

of G on $X \times \mathbb{R}/\mathbb{Z}$, where β is a fixed irrational number. Denote by λ_1 the Lebesgue measure on $\mathbb{R}/\mathbb{Z}$ and choose a nontrivial a.i. sequence $(C_n, n\geq 1)$ of Borel subsets of $\mathbb{R}/\mathbb{Z}$ for the $\mathbb{Z}$-action $(n,t) \to t+n\beta \pmod 1$ on $(\mathbb{R}/\mathbb{Z}, \lambda_1)$. Define $B_n \subset X \times \mathbb{R}/\mathbb{Z}$ by $B_n = X \times C_n$, $n\geq 1$, and observe that $(B_n, n\geq 1)$ is a nontrivial a.i. sequence for the action (1.6) of G on $(X \times \mathbb{R}/\mathbb{Z}, \mu \times \lambda_1)$. The action (1.6) is thus not strongly ergodic, irrespective of whether the original action $(g,x) \to gx$ of G on (X,S,μ) is strongly ergodic. The action (1.6) has another interesting property: consider the map $\psi : X \times \mathbb{R}/\mathbb{Z} \to \mathbb{R}/\mathbb{Z}$ with $\psi(x,t) = t$ for every $(x,t) \in X \times \mathbb{R}/\mathbb{Z}$, and note that

$$\psi(G(x,t)) = \mathbb{Z}\psi(x,t) \quad \mu\times\lambda_1\text{-a.e.}, \tag{1.7}$$

where $\mathbb{Z}\psi(x,t) = \{t+n\beta : n \in \mathbb{Z}\}$ denotes the orbit of $\psi(x,t) = t$ under the $\mathbb{Z}$-action $(n,t) \to t+n\beta \pmod 1$ on $(\mathbb{R}/\mathbb{Z},\lambda_1)$. We can thus find a $\mathbb{Z}$-action which is, in a certain sense, a quotient of the G-action on $(X \times \mathbb{R}/\mathbb{Z}, \mu\times\lambda_1)$. The following result shows that the picture described by (1.7) is typical (cf. [5]).

1.2. <u>Theorem</u>

Let G be a countable group, and let $(g,x) \to gx$ be a nonsingular, ergodic action of G on a nonatomic Lebesgue probability space (X,S,μ). The following conditions are equivalent.

(1) The action of G on (X,S,μ) is not strongly ergodic;

(2) There exists a nonatomic Lebesgue probability space (Y,T,ν), a measure preserving map $\psi : X \to Y$, and a nonsingular, ergodic automorphism T of (Y,T,ν) with

$$\psi(Gx) = \{T^n\psi(x) : n \in \mathbb{Z}\} \quad \mu\text{-a.e.}. \tag{1.8}$$

1.3. <u>Remark</u>

Even if G preserves μ, the automorphism T in (1.8) need not preserve any σ-finite measure equivalent to μ. From W. Krieger's work [6] it follows, however, that one can always find ψ and T in (1.8) such that T preserves a probability measure equivalent to ν.

Theorem 1.2 can also be expressed in terms of ergodic equivalence relations (cf. [4] for the terminology). If R is a countable, nonsingular, ergodic equivalence relation on (X,S,μ) we say that R is strongly ergodic if R has no nontrivial a.i. sequences in S (in this context a sequence $(B_n) \subset S$ is called a.i. if $\lim_n \mu(B_n \Delta V B_n) = 0$ for every automorphism V of $(X,\mathcal{S},\mu)$ with $(x,Vx) \in R$ for μ-a.e. $x \in X$, i.e. for every automorphism V in the full group $[R]$ of R). In this terminology Theorem 1.2 reads as follows.

1.4. Theorem

Let R be a countable, nonsingular, ergodic equivalence relation on $(X,\mathcal{S},\mu)$. Then R is not strongly ergodic if and only if there exists an approximately finite, countable,nonsingular, ergodic equivalence relation R_1 on a nonatomic Lebesgue probability space $(Y,\mathcal{T},\nu)$ and a measure preserving map $\psi:X \to Y$ with

$$\psi(R) = R_1 .$$

In other words, R is not strongly ergodic if and only if it has an approximately finite homomorphic image.

The two equivalent Theoresm 1.2 and 1.4 are in turn equivalent to the fact that the existence of nontrivial a.i. sequences is closely connected with the existence of cocycles $c:G\times X \to \mathbb{Z}$ with $\bar{E}(c) = \{0,\infty\}$ (cf. Theorem 1.1).

Outline of proof of Theorem 1.2 (cf. [5])

First assume that (1.8) is satisfied. It is not difficult to verify that every nontrivial a.i. sequence $(C_n, n\geq 1)$ for T in $\mathcal{T}$ gives rise to a nontrivial a.i. sequence $(\psi^{-1}(C_n), n\geq 1)$ for G in $\mathcal{S}$. We note in passing that (1.8) also defines a cocycle $c:G\times X \to \mathbb{Z}$ with $\bar{E}(c) = \{0,\infty\}$ by setting, for every $g \in G$, $x \in X$,

$$c(g,x) = n \tag{1.9}$$

whenever $\psi(gx) = T^n\psi(x)$.

To prove the converse we assume that the action of G is not strongly ergodic. By Theorem 1.1 we can find a cocycle $c:G\times X \to \mathbb{Z}$ with $\bar{E}(c) = \{0,\infty\}$, defined as in (1.5). Let $\mathcal{T}$ denote the σ-algebra generated by the sets $\{C_n, n\geq 1\}$ and let Y be the space of atoms of $\mathcal{T}$. After ignoring null-sets we may identify Y with the infinite cartesian product $\{0,1\}^{\mathbb{N}}$, $\mathbb{N} = \{1,2,\ldots\}$, by making every $y \in Y$ correspond to the point $(y^{(1)},y^{(2)},\ldots) \in \{0,1\}^{\mathbb{N}}$, where, for every $n \geq 1$, $y^{(n)} = 0$ if $y \subset C_n$, and $y^{(n)} = 1$ if $y \subset X \setminus C_n$. The fact that the cocycle (1.5) is well defined is equivalent to saying that, for every $g \in G$ and a.e. $x \in X$, $\psi(gx)$ and $\psi(x)$ differ in only finitely many coordinates. Apart from technicalities, the proof is now completed by observing that the homomorphic image under ψ of the equivalence relation R defined by G is a subrelation of the odometer equivalence relation on $\{0,1\}^{\mathbb{N}}$, and therefore approximately finite (cf. [9]). □

Theorem 1.2 has a simple, but useful consequence for the cohomology of G :

Corollary 1.5.

Let G be a countable group and let $(g,x) \to gx$ be a nonsingular, ergodic, but not strongly ergodic action of G on $(X,\mathcal{S},\mu)$. Then there exists, for every locally compact, second countable, abelian group A , a cocycle $c \in Z^1(G,U(X,A))$ with $E(c) = A$. In fact, the set $\{c \in \overline{B^1(G,U(X,A))} : E(c) = A\}$ is a dense G_δ in $\overline{B^1(G,U(X,A))}$.

Proof.

The category argument described in [9] shows that it is enough to find a cocycle $c \in \overline{B^1(G,U(X,\mathbb{Z}))}$ with $E(c) = \mathbb{Z}$. The existence of such a cocycle, however, follows immediately from Theorem 1.2: choose a cocycle $d:\mathbb{Z}\times Y \to \mathbb{Z}$ for the $\mathbb{Z}$-action given by the automorphism T of $(Y,\mathcal{T},\nu)$ appearing in Theorem 1.2 with $E(d) = \mathbb{Z}$, and define $c:G\times X \to \mathbb{Z}$ by

$$c(g,x) = d(n,\psi(x))$$

whenever $T^n\psi(x) = \psi(gx)$, $x \in X$, $g \in G$, $n \in \mathbb{Z}$. This cocycle c will satisfy $E(c) = \mathbb{Z}$. □

§2. Concluding remarks

Let G be a countable group, and let $(X,\mathcal{S},\mu)$ be a nonatomic Lebesgue probability space. In [3] it was shown that G has Kazhdan's property T if and only if every ergodic, measure preserving action of G on $(X,\mathcal{S},\mu)$ is strongly ergodic. One can characterize amenable groups in a similar manner: G is amenable if and only if no nonsingular, ergodic action of G on $(X,\mathcal{S},\mu)$ is strongly ergodic ([1,8,11]). Theorem 1.2 is thus of interest only if G is neither amenable, nor a group with property T. This leads us to an important question left unanswered by Theorem 1.2. Assume that $(g,x) \to gx$ is a nonsingular ergodic action of G on $(X,\mathcal{S},\mu)$ which is neither approximately finite nor strongly ergodic. By Theorem 1.2 we can find a Lebesgue probability space $(Y,\mathcal{T},\nu)$ and a measure preserving map $\psi: X \to Y$ with $\psi(Gx) = \{T^n\psi(x) : n \in \mathbb{Z}\}$ for some nonsingular, ergodic automorphism T of $(Y,\mathcal{T},\nu)$. Put

$$\Gamma = \{V \in [G] : \psi(Vx) = \psi(x) \ \mu\text{-a.e.}\} .$$

The group Γ is obviously not ergodic, and it is not difficult to verify that a.e. measure in the ergodic decomposition of μ with respect to Γ is nonatomic (this is the same as saying that there exists no Γ-invariant set of positive measure in $\mathcal{S}$ on which Γ is of type I) (cf. [9, Corollary 7.5 and Theorem 8.7]).

2.1. Problem

Is it possible to choose ψ and T in Theorem 1.2 in such a way that Γ has no nontrivial a.i. sequences (i.e. that Γ is strongly ergodic) on a.e. ergodic component?

We now turn to the existence of nonsingular, ergodic, but not strongly ergodic actions for *every* infinite countable group G. Recall that a nonsingular, ergodic action $(g,x) \to gx$ of G on $(X,\mathcal{S},\mu)$ is said to be of *type* III_1 if the Radon-Nikodym derivative $\rho(g,\cdot) = d\mu g/d\mu$ satisfies $E(\log \rho) = \mathbb{R}$.

2.2. Lemma

If G is an infinite, countable group, then G has an ergodic, free action on $(X,\mathcal{S},\mu)$ which is of type III_1.

Proof

G certainly has a measure preserving, ergodic, free action on (X,S,μ) : take, for example, a Bernoulli action. Now choose an automorphism $V \in [G]$ which is ergodic [9, Theorem 8.22], and use [7] to find a probability measure ν on (X,S) which is quasi-invariant and ergodic for the automorphism V, and such that (the $\mathbb{Z}$-action given by) V is of type III_1 on (X,S,ν). We choose an enumeration $(g_1,g_2,\dots)$ of G, put $\tilde{\mu} = \sum_{k=1}^{\infty} 2^{-k} \nu g_k$, and observe that $\tilde{\mu}$ is nonatomic, quasi-invariant and ergodic under G, and that the action of G on $(X, ,\tilde{\mu})$ is of type III_1. If this action is not free, take its cartesian product with a free Bernoulli action: this new action will be ergodic, free, and of type III_1 (cf. [12]). □

2.3. Theorem

Let G be an infinite, countable group, and let (X,S,μ) be a non-atomic Lebesgue probability space. Then G has a nonsingular, ergodic, free action on (X,S,μ) which is not strongly ergodic.

Proof

Find a nonsingular, ergodic, free action $(g,y) \to gy$ of G on a non-atomic Lebesgue probability space (Y,T,ν) which is of type III_1, and define a cocycle $c: G\times Y \to \mathbb{R}$ by $c(g,\cdot) = \log(d\nu g/d\nu)$. The skew product action (1.1) of G on $(Y\times\mathbb{R},\mu\times\lambda)$ is ergodic, where λ denotes Lebesgue measure, and we can define an ergodic action of G on $(Y\times(\mathbb{R}^2/\mathbb{Z}^2),\nu\times\lambda_2)$ (with λ_2 equal to the Lebesgue measure on $\mathbb{R}^2/\mathbb{Z}^2$) by setting

$$g(y,s,t) = (gy, s+c(g,y) (\mathrm{mod}\ 1), t+\alpha c(g,y) (\mathrm{mod}\ 1))$$

for every $y \in Y$, $(s,t) \in \mathbb{R}^2/\mathbb{Z}^2$, where $\alpha \in \mathbb{R}$ is irrational. This action is easily seen to have nontrivial a.i. sequences. □

2.4. Corollary

Let G be an infinite, countable group, and let (X,S,μ) be a nonatomic probability space. Then G has a nonsingular, ergodic, free action on (X,S,μ) such that the following is true for every locally compact, second countable, abelian group A.

(1) $H^1(G,U(X,A)) \neq \{0\}$.

(2) There exists a cocycle $c \in Z^1(G,U(X,A))$ with $E(c) = A$.

(3) If A is noncompact, there exists a cocycle $c \in Z^1(G,U(X,A))$ with $\bar{E}(c) = \{0,\infty\}$.

References

1. Connes, A., Feldman, J., and Weiss, B. : An amenable equivalence relation is generated by a single transformation. Ergod. Th. and Dynam. Sys. 1 (1981), 431-450.

2. Connes, A., and Krieger, W. : Measure space automorphisms, the normalizers of their full groups, and hyperfiniteness. J. Functional Analysis 24 (1977), 336-352.

3. Connes, A., and Weiss, B. : Property T and asymptotically invariant sequences. Israel J. Math. 37 (1980), 209-210.

4. Feldman, J., and Moore, C.C. : Ergodic equivalence relations, cohomology, and von Neumann algebras. I. Trans. Amer. Math. Soc. 234 (1977), 289-324.

5. Jones, V.F.R., and Schmidt, K. : Asymptotically invariant sequences and hyperfiniteness. Preprint.

6. Krieger, W. : On ergodic flows and the isomorphism of factors. Math. Ann. 223 (1976), 19-70.

7. Krieger, W. : On Borel automorphisms and their quasi-invariant measures. Math. Z. 151 (1976), 19-24.

8. Losert, V., and Rindler, H. : Almost invariant sets. Bull. London Math. Soc. 41 (1981), 145-148.

9. Schmidt, K. : Cocycles of ergodic Transformation Groups. McMillan (India), 1977.

10. Schmidt, K. : Asymptotically invariant sequences and an action of SL(2,$\mathbb{Z}$) on the 2-sphere. Israel J. Math. 37 (1980), 193-208.

11. Schmidt, K. : Amenability, Kazhdan's property T , strong ergodicity and invariant means for ergodic group actions. Ergod. Th. and Dynam. Sys. 1 (1981), 223-236.

12. Schmidt, K., and Walters, P. : Mildly mixing actions of locally compact groups. Proc. London Math. Soc. 45 (1982), 506-518.

Coding of Markov Shifts

Klaus Schmidt

The purpose of this note is to give a brief survey of the problem of finding isomorphisms of Markov shifts which take into account the sequential structure of these processes, and which can be interpreted as realistic 'coding' or 'translation' procedures. The interpretation of such isomorphisms as translation procedures is suggested by the fact that Markov shifts resemble (very idealized) languages: a Markov shift is determined by a finite alphabet, a finite set of 'grammatical rules' which excludes certain combinations of letters (or symbols) in the alphabet, and an assignment of 'likelihood' to the various allowed combinations of symbols, just as ordinary languages distinguish between rare and frequently used words.

In mathematical terms, a Markov shift is completely determined by an irreducible, stochastic matrix $P = (P(i,j)\ ,\ 1 \le i\ ,\ j \le k)\ ,\ k \ge 2$, where <u>stochastic</u> means that $P(i,j) \ge 0$ and $\sum_{r=1}^{k} P(i,r) = 1$ for every $i,j = 1,\ldots,k$, and <u>irreducible</u> indicates that we can find, for every i,j , a positive integer n for which the entry $P^n(i,j)$ of the n-th matrix power of P is positive. The matrix P has a unique left eigenvector $\bar{p} = (\bar{p}(1),\ldots,\bar{p}(k))$ with eigenvalue 1 , positive entries, and $\sum_{r=1}^{k} \bar{p}(r) = 1$. We denote by X_P the space of sequences $x = (x_n, n \in \mathbb{Z})$ in $\{1,\ldots,k\}^{\mathbb{Z}}$ with $P(x_n,x_{n+1}) > 0$ for all $n \in \mathbb{Z}$ and write $T_P : X_P \to X_P$ for the shift transformation $(T_P x)_n = x_{n+1}$. X_P is a compact T_P-invariant subspace of $\{1,\ldots,k\}^{\mathbb{Z}}$, and we define a T_P-invariant probability measure m_P on X_P by setting, for every cylinder set $C = [i_0,\ldots,i_s]_r = \{x \in X_P : x_{r+j} = i_j$ for $j = 0,\ldots,s\}$,

$$m_P(C) = \bar{p}(i_0)P(i_0,i_1)\ldots P(i_{s-1},i_s) .$$

The automorphism T_P of the probability space (X_P,m_P) is called an ergodic Markov shift. In terms of our earlier interpretation the set $\{1,\ldots,k\}$ is the alphabet or set of symbols (including blanks, punctuation, etc.) of a 'language', the set of zero entries of the matrix P corresponds to the grammatical rules of the language by excluding certain combinations of symbols (the restriction to rules governing allowed pairs of letters is completely irrelevant - any set of rules excluding only finitely many strings of finite lengths can be brought into this form by changing the alphabet, if necessary), and m_P describes the frequency of words.

If P and Q are irreducible stochastic matrices, the associated ergodic Markov shifts T_P and T_Q are called isomorphic if there exists a measurable, measure preserving and a.e. bijective map $\phi:X_P \to X_Q$, with $\phi T_P = T_Q\phi$ m_P - a.e.. It is known that T_P and T_Q are isomorphic if and only if they have the same entropy and period, where the entropy $h(T_P)$ of T_P is defined by

$$h(T_P) = - \sum_{\{(i,j):P(i,j)>0\}} \bar{p}(i)P(i,j) \log P(i,j) ,$$

and its period is the greatest common divisor of the set $\{n \geq 1:P^n(1,1) > 0\}$.

This definition of isomorphism has some very serious disadvantages, since it does not reflect the sequential structure of the Markov chains. To see this, think of the two Markov processes (T_P,X_P,m_P) and (T_Q,X_Q,m_Q) as languages, and of the map ϕ in terms of a translator who associates with every two-sided infinite string of letters (i.e. every infinitely long novel) in the first language its translation $\phi(x)$, which is, of course, again an infinitely long novel in the second language. However, in order to write down a single letter of the translation, e.g. $\phi(x)_0$, the translator has to know x completely, i.e. has to have read the complete novel. Assuming finite reading speed, the translator can never start his translation. Another problem lies in the sensitivity of the translation to errors: assume that the translator has read all of x but has misread a single letter, x_0, say. If x_0' denotes what he has read instead of x_0 , denote by x' the point $(\ldots,x_{-2},x_{-1},x_0',x_1,x_2,\ldots)$. Although x and x' differ by a single letter, their translation $\phi(x)$ and $\phi(x')$ need have nothing whatsoever in common!

Substantial impetus towards overcoming these difficulties was given by a emarkable result due to M. Keane and M. Smorodinsky [4], who proved that any two somorphic Markov chains are, in fact, <u>finitarily isomorphic</u>. This means that there xists an isomorphism $\phi: X_P \to X_Q$ of T_P and T_Q and null sets $E_P \subset X_P$, $_Q \subset X_Q$ such that the restrictions of ϕ to $X_P \setminus E_P$ and of ϕ^{-1} to $X_Q \setminus E_Q$ are ontinuous. Another way of expressing this is by saying that there exist measurable, onnegative, integer valued functions a_ϕ, m_ϕ on X_P and $a_{\phi^{-1}}$ and $m_{\phi^{-1}}$ on X_Q ith $\phi(x)_0 = \phi(x')_0$ whenever $x, x' \in X_P \setminus E_P$ satisfy $x_i = x'_i$ for $m_\phi(x) \le i \le a_\phi(x)$, and with ϕ^{-1} satisfying the analogous condition on $X_Q \setminus E_Q$.

The existence of finitary isomorphisms resolves the first difficulty of our ranslator: he can write down $\phi(x)_0$, having read only a finite part of the novel to be translated. The possible effect of misreading a single letter cannot be ealt with so easily, and we shall have to return to this problem later. Another roblem raised by the notion of finitary isomorphisms is whether the functions $_\phi, m_\phi$ can be chosen to be integrable: if not, the translator will have an in-inite expected amount of reading to do for each letter $\phi(x)_i$ of his translation nd will thus proceed with zero expected speed with his translation. Motivated by hese considerations we introduce the following definition (cf. [7,8,9,13]):

. <u>Definition</u> A finitary isomorphism $\phi: X_P \to X_Q$ of two ergodic Markov shifts $_P$ and T_Q is said to satisfy condition F.E. (= finite expected code lengths) f the functions $a_\phi, m_\phi, a_{\phi^{-1}}, m_{\phi^{-1}}$ are all integrable.

Although isomorphic Markov shifts always have finitary isomorphisms, there eed not exist any isomorphisms satisfying condition F.E.. Before we turn to a ist of obstructions to the existence of such isomorphisms I would like to give a imple example of two ergodic Markov shifts which are nontrivially isomorphic via n isomorphism with finite expected code lengths. Consider the matrices

$$P = \begin{pmatrix} 0 & \frac{1}{2} & \frac{1}{2} \\ \frac{1}{2} & 0 & \frac{1}{2} \\ \frac{1}{2} & \frac{1}{2} & 0 \end{pmatrix} \qquad Q = \begin{pmatrix} \frac{1}{2} & \frac{1}{2} \\ \frac{1}{2} & \frac{1}{2} \end{pmatrix}$$

nd the associated Markov shifts T_P and T_Q. Both are aperiodic (i.e. have eriod 1) and have entropy log 2. We denote the alphabet of P by $\{1,2,3\}$ nd that of Q by $\{a,b\}$. The space X_P consists of all possible sequences of the symbols 1,2,3 with no two adjacent symbols being equal, and X_Q is the set of all possible sequences of a's and b's. Clearly the corresponding shifts are quite different: T_P has no fixed points, while T_Q has two. In order to construct an isomorphism one can first try the naive approach: consider the map $\phi : X_P \to X_Q$ determined by the diagram

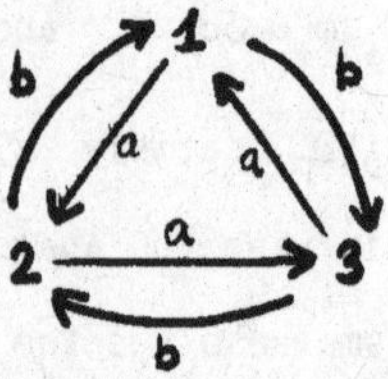

(1)

in the following manner: in order to find $\phi(x)_i$ we read off x_i, x_{i+1} and write down a or b depending on the letter associated with the arrow leading from the symbol x_i to the symbol x_{i+1} in the diagram. The point

$$x = (.....3\ 2\ 3\ 1\ \dot{2}\ 3\ 2\ 3\ 1\ 2\ .....) \qquad (2)$$

thus gets translated as

$$(x) = (.....b\ a\ a\ a\ \dot{a}\ b\ a\ a\ a.....)$$

where the dot indicates the zero coordinate. Unfortunately, the map ϕ is three to one: the strings

$$x' = (.....2\ 1\ 2\ 3\ \dot{1}\ 2\ 1\ 2\ 3\ 1.....)\ ,$$

$$x'' = (.....1\ 3\ 1\ 2\ \dot{3}\ 1\ 3\ 1\ 2\ 3.....)$$

satisfy $\phi(x) = \phi(x') = \phi(x'')$, since we have no way of finding a correct 'starting point' for our translation. If we knew <u>one</u> letter of the original word, we could decide whether x, x', or x" is the 'correct' pre-image of $\phi(x)$. In terms of the diagram (1), the string $\phi(x)$ defines a journey on that diagram by reading off successive coordinates of $\phi(x)$ and choosing the road (= arrow) indicated at each stage. Depending on our position in the diagram at a given time, we could be on three different journeys, all governed by the same set of instructions $\phi(x)$. In order to overcome this problem one can try to modify the diagram (1) to obtain a 'magic set of instructions' which would make us end up in the same symbol, irrespective of our starting point. This is a special case of the road problem in which one has a 'map' of cities connected by one-way-streets and tries to find suitable labels for these streets in such a way that there exists a finite string of instructions which is meaningful for cars starting from any city, and which makes them all end up in the capital at the end of that sequence of instructions, irrespective of their starting point. For more details on this problem cf. [1,11].

In our simple example, the following diagram provides a solution:

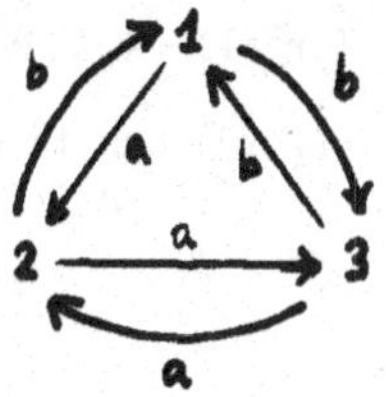

(3)

The magic instruction (a,b) lets us end up in 1, irrespective of our position in the diagram. If we define the map ϕ using (3) instead of (1), we obtain

$$\phi(x) = (\ldots\ldots a\ a\ b\ a\ \dot{a}\ a\ a\ b\ a \ldots\ldots)$$

for x given by (2). Conversely, if we are given $\phi(x)$ and try to determine x_i, we have to look to the left of the coordinate $\phi(x)_i$ until we see the pair $(a,b) = (\phi(x)_{i-j-1}, \phi(x)_{i-j})$, say, with $j \geq 1$. From (3) we now know that $x_{i-j+1} = 1$, and now we can work our way forward again to determine x_i. This procedure is easily seen to have finite expected code lengths ($a_\phi \equiv 1$, $m_\phi \equiv 0$, $a_{\phi^{-1}} = 0$, and $m_{\phi^{-1}}$ is integrable), and has the added advantage of being 'error correcting' in both directions in the following manner: if x and x' differ in only finitely many coordinates then the same is true for $\phi(x)$ and $\phi(x')$ and, conversely, if $\phi(x)$ and $\phi(x')$ differ in finitely many places and lie outside a certain set of measure zero, then x and x' coincide in all but finitely many coordinates. This kind of error correction is different from the usual notion in the sense that translation errors may occur, but with probability one they will affect only a finite part of the translation, so that the transformation will eventually become correct again.

In order to discuss this phenomenon in greater generality we consider an ergodic Markov shift T_P on (X_P, m_P) and define an equivalence relation $\sim$ on X_P by setting $x \sim x'$ whenever x and x' differ in only finitely many coordinates. The following result, due to W. Krieger [7], shows that all isomorphisms satisfying F.E. have the 'correction' property mentioned above.

2. <u>Proposition</u> Let T_P and T_Q be ergodic Markov shifts on the shift spaces (X_P, m_P) and (X_Q, m_Q), respectively, and let $\phi: X_P \to X_Q$ be an isomorphism of T_P and T_Q satisfying condition F.E.. Then there exists a null set $E \subset X_P$ with $x \sim x'$ if and only if $\phi(x) \sim \phi(x')$, for all $x, x' \in X_P \backslash E$.

For the proof of Proposition 2 (cf. [7,9]) one introduces the two functions

$$a^*_\phi(x) = \sup_{n \geq 0} (a_\phi(T_P^{-n}x) - n) \tag{4}$$

and

$$m^*_\phi(x) = \sup_{n \geq 0} (m_\phi(T_P^{n}x) - n) , \tag{5}$$

observes that they are finite m_P-a.e., and notes the existence of a null set $F \subset X_P$ such that, for every $x, x' \in X_P \backslash F$ with $x_i = x'_i$ for all $i \in \mathbb{Z}$ with $-\infty \leq i \leq a^*_\phi(x)$ $(-m^*_\phi(x) \leq i < \infty)$, we have $\phi(x)_i = \phi(x')_i$ for all $i \leq 0$ $(i \geq 0)$. An analogous statement holds for ϕ^{-1}. Combined with ergodicity, this is easily seen to lead to a proof of Proposition 2.

If one denotes by F_P (F_Q) the group of all non-singular automorphisms V of (X_P, m_P) $((X_Q, m_Q))$ satisfying $Vx \sim x$ for a.e.x, Proposition 2 can be restated as

$$\phi F_P \phi^{-1} = F_Q , \tag{6}$$

and equation (6) leads to our first obstruction to the existence of isomorphisms satisfying condition F.E.. If P is an irreducible, stochastic $k \times k$ matrix, denote by Δ_P the countable subgroup of the multiplicative group of real numbers given by

$$\Delta_P = \{\frac{P(i_0,i_1)\dots P(i_{n-1},i_0)}{P(i_0,j_1)\dots P(j_{n-1},i_0)} \; : \; n \geq 1 \; , \; 1 \leq i_r, j_r \leq k \; , \text{ and}$$
$$\text{all occurring terms are non-zero}\}. \qquad (7)$$

3. Theorem [7] Under the assumption of Proposition 2, $\Delta_P = \Delta_Q$.

Proof ([7,9]): Without worrying about nullsets, we note that

$$\Delta_P = \{\frac{dm_P V}{dm_P}(x) : x \in X , V \in F_P\}$$
$$= \{\frac{dm_Q \phi V \phi^{-1}}{dm_Q}(\phi(x)) : x \in X , V \in F_P\} = \Delta_Q . \qquad \square$$

The first explicit example of a nontrivial isomorphism between distinct Bernoulli shifts was due to Meshalkin (Bernoulli shifts are Markov shifts for which the defining matrix P has identical rows): if $P = \begin{pmatrix} \frac14 & \frac14 & \frac14 & \frac14 \\ \vdots & \vdots & \vdots & \vdots \end{pmatrix}$ and $Q = \begin{pmatrix} \frac12 & \frac18 & \frac18 & \frac18 & \frac18 \\ \vdots & \vdots & \vdots & \vdots & \vdots \end{pmatrix}$, the associated Bernoulli shifts T_P and T_Q have equal entropy (log 4) and period (1) and are thus finitarily isomorphic. However, since $\Delta_P = \{1\} \neq \{4^n : n \in \mathbb{Z}\} = \Delta_Q$, there exists no isomorphism between them satisfying F.E..

In order to describe the last two invariants in our list, consider again an irreducible, stochastic $k \times k$ matrix P . We define

$$\Gamma_P = <\{P(i_0,i_1)\dots P(i_{n-1},i_0) : n \geq 1 \; , \; 1 \leq i_r \leq k \; , \text{ and}$$
$$\text{all occurring terms are non zero}\}> \qquad (8)$$

to be the countable subgroup of the multiplicative group of real numbers generated by the set of all non-zero products of the form $P(i_0,i_1)\dots P(i_{n-1},i_0)$, $n \geq 1$. Finally consider, for every $t \in \mathbb{R}$, the nonnegative matrix $(P(i,j)^t, 1 \leq i,j \leq k)$ (where $0^t = 0$ for all $t \in \mathbb{R}!$) , and define (cf. [10,14])

$$\beta_P(t) = \text{maximal eigenvalue of } (P(i,j)^t, 1 \leq i,j \leq k) . \qquad (9)$$

4. Theorem [9] Under the assumptions of Proposition 2, $\Gamma_P = \Gamma_Q$.

5. Theorem [13] Under the assumptions of Proposition 2, $\beta_P(t) = \beta_Q(t)$ for all $t \in \mathbb{R}$.

Before we turn to the proofs of Theorems 4 and 5, note that, for a Bernoulli matrix $P = \begin{pmatrix} p_1 \cdots p_k \\ \vdots \quad \vdots \\ p_1 \cdots p_k \end{pmatrix}$, we have

$$\beta_P(t) = p_1^t + \ldots + p_k^t , \quad t \in \mathbb{R} . \tag{10}$$

One immediately concludes the following corollary.

6. Corollary [13] Let T_P and T_Q be Bernoulli shifts based on the probability vectors $(p_1,\ldots,p_k)$ and $(q_1,\ldots,q_\ell)$. If T_P and T_Q are isomorphic via an isomorphism satisfying condition F.E., then $k = \ell$ and $(q_1,\ldots,q_k)$ is a permutation of $(p_1,\ldots,p_k)$.

The proofs of Theorems 4 and 5 are somewhat technical, and I shall restrict myself to giving the briefest possible outline. Consider the functions

$$I_P(x) = -\log P(x_0,x_1) , \quad x \in X_P , \tag{11}$$

and

$$I_Q(y) = -\log Q(y_0,y_1) , \quad y \in X_Q . \tag{12}$$

In [8] W. Parry proved that, for every finitary isomorphism $\phi : X_P \to X_Q$ of the shifts T_P and T_Q with m_ϕ and $m_{\phi^{-1}}$ integrable, there exists a measurable function $f : X_P \to \mathbb{R}$ such that

$$I_P(x) = I_Q(\phi(x)) + f(T_P x) - f(x) \quad \text{a.e.} \tag{13}$$

If the function f were essentially bounded, one could immediately conclude Theorem 5, since (cf. [10, Proposition 4])

$$\beta_P(t) = \lim_{n\to\infty}\left(\int \exp[(1-t)(I_P(x)+I_P(T_Px)+..+I_P(T_P^{n-1}x))]dm_P(x)\right)^{1/n}$$

$$= \lim_{n\to\infty}\left(\int \exp[(1-t)(I_P(x)+I_P(T_Px)+..+I_P(T_P^{n-1}x)-f(T_P^nx)+f(x))]dm_P(x)\right)^{1/n}$$

$$= \beta_Q(t) .$$

Unfortunately, the function f will in general not be bounded, or even integrable. The following procedure overcomes this problem. Using [2] one extends the function I_P to a function $J_P:G_P \times X_P \to \mathbb{R}$, where G_P denotes the group generated by T_P and F_P, with the following properties:

$$J_P(V_1,V_2x) + J_P(V_2,x) = J_P(V_1V_2,x) \text{ a.e.,} \tag{14}$$

for every $V_1,V_2 \in G_P$,

$$J_P(T_P,x) = I_P(x) , \tag{15}$$

and

$$J_P(V,x) = J_Q(\phi V\phi^{-1},\phi(x)) + f(Vx)-f(x) \tag{16}$$

a.e., for every $V \in G_P$, where J_Q is the analogous function on $G_Q \times X_Q$, and where $f:X_P \to \mathbb{R}$ is given by (13). Using the subgroup $F_P \subset G_P$, and the functions a^*_ϕ and m^*_ϕ in (4) and (5) one can show that f is constant on a certain set $D \subset X_P$ of positive measure ([9, Proposition 4.4]). An application of some cohomological ideas (the asymptotic range, cf. [3,5,12]) now implies Theorem 4.

To prove Theorem 5 one has to know a bit more about the 'shape' of the set on which f is constant: if $f \equiv \alpha$ a.e. on D, put $A = \{x \in X_P : f(x) = \alpha\}$. This set A turns out to have the following property ([13, Theorem 4.1]):

$$\beta_P(t) = \lim_n \left(\int \exp((1-t)J_P(T_P^n,x))dm_P(x)\right)^{1/n}$$

$$= \limsup_n \left(\int_{A\cap T_P^{-n}A} \exp((1-t)J_P(T_P^n,x))dm_P(x)\right)^{1/n}$$

$$= \limsup_n \Big(\int_{\phi(A)\cap T_Q^{-n}\phi(A)} \exp((1-t)J_Q(T_Q^n,x))\,dm_Q(x)\Big)^{1/n}$$

$$\le \beta_Q(t) .$$

Since the problem is symmetric in P and Q, Theorem 5 is also proved.

The group Γ_P is intimately related to a notion of dimension associated with the Markov shift space (X_P,m_P) (cf. [6,15]), and can also be used to strengthen the invariant β_P slightly. We quote the following lemma from [9]:

7. <u>Lemma</u> Let $P = (P(i,j):1 \le i,j \le k)$ be an irreducible stochastic matrix. There exist positive real numbers $r(i)$, $i = 1,\dots,k$, such that, for every $1 \le i,j \le k$,

$$P(i,j)r_j/r_i = P^*(i,j) \in \Gamma_P^{1/d} \cup \{0\} \tag{17}$$

where d is the period of P and $\Gamma_P^{1/d} = \{\alpha \in \mathbb{R}^+ : \alpha^d \in \Gamma_P\}$.

If $\beta_{P^*}(t)$ denotes the maximal eigenvalue of the matrix $(P^*(i,j)^t:1\le i,j\le k)$ $(0^t = 0)$, we clearly have $\beta_{P^*}(t) = \beta_P(t)$ for every $t \in \mathbb{R}$. Now consider the group $\mathrm{Hom}(\Gamma_P^{1/d}, \mathbb{R}_+^*)$ of all homomorphisms of the countable group $\Gamma_P^{1/d}$ to the multiplicative group of positive real numbers and denote, for every $\eta \in \mathrm{Hom}(\Gamma_P^{1/d}, \mathbb{R}_+^*)$, by $\eta(P^*)$ the matrix with entries

$$\eta(P^*)(i,j) = \begin{cases} \eta(P^*(i,j)) & \text{if } P^*(i,j) \ne 0 \\ 0 & \text{otherwise.} \end{cases}$$

Finally we set, for every $\eta \in \mathrm{Hom}(\Gamma_P^{1/d}, \mathbb{R}_+^*)$,

$$\beta_P(\eta) = \text{maximal eigenvalue of } \eta(P^*) . \tag{18}$$

Note that $\beta_P(\eta)$ does not depend on the particular choice of $r(1),\dots,r(k)$ in (17). A minor modification of the proof of Theorem 5 yields the following corollary.

8. <u>Corollary</u> Under the assumption of Proposition 2 we have

$$\beta_P(\eta) = \beta_Q(\eta) \tag{19}$$

for every $\eta \in \mathrm{Hom}(\Gamma_P^{1/d}, \mathbb{R}_+^*) = \mathrm{Hom}(\Gamma_Q^{1/d}, \mathbb{R}_+^*)$.

I would like to conclude this note with a few comments on the nature of the invariants $\Gamma_P, \Delta_P, \beta_P$. Although the isomorphism ϕ in Proposition 2 is only defined a.e., and could, in particular, destroy all information about periodic orbits of T_P , these invariants are, in fact, linked in an obvious manner to the set of periodic orbits of T_P . The beta-function $\beta_P(t)$, $t \in \mathbb{R}$, was introduced by S. Tuncel as an invariant of finite equivalence of Markov shifts [14], and conjectured to be a complete invariant for this notion of equivalence in [10]. If this conjecture is true, one could conjecture further that the existence of a finitary isomorphism ϕ with finite expected code lengths between Markov shifts T_P and T_Q implies the existence of an irreducible stochastic matrix R with associated shift T_R and of bounded-to-one, a.e. one-to-one, continuous maps $\phi_1 : X_R \to X_P$, $\phi_2 : X_R \to X_Q$ such that $\phi_1 T_R = T_P \phi_1$ and $\phi_2 T_R = T_Q \phi_2$.

REFERENCES

1. R.L. Adler and B. Marcus: Topological entropy and equivalence of dynamical systems. Mem. A.M.S. 219 (1979).
2. R.A.R. Butler and K. Schmidt: An information cocycle for groups of non-singular transformations. Preprint (Warwick 1982).
3. J. Feldman and C.C. Moore: Ergodic equivalence relations, cohomology and von Neumann algebras I. T.A.M.S. 234 (1977), 289-324.
4. M. Keane and M. Smorodinsky: Finitary isomorphisms of irreducible Markov shifts. Israel J. Math 34 (1979), 281-286.
5. W. Krieger: On the Araki-Woods asymptotic ratio set and nonsingular transformations of a measure space. Lecture Notes in Mathematics, Vol.160: Contributions to Ergodic Theory and Probability, 158-177. Springer, 1970.
6. W. Krieger: On dimension functions and topological Markov chains. Invent. Math. 56 (1980), 239-250.
7. W. Krieger: On the finitary isomorphisms of Markov shifts that have finite expected coding time. Z. Wahrscheinlichkeits theorie verw. Geb.(to appear).

8. W. Parry: Finitary isomorphisms with finite expected code lengths. Bull. L.M.S. 11 (1979), 170-176.

9. W. Parry and K. Schmidt: Natural coefficients and invariants for Markov shifts. Preprint (Warwick 1982).

10. W. Parry and S. Tuncel: On the classification of Markov chains by finite equivalence. Ergod. Th. Dynam. Sys. 1 (1981), 303-335.

11. W. Parry and S. Tuncel: Classification problems in Ergodic Theory. London Math. Soc. Lecture Notes, Vol. 67. C.U.P. 1982.

12. K. Schmidt: Cocycles of Ergodic Transformation groups. MacMillan (India), 1977.

13. K. Schmidt: Invariants for finitary isomorphisms with finite expected code lengths. Preprint (UCLA, 1983).

14. S. Tuncel: Conditional pressure and coding. Israel J. Math. 39 (1981), 101-112.

15. S. Tuncel: A dimension, dimension modules, and Markov chains. Proc. L.M.S. 46 (1983), 100-116.

C*-ALGEBRAS OF ANOSOV FOLIATIONS

Hiroshi Takai(*)

Tokyo Metropolitan University
Setagaya, Tokyo, JAPAN

1. Introduction In differentiable dynamical systems, Anosov maps are one of the most important notions to understand the phenomena and problems of the qualitative theory of ordinary differential equations. Among many results concerning them, Bowen [1] showed that the foliations constructed from transitive Anosov maps give rise to hyperfinite and ergodic equivalence relations. This result was interpreted by Connes [2] that the von Neumann algebras associated to (transitive) Anosov foliations are hyperfinite, so that their associated C*-algebras are nuclear. Moreover he found a Thom isomorphism of K-theory for the C*-algebra of an Anosov foliation on the unit tangent bundle of a compact Riemann surface with genus greater than one. In this respect, they could be quite useful models for algebraic topology and noncommutative differential geometry. They also might give a good chance to find a new algebraic invariant in C*-algebra theory.

In this note, we shall report some algebraic interpretation and K-theoretic meaning of the C*-algebras of Anosov foliations on infrahomogeneous spaces.

2. Anosov foliations Let G be a connected Lie group with Lie algebra G, K a compact subgroup of G with Lie algebra k and Γ a torsion free uniform discrete subgroup of G. Given a $X \in G$ with $(\exp tX)K(\exp -tX)=K$ for all $t \in \mathbb{R}$, the flow θ_t: $\Gamma gK \to \Gamma g(\exp tX)K$ is *Anosov* if $\mathrm{Ker}\,(\mathrm{ad}X) = R + \mathbb{R}X$ and $\mathrm{Sp}_{\mathbb{C}}(\mathrm{ad}X) \cap (i\mathbb{R}\setminus\{0\}) = \phi$. Let E^+, E^- be the invariant subspaces in the primary decomposition of adX corresponding to eigenspaces of positive and negative real parts respectively. Then $G = R + \mathbb{R}X + E^+ + E^-$, so the tangent space $T_k(G/K)$ of G/K at K is $\mathbb{R}X + E^+ + E^-$. This implies that θ_t has a hyperbolic splitting on the tangent bundle $T(\Gamma\backslash G/K)$ of $\Gamma\backslash G/K$ because of considering the differential of the projection from $T(G/K)$ to $T(\Gamma\backslash G/K)$. Let W_x^s (resp. W_x^u) be a stable (resp. unstable) submanifold of $\Gamma\backslash G/K$ at x with respect to θ_t. Then the collection of

---

(*) On leave at Mathematics Institute, The University of Warwick.
The work is partially supported by S.E.R.C.

all those sets gives rise to a foliation F_h of $\Gamma\backslash G/K$ which is called *horocycle*. Similarly the set of all θ_t-orbits of W_x^s (resp. W_x^u) becomes a foliation F_A of $\Gamma\backslash G/K$ which is called *Anosov*. According to Tomter's paper [9] of Anosov flows on infra-homogeneous spaces, it is assumed basically that G is one of the following groups: (I) a noncompact semi-simple Lie group of real rank one, (II) the semidirect product $N\times_s \mathbb{R}$ of a nilpotent (simply connected) Lie group N by a hyperbolic action of $\mathbb{R}$, or (III) the semidirect product $R\times_s S$ of a nilpotent radical R of G by a simple group S of real rank one. In what follows, we shall study Anosov foliation structure of $\Gamma\backslash G/K$ case by case.

Case (I): Let L be a maximal compact subgroup of G containing K. Then it is easy to see that $\Gamma\backslash G/K$ is a principal L/K-bundle over $\Gamma\backslash G/L$. Set F_A (resp. F_h) be an Anosov (resp. a horocycle) foliation of $\Gamma\backslash G/K$. By definition, each leaf of F_A is nothing but $\Gamma\backslash(G/L \times \{\ell K\})$, $\ell \in L$ (cf: [6]). So we have the following lemma:

Lemma 2.1 $(\Gamma\backslash G/K, F_A)$ is a foliated L/K-bundle over $\Gamma\backslash G/L$.

Let M be the centralizer of exptX in L. Then the Lie algebra of M is equal to R. So $M \subset K$ and K/M is finite. Let $T_1(\Gamma\backslash G/L)$ be the unit tangent bundle over $\Gamma\backslash G/L$. Considering the differential of the mapping $\Gamma gL \to \Gamma g(\exp tX)L$, we can show that $\Gamma\backslash G/M$ is diffeomorphic to $T_1(\Gamma\backslash G/L)$. Then there is an Anosov foliation $\tilde{F}_A$ of $T_1(\Gamma\backslash G/L)$ whose image under the finite covering map to $\Gamma\backslash G/K$ is just F_A. Therefore we have the followin lemma:

Lemma 2.2 $(T_1(\Gamma\backslash G/L), \tilde{F}_A)$ is a foliated S^n-bundle over $\Gamma\backslash G/L$ whose imag under the finite covering map to $\Gamma\backslash G/K$ is F_A where $n = \dim G/L - 1$.

For a horocycle foliation F_h of $\Gamma\backslash G/K$, we observe the following thing. Let $G = L(\exp\mathbb{R}X)N^+$ be the Iwasawa decomposition of G with respect to L and $\exp\mathbb{R}X$. If K is connected, then $\Gamma\backslash G$ is a principal K-bundle over $\Gamma\backslash G/K$ and lifting F_h to a foliation $\tilde{F}_h$ of Γ/G, we can show that each leaf of $\tilde{F}_h$ is KN^+-orbital. So we have the following:

Lemma 2.3 If K is connected, then $\Gamma\backslash G$ is a principal K-bundle over $\Gamma\backslash G/K$ and the pull back foliation $\tilde{F}_h$ of F_h to $\Gamma\backslash G$ is KN^+-orbital.

By the similar way as before, if M is connected, in other words G is not isomorphic to the Lie algebra $SL(2,\mathbb{R})$ of $SL(2,\mathbb{R})$, then we have the following:

Lemma 2.4 $\Gamma\backslash G$ is a principal M-bundle over $T_1(\Gamma\backslash G/L)$ and the pull back foliation of a horocycle foliation F_h of $T_1(\Gamma\backslash G/L)$ to $\Gamma\backslash G$ is MN^+-orbital.

Remark 2.1 If K is the normalizer of $\exp\mathbb{R}X$ in L, K/M has order two because it is a Weyl group of a semisimple Lie group G of real rank one.

Remark 2.2 Case (I) contains all the unit sphere bundles over hyperbolic n-manifolds with geodesic flows.

Case (II) : In this case, it follows from Tomter [9] that K is trivial and $(\Gamma\backslash G, \exp tX)$ is the suspension of a hyperbolic diffeomorphism of $N \cap \Gamma\backslash N$. Let F_h be a horocycle foliation of $\Gamma\backslash G$ with respect to $\exp tX$. Then each (stable) lead of F_h has a form $\Gamma\backslash(N^+ \times \{t\})$, $t \in \mathbb{R}$ where $N^+ = \exp E^+$. Similarly each stable leaf of a horocycle foliation F_h of $N \cap \Gamma\backslash N$ with respect to σ is diffeomorphic to $N \cap \Gamma\backslash N^+$. Since $\Gamma = \mathbb{Z}\cdot(N \cap \Gamma) = (N \cap \Gamma)\cdot\mathbb{Z}$ where $\mathbb{Z}$ is the infinite cyclic group coming from the suspension equivalence, we see that each leaf of F_h is diffeomorphic to one of F_h' and vice versa. Summing up the above argument, we have the following lemma:

Lemma 2.5 Let F_h be a horocycle foliation of $\Gamma\backslash G$ with respect to $\exp tX$ and F_h' the horocycle foliation of $N \cap \Gamma\backslash N$ with respect to a hyperbolic mapping σ corresponding to $\exp tX$. Then F_h' is the image of F_h under the projection π of $\Gamma\backslash G$ onto $N \cap \Gamma\backslash N$.

Since $\Gamma\backslash G$ is a S^1-bundle over $N \cap \Gamma\backslash N$ with π, $(N \cap \Gamma\backslash N, F_h')$ is the push forward foliation of $(\Gamma\backslash G, F_h)$ under π. In the case of an Anosov foliation F_A of $\Gamma\backslash G$ with respect to $\exp tX$, each leaf of F_A is just $\Gamma\backslash N^+ \times_s \mathbb{R}$ up to diffeomorphisms. This means that F_A is the pull back foliation of F_h' under π. So we have the following lemma:

Lemma 2.6 $(\Gamma\backslash G, F_A)$ is the pull back foliation of $(N \cap \Gamma\backslash N, F_h')$ under the projection π of $\Gamma\backslash G$ onto $N \cap \Gamma\backslash N$.

Remark 2.3 (Case (II) contains all infra-solvmanifolds with hyperbolic automorphisms.

Case (III) : In this case, $(\Gamma\backslash S/K, \exp tX)$ is as in Case (I) and $(\Gamma\backslash R \times_s \exp\mathbb{R}X, \exp tX)$ is as in Case (II). Hence both systems have simply connected nilpotent Lie subgroups S^+, R^+ of S, R corresponding to the stable subsapces E_S^+, E_R^+ of the Lie algebras $\mathcal{S}, \mathcal{R}$ with respect to $\exp tX$ respectively. Similarly S^-, R^- are also defined for unstable subspaces.

Let F_h be a horocycle foliation of $\Gamma\backslash R \times_s S/K$. By the same way as we have seen in Lemma 2.3, we can show the following lemma:

Lemma 2.7 If K is connected, $\Gamma\backslash R \times_s S$ is a principal K-bundle over $\Gamma\backslash R \times_s S/K$ and the pull back foliation $\tilde{F}_h$ of F_h to $\Gamma\backslash R \times_s S$ is $R^j \times_s KS^j$-orbital where j = + or −.

Similarly we easily see leaf structure for the pull back foliation $\tilde{F}_A$ of an Anosov foliation F_A of $\Gamma\backslash R \times_s S/K$ to $\Gamma\backslash R \times_s S$ as follows:

Lemma 2.8 If K is connected, $\tilde{F}_A$ is $R^j \times_s K(\exp \mathbb{R}X)S^j$-orbital where $j = +$ or $-$.

Remark 2.4 In the above two lemmas, suppose K is not connected and K_o its connected component of the identity. Then $\Gamma\backslash T \times_s S/K_o$ is a K/K_o-covering manifold of $\Gamma\backslash R \times_s S/K$. Since K is compact, K/k_o is finite.

3. Foliation C*-algebras Throughout this section, we use the same notions and notations as in Connes' note [3]. Let (M,F) be a foliated manifold, C*(M,F) its associated C*-algebra. Suppose (M,F) is a foliated F-bundle over B, namely F is transverse to F. Let θ be the total holonomy homomorphism of the fundamental group $\pi_1(B)$ of B on F, and let H be the groupoid whose regular C*-algebra is the reduced crossed product $(C(F) \times_\theta \pi_1(B))_r$ of C(F) by the action θ of $\pi_1(B)$. Looking at the holonomy groupoid Hol(F) of (M,F), it is Borel isomorphic to $H \times B \times B$. Finding a suitable imprimitivity bimodule with respect to C*(M,F) and $(C(F) \times_\theta \pi_1(B))_r$ in the sense of Rieffel, we have the following proposition:

Proposition 3.1 ([6]) Let (M,F) be a foliated F-bundle over B. Suppose the holonomy groupoid of (M,F) is Hausdorff, then C*(M,F) is isomorphic to the tensor product $(C(F) \times_\theta \pi_1(B))_r \otimes \mathcal{C}(L^2(B))$ of $(C(F) \times_\theta \pi_1(B))_r$ and the C*-algebra $\mathcal{C}(L^2(B))$ of all compact operators on $L^2(B)$.

Let F_A be an Anosov foliation of $\Gamma\backslash G/K$ where G is a Case (I). By Lemma 2.1, $(\Gamma\backslash G/K, F_A)$ is a foliated L/K-bundle over $\Gamma\backslash G/L$ where L is a maximal compact subgroup of G containing K. Let $G = L(\exp \mathbb{R}X)N^+$ be the Iwasawa decomposition of G with respect to L and $\exp \mathbb{R}X$. Then L/K is diffeomorphic to $G/K(\exp \mathbb{R}X)\, N^+$. So $(\Gamma\backslash G/K, F_A)$ can be viewed as a foliated $G/K(\exp \mathbb{R}X)N^+$-bundle over $\Gamma\backslash G/L$. Since G/L is simply connected (actually contractible), $\pi_1(\Gamma\backslash G/L) = \Gamma$. Then it follows from Proposition 3.1 that $C^*(\Gamma\backslash G/K, F_A)$ is stably isomorphic to the reduced crossed product $(C(G/K(\exp \mathbb{R}X)N^+) \times_\lambda \Gamma)_\gamma$ of $C(G/K(\exp \mathbb{R}X)N^+)$ by the left translation λ of Γ. Since E^+ is K-invariant, $N^+ = \exp E^+$ is Ad(K)-invariant. So $K(\exp \mathbb{R}X)N^+ = N^+ \times_s K(\exp \mathbb{R}X)$ is an amenable Lie subgroup of G. Due to Rieffel [7], the latter C*-algebra is stably isomorphic to the reduced crossed product $(C(\Gamma\backslash G) \times_\rho K(\exp \mathbb{R}X)N^+)_\gamma$ of $G(\Gamma\backslash G)$ by the right translation ρ of $K(\exp \mathbb{R}X)N^+$. Therefore we have the following theorem:

Theorem 3.2 $C^*(\Gamma\backslash G/K, F_A)$ is stably isomorphic to the crosssed product $C(\Gamma\backslash G) \times_\rho K(\exp \mathbb{R}X)N^j$ of $C(\Gamma\backslash G)$ by the right translation ρ of $K(\exp \mathbb{R}X)N^j$, where $j = +$ or $-$.

Putting K = M the centralizer of $\exp \mathbb{R}X$ in L, we have from Lemma 2.2 the following corollary:

Corollary 3.3 $C^*(T_1(\Gamma\backslash G/L), F_A)$ is stably isomorphic to the crossed product $(C(S^n) \times_\theta \Gamma)_\gamma$ of $C(S^n)$ by the total holonomy action θ of Γ where $n = \dim G/L-1$.

Let F be a connected space and M be a F-bundle over B. Suppose $\mathcal{F}$ is a foliation of B and $\tilde{\mathcal{F}}$ is the pull back foliation of $\mathcal{F}$ to M. Let $Hol(\tilde{\mathcal{F}})$, $Hol(\mathcal{F})$ be the holonomy groupoids of $\tilde{\mathcal{F}}$, $\mathcal{F}$ respectively. By the definition of $\tilde{\mathcal{F}}$, we can easily see that the local charts of $Hol(\tilde{\mathcal{F}})$ are divided into both $(F \times F) \times (\mathcal{F} \times \mathcal{F})$-direction and the transverse direction of $\mathcal{F}$. Therefore $Hol(\tilde{\mathcal{F}})$ is (Borel) isomorphic to $Hol(\mathcal{F}) \times F \times F$. Hence using Rieffel's machine, we have the following proposition:

Proposition 3.3 Let F be a connected manifold and M be a F-bundle over B. Let $\mathcal{F}$ be a foliation of B and $\tilde{\mathcal{F}}$ be the pull back foliation of $\mathcal{F}$ to M. Then $C^*(M,\tilde{\mathcal{F}})$ is isomorphic to $C^*(B,\mathcal{F}) \otimes \mathfrak{C}(L^2(F))$.

Let $\mathcal{F}_h$ be a horocycle foliation of $\Gamma\backslash G/K$ and suppose K is connected. Then it follows from Lemma 2.3 that $\Gamma\backslash G$ is a principal K-bundle over $\Gamma\backslash G/K$ and the pull back foliation $\tilde{\mathcal{F}}_h$ of $\mathcal{F}_h$ to $\Gamma\backslash G$ is KN^+-orbital or KN^--orbital where $N^- = \exp E^-$. By Proposition 3.3, $C^*(M,\tilde{\mathcal{F}})$ is isomorphic to $C^*(\Gamma\backslash G/K, \mathcal{F}_h) \otimes \mathfrak{C}(L^2(K))$. Since Γ is torsion free and uniform discrete, $\Gamma \cap KN^+ = \{e\}$ and $\Gamma \cap KN^- = \{e\}$. So the right translation ρ of KN^+ or KN^- on $\Gamma\backslash G$ is free. Therefore $C^*(M,\tilde{\mathcal{F}})$ is isomorphic to the crossed product $C(\Gamma\backslash G) \times_\rho KN^+$ or $C(\Gamma\backslash G) \times_\rho KN^-$. Then we have the following theorem:

Theorem 3.4 If K is connected, then $C^*(\Gamma\backslash G/K, \mathcal{F}_h)$ is stably isomorphic to $C(\Gamma\backslash G) \times_\rho KN^j$ where $j = +$ or $-$.

Corollary 3.5 If G is not isomorphic to $SL(2,\mathbb{R})$, then $C^*(T_1(\Gamma\backslash G/L), \mathcal{F}_h)$ is stably isomorphic to $C(\Gamma\backslash G) \times_\rho MN^j$ where $j = +$ or $-$.

Remark 3.1 If G is connected and simply connected, we may assume that K is connected.

Now let $\mathcal{F}_A$ be an Anosov foliation of $\Gamma\backslash G$ where G is of Case (II). By Lemma 2.6, $Hol(\mathcal{F}_A)$ is isomorphic to the skew product $Hol(\mathcal{F}_h) \times_s \mathbb{R}$ of $Hol(\mathcal{F}_h)$ by $\mathbb{R}$ in the sense of Renault [11]. So we have the following proposition:

Proposition 3.6 $C^*(\Gamma\backslash G, F_A)$ is isomorphic to $C^*(\Gamma\backslash G, F_h) \times_s \mathbb{R}$.
Let F_h' be a horocycle foliation of $N \cap \Gamma\backslash N$ defined in Case (II). Then it is N^j-orbital where $N^j = \exp E^j$ (j = + or -). Since each leaf has a trivial holonomy, Hol $(F_h') = (N \cap \Gamma\backslash N) \times N^j$ up to isomorphism.
As we have seen in the preceding section, $\Gamma\backslash G$ is a S^1-bundle over $N \cap \Gamma\backslash N$ and F_A is the pull back foliation of F_h'.
Let F_h be a horocycle foliation of $\Gamma\backslash G$. By Lemma 2.5, F_h' is the image of F_h under the projection of $\Gamma\backslash G$ to $N \cap \Gamma\backslash N$. Since $\Gamma\backslash G$ is a S^1-bundle over $N \cap \Gamma\backslash N$, codim F_h = codim $F_h' + 1$. Comparing Hol (F_h) with Hol (F_h') we can see that Hol (F_h) is isomorphic to Hol $(F_h') \times S^1$. Therefore we get the following theorem combining the discussion before:

Theorem 3.7 $C^*(\Gamma\backslash G, F_h)$ is isomorphic to $(C(N \cap \Gamma\backslash N) \times_\rho N^j) \otimes C(S^1)$ where j = + or -.

We now consider a horocycle foliation F_h of $\Gamma\backslash G/K$ where G is of Case (III). By Lemma 2.7, if K is connected, the pull back foliation $\tilde{F}_h$ of F_k to $\Gamma\backslash R\times_s S$ is $R^j \times_s KS^j$-orbital where j = + or -. Since Γ is a torsion free uniform discrete subgroup of $R\times_s S$, $\Gamma \cap KS^j = \{e\}$ for j = + or -. By the similar way as we have seen in the case (II), each leaf of $\tilde{F}_h$ has a trivial holonomy. So Hol $(\tilde{F}_h) = (\Gamma\backslash R\times_s S) \times (R^j \times_s KS^j)$ up to isomorphism. Due to Proposition 3.3, we have the following theorem:

Theorem 3.8 If K is connected, then $C^*(\Gamma\backslash R\times_s S/K, F_h)$ is stably isomorphic to the crossed product $C(\Gamma\backslash R\times_s S) \times_\rho (R^j \times_s KS^j)$ of $C(\Gamma\backslash R\times_s S)$ by the right translation ρ of $R^j \times_s KS^j$ where j = + or -.

We finally consider an Anosov foliation F_A of $\Gamma\backslash R\times_s S/K$. By Lemma 2.8, if K is connected, the pull back foliation $\tilde{F}_A$ of F_A to $\Gamma\backslash R\times_s S$ is $R^j \times_s K(\exp \mathbb{R}X)S^j$-orbital where j = + or -. Then $R^j \times_s K(\exp \mathbb{R}X)S^j = (R^j \times_s KS^j) \times_s \exp \mathbb{R}X$. Since each $(R^j \times_s KS^j)$-orbit has a trivial holonomy, so does each $R^j \times_s K(\exp \mathbb{R}X)S^j$-orbit. Therefore Hol $(\tilde{F}_A) = (\Gamma\backslash R\times_s S) \times (R^j \times_s K(\exp \mathbb{R}X)S^j)$ up to isomorphism. So we have the following theorem using Proposition 3.3:

Theorem 3.9 If K is connected, then $C^*(\Gamma\backslash R\times_s S/K, F_A)$ is stably isomorphic to $C(\Gamma\backslash R\times_s S) \times_\rho (R^j \times_s K(\exp \mathbb{R}X)S^j)$ where j = + or -.

Remark 3.2 When S is simply connected, K can be chosen as a compact connected Lie subgroup of S.

4. <u>KK-theory</u> In this section, we study KK-theory for $C^*(M,F)$ of an Anosov (or a horocycle) foliation F on an infra-homogeneous space M. Before going into the main part, we announce the following crucial lemma due to Kasparov [5]:

<u>Lemma 4.1</u> Let (A,G,α) be a C^*-dynamical system where A is separable and G is a connected amenable Lie group. Let G_c be its maximal compact subgroup of G, V be the cotangent space of G/G_c at G_c and C_V the Clifford algebra associated to V. Then $KK(A \times_\alpha G)$ is isomorphic to $KK(A \otimes C_V) \times_{\alpha \otimes \sigma} G_c)$ where $KK(\cdot)$ means KK-homology and KK-cohomology at the same time, and σ is the action of G_c on C_V associated to the coadjoint action Ad_* of G_c on V.

Let $F = F_h$ or F_A of $\Gamma\backslash G/K$ where G is of Case (I). By Theorem 3.2, $KK(C^*(\Gamma\backslash G/K,\ F_A)) = KK(C(\Gamma\backslash G) \times_\rho K(\exp \mathbb{R}X)N^j)$ where j = + or −. Suppose G is simply connected, then K is connected. Since $K(\exp \mathbb{R}X)N^j$ is amenable, the right hand side of the equality is $KK((C(\Gamma\backslash G) \otimes_{\mathbb{R}} C_V) \times_{\rho \otimes \sigma} K)$ due to Lemma 4.1 where $V = (\mathbb{R}X + E^j)^*$. If $\dim V = 2n$, $C_V = M_{2n}(\mathbb{R})$. Since $\rho \otimes \sigma$ is free on $\Gamma\backslash G$, it follows from Green [4] that $(C(\Gamma\backslash G) \otimes_{\mathbb{R}} C_V) \times_{\rho \otimes \sigma} K$ is isomorphic to $C(\Gamma\backslash G/K) \otimes \mathcal{C}(L^2(K) \otimes_{\mathbb{R}} \mathbb{R}^{2n})$. So $KK(C^*(\Gamma\backslash G/K,\ F_A)) = KK(\Gamma\backslash G/K)$. If $\dim V = 2n+1$, $\dim E^j = 2n$. By Theorem 3.4, $KK(C^*(\Gamma\backslash G/K,\ F_h)) = KK(C(\Gamma\backslash G) \times_\rho KN^j)$. By the same way as before, $KK(C^*(\Gamma\backslash G) \times_\rho KN^j) = KK(\Gamma\backslash G/K)$. Since $KK(C^*(\Gamma\backslash G/K,\ F_A)) = KK^1(C^*(\Gamma\backslash G/K,\ F_h))$, we conclude that $KK(C^*(\Gamma\backslash G/K,\ F)) = KK^{\dim F}(\Gamma\backslash G/K)$ for $F = F$ or F_h. Now suppose G is of Case (II), and $F = F$ or F_h of $\Gamma\backslash G$. By Theorem 3.7, $KK(C^*(\Gamma\backslash G,\ F_h)) = KK(C(N \cap \Gamma\backslash N) \times_\rho N^j) \otimes C(S^1)$. By the nilpotency of N^j and Fack-Skandaris [10], the right hand side is $KK^{\dim N^j}((N \cap \Gamma\backslash N) \times S^1)$. Since $\Gamma\backslash G$ is diffeomorphic to $S^1 \times (N \cap \Gamma\backslash N)$, we obtain that $KK^{\dim N^j}(\Gamma\backslash G) = KK^{\dim N^j}((N \cap \Gamma\backslash N) \times S^1)$. By Proposition 3.6, $KK(C^*(\Gamma\backslash G,\ F_A)) = KK^1(C^*(\Gamma\backslash G,\ F_h) = KK^{1+\dim N^j}(S^1 \times (N \cap \Gamma\backslash N)) = KK^{\dim N^j+1}(\Gamma\backslash G)$. Summing up the above argument, we have that $KK(C^*(\Gamma\backslash G, F)) = KK^{\dim F}(\Gamma\backslash G)$ for $F = F_A$ or F_h. Finally suppose G is of case (III) and K is connected. By Theorem 3.8, $KK(C^*(\Gamma\backslash R \times_s S/K, F_h)) = KK(C(\Gamma\backslash R \times_s S) \times_\rho (R^j \times_s KS^j))$. Since $R^j \times_s KS^j$ is connected amenable, it follows from Lemma 4.1 that the right hand side of the equality is $KK((C(\Gamma\backslash R \times_s S) \otimes C_V) \times_{\rho \otimes \sigma} K)$ where $V = (E_k^j + E_s^j)^*$.

By Theorem 3.9, $KK(C^*(\Gamma\backslash R \times_s S/K,\ F_A)) = KK(C(\Gamma\backslash R \times_s S) \times_\rho (R^j \times_s K(\exp \mathbb{R}X)S^j))$. Since $R^j \times_s K(\exp \mathbb{R}X)S^j$ is connected amenable, it also follows from Lemma 4.1 that the right hand side of the above equality is $KK((C(\Gamma\backslash R \times_s S) \otimes C_W) \times_{\rho \otimes \sigma} K)$ where $W = V + (\mathbb{R}X)^*$. By the same way as Case (I), we conclude that $KK(C^*(\Gamma\backslash R \times_s S/K, F)) = KK^{\dim F}(\Gamma\backslash R \times_s S/K)$ for $F = F_A$ or F_h.

Consequently we have the following theorem:

Theorem 4.2 Let (M,F) be a foliated manifold. Suppose F is an Anosov (or a horocycle) foliation of an infra-homogeneous space $M = \Gamma\backslash G/K$ where Γ is a torsion free uniform discrete subgroup of a connected Lie group G and K is a compact connected Lie subgroup of G, then $KK(C^*(M,F)) = KK^{\dim F}(M)$.

Corollary 4.3 ([8]) Let M be a compact Riemannian locally symmetric space of strictly negative curvature and F be an Anosov (or a horocycle) foliation of the unit tangent bundle $T_1(M)$ over M. Then $KK(C^*(T_1(M),F)) = KK^{\dim F}(T_1(M))$.

Acknowledtement: The author would like to thank Professor D.E. Evans for inviting him to the Mathematical Institute, University of Warwick where this work was completed.

References

[1] R. Bowen, Anosov foliations are hyperfinite, Ann. Math., 106 (1977), 547-565.

[2] A. Connes, The von Neumann algebra of a foliation, Springer Lecture Notes, Physics 80 (1978), 145-151.

[3] A. Connes, A survey of foliations and operator algebras, Proc. Symp. Pure Math., 38 (1982) Part 1, 521-628.

[4] P. Green, C*-algebras of transformation groups with smooth orbit space, Pacific J. Math., 72 (1977), 71-97.

[5] G.G. Kasparov, K-theory, group C*-algebras and higher signatures (Part 1), Preprint (1981).

[6] T. Natsume and H. Takai, Connes algebras associated to foliated bundles, Preprint, Univ. Rome (1982).

[7] M.A. Rieffel, Morita equivalence of certain transformation group C*-algebras, Math. Ann., 222 (1976), 7-22.

[8] H. Takai, KK-theory for the C*-algebras of Anosov foliations, Proceeding of the third Japan - U.S. Seminar on C*-algebras (1983).

[9] P. Tomter, Anosov flows on infra-homogeneous spaces, Proc. Symp. Pure Math., 14 (1970), 299-327.

[10] T.Fack and G. Skandaris, Connes' analogue of the Thom isomorphism for the Kasparov groups, Invent. Math., 64 (1981), 7-14.

[11] J. Renault, A groupoid approach to C*-algebras, Lecture Notes in Math., Springer 793 (1980).

A LATTICE-THEORETIC CHARACTERIZATION OF CHOQUET SIMPLEXES

Silviu Teleman

Let E be a Hausdorff locally convex topological real vector space and let $K \subset E$ be a (non-empty) compact convex set. The set K is said to be a *Choquet simplex* if for any $x \in K$ there is a *unique* Choquet maximal Radon probability measure μ_x on K, representing x; i.e. such that $b(\mu_x)=x$.

The set K is said to be a *Bauer simplex* if it is a Choquet simplex and if ex K is closed in K. For Bauer simplexes there is an elegant lattice-theoretic characterization, due to H.Bauer (see [1], Satz 1):

The compact convex set K *is a Bauer simplex if, and only if, the ordered real vector space* A(K) *is a (vector) lattice.*

Here A(K) denotes the real vector space of all continuous affine real functions on K, equipped with the usual linear and order structures.

There is also a lattice-theoretic characterization for Choquet simplexes, due to G.Choquet and P.A.Meyer (see [6], [7] and [8]; and also [11], Ch.9;[4], Ch.III;[9], Ch.XI, Theorem T 29). Namely, we can assume, without any loss of generality, that K is *regularly embedded* in E; i.e., the affine subspace of E, generated by K, does not contain 0. Then the cone $\tilde{K}=\{\alpha x;\alpha\in\mathbb{R}_+,x\in K\}$ induces a real vector space order relation in the real vector subspace $\tilde{K}-\tilde{K} \subset E$, in the usual manner. Then the G.Choquet-P.A.Meyer Theorem says that:

The compact convex set K *is a Choquet simplex if, and only if,* $\tilde{K}-\tilde{K}$ *is a vector lattice.*

REMARK. Since $\tilde{K}-\tilde{K}$ is isomorphic with the Banach dual space $A(K)^*$ by an order preserving linear isomorphism, the Choquet-Meyer Theorem can be given also the more eloquent formulation:

The compact convex set K *is a Choquet simplex if, and only if,*

A(K)* *is a vector lattice.*

There are also characterizations of Choquet simplexes in terms of extension properties.

We refer to a recent paper by C.J.K.Batty (see [3]), where a general result of this kind is obtained and where references to previous results are given (see also [4], Ch.III).

In this Note we shall give a lattice-theoretic characterization for Choquet simplexes, similar to that of H.Bauer, involving the space of the (bounded) strongly universally measurable affine real functions on K, which where introduced in [16].

We shall frequently refer to results obtained in [13], [14], [15] and [16] (see, also, [2]).

I. Let E be a Hausdorff locally convex topological real vector space and let $K \subset E$ be a (non empty) compact convex subset. Let ex K be the set of the extreme points of K, and equip ex K with the Choquet topology, denoted by C (see [13], p.27; and also [4], Ch.II, §2). Let A(K) be the real vector space of all continuous affine real functions on K; denote by $M_+^1(K)$ the convex set of all Radon probability measures on K, and by $M_{x_o}^1(K)$ the convex set of all $\mu \in M_+^1(K)$, such that the barycenter $b(\mu)=x_o$.

Let $\mathcal{B}_o(K)$ denote the σ-algebra of all Baire measurable subsets of K; $\tilde{\mathcal{B}}_o(\text{ex } K)$ the σ-algebra of subsets of ex K, given by

$\tilde{\mathcal{B}}_o(\text{ex } K)=\{D \cap (\text{ex } K);\ D \in \mathcal{B}_o(K)\}$,

and denote by $\mathcal{B}(\text{ex } K;C)$ the σ-algebra of all Borel measurable subsets of ex K, with respect to the Choquet topology.

By virtue of results obtained in [2],[13], [14] and [15], for any Choquet maximal measure $\mu \in M_+^1(K)$ there exists a *regular* probability measure $\tilde{\mu}:\mathcal{B}(\text{ex } K;C) \to [0,1]$, such that

1) $\tilde{\mathcal{B}}_o(\text{ex } K) \subset \mathcal{B}(\text{ex } K;\ C)^{\sim}_{\tilde{\mu}}$;

2) $\tilde{\mu}(D \cap (\text{ex } K))=\mu(D),\ \forall D \in \mathcal{B}_o(K)$;

3) $\tilde{\mu}(F \cap (\text{ex } K))=\mu(F)$, for any compact extremal subset $F \subset K$.

REMARKS. 1. The measure $\tilde{\mu}$, which is uniquely determined by the regularity property and by properties 1), 2) and 3) above, was called the *extended boundary measure* associated to μ, and denoted by $\tilde{\mu}_1$ in [2] and [15]. We now prefer to drop the subscript 1, which had in [2] and [15] the unique rôle to help distinguish the measure $\tilde{\mu}_1$ from its restriction $\tilde{\mu}_o$ to $\tilde{\mathcal{B}}_o(\text{ex } K)$.

2. The measure $\tilde{\mu}_o$ has now only a limited importance. This fact is demonstrated by Examples in ([2], §7) and by Theorem 1 from [16]. The key moment in the proof of this theorem, in which $\tilde{\mu}_1$ does the job, whereas $\tilde{\mu}_o$ fails to do it, is the obtention of the first inequality in formula (1), from the proof of the theorem (see [16], p.13), theorem which will play an important rôle in the sequel.

In [16] we have introduced the *strongly universally measurable* bounded affine functions $h:K\to\mathbb{R}$, by extending the notion of a *universally measurable element* given by G.E.Pedersen for the case of C*-algebras (see [10], p.104). We recall the definition given in [16]:

A bounded affine function $h:K\to\mathbb{R}$ is said to be *strongly universally measurable* if for any $x_o\in K$ and any $\varepsilon>0$ there exist bounded affine lower semicontinuous functions $k,\ell:K\to\mathbb{R}$, such that

$$-k\leq h\leq \ell, \text{ on } K, \qquad \text{and } \ell(x_o)+k(x_o)<\varepsilon.$$

If we denote by $A^b(K)$ the real vector space of all bounded affine real functions on K, and by U(K) the set of all strongly universally measurable (bounded affine) real functions on K, then

$$A(K)\subset U(K)\subset A^b(K),$$

and U(K) is a Banach subspace of $A^b(K)$, the latter being endowed with the sup-norm.

Moreover, U(K) obviously contains all the bounded affine semicontinuous real functions on K. We note that the definition of the strongly universally measurable bounded affine functions on K involves only topological properties of K.

We shall say that a function $f:\text{ex } K\to\mathbb{R}$ is *universally measurable* if f is $\tilde{\mu}$-measurable for the extended boundary measure $\tilde{\mu}$, corresponding to *any* Choquet maximal measure $\mu\in M_+^1(K)$. We shall denote by M(ex K) the set of all universally measurable functions $f:\text{ex } K\to\mathbb{R}$, and by $M^b(\text{ex } K)$ the subset of all the *bounded* universally measurable functions $f:\text{ex } K\to\mathbb{R}$.

It is obvious that $M^b(\text{ex } K)$ is a Banach space, if it is endowed with the usual algebraic operations and the sup-norm.

According to results obtained in ([16], Theorem 3), for any $h\in U(K)$ we have $h|\text{ex } K\in M^b(\text{ex } K)$, whereas from ([16], Theorem 4), we immediately infer that the restriction mapping

$$r:U(K)\ni h\to r(h)=h|\text{ex } K\in M^b(\text{ex } K)$$

is an *isometry*. Moreover, by ([16], Theorem 3), we have

$$h(b(\mu))=\int_{\text{ex } K} r(h)\,d\tilde{\mu}, \qquad h\in U(K),$$

for any Choquet maximal measure $\mu\in M_+^1(K)$.

II. We shall now assume that K is a Choquet simplex. Then, for any $x\in K$, there exists a *unique* Choquet maximal measure $\mu_x \in M_+^1(K)$, such that $b(\mu_x)=x$. Since the set of all Choquet maximal measures in $M_+^1(K)$ is a convex subset of $M_+^1(K)$, and since the mapping $b:M_+^1(K)\to K$ is affine, we infer that the mapping $K\ni x\to\mu_x \in M_+^1(K)$ is affine.

Since the Radon-Nikodym derivatives have Baire measurable representatives, we immediately infer that for any Choquet maximal measures $\mu,\nu\in M_+^1(K)$, and any $t\in[0,1]$ we have

$$(t\mu+(1-t)\nu)^{\sim}=t\tilde{\mu}+(1-t)\tilde{\nu}.$$

For any $f\in M^b(\text{ex } K)$ we shall define $T(f):K\to\mathbb{R}$ by

$$T(f)(x)=\tilde{\mu}_x(f), \qquad x\in K.$$

It is obvious that $T(f)\in A^b(K)$, for any $f\in M^b(\text{ex } K)$.

For any $x\in \text{ex } K$ we have $\mu_x=\varepsilon_x$ (the Dirac measure at x) and, therefore, $\tilde{\mu}_x=\varepsilon_x$. It follows that

$$T(f)(x)=f(x), \qquad x\in \text{ex } K,\ f\in M^b(\text{ex } K).$$

We infer that any $f\in M^b(\text{ex } K)$ has a bounded affine real extension to K.

Let now $\tilde{F}\subset \text{ex } K$ be any C-closed subset. This means that there exists a compact extremal subset $F\subset K$, such that $F\cap(\text{ex } K)=\tilde{F}$, and, by virtue of the extremality property of F, the function χ_F is convex on K, and upper semicontinuous on K, since F is closed. We have

$$\tilde{\mu}_x(\tilde{F})=\mu_x(F)=\mu_x(\chi_F)=\sup\{\mu(\tilde{\chi}_F);\mu\sim\varepsilon_x\}=\bar{\chi}_F(x), \qquad x\in K,$$

where we have taken into account the fact that there is a unique Choquet maximal measure $\mu_x\in M_+^1(K)$, representing x, the convexity of the function $\chi_{\tilde{F}}$, as well as the fact that χ_F is upper semicontinuous (see [14], Proposition 1).

We have proved the following

PROPOSITION 1. *If* K *is any Choquet simplex, then for any* C-*closed subset* $\tilde{F}\subset$ ex K, *the function* $T(\chi_{\tilde{F}})$ *is affine and upper semicontinuous.*

Let now $f\in M^b(\text{ex } K)$ be such that $f\geq 0$. Let $x_o\in K$ and $\varepsilon>0$ be given. Let $0=t_o<t_1<\dots<t_n=\|f\|$ be such that $t_i-t_{i-1}<\frac{\varepsilon}{2}$, $i=1,2,\dots,n$, and define

$$A_i=\{x\in \text{ex } K;\ t_{i-1}<f(x)\leq t_i\}, \qquad i=1,2,\dots,n.$$

Let $g=\sum_{i=1}^{n} t_{i-1}\chi_{A_i}$. Then we have $g\leq f$ and

$$\int_{\text{ex } K}(f-g)\,d\tilde{\mu}_{x_o}<\frac{\varepsilon}{2}. \tag{1}$$

Since $\tilde{\mu}_{x_o}$ is regular, we can find a C-closed subset $\tilde{F}_i\subset A_i$ such that

$$\tilde{\mu}_{x_c}(A_i\setminus\tilde{F}_i)<\frac{\varepsilon}{2nt_{i-1}}, \qquad i=1,2,\ldots,n.$$

If we define $h=\sum_{i=1}^{n} t_{i-1}\chi_{\tilde{F}_i}$ on ex K, we have

(2) $\qquad \int_{\text{ex } K}(g-h)\,d\tilde{\mu}_{x_o}<\frac{\varepsilon}{2},$

and, therefore, from (1) and (2) we infer that

(3) $\qquad T(f)(x_o)-T(h)(x_o)=\int_{\text{ex } K}(f-h)\,d\tilde{\mu}_{x_o}<\varepsilon$.

On the other hand, from $h\le f$ on ex K, we infer that

(4) $\qquad T(h)\le T(f)$, $\qquad$ on K.

From (3), (4) and from Proposition 1 we immediately infer the following

THEOREM 1. *If* K *is any Choquet simplex, then for any* $f\in M^b(\text{ex } K)$, *the function* T(f) *is the unique strongly universally measurable bounded affine real function on* K, *which extends* f.

We immediately obtain the following:

COROLLARY 1. *For any Choquet simplex* K, *the restriction mapping* $r:U(K)\to M^b(\text{ex } K)$ *is an isomorphism of Banach spaces.*

We can now prove

COROLLARY 2. *For any Choquet simplex* K *the space* U(K) *is a vector lattice.*

PROOF. Let $h_1,h_2\in U(K)$ and define $h_o=T(r(h_1)\vee r(h_2))$, where $\vee$ denotes the "join operation" in $M^b(\text{ex } K)$. We have

$r(h_o)=r(h_1)\vee r(h_2)\ge r(h_i)$, $\qquad i=1,2,$

and, therefore, $h_o\ge h_i$, on K, i=1,2 (see [16], Theorem 4). On the other hand, if $h\in U(K)$ and if $h\ge h_i$, i=1,2, we have

$r(h)\ge r(h_1)\vee r(h_2)$,

and, therefore, by ([16], Theorem 3):

$h=T(r(h))\ge T(r(h_1)\vee r(h_2))=h_o$;

i.e., h_o is the l.u.b. of the set $\{h_1,h_2\}$ in the ordered vector space U(K). Q.E.D.

REMARKS.1. In view of Corollary 1, since $M^b(\text{ex } K)$ in endowed with an obvious structure of a (commutative) real C*- algebra the same can be carried over to U(K), extending its natural Banach space structure.

2. We have seen that if K is a Choquet simplex, then there exists a linear mapping $T:M^b(\text{ex } K)\to U(K)$, such that $r\circ T=id$ and $T\circ r=id$.

Assume now that K is a compact convex set such that there exists a mapping $S:M^b(\text{ex } K)\to U(K)$, such that $r\circ S=id$; equivalently, any $f\in M^b(\text{ex } K)$

can be extended as a function in U(K). Then r is bijective and $S=r^{-1}$ is linear.

III. We shall now prove that the preceding results have a converse. More precisely, we shall now prove the main theorem (cf.[7], Théorème 11 and Critère 6, p.147).

THEOREM 2. *For any compact convex set* K, *the following statements are equivalent:*

1) K *is a Choquet simplex;*
2) U(K) *is a vector lattice;*
3) *The mapping* $r:U(K)\to M^b(\mathrm{ex}\,K)$ *is surjective.*

PROOF. The implication 1) $\Longrightarrow$ 3) has already been proved.
3) $\Longrightarrow$ 2) Indeed, let $h_1,h_2\in U(K)$. Then $r(h_1)\vee r(h_2)\in M^b(\mathrm{ex}\,K)$. Let $h_o=r^{-1}(r(h_1\vee r(h_2))$. Then we have

$$r(h_o)=r(h_1)\vee r(h_2)\geq r(h_i),\qquad i=1,2;$$

hence $h_o\geq h_i$ on K, i=1,2. On the other hand, if $h\in U(K)$, and $h\geq h_i$, i=1,2, then

$$r(h)\geq r(h_1)\vee r(h_2),$$

and, therefore,

$$h=r^{-1}(r(h))\geq r^{-1}(r(h_1)\vee r(h_2))=h_o;$$

i.e., h_o is the l.u.b. of the set $\{h_1,h_2\}$ in U(K).
2) $\Longrightarrow$ 1). Let us denote by $h_1\vee h_2$ the join of h_1,h_2 in $C(K;\mathbb{R})$, and by $h_1\sqcup h_2$ the join of $h_1,h_2\in U(K)$ in U(K) itself (U(K) is not assumed to be a sub-lattice of $C(K;\mathbb{R})$).

For $h_1,h_2,\dots,h_n\in A(K)$, the function $h_1\vee h_2\vee\dots\vee h_n$ is continuous and convex on K, whereas $h_1\sqcup h_2\sqcup\dots\sqcup h_n$ is affine, and belongs to U(K).

We have

$$h_1\vee h_2\vee\dots\vee h_n\leq h_1\sqcup h_2\sqcup\dots\sqcup h_n\leq h,$$

for any $h\in A(K)$, such that $h_i\leq h$, i=1,2,...,n. We infer that

$$h_1\vee h_2\vee\dots\vee h_n\leq h_1\sqcup h_2\sqcup\dots\sqcup h_n\leq (h_1\vee h_2\vee\dots\vee h_n)^-,$$

where $\bar{f}$ denotes the concave upper semicontinuous hull of $f\in C(K;\mathbb{R})$.

We infer that for any $\mu\in M_+^1(K)$ we have

$$\mu(h_1\vee h_2\vee\dots\vee h_n)\leq (h_1\sqcup h_2\sqcup\dots\sqcup h_n)(b(\mu))\leq$$
$$\leq\mu((h_1\vee h_2\vee\dots\vee h_n)^-),$$

since the barycentric calculus holds for any function in U(K) (see [16], Theorem 3).

We infer that for any Choquet maximal measure $\mu\in M_+^1(K)$ we have

$$\mu(h_1\vee h_2\vee\dots\vee h_n)=(h_1\sqcup h_2\sqcup\dots\sqcup h_n)(b(\mu)),\tag{1}$$

(see [13], Lemma 1.2, b)).

We infer from (1) that for any Choquet maximal measures $\mu_1,\mu_2 \in M_+^1(K)$, such that $b(\mu_1)=b(\mu_2)$, we have

(2) $\mu_1(h_1 \vee h_2 \vee \dots \vee h_n)=\mu_2(h_1 \vee h_2 \vee \dots \vee h_n)$,

for any $h_1,h_2,\dots,h_n \in A(K)$. Since the set

$\{h_1 \vee h_2 \vee \dots \vee h_n : h_i \in A(K), \quad i=1,2,\dots,h, \ n \in \mathbb{N}^*\}$

is total in $C(K;\mathbb{R})$ (by virtue of the Stone-Weierstrass Theorem), we infer from (2) that $\mu_1=\mu_2$. It follows that K is a Choquet simplex, and the Theorem is proved.

REMARK. Let A be any C*-algebra with a unit element and let E(A) be the state space of A, endowed with the $\sigma(A^*;A)$-topology. Then, in view of H.Bauer's Therorem, the Theorem of Sherman can be stated as follows:

E(A) *is a Bauer simplex if, and only if,* A *is commutative.* (See [12], Theorem 16). The same proof, as that given in [12]to Sherman's Theorem, yields the following result:

E(A) *is a Choquet simplex if, and only if,* A *is commutative.*

IV. In view of the preceding results, it appears that the Banach space $M^b(\text{ex } K)/\text{im}\, r$ gives a measure of the extent to which the compact convex set K differs from a Choquet simplex.

Also, it would be interesting to characterize those $f \in M^b(\text{ex } K)$, which admit of an extension in $A^b(K)$. Equivalently, we can consider the restriction mapping $r^b: A^b(K) \to \mathbb{R}^{\text{ex } K}$, and then characterize $(r^b)^{-1}(M^b(\text{ex } K))$.

This question has a relevance for Naimark's Problem (see [13], §7, where a detailed discussion can be found). Indeed let H be any Hilbert space and let K(H) be the (elementary) C*-algebras of all compact operators on H. Let $E_o(K(H))$ be the compact convex set of all quasi-states of K(H). Then the self-adjoint part $K(H)_h$ of K(H) can be identified, as a Banach space, with the set $A_o(E_o(K(H))) \subset A(E_o(K(H)))$ of all continuous affine real functions on $E_o(K(H))$, vanishing at 0, whereas the self-adjoint part $K(H)_h^{**}=L(H)_h$ of the von Neumann envoloping algebra $K(H)^{**}$ of K(H) can be identified, as a Banach space, with the set $A_o^b(E_o(K(H))) \subset$ $\subset A^b(E_o(K(H)))=A(E_o(K(H)))^{**}$ of all bounded affine real functions on $E_o(K(H))$, vanishing at 0.

Since any minimal projection $e \in K(H)^{**}$ is continuous as an element in $A^b(E_o(K(H)))$, it easily follows that

$A^b(E_o(K(H)))=U(E_o(K(H)))$;

we immediately infer that

(1) $r_o^b(A_o^b(E_o(K(H)))) \subset M_\Omega^b(P(K(H)))$,

where by r_o^b we have denoted the restriction mapping

$A_o^b(E_o(K(H))) \ni f \to f|P(K(H))$,

whereas $M^b_\Omega(P(K(H)))$ is the set of all bounded universally measurable real functions on $P(K(H))=(\text{ex } E_o(K(H)))\setminus\{0\}$, the pure states set of K(H). Here "universally measurable" should be understood in the broader sense "with respect to any extended boundary measure on P(K(H)), corresponding to any maximal orthogonal measure on $E_o(K(H))$, whose barycenter is a state of K(H)" (see [13], for details).

The solution of Naimark's problem is related to the inclusion (1). Namely, let A be a C*-algebra, having a one-point spectrum, i.e., card $\hat{A}=1$. The problem is to show that A is elementary, i.e., that there exists a Hilbert space H, such that A be isomorphic to K(H).

In any case, since any two irreducible representations of A are unitarily equivalent, it follows that one has an irreducible embedding $A \subset L(H_A)$, where the Hilbert space H_A is determined up to an isomorphism.

As above, we can consider the restriction mapping

$$r^b_o : A^b_o(E_o(A)) \to \mathbb{R}^{P(A)}$$

and the subspace $M^b_\Omega(P(A)) \subset \mathbb{R}^{P(A)}$.

We recall that a cardinal α is sometimes said to be "measurable" if on (any) set S, whose cardinal card $S=\alpha$, there exists a probability measure $\lambda: \mathcal{P}(S)\to[0,1]$, such that $\lambda(\{s\})=0$, for any $s\in S$.

We can state the following results

a) *If* A *is an elementary* C*-*algebra, then*

$$(*)\quad r^b_o(A^b_o(E_o(A))) \subset M^b_\Omega(P(A)).$$

b) *Conversely, if* A *is a* C*-*algebra, such that card* $\hat{A}=1$, *if the inclusion* (*) *holds and if the Hilbert dimension of* H_A *is a non-measurable cardinal, then* A *is elementary.*

The details of the proof will appear elsewhere.

BIBLIOGRAPHY

1. Bauer, H.: Kennzeichnung kompakter Simplexe mit abgeschlossener Extremalpunktmenge, *Arkiv der Mathematik*, 14(1963), 415-421.
2. Batty, C.J.K.: Some properties of maximal measures on compact convex sets, preprint, 1981. A revised version is to the appear in *Math.Proc.Cambridge Phil.Soc.*
3. Batty, C.J.K.: A characterization of simplexes by an extension property, preprint 1982. To appear in *Quarterly Journal of Mathematics (Oxford)*, in a revised form.
4. Boboc, N.; Bucur, Gh.: *Conuri convexe de funcţii continue pe spaţii compacte*, Ed.Acad.R.S.R., Bucureşti, 1976.
5. Choquet, G.: Le théorème de représentation intégrale dans les ensembles convexes compacts, *Ann.Inst.Fourier*, 10(1960), 333-344.

6. Choquet, G.: Remarques à propos de la démonstration d'unicité de P.A.Meyer, *Séminaire Brelot-Choquet-Deny* (Théorie du Potentiel), 6e année, 1962, exposé 8.

7. Choquet, G.; Meyer, P.A.: Existence et unicité des représentations intégrales dans les ensembles convexes compacts quelconques, *Ann.Inst.Fourier*, 13(1963), 139-154.

8. Meyer, P.A.: Sur les démonstrations nouvelles du théorème de Choquet, *Seminaire Brelot-Choquet-Deny* (Théorie du Potentiel), 6e année, 1962, exposé 7.

9. Meyer P.A.: *Probability and Potentials*, Blaisdell Publ.Comp., Waltham, Toronto, London, 1966.

10. Pedersen, G.E.: *C*-algebras and their automorphism groups*, Academic Press, London-New York-San Francisco, 1979.

11. Phelps, R.R.: *Lectures on Choquet's Theorem*, D.van Nostrand Co., Princeton-Toronto-New York-London, 1966.

12. Skau, C.F.: Orthogonal measures on the state space of a C*-algebra, in *Algebras in Analysis*, ed. by W.Williamson, Academic Press, London-New York-San Fancisco, 1975.

13. Teleman, S.: An introduction to Choquet Theory with applications to Reduction Theory, *INCREST Preprint series in Mathematics*, 71(1980).

14. Teleman, S.: On the regularity of the boundary measures, *INCREST Preprint series in Mathematics*, 30(1981).

15. Teleman, S.: Measure-theoretic properties of the Choquet and of the maximal topologies, *INCREST Preprint series in Mathematics*, 33(1982).

16. Teleman, S.: On the non-commutative extension of the theory of Radon measure, *INCREST Preprint series in Mathematics*, 1(1983).

S.Teleman
Department of Mathematics, INCREST
Bdul Păcii 220, 79622 Bucharest
Romania.

DIRAC INDUCTION FOR SEMI-SIMPLE LIE GROUPS HAVING ONE CONJUGACY CLASS OF CARTAN SUBGROUPS

Alain Valette(*)

0. Introduction

This paper deals with a particular case of a conjecture due to Connes and Kasparov, which relates K-theory, index theory, and group representations. To explain this conjecture in more details, we immediately introduce some important notations.

Let G be a connected Lie group; we denote by $C^*_r(G)$ the reduced C*-algebra of G, i.e. the norm closure of $L^1(G)$ acting by left convolution on $L^2(G)$. The spectrum of $C^*_r(G)$, endowed with the Jacobson topology, is the reduced dual of G, denoted by $\hat{G}_r$. If G is a type I group, $\hat{G}_r$ coincides with the support of the Plancherel measure. For the basic results about (group) C*-algebras, we refer to Dixmier's book [6].

In order to describe simply the Connes-Kasparov conjecture, we assume that G is a connected semi-simple Lie group with finite centre. Let K be a maximal compact subgroup of G. We denote by $\mathfrak{g}, \mathfrak{k}$ the Lie algebras of G, K respectively. Let $\mathfrak{g} = \mathfrak{k} + \mathfrak{p}$ be a Cartan decomposition of $\mathfrak{g}$ (for basic structure of real semi-simple Lie algebras, see [7]). The Killing form B on $\mathfrak{g}$ induces a positive definite scalar product on $\mathfrak{p}$; in this way, we get from the adjoint action of K on $\mathfrak{p}$ a homomorphism $K \to SO(\mathfrak{p})$. We say that G/K has a G-invariant spin structure if this homomorphism lifts to the double covering $Spin(\mathfrak{p})$ of $SO(\mathfrak{p})$. This can always be realized by passing to a suitable double covering of G. We will make this assumption in the sequel (see [18] for a discussion of the case without G-invariant spin structure). Let S be the spin module of $Spin(\mathfrak{p})$: if q is the dimension of $\mathfrak{p}$, this is a representation of degree $2^{[q/2]}$; if q is odd, this is an irreducible representation, while if q is even, this is the direct sum of two inequivalent irreducible representations. S also is a module over the Clifford algebra of $\mathfrak{p}$; this module is irreducible (resp. $\mathbb{Z}/2$-graded irreducible) if q is odd (resp. even). Since we have a G-invariant spin structure on G/K, we may view S as a module over K: let χ be the representation of K obtained in this way. Let ρ be an arbitrary finite dimensional representation of K. The Dirac operator D_ρ with coefficients in ρ

is a first order G-invariant differential operator acting on the space of C^∞-sections with compact support of the homogeneous vector bundle over G/K induced from $\chi \otimes \rho$. This space of sections is of course equal to

$$(C_c^\infty(G) \otimes \chi \otimes \rho)^K = \{\xi : G \to \chi \otimes \rho : \xi \text{ is } C^\infty \text{ with compact support and, for any } g \in G,\ k \in K: \xi(gk) = (\chi \otimes \rho)(k^{-1})\xi(g)\}.$$

With this formalism, the Dirac operator D_ρ is obtained by taking an orthonormal basis $(Y_i)_{1\le i\le q}$ of $\mathfrak{p}$, and defining:

$$D_\rho = \sum_{i=1}^{q} Y_i \otimes c(Y_i) \otimes 1$$

where $c(Y)$ denotes the Clifford multiplication by the vector Y (for more details, see [16,§1]). This operator is elliptic, G-invariant and formally self-adjoint. Moreover, for q even, it anticommutes with the $\mathbb{Z}/2$-grading on $(C_c^\infty(G) \otimes \chi \otimes \rho)^K$. Hence, by Kasparov's theorem 2 in [8], it defines an unbounded Kasparov bimodule (in the sense of Baaj-Julg [3]), of degree q (mod.2) on the C*-module completion of $(C_c^\infty(G) \otimes \chi \otimes \rho)^K$ over $C_r^*(G)$. So, by Proposition 2.2 in [3], we get an element of $K_q(C_r^*(G))$, called the <u>analytical index</u> of D_ρ, and denoted by $\mathrm{Ind}_a(D_\rho)$. The map $\mathrm{Ind}_a(D_\rho)$ extends to a group homomorphism Δ from the representation ring R(K) of K into $K_q(C_r^*(G))$. This map Δ is the <u>Dirac induction</u>. We may now state the <u>Connes-Kasparov conjecture</u>:

<u>Conjecture</u>: Under the above assumptions on G, the Dirac induction $\Delta: R(K) \to K_q(C_r^*(G))$ is a group isomorphism, and $K_{q+1}(C_r^*(G)) = 0$.

For a discussion of this conjecture in the case where G is not necessarily semi-simple, see [4], [10], [17], [18].

It is known, from Kasparov's result [8, Theorem 2], that any first order G-invariant differential elliptic operator D over a proper G-manifold defines an element $\mathrm{Ind}_a(D)$ of one of the K-groups of $C_r^*(G)$. Thus the meaning of the conjecture is that, to generate $K_*(C_r^*(G))$, it suffices to consider one operator, namely the Dirac operator (twisted by representations of K), on one proper G-manifold, namely the symmetric space G/K. This must be related to the fact, proved by Baum-Connes [4], that G/K is a final object in the category of proper G-manifolds and G-maps. Also, we see that the analytical index of a first order G-invariant differential elliptic operator on G/K only depends on the underlying homogeneous vector bundles.

In the case of semi-simple Lie groups, the above conjecture

was proved by Penington-Plymen [17] for complex groups; they used an explicit description of the Dirac induction. For the Lorentz groups SO(n, 1) (and their double covering Spin(n, 1), $n \geq 2$), the conjecture was proved abstractly by Kasparov [9]; soon after, a different proof was given by the author [18], using an explicit computation of Δ. In this paper, we extend the techniques of Penington-Plymen [17] to cover the case of semi-simple Lie groups with just one conjugacy class of Cartan subgoups. We recall that a Lie subalgebra of $\mathfrak{g}$ is a <u>Cartan subalgebra</u> if it is a maximal abelian subalgebra consisting of semi-simple elements, and that a subgroup of G is a <u>Cartan subgroup</u> if it is the centralizer in G of some Cartan subalgebra in $\mathfrak{g}$. So, we demand that any two Cartan subgroups of G are conjugate. Examples of such groups are compact groups, complex groups, the Lorentz groups SO(2n+1,1), the groups $SL_n(\mathbb{H})$ of $n \times n$ matrices over the quaternions with Dieudonné determinant 1,...

<u>Remark</u>: While I was completing this paper, I received a manuscript of A. Wassermann [22], announcing a proof of the conjecture for any connected linear reductive Lie group (so this takes care of $SL_n(\mathbb{R})$, but not of its universal covering). This completely supersedes the above results, since all the groups just mentioned are linear. Wassermann's proof uses some deep machinery from representation theory of semi-simple Lie groups: Plancherel theorem, limits of discrete series, intertwining operators, theory of K-types, classification of tempered irreducible representations. However, I think that the content of the present paper still remains interesting as an example in which the Plancherel theorem is essentially sufficient to get the result (there are no non-trivial intertwining operators, and the K-types are quite easy to determine).

This paper is organized as follows: in the first section, we recall a result of Parthasarathy [16] which relates the square of the Dirac operator D_ρ (for $\rho \in \hat{K}$) to the Casimir operator on G. As a consequence of this formula, the operator D_ρ^2 acts by a scalar in any irreducible representation of G. We compute this scalar for representations belonging to the principal series of G.

In section 2, we collect information on the structure of

$C_r^*(G)$. To this end, we need some basic results from Harish-Chandra's theory (available in [14], [21]); one of these results essentially says that all the representations in $\hat{G}_r$ are obtained by inducing up representations of certain distinguished subgroups of G, the so-called cuspidal parabolic subgroups, and then decomposing these induced representations into irreducible components. We will prove that, if any such representation (induced from a cuspidal parabolic subgroup) is irreducible, then $\hat{G}_r$ is a Hausdorff locally compact space, and $C_r^*(G)$ is isomorphic to $C_0(\hat{G}_r, \mathcal{K})$, the algebra of norm continuous functions, vanishing at infinity, from $\hat{G}_r$ to the algebra $\mathcal{K}$ of compact operators on the usual Hilbert space. The assumptions of this theorem are satisfied if G has one conjugacy class of Cartan subgroups, as shown by a result of Wallach [19]. Another group to which the result applies is $SL_3(\mathbb{R})$; we will treat this example in some details (in particular, we will compute abstractly the K-groups of the reduced C*-algebra of $SL_3(\mathbb{R})$).

In the third section, we put together the results of the previous sections to determine the Dirac induction in the case where G has one conjugacy class of Cartan subgroups. The other ingredients that we need are some lemmas about Weyl groups, and the concept of Fourier transform of an induced C*-module over $C_r^*(G)$; this rests on an idea of Atiyah-Schmid [2], and was already exploited in [17] and [18].

1. The Dirac operator and the principal series

Let G be a connected semi-simple Lie group with finite centre, and let K be a maximal compact subgroup. We assume throughout that G/K has a G-invariant spin structure. In this paragraph, we first recall a formula of Parthasarathy [16] which relates the square of the Dirac operator to the Casimir operator on G. As a consequence of this formula, the square of Dirac acts by a scalar in any irreducible representation of G. We compute this scalar for representations in the principal series of G.

The restriction of the Killing form B of $\mathfrak{g}$ to $\mathfrak{k}$ is negative definite. So let $(X_j)_{1 \leq j \leq \dim K}$ be a basis of $\mathfrak{k}$ such that

$$B(X_i, X_j) = -\delta_{ij}$$

The <u>Casimir operator</u> on K, resp. G, is defined by:

$$\Omega_K = -\sum_{j=1}^{\dim K} X_j^2$$

$$\text{resp.} \quad \Omega_G = \Omega_K + \sum_{i=1}^{q} Y_i^2$$

It turns out that Ω_K, resp. Ω_G, is independent of the choice of an orthonormal basis in $\mathfrak{k}$, resp. $\mathfrak{g}$, and defines an element in the centre of the enveloping algebra of $\mathfrak{k}$, resp. $\mathfrak{g}$ (for all these assertions, see [21]).

Let T be a maximal torus in K, with Lie algebra $\mathfrak{t}$. The Killing form B induces a positive definite scalar product <.,.> on $i\mathfrak{t}^*$. Let ρ_K be half the sum of the positive roots, for a given ordering on the root system of K with respect to T; let $\overline{C_K}$ be the associated closed Weyl chamber in $i\mathfrak{t}^*$. For any weight τ in $\overline{C_K}$, we denote by E_τ the irreducible representation of K with highest weight τ. The following result is due to Parthasarathy [16, lemma 2.2 and prop. 3.1].

<u>1.1. Theorem</u>: <u>For</u> $\rho = E_\tau$:

i) $D_\rho^2 = -\Omega_G + \langle\tau + 2\rho_K, \tau\rangle - \chi(\Omega_K)$

ii) $\chi(\Omega_K)$ <u>is a scalar operator</u>

In his paper, Parthasarathy proves ii) only in the case where G has discrete series representations, and states without proof that it holds in full generality. Since ii) is of fundamental importance for us (after all, this says that D_ρ^2 is just $-\Omega_G$, up to a constant), we shall device in the Appendix a proof of this fact in the case where G has one conjugacy class of Cartan subgroups (this is the only case in which we use Theorem 1.1).

We now turn to the principal series representations of G. Before that, we recall the definition of a minimal parabolic subgroup. Let $\mathfrak{a}_0$ be a maximal abelian subalgebra of $\mathfrak{p}$. An element $\lambda \in \mathfrak{a}_0^*$ is a <u>restricted root</u> if $\lambda \neq 0$ and if

$$\mathfrak{g}^\lambda = \{X \in \mathfrak{g} : [H,X] = \lambda(H)X, \text{ for all } H \in \mathfrak{a}_0\}$$

is not zero. The set Σ of all restricted roots is a root system (in general not reduced). Choose some ordering on Σ. Then

$$\mathfrak{n}_0 = \sum_{\lambda\in\Sigma, \lambda>0} \mathfrak{g}^\lambda$$

is a nilpotent subalgebra of $\mathfrak{g}$, and

$$\mathfrak{g} = \mathfrak{k} + \mathfrak{a}_0 + \mathfrak{n}_0$$

is the Iwasawa decomposition of $\mathfrak{g}$. If A_0 (resp. N_0) is the analytic subgroup of G with Lie algebra $\mathfrak{a}_0$ (resp. $\mathfrak{n}_0$), then

$$G = KA_0N_0$$

is the Iwasawa decomposition of G. Now let M_0 (resp. $M_0{}^*$) be the centralizer (resp. normalizer) of A_0 in K. Then

$$P_0 = M_0A_0N_0$$

is the (standard) minimal parabolic subgroup, and

$$W_0 = M_0{}^*/M_0$$

is the Weyl group of P_0. Fix $\delta \in \hat{M}_0$, $\lambda \in \hat{A}_0$. The representations of G of the form

$$\pi(\delta, \lambda) = \mathrm{Ind}_{P_0}^G\ \delta \otimes \lambda \otimes 1$$

(obtained by Mackey induction) are the principal series representations. We recall that these representations are obtained as follows: the Hilbert space $\mathcal{H}_{\pi(\delta,\lambda)}$ is the set of functions $f\colon G \to \delta$ such that

$$f(gman) = \delta(m^{-1})\Delta_0(a)^{-\frac{1}{2}}e^{-i\lambda(\log a)}f(g)$$

$$\int_{K/M} \|f(\dot{k})\|^2 d\dot{k}$$

where $d\dot{k}$ denotes the unique K-invariant probability measure on K/M, and Δ_0 denotes the modular function on P_0. The Weyl group W_0 acts on $\hat{M}_0 \times \hat{A}_0$. We have a famous result of Bruhat (see [14, pp. 22-23]):

1.2. Theorem: Let $\pi(\delta,\lambda)$, $\pi(\delta',\lambda')$ be two representations in the principal series.

i) $\pi(\delta,\lambda)$ is equivalent to $\pi(\delta',\lambda')$ iff there exists $w \in W_0$ such that $w(\delta,\lambda) = (\delta',\lambda')$

ii) The dimension of the space of intertwining operators between $\pi(\delta,\lambda)$ and $\pi(\delta',\lambda')$ is less or equal to

$$\mathrm{card}\{w \in W_0 : w(\delta,\lambda) = (\delta',\lambda')\}$$

From this theorem, we see that, for a fixed $\delta \in \hat{M}_0$, almost all the $\pi(\delta,\lambda)$'s are irreducible. On these representations, the Casimir operator acts by a scalar, but since it is possible to realize coherently all the $\pi(\delta,\lambda)$'s on the same Hilbert space (see [14, p.17]), we see that by continuity Ω_G acts everywhere by a scalar. We want to compute this scalar. To do that, we will need some more information about relations between root systems, which can be found in [21].

Let $\mathfrak{h}_0$ be a maximal abelian subalgebra of $\mathfrak{g}$, containing $\mathfrak{a}_0$;

put $\mathfrak{b}_0 = \mathfrak{h}_0 \cap \mathfrak{k}$. Then $\mathfrak{a}_0 = \mathfrak{h}_0 \cap \mathfrak{p}$ and $\mathfrak{h}_0 = \mathfrak{a}_0 + \mathfrak{b}_0$. By conjugating if necessary, we may assume that $\mathfrak{b}_0$ is a Cartan subalgebra of $\mathfrak{m}_0$. The complexification $(\mathfrak{h}_0)_{\mathbb{C}}$ is a Cartan subalgebra of $\mathfrak{g}_{\mathbb{C}}$, and we denote by Φ the associated root system. These roots are real on $(\mathfrak{a}_0 + i\mathfrak{b}_0)^*$, and B induces a positive definite scalar product on $(\mathfrak{a}_0 + i\mathfrak{b}_0)^*$.

The conjugation s of $\mathfrak{g}_{\mathbb{C}}$ with respect to $\mathfrak{g}$ induces an involution, also denoted by s, on $(\mathfrak{a}_0 + i\mathfrak{b}_0)^*$: s is 1 on $\mathfrak{a}_0^*$ and -1 on $i\mathfrak{b}_0^*$; s acts on Φ. We denote by p the orthogonal projection from $(\mathfrak{a}_0 + i\mathfrak{b}_0)^*$ to $\mathfrak{a}_0^*$. Then $p(\Phi) = \Sigma$ (see [21, 1.1.3.]). Define

$$\Phi^- = \{\alpha \in \Phi : s\alpha = \alpha\}$$
$$\Phi^+ = \Phi \setminus \Phi^-$$

Let $\mathfrak{m}_0$ be the Lie algebra of M_0. By [20, 7.5.11], Φ^- coincides with the root system of $(\mathfrak{m}_0)_{\mathbb{C}}$ with respect to $(\mathfrak{b}_0)_{\mathbb{C}}$. It is possible to put an ordering on Φ in such a way that, if α is a positive root in Φ^+, then $s\alpha$ also is positive. This can be done by taking the lexicographic order associated to a basis $\{A_1, \ldots, A_m, B_1, \ldots, B_n\}$ of $\mathfrak{a}_0 + i\mathfrak{b}_0$, where $\{A_1, \ldots, A_m\}$ is a basis of $\mathfrak{a}_0$, and $\{B_1, \ldots, B_n\}$ is a basis of $i\mathfrak{b}_0$. If moreover Σ is ordered lexicographically with respect to the same basis of $\mathfrak{a}_0$, then $p: \Phi \to \Sigma$ preserves the order.

Any root $\lambda \in \Sigma$ has a <u>multiplicity</u> $m(\lambda)$: it is the cardinal of $p^{-1}(\lambda) \cap \Phi^+$; it is also the dimension of $\mathfrak{g}^\lambda$. Put

$$\rho_\Sigma = \frac{1}{2} \sum_{\lambda \in \Sigma,\ \lambda > 0} m(\lambda)\lambda$$

The following lemma is classical (see [14, p.16]).

<u>1.3. Lemma</u>: <u>The modular function of</u> P_0 <u>is</u>

$$\Delta_0(man) = e^{2\rho_\Sigma(\log a)}$$

To compute the action of Ω_G on the principal series, we will need one more lemma. The Killing form B is non-degenerate on $\mathfrak{a}_0$. So for any root $\lambda \in \Sigma$, there exists a $Q_\lambda \in \mathfrak{a}_0$ such that, for any $H \in \mathfrak{a}_0$, we have $\lambda(H) = B(H, Q_\lambda)$.

<u>1.4. Lemma</u>: For any $\lambda \in \Sigma$, there exist $m(\lambda)$ $\mathbb{C}$-linearly independent elements X_i^λ $(1 \le i \le m(\lambda))$ in $(\mathfrak{g}^\lambda)_{\mathbb{C}}$, such that:

$$B(X_i^\lambda, X_j^\mu) = 0 \text{ if } \lambda + \mu \neq 0$$

$$B(X_i^{\lambda}, X_j^{-\lambda}) = \delta_{ij}$$
$$[X_i^{\lambda}, X_j^{-\lambda}] = Q_{\lambda}$$

Proof: We recall that, for any $\alpha \in \Phi$, there exists $H_\alpha \in (\mathfrak{h}_0)_{\mathbb{C}}$ such that, for any $H \in (\mathfrak{h}_0)_{\mathbb{C}}$, one has $\alpha(H) = B(H, H_\alpha)$. Also, there exist $X_\alpha \in \mathfrak{g}_\alpha$, $X_{-\alpha} \in \mathfrak{g}_{-\alpha}$ such that $B(X_\alpha, X_{-\alpha}) = 1$ and $[X_\alpha, X_{-\alpha}] = H_\alpha$ (see [14]). Clearly

$$Q_\lambda = \tfrac{1}{2}(H_\alpha + sH_\alpha) \text{ if } p(\alpha) = \lambda.$$

Assume $\alpha \neq s\alpha$. Then

$$B(2^{-\frac{1}{2}}(X_\alpha + sX_\alpha), 2^{-\frac{1}{2}}(X_{-\alpha} + sX_{-\alpha})) = 1 = B(2^{-\frac{1}{2}}(X_\alpha - sX_\alpha), 2^{-\frac{1}{2}}(X_{-\alpha} - sX_{-\alpha}))$$
$$[2^{-\frac{1}{2}}(X_\alpha + sX_\alpha), 2^{-\frac{1}{2}}(X_{-\alpha} + sX_{-\alpha})] = Q_\lambda = [2^{-\frac{1}{2}}(X_\alpha - sX_\alpha), 2^{-\frac{1}{2}}(X_{-\alpha} - sX_{-\alpha})]$$

One concludes easily using these formulas plus the fact that

$$B(X_\alpha + sX_\alpha, X_\alpha - sX_\alpha) = 0$$

Let ρ_{M_0} be half the sum of the positive roots of M_0; let σ be the highest weight of $\delta \in \hat{M}_0$ in the closed Weyl chamber $\overline{C_{M_0}}$.

1.5. Theorem: On the principal series representation $\pi(\delta, \lambda)$, the Casimir operator Ω_G acts by the scalar

$$\pi(\delta, \lambda)(\Omega_G) = \langle \sigma + 2\rho_{M_0}, \sigma \rangle - \langle \lambda, \lambda \rangle - \langle \rho_\Sigma, \rho_\Sigma \rangle$$

Proof: By [20, 7.3.4], one has a direct sum decomposition

$$\mathfrak{g} = \mathfrak{m}_0 + \mathfrak{a}_0 + \mathfrak{n}_0 + \vartheta\mathfrak{n}_0$$

where $\vartheta\mathfrak{n}_0 = \sum_{\alpha \in \Sigma,\ \alpha > 0} \mathfrak{g}^\alpha$

Moreover, $\mathfrak{m}_0$, $\mathfrak{a}_0$, $\mathfrak{n}_0 + \vartheta\mathfrak{n}_0$ are pairwise orthogonal with respect to B. Let (M_i) be an orthonormal basis of $\mathfrak{m}_0$, (A_j) an orthonormal basis of $\mathfrak{a}_0$, and (X_k^α) the basis of $(\mathfrak{n}_0 + \vartheta\mathfrak{n}_0)_{\mathbb{C}}$ introduced in lemma 1.4. Since the Casimir operator does not depend on the choice of the basis of $\mathfrak{g}_{\mathbb{C}}$ (see [21, 2.3.3.]), it can be written:

$$\Omega_G = -\sum_i M_i^2 + \sum_j A_j^2 + \sum_{\alpha > 0} \sum_{k=1}^{m(\alpha)} (X_k^\alpha X_k^{-\alpha} + X_k^{-\alpha} X_k^\alpha)$$

Note that $\sum_i M_i^2$ is just the Casimir operator on M_0, denoted by Ω_0. Now, we let Ω_G act on the left on the space of C^∞-functions f on G such that

$$f(gman) = \delta(m)^{-1} e^{-(\rho_\Sigma + i\lambda)(\log a)} f(g)$$

(here, we have used lemma 1.3). Since we a priori know that Ω_G acts by a scalar, it is enough to compute $(\pi(\delta, \lambda)\Omega_G)f(e)$, for such a function f. Then clearly one has:

$\pi(\delta,\lambda)(\Omega_0)f(e) = \delta(\Omega_0)f(e) = <\sigma + 2\rho_{M_0}, \sigma> f(e)$

$\pi(\delta,\lambda)(A_j)f(e) = (\rho_\Sigma + i\lambda)(A_j)f(e)$

$\pi(\delta,\lambda)(N)f(e) = 0 \qquad (\text{for } N \in \mathfrak{n}_0)$

Since, for $\alpha > 0$, we have

$$X_k^\alpha X_k^{-\alpha} + X_k^{-\alpha}X_k^\alpha = [X_k^\alpha, X_k^{-\alpha}] + 2X_k^{-\alpha}X_k^\alpha$$

then, by lemma 1.4:

$\pi(\delta,\lambda)(X_k^\alpha X_k^{-\alpha} + X_k^{-\alpha}X_k^\alpha)f(e) = \pi(\delta,\lambda)(Q_\alpha)f(e)$

since X_k^α belongs to $\mathfrak{n}_0$. So

$$\begin{aligned}\pi(\delta,\lambda)\Omega_G &= <\sigma + 2\rho_{M_0}, \sigma> + \sum_j (\rho_\Sigma + i\lambda)(A_j)^2 - \sum_{\alpha\in\Sigma,\ \alpha>0}\ \sum_{k=1}^{m(\alpha)} (\rho_\Sigma + i\lambda)(Q_\alpha)\\ &= <\sigma + 2\rho_{M_0}, \sigma> + <\rho_\Sigma + i\lambda, \rho_\Sigma + i\lambda> - \sum_{\alpha\in\Sigma,\ \alpha>0} m(\alpha)<\rho_\Sigma + i\lambda, \alpha>\\ &= <\sigma + 2\rho_{M_0}, \sigma> + <\rho_\Sigma + i\lambda, \rho_\Sigma + i\lambda> - <\rho_\Sigma + i\lambda, 2\rho_\Sigma>\\ &= <\sigma + 2\rho_{M_0}, \sigma> - <\lambda, \lambda> - <\rho_\Sigma, \rho_\Sigma>\end{aligned}$$

which is the desired formula.

Putting together 1.1 and 1.5, we immediately obtain:

<u>1.6. Corollary</u>: <u>For</u> $\rho = E_\tau$:

$$\pi(\delta,\lambda)(D_\rho{}^2) = -<\sigma + 2\rho_{M_0}, \sigma> + <\lambda, \lambda> + <\rho_\Sigma, \rho_\Sigma> + <\tau + 2\rho_K, \tau> - \chi(\Omega_K)$$

2. $C_r^*(G)$ when the reduced dual is Hausdorff

In this section, we prove that, if the reduced dual $\hat{G}_r$ of a semi-simple Lie group G with finite centre is Hausdorff, then the reduced C*-algebra $C_r^*(G)$ has a particularly simple form:

$$C_r^*(G) = C_0(\hat{G}_r, \mathcal{K})$$

We will also give some examples of application of this result.

We begin with some background on Harish-Chandra theory (a more detailed exposition can be found in [14], [21]).

<u>2.1. Definition</u>: A closed subgroup P of G is <u>parabolic</u> if P is the normalizer of its Lie algebra, and if P contains a conjugate of the standard minimal parabolic subgroup P_0.

A parabolic subgroup P has a <u>Langlands decomposition</u> P = MAN obtained as follows: N is the maximal connected normal subgroup of P consisting of unipotents elements; there exists then a closed reductive subgroup L of P such that P = L.N (a semi-direct product). A is a maximal connected split abelian subgroup in the centre of L; L is then equal to the centralizer of A in G (see [21, 1.2.4.2]). Finally,

$$M = \bigcap_{\chi} \operatorname{Ker} |\chi|$$

where χ runs over the set of homomorphisms from L to the multiplicative group of real numbers. Then L is the direct product of M and A. We denote by Z(A) (resp. N(A)) the centralizer (resp. normalizer) of A in G.

2.2. Definition: The group $W_P = N(A)/Z(A)$ is the Weyl group of P.

It is easy to see that W_P acts on $\hat{A}$ and $\hat{M}$. We need one more fact about Cartan subalgebras; as earlier, let $\mathfrak{g} = \mathfrak{k} + \mathfrak{p}$ be a Cartan decomposition of $\mathfrak{g}$, with associated Cartan involution ϑ. As in §1, let $\mathfrak{a}_0$ be a maximal abelian subalgebra of $\mathfrak{p}$, and let $\mathfrak{h}_0$ be a Cartan subalgebra extending $\mathfrak{a}_0$. Then $\mathfrak{h}_0$ is automatically ϑ-stable. If $\mathfrak{h}$ is any Cartan subalgebra, then by conjugating if necessary we may assume that $\mathfrak{h}$ is ϑ-stable and moreover:

$$\mathfrak{h} \cap \mathfrak{p} \subseteq \mathfrak{a}_0$$

$$\mathfrak{h} \cap \mathfrak{k} \supseteq \mathfrak{b}_0 = \mathfrak{h}_0 \cap \mathfrak{k}$$

(for all this, see [21, 1.3.1.]). Let P be a parabolic subgroup; by conjugating, it is possible to obtain $MA = P \cap \vartheta P$ (see [14, p.28]). We make this assumption from now on.

2.3. Definition: A parabolic subgroup P = MAN is cuspidal if there exists a ϑ-stable Cartan subalgebra $\mathfrak{h}$ such that $\mathfrak{h} \cap \mathfrak{p}$ is the Lie algebra of A. The centralizer of $\mathfrak{h}$ in G is a Cartan subgroup which is said to be compatible with P. In this case, $\mathfrak{b} = \mathfrak{h} \cap \mathfrak{k}$ is a Cartan subalgebra in the Lie algebra of M.

We see that $\mathfrak{h}_0$ is compatible with P_0, so P_0 is cuspidal. At the other extreme, G is cuspidal if and only if G contains a compact Cartan subgroup. Note that this is precisely Harish-Chandra's condition for the existence of discrete series representations of G. More generally, there is the following result of Harish-Chandra (see [21]):

2.4. Theorem: Let P = MAN be a parabolic subgroup. The following properties are equivalent:

i) P is cuspidal

ii) M admits discrete series representaions

2.5. Definition: Let $P_1 = M_1A_1N_1$ and $P_2 = M_2A_2N_2$ be two parabolic subgroups of G. Then P_1 is associated to P_2 if A_1, A_2 are conjugate in G.

It is easily checked that, if $g \in G$ conjugates A_1 and A_2, then g also conjugates M_1 and M_2. Moreover, P_1 and P_2 are associated if and only if the corresponding compatible Cartan subgroups H_1, H_2 are conjugate .

Let P = MAN be a cuspidal parabolic subgroup; let σ be an element of the set $\hat{M}_d$ of discrete series representations of M; let τ be a unitary character of A. The representations

$$\pi(\sigma,\tau) = \mathrm{Ind}_P^G \sigma \otimes \tau \otimes 1$$

are called representations of the P-series (for $P = P_o$, these are exactly the principal series representations). The following result is due to Lipsman [13].

2.6. Proposition: i) Let $P_1 = MAN_1$, $P_2 = MAN_2$ be two associated cuspidal parabolic subgroups. For σ in $\hat{M}_d$, τ in $\hat{A}$, the representations $\mathrm{Ind}_{P_1}^G \sigma \otimes \tau \otimes 1$ and $\mathrm{Ind}_{P_2}^G \sigma \otimes \tau \otimes 1$ are unitarily equivalent;

ii) Let P_1, P_2 be two non-associated cuspidal parabolic subgroups. Let π_1 (resp. π_2) be an element of the P_1-series (resp. P_2-series). Then π_1, π_2 are disjoint.

It turns out that Theorem 1.2 applies without change to the study of equivalence and irreducibility in the P-series. We may now state Harish-Chandra's famous result on the structure of the reduced dual of G (see [14]).

2.7. Theorem: Let $\mathcal{P}$ be the set of associativity classes of cuspidal parabolic subgroups of G. Then $\hat{G}_r$ is the disjoint union, over $P \in \mathcal{P}$, of the irreducible components of the P-series representations.

To study $C_r^*(G)$, we will need the following results, also due to Harish-Chandra (see [14]).

2.8. Theorem: Let π be an irreducible representation of G, and f be a function in $C_c^\infty(G)$.

i) $\pi(f)$ is a trace class operator;

ii) The linear form $f \to \mathrm{Tr}\,\pi(f)$ is a distribution on $C_c^\infty(G)$;

iii) There exists a function Θ_π, locally integrable and real analytic on some open dense subset of G, such that:

$$\mathrm{Tr}\,\pi(f) = \int_G \Theta_\pi(g) f(g)\,dg$$

With this result in hand, we may state our structure theorem for $C_r^*(G)$ when $\hat{G}_r$ is Hausdorff (actually, the result is a little bit more precise).

2.9. Theorem: Assume that, for any cuspidal parabolic subgroup P of G, any representation $\pi(\sigma,\tau)$ is irreducible. Then

i) $\hat{G}_r = \coprod_{P\in\mathcal{P}} (\hat{M}_d \times \hat{A})/W_P$

ii) Endow $\hat{G}_r$ with the natural topology (obtained by viewing $\hat{M}_d$ as a discrete space, and $\hat{A}$ as a vector space). Then

$$C_r^*(G) = C_0(\hat{G}_r, \mathcal{K})$$

where $\mathcal{K}$ denotes the algebra of usual compact operators.

iii) The natural topology and the Jacobson topology coincide on $\hat{G}_r$.

Proof: i) is an immediate consequence of Theorem 2.8.

ii) Fix a cuspidal parabolic subgroup $P = MAN$, and $\sigma \in \hat{M}_d$. Recall from [14, p.17] that there exists a Hilbert space $\mathcal{H}_\sigma$ on which all the $\pi(\sigma,\tau)$'s ($\tau \in \hat{A}$) are coherently represented. Denote by C_σ the connected component of $\hat{G}_r$ associated to σ. Define the partial Fourier transform by

$$\alpha_{P,\sigma}: C_r^*(G) \to \ell^\infty(C_\sigma, \mathcal{K}(\mathcal{H}_\sigma)): x \to (\pi(\sigma,\tau) \to \pi(\sigma,\tau)(x))$$

and the global Fourier transform $\alpha: C_r^*(G) \to \ell^\infty(\hat{G}_r, \mathcal{K})$ by

$$\alpha = \bigoplus_{P\in\mathcal{P}} \bigoplus_{\sigma\in\hat{M}_d/W_P} \alpha_{P,\sigma}$$

Clearly, α is pointwise a *-homomorphism. We have to show that maps $C_r^*(G)$ to $C_0(\hat{G}_r, \mathcal{K})$. So let $(\pi_n)_{n\in\mathbb{N}}$ be a sequence in $\hat{G}_r$ tending to ∞ in the natural topology. We have to show that

$$\lim_{n\to\infty} \|\pi_n(x)\| = 0 \qquad (x \in C_r^*(G))$$

Since there are but finitely many associativity classes of cuspidal parabolic subgroups (see [21, 1.3.1.11]), we may as well assume that there exists a cuspidal parabolic subgroup $P = MAN$ such that π_n is of the form $\pi(\sigma_n, \tau_n)$, where $\sigma_n \in \hat{M}_d$ and $\tau_n \in \hat{A}$. We may also assume that x belongs to $C_c^\infty(G)$. The Riemann-Lebesgue lemma for P then shows that:

$$\lim_{n\to\infty} \mathrm{Tr}\, \pi(\sigma_n, \tau_n)(x) = 0$$

This was proved by Lipsman [12] (see also [21, 5.5.4.1]), in the case where P is minimal parabolic; a general proof can be found in Arthur [1](see in particular Theorem 3.1 of [1]).(*) Since

$$\|\pi(\sigma_n, \tau_n)(x)\|^2 \leq \mathrm{Tr}\, \pi(\sigma_n, \tau_n)(x^*x)$$

we see that $\lim_{n\to\infty} \|\pi(\sigma_n, \tau_n)(x)\| = 0$

It remains to prove the norm continuity of $\alpha(x)$. So take a convergent sequence $\pi_n \to \pi$ in $\hat{G}_r$. So there exist a cuspidal parabolic $P = MAN$ and a $\sigma \in \hat{M}_d$ such that $\pi_n = \pi(\sigma, \tau_n)$ for n big enough, and $\pi = \pi(\sigma,\tau)$. As above, we assume that x belongs to $C_c^\infty(G)$. Then

$$\lim_{n\to\infty} \Theta_{\pi(\sigma, \tau_n)}(g) = \Theta_{\pi(\sigma,\tau)}(g) \quad \text{almost everywhere on } G$$

(see [13]). By Theorem 2.8 and the dominated convergence theorem

(1) $$\lim_{n\to\infty} \mathrm{Tr}\, \pi(\sigma, \tau_n)(x) = \mathrm{Tr}\, \pi(\sigma,\tau)(x)$$

By lemma 4.5 of [15], we deduce from this:

$$\lim_{n\to\infty} \|\pi(\sigma, \tau_n)(x) - \pi(\sigma,\tau)(x)\| = 0$$

So the image of α is contained in $C_0(\hat{G}_r, \mathcal{K})$. Clearly, by Theorem 2.7, the map α is one-to-one. Moreover, the range of α separates the pure states of $C_0(\hat{G}_r, \mathcal{K})$. Indeed, these are of the form

$$f \to \langle f(\pi)\xi, \xi\rangle \qquad (\pi \in \hat{G}_r,\ \xi \in \mathcal{H}_\pi,\ \|\xi\| = 1)$$

If we have for any x in $C_r^*(G)$:

$$\langle \pi(x)\xi, \xi\rangle = \langle \pi'(x)\xi', \xi'\rangle$$

then by Proposition 2.6, the representations π and π'come from the same parabolic P, i.e. we may write $\pi = \pi(\sigma,\tau)$ and $\pi' = \pi(\sigma',\tau'$
Now because of Theorem 1.2 (valid for any cuspidal parabolic !) we find an element of W_P which conjugates (σ,τ) and (σ', τ'), so that we actually have $\pi = \pi'$ in $\hat{G}_r$. So we have, for any $x \in C_r^*(G)$:

$$\langle \pi(x)\xi, \xi\rangle = \langle \pi(x)\xi', \xi'\rangle$$

But since $\pi(x)$ can be any compact operator on $\mathcal{H}_\pi$, we must have $\xi = \xi'$. So by the Stone-Weierstrass theorem (see [6, 11.1.8]), the range of α is the whole of $C_0(\hat{G}_r, \mathcal{K})$.

iii) is an immediate consequence of ii). This concludes the proof.

(*) According to the reviewer of Lipsman's paper [12] (see MR 42#4673), the Riemann-Lebesgue lemma for arbitrary cuspidal parabolic subgroups is an unpublished work of Lipsman. A self-contained proof of this lemma was recently given by Penington,"Semi-simple Lie groups", Thesis, Oxford 1983.

Remarks: i) The fact that the Jacobson topology and the natural topology coincide on the set of irreducible principal series representations was first proved by Lipsman [12].

ii) It is possible to prove ii) in Theorem 2.9 in a somewhat different way. Assume that we a priori know that the natural topology on $\hat{G}_r$ coincides with the Jacobson topology. Then formula (1) shows that $C^*_r(G)$ is a continuous trace C^*-algebra. $C^*_r(G)$ is then determined, up to isomorphism, by its Dixmier-Douady invariant in $H^3(\hat{G}_r, \mathbb{Z})$(see [6, 10.9.5]). But since $\hat{G}_r$ is a disjoint union of contractible spaces (namely linear varieties and closed cones), $H^3(\hat{G}_r, \mathbb{Z})$ is zero. So $C^*_r(G) = C_0(\hat{G}_r, \mathcal{K})$. However, we think that this proof misses the important fact that the isomorphism actually comes from the global Fourier transform.

iii) From the point of view of K-theory, the interest of Theorem 2.9 is that it reduces the computation of $K_*(C^*_r(G))$ to the computation of the K-theory of the locally compact space $\hat{G}_r$.

We are now going to give some examples of applications of Theorem 2.9. The next result was obtained by Penington-Plymen in the case of complex groups (see [17]).

2.10. Proposition: Let G be a semi-simple Lie group with just one conjugacy class of Cartan subgroups. Then $\hat{G}_r = (\hat{M}_0 \times \hat{A}_0)/W_0$ and

$$C^*_r(G) = C_0(\hat{G}_r, \mathcal{K})$$

Proof: From the remarks following 2.2, it follows that there is, up to associativity, just one cuspidal parabolic subgroup, namely the minimal parabolic P_0; so the representations in $\hat{G}_r$ are precisely the principal series representations. A result of Wallach [19](see also [21,p.463]) shows that in this case any principal series representation is irreducible. So Theorem 2.9 applies.

We now turn to the case of Lorentz groups.

2.11. Proposition: For $G = SO(n,1)$ $(n \geq 2)$, $\hat{G}_r$ is Hausdorff and

$$C^*_r(G) = C_0(\hat{G}_r, \mathcal{K})$$

Proof: For n odd, SO(n,1) has one conjugacy class of Cartan subgroups, so 2.10 applies. For n even, SO(n,1) has, up to conjugacy, two Cartan subgroups: a compact one and a non-compact one. So the reduced dual consists of $\hat{G}_d$, the set of discrete series representations, and the irreducible components of the principal series representations. But Knapp and Stein [11] have shown that any such representation is irreducible; so $\hat{G}_r = \hat{G}_d \amalg (\hat{M}_o \times \hat{A}_o)/W_o$ and 2.9 applies.

Note that the case n=2 already shows that the reduced dual of the double covering Spin(n,1) of SO(n,1) needs not be Hausdorff. (Actually, Knapp-Stein proved in [11] that the reduced dual of Spin(2n,1) is never Hausdorff). That the Dirac induction is an isomorphism for Spin(2n,1) was proved in an abstract way in [9], and in a computational manner in [18].

Our final example will be $SL_3(\mathbb{R})$. Up to associativity, this group has two cuspidal parabolic subgroups, namely

$$P_o = \begin{pmatrix} * & * & * \\ 0 & * & * \\ 0 & 0 & * \end{pmatrix} \qquad P = \begin{pmatrix} * & * & * \\ * & * & * \\ 0 & 0 & * \end{pmatrix}$$

Their Langlands decompositions are $P_o = M_oA_oN_o$ and $P = MAN$, where

$$M_o = \left\{ \begin{pmatrix} \varepsilon_1 & 0 & 0 \\ 0 & \varepsilon_2 & 0 \\ 0 & 0 & \varepsilon_3 \end{pmatrix}, \varepsilon_1\varepsilon_2\varepsilon_3 = 1,\ \varepsilon_i = \pm 1 \right\} ; \ M = \left\{ \begin{pmatrix} a & b & 0 \\ c & d & 0 \\ 0 & 0 & e \end{pmatrix}, e = \pm 1,\ (ad-bc)e = 1 \right\}$$

$$A_o = \left\{ \begin{pmatrix} a_1 & 0 & 0 \\ 0 & a_2 & 0 \\ 0 & 0 & a_3 \end{pmatrix}, a_1a_2a_3 = 1,\ a_i > 0 \right\} ; \ A = \left\{ \begin{pmatrix} a & 0 & 0 \\ 0 & a & 0 \\ 0 & 0 & a^{-2} \end{pmatrix}, a > 0 \right\}$$

$$N_o = \left\{ \begin{pmatrix} 1 & * & * \\ 0 & 1 & * \\ 0 & 0 & 1 \end{pmatrix} \right\} \qquad N = \left\{ \begin{pmatrix} 1 & 0 & * \\ 0 & 1 & * \\ 0 & 0 & 1 \end{pmatrix} \right\}$$

<u>2.12. Proposition</u>: <u>Let G be</u> $SL_3(\mathbb{R})$. <u>Then</u>

i) $\hat{G}_r = (\hat{M}_o \times \hat{A}_o)/W_o \amalg (\hat{M}_d \times \hat{A})$

ii) $C^*_r(G) = C_o(\hat{G}_r, \mathcal{K})$

iii) $K_o(C^*_r(G)) = 0$, <u>and</u> $K_1(C^*_r(G))$ <u>is free abelian with infinite rank, with one generator for any element of</u> $\hat{M}_d$.

<u>Proof</u>: i) First of all, any representation in the principal series is irreducible: this is a result of Wallach [19], which is even valid for $SL_{2n+1}(\mathbb{R})$. On the other hand, any representation in the P-series is irreducible; this will follow from 1.2 and the fact that W_P is zero. To prove this last statement, let w be an element in N(A). Then w induces an automorphism of A, i.e. there exists $s \in \mathbb{R}$ such that $waw^{-1} = a^s$ (since $A \simeq \mathbb{R}_o^+$). By computing traces, we get:

$2a^s + a^{-2s} = Tr(waw^{-1}) = Tr\ a = 2a + a^{-2}$ (for any $a \in A$)

This clearly implies s = 1, i.e. w belongs to Z(A); so $W_P = 0$.

ii) follows immediately from i) and 2.9.

iii) It suffices to show that $K^i((\hat{M}_o \times \hat{A}_o)/W_o)$ is zero both in degree 0 and in degree 1. But clearly W_0 = Sym 3, and $\hat{M}_0 = (\mathbb{Z}/2)^2$; so W_0 has two orbits on $\hat{M}_0$, each of them having non-trivial isotropy. This shows that $(\hat{M}_0 \times \hat{A}_0)/W_0$ is a disjoint union of two 2-dimensional closed convex cones; and cones have trivial K-theory.

Remark: The group M is the semi-direct product of $SL_2(\mathbb{R})$ by the subgroup generated by $\sigma = \begin{pmatrix} 0 & 1 \\ 1 & 0 \end{pmatrix}$. So the representation theory of M can be obtained from the Mackey machine. In particular, by [14, p.76], if γ belongs to the full dual $\hat{M}$ of M, there is a π in $SL_2(\mathbb{R})$ such that one of the following cases happens:

i) $\pi^\sigma = \pi$ and $\gamma|_{SL_2(\mathbb{R})}$ is a multiple of π;

ii) $\pi^\sigma \neq \pi$ and $\gamma = \mathrm{Ind}_{SL_2(\mathbb{R})}^{M} \pi$.

(Here $\pi(.) = \pi(\sigma . \sigma)$). Note that, if π is in the discrete series of $SL_2(\mathbb{R})$, then $\pi^\sigma \neq \pi$; indeed, such representations either have a higher K-type or a lower K-type (K = SO(2)). A simple computation shows that σ exchanges the higher K-type representations and the lower K-type representations. This shows that, if γ belongs to $\hat{M}_d$, then case i) may not occur. So the discrete series of M is just the discrete series of $SL_2(\mathbb{R})$ divided by the free action of σ.

3. Groups with one conjugacy class of Cartan subgroups

In this section, we completely determine the Dirac induction for semi-simple Lie groups with just one conjugacy class of Cartan subgroups, and show that it is an isomorphism.

From now on, G will always denote a connected semi-simple Lie group having one conjugacy class of Cartan subgroups. We first collect some results on the structure of G, which can be found e.g. in Wallach's book [20, §7.9]. We keep the notations of the preceding paragraphs.

3.1. Proposition: i) $\mathfrak{b}_0$ is a Cartan subalgebra of $\mathfrak{k}$;

ii) Let R be the set of restrictions to $i\mathfrak{b}_0$ of elements of Φ^+. Then $R \cup \Phi^-$ is precisely the root system of $\mathfrak{k}_{\mathbb{C}}$ with respect to $(\mathfrak{b}_0)_{\mathbb{C}}$;

iii) G is linear.

iv) K is semi-simple;

v) M_0 is connected.

From this, we immediately deduce:

3.2. Corollary: i) G has finite centre;

ii) For any $\lambda \in \Sigma$: the multiplicity $m(\lambda)$ is even;

iii) Σ is a reduced root system;

iv) N_0 is even-dimensional.

Proof: i) Obvious from 3.1.iii).

ii) If there is a λ with odd multiplicity, then for some $\alpha \in \Phi$ we have $\alpha = s\alpha$. This means that the restriction of α to $i\mathfrak{b}_0$ is zero, contradicting 3.1.ii).

iii) Fix λ in Σ, and assume $2\lambda \in \Sigma$. Then by [21, p.33], $m(2\lambda)$ is odd, a contradiction.

iv) $\dim N_0 = \sum_{\lambda\in\Sigma,\ \lambda>0} \dim \mathfrak{g}^{\lambda} = \sum_{\lambda\in\Sigma,\ \lambda>0} m(\lambda)$

We now have to consider Weyl groups. We recall that W_0 is defined to be M_0^*/M_0; it also coincides with the Weyl group of Σ (see [21, p.28]). We define $W(K)$ (resp. $W(M_0)$) to be the Weyl group of the root system of $\mathfrak{k}_{\mathbb{C}}$ (resp. $(\mathfrak{m}_0)_{\mathbb{C}}$) with respect to $(\mathfrak{b}_0)_{\mathbb{C}}$. Note that $W(M_0)$ is also the Weyl group of the root system Φ^-.

3.3. Lemma: i) R is a root system in $i\mathfrak{b}_0^*$;

ii) There is a short exact sequence:

$$0 \to W(M_0) \to W(K) \to W_0 \to 0$$

Proof: i) We verify that R satisfies the conditions of [5, p.142]. It is clear from 3.1 that R is finite and does not contain 0. Now let α be an element of Φ^+. The reflection with respect to the orthogonal of α, resp. $s\alpha$, maps $\beta \in (\mathfrak{a}_0 + i\mathfrak{b}_0)^*$ to

$$\beta - \frac{2\langle\beta, \alpha\rangle}{\langle\alpha, \alpha\rangle}\alpha$$

resp. $$\beta - \frac{2\langle\beta, s\alpha\rangle}{\langle\alpha, \alpha\rangle}s\alpha$$

The product s_α of these two reflections maps β to

$$\beta - \frac{2\langle\beta, \alpha\rangle}{\langle\alpha, \alpha\rangle} - \frac{2\langle\beta, s\alpha\rangle}{\langle\alpha, \alpha\rangle}s\alpha$$

(because $\langle\alpha, s\alpha\rangle = 0$; this is proved in [21, p.33]). When restricted to $i\mathfrak{b}_0^*$, s_α is precisely the reflection with respect to the orthogonal of the element of R obtained by restricting α. Next, we have to show that, for $\sigma, \tau \in R$, then

$$2\frac{\langle\sigma, \tau\rangle}{\langle\sigma, \sigma\rangle} \text{ is an integer.}$$

But we may write $\sigma = \frac{1}{2}(\alpha - s\alpha)$, $\tau = \frac{1}{2}(\beta - s\beta)$, for some $\alpha, \beta \in \Phi^*$.

Then

$$2\frac{\langle\sigma, \tau\rangle}{\langle\sigma, \sigma\rangle} = \frac{2}{\langle\alpha, \alpha\rangle}(\langle\alpha, \beta\rangle - \langle\alpha, s\beta\rangle)$$

and this is clearly an integer. Finally, we have to show that R spans $i\mathfrak{b}_0^*$ linearly. But, if $\{\beta_1, \ldots, \beta_n\}$ is a fundamental system of roots in Φ^-, we find by [21, p.23] a root $\alpha \in \Phi^+$ such that

$$\langle\alpha, \beta_i\rangle < 0 \quad (1 \leq i \leq n)$$

So $\alpha + \beta_i$ is a root in Φ^+, and if β is the restriction of α to $i\mathfrak{b}_0$, then the family $\{\beta, \beta + \beta_1, \ldots, \beta + \beta_n\}$ generates $i\mathfrak{b}_0^*$.

ii) Let W(G) be the Weyl group of Φ, and let W_s be the subgroup

$$W_s = \{w \in W(G) : ws = sw\}$$

Clearly $W(M_0)$ is normal in W_s. Moreover, by [21, 1.1.3.3.], the restriction of Φ to $\mathfrak{a}_0$ induces an homomorphism of W_s onto W_0, the kernel of which is $W(M_0)$. Thus, we have the short exact sequence:

$$0 \to W(M_0) \to W_s \to W_0 \to 0$$

Now the restriction of Φ to $i\mathfrak{b}_0$ induces an isomorphism between W_s and W(K). Indeed, if σ is a root in R such that $\sigma = \frac{1}{2}(\alpha - s\alpha)$ for some $\alpha \in \Phi^+$, then, as we saw earlier, s_α belongs to W_s and restricts to the reflection with respect to the orthogonal of σ; so the canonical map $W_s \to W(K)$ is onto. It is also one-to-one because otherwise there would exist a root $\alpha \in \Phi^+$ such that $\alpha = s\alpha$, contradicting 3.2.ii).

Let $\{B_1, \ldots, B_n\}$ be a basis of $i\mathfrak{b}_0$. The roots in Φ^-, R, $R \cup \Phi^-$ are ordered lexicographically with respect to this basis. The next lemma will show that the exact sequence in 3.3.ii) is actually split. Note that this part of lemma 3.3 also shows that W(K) acts on Φ^- as an automorphism group.

3.4. Lemma: Let W be the subgroup of W(K) which globally preserves the positive roots in Φ^-. Then:

i) W(K) is the semi-direct product of $W(M_0)$ and W;

ii) $W \simeq W_0$.

Proof: i) Follows immediately from [5, p.155].

ii) $W = W(K)/W(M_0) = W_0$ by 3.3.

Lemma 3.2 allows us to assume without loss of generality that G is simply connected. With this assumption, half the sums of the positive roots of the various root systems attached to $\mathfrak{b}_0$ will actually be weights of $\mathfrak{b}_0$. We denote by $\overline{C_K}$ (resp. $\overline{C}$, $\overline{C_0}$) the closed fundamental Weyl chamber associated to the positive roots of $R \cup \Phi^-$ (resp. R, Φ^-). Clearly $\overline{C_K} = \overline{C} \cap \overline{C_0}$.

3.5. Lemma: i) $\overline{C_K}$ is a fundamental domain for the action of W in $\overline{C_0}$.

ii) Let σ be a W-regular element in $\overline{C_0}$. Then $\sigma + \rho_{M_0}$ is W(K)-regular.

iii) Let σ be a weight in $\overline{C_K}$. σ is W-regular if and only if there exists a weight τ in $\overline{C_K}$ such that $\sigma + \rho_{M_0} = \tau + \rho_K$.

Proof: i) follows immediately from 3.4.

ii) Let k be an element in the stabilzer of $\sigma + \rho_{M_0}$ in $W(K)$. By 3.4, we may write $k = mw$, where $m \in W(M_0)$ and $w \in W$. Then $w(\sigma + \rho_{M_0})$ belongs to the open Weyl chamber C_0. Since $m(w(\sigma + \rho_{M_0}))$ belongs to C_0, m is equal to 1. So we have

$$\sigma + \rho_{M_0} = w(\sigma + \rho_{M_0}) = w\sigma + \rho_{M_0}$$

i.e. $\sigma = w\sigma$. But since σ is W-regular, w has to be 1.

iii) First, we prove that ρ_{M_0} belongs to $\overline{C_K}$. Since $\overline{C_K}$ is a fundamental domain for $W(K)$, there is a $k \in W(K)$ such that $k\rho_{M_0}$ belongs to $\overline{C_K}$. It is enough to show that k stabilizes ρ_{M_0} ; to see that, write $k = mw$ as above. Then $k\rho_{M_0} = m\rho_{M_0}$ has to belong to $\overline{C_K} \subseteq \overline{C_0}$. This implies $m=1$, i.e. $k=w$. Now assume that σ is W-regular; then $\sigma + \rho_{M_0}$ belongs to $\overline{C_K}$, and is W(K)-regular by part i). But any W(K)-regular weight in $\overline{C_K}$ is of the form $\tau + \rho_K$, for some weight τ in $\overline{C_K}$. Conversely, if $w \in W$ stabilizes σ, we have

$$\tau + \rho_K = \sigma + \rho_{M_0} = w\sigma + \rho_{M_0} = w(\tau + \rho_K)$$

which implies $w = 1$, since $\tau + \rho_K$ is W(K)-regular.

With this lemma in hand, we may compute the K-theory of $C_r^*(G)$.

3.6. Proposition: Let q be the (mod.2) dimension of G/K.

i) $K_{q+1}(C_r^*(G)) = 0$

ii) $K_q(C_r^*(G))$ is free abelian of infinite rank; an element $\delta \in \hat{M}_0$ provides a copy of $\mathbb{Z}$ to $K_q(C_r^*(G))$ if and only if, for some $w \in W_0$, the highest weight of $w\delta$ is of the form $\tau + \rho_K - \rho_{M_0}$, where τ is a weight in $\overline{C_K}$.

Proof: i) By 2.10, we have

$$K_i(C_r^*(G)) = K^i((\hat{M}_0 \times \hat{A}_0)/W_0) \qquad (i = 0, 1)$$

and $(\hat{M}_0 \times \hat{A}_0)/W_0$ is a disjoint union of closed convex cones and copies of $\hat{A}_o$. Since, by 3.2.iv), q is also the (mod.2) dimension of $\hat{A}_o$, we have $K^{q+1}((\hat{M}_o \times \hat{A}_o)/W_o) = 0$.

ii) An element $\delta \in \hat{M}_o$ contributes a copy of $\mathbb{Z}$ to $K^q((\hat{M}_o \times \hat{A}_o)/W_o)$ if and only if δ is W_o-regular. To examine this question, we may assume by 3.5.i) that the highest weight σ of δ belongs to $\overline{C_K}$. The conclusion follows then from 3.5.iii).

We consider now the Dirac induction on G. Since G is assumed to be simply connected, there is a G-invariant spin structure on G/K. We need some information about the representation χ of K.

<u>3.7. Lemma</u>: <u>The highest weight of K in</u> χ <u>is</u> $\rho_K - \rho_{M_o}$. <u>Its multiplicity</u> is $2^{[\frac{1}{2}\dim A_o]}$.

<u>Proof</u>: We begin by showing that the non-zero weights of the adjoint representation of K on $\mathfrak{p}_{\mathbb{C}}$ are precisely the elements of R. For any $\alpha \in \Phi^+$, $H \in \mathfrak{b}_o$, the element $X_\alpha - \vartheta X_\alpha$ belongs to $\mathfrak{p}_{\mathbb{C}}$ and

$$[H, X_\alpha - \vartheta X_\alpha] = \alpha(H)(X_\alpha - \vartheta X_\alpha)$$

The $X_\alpha - \vartheta X_\alpha$'s ($\alpha \in \Phi^+$, $\alpha > 0$) are $\mathbb{C}$-linearly independent and span a subspace of $\mathfrak{p}_{\mathbb{C}}$ of dimension dim N_o. On the other hand, if X belongs to $(\mathfrak{a}_o)_{\mathbb{C}}$, we have $[H,X] = 0$ for $H \in \mathfrak{b}_o$. As dim $\mathfrak{p}$ = dim A_o + dim N_o, we have the list of all weights of K in $\mathfrak{p}_{\mathbb{C}}$. If we denote by $\alpha_1, \dots, \alpha_p$ the set of positive roots in R, the adjoint representation of $\mathfrak{b}_o$ in the standard Cartan subalgebra of $\mathfrak{so}(\mathfrak{p})$ has the form

$$H \to \begin{pmatrix} 0 & -\alpha_1(H) & 0 & \cdots & & & & 0 \\ \alpha_1(H) & 0 & & & & & & \vdots \\ 0 & & \ddots & & & & & \vdots \\ \vdots & & & 0 & -\alpha_p(H) & & & \vdots \\ \vdots & & & \alpha_p(H) & 0 & & & \vdots \\ \vdots & & & & & 0 & & \vdots \\ \vdots & & & & & & \ddots & \vdots \\ 0 & \cdots & & & & & \cdots & 0 \end{pmatrix}$$

(We recall that the standard Cartan subalgebra of $\mathfrak{so}(\mathfrak{p})$ is

$$\begin{pmatrix} 0 & -\lambda_1 & \cdot & \cdot & \cdot & 0 \\ \lambda_1 & 0 & & & & \cdot \\ \cdot & & \cdot & & & \cdot \\ \cdot & & & & 0 & -\lambda_m \\ \cdot & & & & & \\ 0 & \cdot & \cdot & \cdot & \lambda_m & 0 \end{pmatrix} \quad \text{or} \quad \begin{pmatrix} 0 & -\lambda_1 & \cdot & \cdot & \cdot & 0 \\ \lambda_1 & 0 & & & & \cdot \\ \cdot & & \cdot & & & \cdot \\ \cdot & & & 0 & -\lambda_m & \cdot \\ \cdot & & & \lambda_m & 0 & \cdot \\ 0 & \cdot & \cdot & \cdot & \cdot & 0 \end{pmatrix}$$

according to the parity of q; here $m = [\frac{1}{2}q]$). It is well-known that the weights of the spin representation of Spin($\mathfrak{p}$) are of the form $\frac{1}{2}(\pm\lambda_1 \pm \ldots \pm\lambda_m)$. From this, we see that the weights of χ are of the form $\frac{1}{2}(\pm\alpha_1 \pm \ldots \pm \alpha_p)$, each of them being of multiplicity $2^{[\frac{1}{2}\dim A_0]}$. Finally, since the highest weight of the spin representation of Spin($\mathfrak{p}$) is $\frac{1}{2}(\lambda_1 + \ldots + \lambda_m)$, the highest weight of χ is

$$\tfrac{1}{2}(\alpha_1 + \ldots + \alpha_p) = \rho_K - \rho_{M_0}$$

By 1.1.ii), the Casimir operator of K acts by a scalar operator on χ. By the preceding lemma, it suffices to compute it on $E_{\rho_K - \rho_{M_0}}$. Then clearly

$$\chi(\Omega_K) = \langle 3\rho_K - \rho_{M_0}, \rho_K - \rho_{M_0}\rangle$$

(Note that we will actually prove in the appendix that χ is the sum of $2^{[\frac{1}{2}\dim A_0]}$ copies of $E_{\rho_K - \rho_{M_0}}$).

3.8. Lemma: Let ρ_G be half the sum of the positive roots of Φ. Then $\chi(\Omega_K) = \langle\rho_G, \rho_G\rangle - \langle\rho_K, \rho_K\rangle$

Proof: Of course it suffices to prove that

$$\langle 3\rho_K - \rho_{M_0}, \rho_K - \rho_{M_0}\rangle = \langle\rho_G, \rho_G\rangle - \langle\rho_K, \rho_K\rangle$$

Let $\{B_1, \ldots, B_n\}$ be a basis of $i\mathfrak{b}_0$, $\{A_1, \ldots, A_m\}$ be a basis of $\mathfrak{a}_0$; the set Φ is lexicographically ordered with respect to the basis $\{B_1, \ldots, B_n, A_1, \ldots, A_m\}$ of $i\mathfrak{b}_0 + \mathfrak{a}_0$. This ordering has the property that, if $\alpha \in \Phi^+$ is positive, then $-s\alpha$ is positive. Then

$$\rho_G = \tfrac{1}{2}\sum_{\alpha\in\Phi^-,\ \alpha>0} \alpha + \tfrac{1}{2}\sum_{\alpha\in\Phi^+,\ \alpha>0} \alpha = \rho_{M_0} + 2\rho_R$$

where ρ_R is half the sum of the positive roots of R. But clearly

$$\rho_{M_0} + 2\rho_R = 2\rho_K - \rho_{M_0}$$

The result now follows from an elementary computation.

3.9. Corollary: For $\rho = E_\tau$, $\lambda \in \hat{A}_0$, $\delta \in \hat{M}_0$ with highest weight $\sigma \in \overline{C_{M_0}}$:

$$\pi(\delta, \lambda)(D_\rho^2) = -\langle\sigma + \rho_{M_0}, \sigma + \rho_{M_0}\rangle + \langle\tau + \rho_K, \tau + \rho_K\rangle + \langle\lambda, \lambda\rangle$$

Proof: By 1.6 and 3.8, we have

$$\pi(\delta,\lambda)(D_\rho^2) = -\langle\sigma + \rho_{M_0}, \sigma + \rho_{M_0}\rangle + \langle\lambda, \lambda\rangle + \langle\rho_\Sigma, \rho_\Sigma\rangle + \langle\rho_{M_0}, \rho_{M_0}\rangle$$
$$+ \langle\tau + \rho_K, \tau + \rho_K\rangle - \langle\rho_G, \rho_G\rangle$$

But consider an ordering on Φ such that $\alpha \in \Phi^+$, α positive, implies $s\alpha$ positive (such an ordering was used in the remarks preceding 1.3). The length of ρ_G remains the same in this new ordering, since any two sets of positive roots in Φ are conjugate by an element in W(G). For this chosen ordering, we obviously have $\rho_G = \rho_{M_o} + \rho_\Sigma$, hence

$$\langle\rho_G, \rho_G\rangle = \langle\rho_{M_o}, \rho_{M_o}\rangle + \langle\rho_\Sigma, \rho_\Sigma\rangle$$

since $\langle\rho_{M_o}, \rho_\Sigma\rangle = 0$. This concludes the proof.

We now proceed to define the concept of Fourier transform of an induced C*-module over $C^*_r(G)$. This notion is directly inspired by the notion of Plancherel decomposition of an induced representation of G, due to Atiyah and Schmid [2]. Let ρ be a finite dimensional representation of K, with degree d_ρ. As we saw earlier, the space of C^∞-sections with compact support of the induced homogeneous vector bundle over G/K is

$$\mathcal{E}_\rho = (C_c^\infty(G) \otimes \rho)^K$$

This space is a right module over $C_c^\infty(G)$ by

$$\xi f(x) = \int_G \xi(gx)f(g)dg \qquad (\xi \in \mathcal{E}_\rho,\ f \in C_c^\infty(G))$$

Moreover, it is a pre-C*-module when endowed with the $C_c^\infty(G)$-valued scalar product:

$$\langle\xi, \eta\rangle(x) = \int_G \langle\xi(g), \eta(xg)\rangle dg$$

Fix an element $\delta\ \hat{M}_o$; let $\mathcal{H}_\delta$ be a Hilbert space on which the representations $\pi(\delta,\lambda)$ $(\lambda \in \hat{A}_o)$ are coherently represented. By [14, p.40], we have by Frobenius reciprocity:

$$\pi(\delta,\lambda)_{|K} = \mathrm{Ind}_{M_o}^K\ \delta$$

(this does not depend on λ!). We denote by $\mathcal{H}_\delta{}^o$ the opposite Hilbert space, i.e. the additive group $\mathcal{H}_\delta$ with conjugate operation of the complex numbers, and corresponding inner product. We denote by $\mathrm{Hom}_K(\mathcal{H}_\delta{}^o, \mathcal{H}_\delta{}^o \otimes \rho)$ the space of K-intertwining operators between $\pi(\delta,\lambda)_{|K}$ and $1 \otimes \rho$ (note that any operator in this space has finite rank). Let C_δ be the connected component of $\hat{G}_r$ associated to δ. The space $C_o(C_\delta, \mathrm{Hom}_K(\mathcal{H}_\delta{}^o, \mathcal{H}_\delta{}^o \otimes \rho))$ is a right module over $C_o(C_\delta, \mathcal{K}(\mathcal{H}_\delta))$ by

$$(\xi \cdot f)(\pi) = (f(\pi)^* \otimes 1)\xi(\pi)$$

(the reason for introducing $\mathcal{H}_\delta{}^o$ is that this action so becomes linear; working with $\mathcal{H}_\delta$ it would be antilinear). The space $C_o(C_\delta, \mathrm{Hom}_K(\mathcal{H}_\delta^o, \mathcal{H}_\delta^o \otimes \rho))$ also becomes a C*-module when endowed with the $C_o(C_\delta, \mathcal{K}(\mathcal{H}_\delta))$-valued scalar product:

$$\langle\xi, \eta\rangle(\pi) = (1 \otimes \mathrm{tr})(\eta(\pi)\xi(\pi)^*)$$

where $1 \otimes \mathrm{tr}$ denotes the conditional expectation

$$\mathcal{L}(\mathcal{H}_\delta^o \otimes \rho) \to \mathcal{L}(\mathcal{H}_\delta^o): S \otimes T \to \mathrm{tr}(T)S$$

and tr is the normalized trace

$$\mathrm{tr}(T) = d_\rho^{-1} \sum_{j=1}^{d_\rho} T_{jj}$$

Now, we define the <u>partial Fourier transform</u>:

$$\alpha_{\rho,\delta}: (C_c^\infty(G) \otimes \rho)^K \to \ell^\infty(C_\delta, \mathrm{Hom}_K(\mathcal{H}_\delta{}^\circ, \mathcal{H}_\delta{}^\circ \otimes \rho))$$

$$\xi \to (\pi \to d_\rho^{\frac{1}{2}} \int_G \pi(g) \otimes \xi(g)\, dg)$$

Arguments similar to those used in 2.9 show that $\alpha_{\rho,\delta}$ actually maps $(C_c^\infty(G) \otimes \rho)^K$ to $C_0(C_\delta, \mathrm{Hom}_K(\mathcal{H}_\delta{}^\circ, \mathcal{H}_\delta{}^\circ \otimes \rho))$. With this, we define the global Fourier transform

$$\alpha_\rho: (C_c^\infty(G) \otimes \rho)^K \to \bigoplus_{\delta \in \hat{M}_0/W_0} C_0(C_\delta, \mathrm{Hom}_K(\mathcal{H}_\delta{}^\circ, \mathcal{H}_\delta{}^\circ \otimes \rho))$$

by $\alpha_\rho = \bigoplus_{\delta \in \hat{M}_0/W_0} \alpha_{\rho,\delta}$ (here we view $\bigoplus_{\delta \in \hat{M}_0/W_0} C_0(C_\delta, \mathrm{Hom}_K(\mathcal{H}_\delta{}^\circ, \mathcal{H}_\delta{}^\circ \otimes \rho))$ as a C*-module over $C_0(\hat{G}_r, \mathcal{K})$).

<u>3.10. Lemma</u>: <u>For any</u> $\xi, \eta \in (C_c^\infty(G) \otimes \rho)^K$, $f \in C_c^\infty(G)$:

$$\alpha_\rho(\xi f) = \alpha_\rho(\xi)\alpha(f)$$

$$\langle \alpha_\rho(\xi), \alpha_\rho(\eta) \rangle = \alpha \langle \xi, \eta \rangle$$

<u>where</u> α <u>denotes the global Fourier transform on</u> $C_c^\infty(G)$.

<u>Proof</u>: Fix δ in $\hat{M}_0$, and π in C_δ. Then

$$\alpha_\rho(\xi f)(\pi) = d_\pi^{\frac{1}{2}} \int_G \int_G \pi(g) \otimes \xi(hg) f(h)\, dg\, dh \qquad (h \to h^{-1})$$

$$= d_\pi^{\frac{1}{2}} \int_G \int_G \overline{f(h^{-1})}\, \pi(g) \otimes \xi(h^{-1}g)\, dg\, dh \qquad (g \to hg)$$

$$= d_\pi^{\frac{1}{2}} \int_G \int_G (f^*(h)\pi(h) \otimes 1)(\pi(g) \otimes \xi(g))\, dg\, dh$$

$$= \alpha(f^*)(\pi)\alpha_\rho(\xi)(\pi) = (\alpha_\rho(\xi).\alpha(f))(\pi)$$

(we used at the second step the fact that we are working on $\mathcal{H}_\delta{}^\circ$).

Moreover

$$\langle \alpha_\rho(\xi), \alpha_\rho(\eta) \rangle(\pi) = (1 \otimes \mathrm{tr})(\alpha_\rho(\eta)(\pi)\, \alpha_\rho(\xi)^*(\pi))$$

$$= d_\rho (1 \otimes \mathrm{tr}) \int_G \int_G (\pi(g) \otimes \eta(g))(\pi(h) \otimes \xi(h))^*\, dg\, dh$$

But, for $v \otimes w \in \mathcal{H}_\delta{}^\circ \otimes \rho$, we have

$$(\pi(h) \otimes \eta(h))^*(v \otimes w) = \langle w, \xi(h) \rangle \pi(h^{-1})v$$

hence

$$(\pi(g) \otimes \eta(g))(\pi(h) \otimes \xi(h))^*(v \otimes w) = \pi(gh^{-1})v \otimes \eta(g)\langle \xi(h), w \rangle$$

So the operator we get is the tensor product of $\pi(gh^{-1})$ and a rank one operator. The trace of this rank one operator is $d_\rho^{-1} \langle \xi(h), \eta(g) \rangle$. So we get:

$$<\alpha_\rho(\xi), \alpha_\rho(\eta)>(\pi) = \int_G\int_G \pi(gh^{-1}) <\xi(h), \eta(g)> dg\, dh \qquad (g \to gh)$$

$$= \int_G\int_G \pi(g) <\pi(h), \eta(gh)> dg\, dh$$

$$= (\alpha<\xi, \eta>)(\pi)$$

This finishes the proof.

It follows from this lemma that the C^*-module completion of $(C_c^\infty(G) \otimes \rho)^K$ over $C_r^*(G)$ is isometrically isomorphic to the module $\bigoplus_{\delta \in \hat{M}_0/W_0} C_0(C_\delta, \mathrm{Hom}_K(\mathcal{H}_\delta^0, \mathcal{H}_\delta^0 \otimes \rho))$ over $C_0(\hat{G}_r, \mathcal{K})$. As a consequence, it is possible to define the Fourier transform of any (bounded or unbounded) C^*-module endomorphism D of $(C_c^\infty(G) \otimes \rho)^K$. In particular, for any $\pi \in C_\delta$, we get an operator $\pi(D)$ acting on $\mathrm{Hom}_K(\mathcal{H}_\delta^0, \mathcal{H}_\delta^0 \otimes \rho)$. But this last space, endowed with the Hilbert-Schmidt norm, is isometric to $\mathcal{H}_\delta^0 \otimes (\mathcal{H}_\delta^{0*} \otimes \rho)^K$. So we have

$$\pi(D) \in \mathrm{End}(\mathcal{H}_\delta^0 \otimes (\mathcal{H}_\delta^{0*} \otimes \rho)^K)$$

But $\pi(D)$, being a module endomorphism, has to commute with the action of $\mathcal{K}(\mathcal{H}_\delta)$ on the first factor. So $\pi(D)$ is actually given by $\pi(D) = 1 \otimes M_\pi$, where $M_\pi \in \mathrm{End}(\mathcal{H}_\delta^{0*} \otimes \rho)^K$; note that $(\mathcal{H}_\rho^{0*} \otimes \rho)^K$ is nothing but the space of K-intertwining operators between $\pi(\delta,\lambda)|_K$ and ρ; so it is finite-dimensional. Moreover, what we did in 3.9 was actually the computation of the Fourier transform of the square of the Dirac operator; in this case the matrix M_π turns out to be a scalar matrix.

If D is a G-invariant elliptic differential operator on $(C_c^\infty(G) \otimes \rho)^K$, the computation of the K-theory element $\mathrm{Ind}_a(D)$ is now reduced to the computation of the K-theory element given on each C_δ by the continuous field of matrices M_π.

We proceed to identify the Dirac induction, starting from an irreducible representation $\rho = E_\tau$ of K. The next lemma (similar to lemma 5.3 in [17]) will help to locate the C_δ's on which $\mathrm{Ind}_a(D_\rho)$ is not zero.

<u>3.11. Lemma</u>: <u>Let</u> δ <u>be an element of</u> $\hat{M}_0$, <u>with highest weight</u> σ <u>in</u> $\overline{C_K}$ <u>(which we may assume by 3.5). For any</u> $\lambda \in \hat{A}_0$, <u>the space of K-intertwining operators between</u> $\pi(\delta,\lambda)|_K$ <u>and</u> $\chi \otimes E_\tau$ <u>is of dimension</u> $2^{[\frac{1}{2}\dim A_0]}$ <u>if</u> $\sigma = \tau + \rho_K - \rho_{M_0}$, <u>and is zero if</u> $\tau + \rho_K - \rho_{M_0} - \sigma$ <u>is not a positive linear combination of positive roots of</u> K.

Proof: As we saw earlier:

$$\pi(\delta,\lambda)|_K = \mathrm{Ind}_{M_0}^K \delta$$

Now

$$\mathrm{Ind}_{M_0}^K \delta = E_\sigma \oplus E$$

where E is a direct sum of irreducible K-modules with highest weights of the form $\sigma + \sum$, where $\sum$ is a non-zero sum of positive roots. On the other hand, we have, by 3.7:

$$\chi \otimes E_\tau = 2^{[\frac{1}{2}\dim A_0]}\, E_{\tau+\rho_K-\rho_{M_0}} \oplus E'$$

where E' is a direct sum of irreducible K-modules with highest weights of the form $\tau + \rho_K - \rho_{M_0} - \sum'$. The conclusion follows easily from Schur's lemma.

We are now in position to prove the main result of this section.

3.12. Theorem: The Dirac induction $\Delta: R(K) \to K_q(C_r^*(G))$ is an isomorphism given as follows: the representation $E_\tau = \rho$ is mapped to the Bott element on the copy of $\hat{A}_0$ in $\hat{G}_r$ associated to the element of $\hat{M}_0$ with highest weight $\tau + \rho_K - \rho_{M_0}$.

Proof: By 3.11, we know that the K-theory element $\mathrm{Ind}_a(D_\rho)$ is located on the C_δ's for which the highest weight σ of δ is of the form $\sigma = \tau + \rho_K - \rho_{M_0} - \sum$, where $\sum$ is a sum of positive roots of K. We first show that $\mathrm{Ind}_a(D_\rho)$ is zero on those C_δ's for which $\sum$ is not zero. Indeed, in such a case, we have by 3.9:

$$\pi(\delta,\lambda)(D_\rho^2) \geq -\langle\sigma+\rho_{M_0},\sigma+\rho_{M_0}\rangle + \langle\tau+\rho_K,\tau+\rho_K\rangle > 0$$

so that, on these C_δ's, the Fourier transform of D_ρ is invertible, and does not give anything in K-theory. It remains to show that, for $\sigma = \tau + \rho_K - \rho_{M_0}$, the element $\mathrm{Ind}_a(D_\rho)$ precisely gives what we call the Bott element on C_δ (note that $C_\delta = \hat{A}_0$, by 3.6). This element, giving the canonical generator of $K_q(C_0(C_\delta,\mathcal{K}(\mathcal{H}_\delta)))$, is described by means of the C*-module $C_0(C_\delta, \mathcal{H}_\delta^0 \otimes S)$, where S denotes the spin module of $\mathrm{Spin}(C_\delta)$. As endomorphism F on this C*-module, we take

$$(F\xi)(\lambda) = (1 \otimes c(\lambda))\xi(\lambda) \qquad (\xi \in C_0(C_\delta, \mathcal{H}_\delta^0 \otimes S))$$

where $c(\lambda)$ denotes the Clifford multiplication by $\lambda \in C_\delta$. This F defines an unbounded Kasparov element (in the sense of [3]), which is the required generator in K-theory. Now, for

$\sigma = \tau + \rho_K - \rho_{M_0}$, we have

$$\pi(\delta,\lambda)(D_\rho{}^2) = \langle\lambda, \lambda\rangle$$

by 3.9. Moreover, if we write

$$\pi(\delta,\lambda)(D_\rho) = 1 \otimes M_{\pi(\delta,\lambda)}$$

like in the remarks following 3.10, we see by 3.11 that $M_{\pi(\delta,\lambda)}$ acts on a space with dimension $S^{[\frac{1}{2}\dim C_\delta]}$. Since the family $\pi(\delta,\lambda)(D_\rho)$ depends linearly on λ (as the Fourier transform of a first order differential operator), we see that $M_{\pi(\delta,\lambda)}$ is precisely the Clifford multiplication by λ on the spin module of $\mathrm{Spin}(C_\delta)$, so that $\mathrm{Ind}_a(D_\rho)$ defines the K-theory element described here above. The fact that the Dirac induction Δ is an isomorphism follows immediately from 3.6.ii). This concludes the proof.

<u>Remark</u>: If G is a compact semi-simple Lie group, then $M_0 = K = G$, so that $\rho_K = \rho_{M_0}$ and one finds the trivial fact that R(G) is isomorphic to $K_0(C_r^*(G))$. If G is a complex semi-simple Lie group, then M_0 is a maximal torus in K (see [7, p.238]), so $\rho_{M_0} = 0$. The preceding result was then obtained by Penington-Plymen [17]. These two cases are opposite of each other and, according to our result, semi-simple Lie groups with one conjugacy class of Cartan subgroups seem to "interpolate" between compact and complex groups.

Appendix: A result of Parthasarathy

In this section, we give a proof of the fact that, if G is a connected simply connected semi-simple Lie group with one conjugacy class of Cartan subgroups, then the Casimir operator Ω_K of K acts by a scalar on the representation χ. As we noticed in the remarks following 1.1, this result was claimed by Parthasarathy [16] for any semi-simple Lie group, but proved only in the case where the group has discrete series representations.

The proof that we give depends on the Cartan classification of simple real Lie algebras; it would be desirable to have a proof avoiding this classification.

A.1. Lemma: Let $\mathfrak{g}$ be a simple real Lie algebra having one conjugacy class of Cartan subalgebras. Then one of the following situations occurs:

i) $\mathfrak{g}$ is compact;

ii) $\mathfrak{g}$ is complex;

iii) $\mathfrak{g} = \mathfrak{so}(2n-1,1) = D_n^{\mathbb{R},1}$

iv) $\mathfrak{g} = \mathfrak{sl}_n(\mathbb{H}) = A_{2n-1}^{\mathbb{H}}$

v) $\mathfrak{g} = E_{6(-26)}$

Proof: Assume that $\mathfrak{g}$ is neither compact nor complex: then $\mathfrak{g}$ has one of the Satake diagrams given in [21, pp.30-32]; these tables also give the multiplicities $m(\lambda)$ and $m(2\lambda)$, where λ runs over a set of simple roots for Σ. By 3.2, we must have $m(\lambda)$ even and $m(2\lambda) = 0$. This singles out the diagrams:

$A_{2n-1}^{\mathbb{R}}$ 

$D_n^{\mathbb{R},1}$ 

$E_{6(-26)}$ 

It is easily verified that the corresponding groups have just one conjugacy class of Cartan subgroups. For the adjoint group of $E_{6(-26)}$, this can be done by identifying it to the automorphism group of the projective plane over the Cayley numbers.

A.2. Proposition: The representation χ is the sum of $2^{[\frac{1}{2}\dim A_0]}$ copies of the irreducible representation $E_{\rho_K-\rho_{M_0}}$ of K.

Proof: We have $\dim\chi = 2^{[\frac{1}{2}q]} = 2^{\frac{1}{2}\dim N_0}.2^{[\frac{1}{2}\dim A_0]}$. By 3.7, it is enough to prove that the dimension d of $E_{\rho_K-\rho_{M_0}}$ is equal to $2^{\frac{1}{2}\dim N_0}$. By the Weyl dimension formula, this dimension is given by

$$d = \prod_{\alpha>0} \langle 2\rho_K - \rho_{M_0}, \alpha\rangle \Big/ \prod_{\alpha>0} \langle \rho_K, \alpha\rangle$$

where α runs over the positive roots of K. Clearly, we may assume that G is almost simple. By A.1, there are five cases to consider.

i) G is compact: obviously $d=1$, and $N_0 = 1$.

ii) G is complex: then M_0 is a maximal torus in K, so $\rho_{M_0} = 0$, and $d = 2^p$ where p is the number of positive roots of K. But since Φ^- is empty, we have $p = \frac{1}{2} \dim N_0$.

iii) If $G = Spin(2n-1,1)$ (the universal covering of $SO(2n-1,1)$), then $K = Spin(2n-1)$ and $\mathfrak{p} = \mathbb{R}^{2n-1}$. A simple matrix computation shows that the adjoint representation of K on $\mathfrak{p}$ is nothing but the vector representation of $Spin(2n-1)$. So χ is irreducible, of degree 2^{n-1}. On the other hand, $\dim N_0 = 2n-2$.

iv) If $G = SL_n(\mathbb{H})$, then $M = Sp(n)$ and $M_0 = (Sp(1))^n$. With the notations of [5, p.254], we denote by $\{\varepsilon_i, 1 \leq i \leq n\}$ the canonical basis of $\mathbb{R}^n$. The positive roots of K ere the $2\varepsilon_i$'s $(1 \leq i \leq n)$, and the $\varepsilon_i \pm \varepsilon_j$'s $(1 \leq i < j \leq n)$, while the positive roots of M_0 are the $2\varepsilon_i$'s $(1 \leq i \leq n)$. So

$$\rho_{M_0} = \sum_{i=1}^{n} \varepsilon_i$$

$$\rho_K = \sum_{i=1}^{n} (n+1-i)\varepsilon_i$$

Two straightforward calculations then show that $d = 2^{n(n-1)}$ and $\dim N_0 = 2n(n-1)$; we leave them to the skeptical reader.

v) If $G = E_{6(-26)}$, we have $K = F_4$ and $M = Spin(8) = D_4$. In the notations of [5, pp.256-272], the positive roots of K are ε_i $(1 \leq i \leq 4)$, $\varepsilon_i \pm \varepsilon_j$ $(1 \leq i < j \leq 4)$, $\frac{1}{2}(\varepsilon_1 \pm \varepsilon_2 \pm \varepsilon_3 \pm \varepsilon_4)$. The positive roots of M_0 are $\varepsilon_i \pm \varepsilon_j$ $(1 \leq i < j \leq 4)$. So

$$\rho_{M_0} = 3\varepsilon_1 + 2\varepsilon_2 + \varepsilon_3$$

$$\rho_K = \tfrac{1}{2}(11\varepsilon_1 + 5\varepsilon_2 + 3\varepsilon_3 + \varepsilon_4)$$

Again, tedious calculations yield $d = 2^{12}$ and $\dim N_0 = 24$. This concludes this (messy) proof.

References

[1] J.Arthur, Harmonic analysis on the Schwartz space of a reductive Lie group II, mimeographed notes, Yale University, 1975.

[2] M.F.Atiyah & W.Schmid, A geometric construction of the discrete series for semi-simple Lie groups, Inventiones Math. 42 (1977), 1-62.

[3] S.Baaj & P.Julg, Théorie bivariante de Kasparov et opérateurs non bornés dans les C*-modules hilbertiens, C.R.Acad.Sc. Paris 296, Sér.I (1983), 875-878.

[4] P.Baum & A.Connes, Geometric K-theory for Lie groups and

foliations, to appear in Proc. 1983 U.S.-Japan seminar, "Geometric methods in Operator Algebras".

[5] N.Bourbaki, Groupes et algèbres de Lie (Chap.IV,V,VI), Hermann (Paris), 1968.

[6] J.Dixmier, C*-algebras, North-Holland (Amsterdam), 1982.

[7] S.Helgason, Differential geomerty and symmetric spaces, Monographs in Pure & Applied Math.12, Academic Press, 1962.

[8] G.G.Kasparov, The index of invariant elliptic operators, K-theory and Lie group representations, Dokl.Akad.Nauk.USSR. 268 (1983), 533-537.

[9] G.G.Kasparov, Lorentz groups: K-theory of unitary representations and crossed products, Preprint Chernogolovka, 1983.

[10] G.G.Kasparov, Operator K-theory and its applications: elliptic operators, group representations, higher signatures, C*-extensions, to appear in Proc. International Congress of Math. Warsaw, 1983.

[11] A.W.Knapp & E.M.Stein, Intertwining operators for semi-simple Lie groups, Ann. of Math. 93 (1971), 489-578.

[12] R.L.Lipsman, The dual topology for the principal and discrete series on semi-simple Lie groups, Trans.A.M.S. 152 (1970), 399-417.

[13] R.L.Lipsman, On the characters and equivalence of continuous series representations, J.Math.Soc.Japan 23(1971), 452-480.

[14] R.L.Lipsman, Group representations, Springer Lect. Notes in Maths. 388, 1974.

[15] D.Milicic', Topological representation of the group C*-algebra of $SL_2(\mathbb{R})$, Glasnik Matem. 6(26)(1971), 231-246.

[16] R.Parthasarathy, Dirac operator and the discrete series, Ann. of Math. 96 (1972), 1-30.

[17] M.G.Penington & R.J.Plymen, The Dirac operator and the principal series for complex semi-simple Lie groups, Journ. of Funct. Anal. 53 (1983), 269-286.

[18] A.Valette, K-theory for the reduced C*-algebra of a semi-simple Lie group with real rank 1 and finite centre, to appear in Quart. Journ. Math. (Oxford).

[19] N.R.Wallach, Cyclic vectors and irreducibility for principal series representations, Trans.Amer.Math.Soc.158(1971), 107-113.

[20] N.R.Wallach, Harmonic analysis on homogeneous spaces, Marcel

Dekker (New-York), 1973.

[21] G.Warner, Harmonic analysis on semi-simple Lie groups I, II, Springer 1972.

[22] A.Wassermann, A proof of the Connes-Kasparov conjecture for linear reductive Lie groups, preprint (University of Liverpool), Dec. 1983.

Author's address: Département de Mathématiques CP 214
Université Libre de Bruxelles
Boulevard du Triomphe
B-1050 Bruxelles.

(*) Research Assistant at the Belgian National Fund for Scientific Research.

SYMMETRIES OF SOME REDUCED FREE PRODUCT C*-ALGEBRAS

Dan Voiculescu

The reduced free product of C*-algebras considered in this paper is a generalization of the construction in [7] with the trace states replaced by arbitrary states. In this context we study symmetries of two types of C*-algebras arising as reduced free products.

One type of C*-algebras are the free group factors. It is known that the canonical anticommutation relations yield a certain automorphic action of the unitary group of a Hilbert space on the hyperfinite II_1 factor. We show that there is an analogoue of this construction on the Fock space for Boltzmann statistics, which gives an action of the orthogonal group of an n-dimensional real Hilbert space on the factor of the free group on n generators. This appears to be related to a noncommutative central limit theorem, where the reduced free products replace the tensor products in the definition of independent random variables. The limit distribution is not the usual gaussian one, but a certain modified arcsine law.

The other type of C*-algebras are extensions of Cuntz-algebras acting on the Fock-space for Boltzmann statistics as in ([17], [16]). The automorphism groups appearing in this situation are the groups U(n,1).

The paper has six sections.

The first section contains generalities and various remarks about reduced free products of C*-algebras with specified states.

In §2 we remark that the realization of certain extensions of the Cuntz-algebras ([34], [32]) on the Fock-space for Boltzmann statistics ([17],[16]), amounts to viewing these algebras as reduced free products of copies of the C*-algebra of an isometry with respect to the corresponding "vacuum" states and we construct a representation of U(n,1) on the Fock-space implementing an automorphic action on these algebras.

In the third section we construct a functor from real Hilbert spaces and contractions to C*-algebras and completely positive maps. The non-commutative C*-algebras arising in this way have unique trace states, which are faithful and the associated von Neumann algebras are the free group factors. The corresponding action of the orthogonal group O(n) on the factor of the free group on n generators restricted to the permutation matrices, yields the action of the symmetric group by permutations of the generators (and is of course different from the automorphic actions considered in [3], [33]). We exhibit a convenient basis for computations with this action of O(n) in terms of certain Gegenbauer polynomials.

The fourth section considers the addition problem for non-commutative random variables independent in the sense of free products and we prove the corresponding central limit theorem (for other types of non-commutative central limit theorems see [24],[18] and references given there).

In §5 we show how the definitions in §1 can be generalized to reduced free products with amalgamation by using Hilbert-modules over C*-algebras.

In section 6 we consider the problem of decomposing the representation of U(n,1) on the tensor algebra of $\mathbb{C}^n$ constructed in §2 into irreducible representations. It appears that this problem is naturally related to a similar problem involving the Lie superalgebras $\ell(n,1)$.

This paper has been circulated as INCREST Preprint no.29/1983. Recently we learned from Mathematical Reviews (MR 83 H:46070) about a paper by D.Avitzour: "Free products of C*-algebras" (Trans.Amer.Math. Soc.271(1982), 423-465), where free product states and reduced free products have also been considered.

§1

This section is devoted to generalities about free products for Hilbert spaces and C*-algebras. The reduced free product of C*-algebras considered here generalizes the construction in [7] from the case of C*-algebras with specified trace-states to the case of C*-algebras with arbitrary specified states.

We discuss commutants of reduced free products, generalizing [7], and commutants modulo the compacts, extending some constructions which appeared in connection with K-theory computations for free groups ([35], [29], [15], [12]).

1.1. The natural frame-work for our considerations is the category having as objects pairs (H,ξ) where H is a complex Hilbert space and $\xi\in H$ is a unit vector and the morphisms are contractions $T:H\to H'$

such that $T\xi=\xi'$. The free product of a family $((H_\iota,\xi_\iota))_{\iota\in I}$ is (H,ξ) with

$$H=\mathbb{C}\xi\oplus\bigoplus_{n\geq 1}\bigoplus_{(\iota_1,\ldots,\iota_n)\in D_n(I)}(H^o_{\iota_1}\otimes\ldots\otimes H^o_{\iota_n})$$

where $H^o_\iota=H_\iota\ominus\mathbb{C}\xi_\iota$ and

$$D_n(I)=\{(\iota_1,\ldots,\iota_n)\mid \iota_j\neq\iota_{j+1},\ 1\leq j\leq n-1,\ \iota_j\in I, 1\leq j\leq n\}.$$

Given morphisms $T_\iota:H_\iota\to H'_\iota$, remark that $T_\iota(H^o_\iota)\subset H'^o_\iota$ so that there corresponds a morphism $T:H\to H'$. We shall write $(H,\xi)=\mathop{\ast}\limits_{\iota\in I}(H_\iota,\xi_\iota)$ and $T=\mathop{\ast}\limits_{\iota\in I}T_\iota$. Also, we shall sometimes abuse notations and write $H=\ast_\iota H_\iota$, $\xi=\ast_\iota\xi_\iota$. Another notation which we shall use is P_o for the projection of H onto $\mathbb{C}\xi$ and $P_{(\iota_1,\ldots,\iota_n)}$ for the projection of H onto $H^o_{\iota_1}\otimes\ldots\otimes H^o_{\iota_n}$. Also, $H(\ell,\iota)$, $H(r,\iota)$ will denote the subspaces of H defined by

$$H(\ell,\iota)=\mathbb{C}\xi\oplus\bigoplus_{n\geq 1}\bigoplus_{\substack{(\iota_1,\ldots,\iota_n)\in D_n(I)\\ \iota_1\neq\iota}}H^o_{\iota_1}\otimes\cdots\otimes H^o_{\iota_n}$$

$$H(r,\iota)=\mathbb{C}\xi\oplus\bigoplus_{n\geq 1}\bigoplus_{\substack{(\iota_1,\ldots,\iota_n)\in D_n(I)\\ \iota_n\neq\iota}}H^o_{\iota_1}\otimes\cdots\otimes H^o_{\iota_n}$$

1.2. For each $\iota\in I$ we define unitary operators $V_\iota:H\to H_\iota\otimes H(\ell,\iota)$ by

$$V_\iota(h_1\otimes\ldots\otimes h_n)=\begin{cases} h_1\otimes(h_2\otimes\ldots\otimes h_n) & \text{if } \iota_1=\iota,\ n\geq 2\\ h_1\otimes\xi & \text{if } \iota_1=\iota,\ n=1\\ \xi_\iota\otimes(h_1\otimes\ldots\otimes h_n) & \text{if } \iota_1\neq\iota,\end{cases}$$

$$V_\iota\xi=\xi_\iota\otimes\xi$$

where $h_j\in H^o_{\iota_j}$, $(\iota_1,\ldots,\iota_n)\in D_n(I)$.

Similarly, there are unitaries $W_\iota:H\to H(r,\iota)\otimes H_\iota$ defined by:

$$W_\iota(h_1\otimes\ldots\otimes h_n)=\begin{cases} (h_1\otimes\ldots\otimes h_{n-1})\otimes h_n & \text{if } \iota_n=\iota,\ n\geq 2\\ \xi\otimes h_1 & \text{if } \iota_1=\iota,\ n=1\\ (h_1\otimes\ldots\otimes h_n)\otimes\xi_\iota & \text{if } \iota_n\neq\iota\end{cases}$$

$$W_\iota\xi=\xi\otimes\xi_\iota$$

where $h_j\in H^o_{ij}$, $(\iota_1,\ldots,\iota_n)\in D_n(I)$.

Using these unitaries we define two representations, λ_ι and respectively ρ_ι of $\mathcal{L}(H_\iota)$ (the bounded operators on H_ι) by the formulae:

$$\lambda_\iota(T)=V_\iota^{-1}(T\otimes I)V_\iota,\quad \rho_\iota(T)=W_\iota^{-1}(I\otimes T)W_\iota$$

1.3. It is easy to check that for $\iota,\iota'\in I$ and $T\in L(H_\iota)$, $T'\in L(H_{\iota'})$ we have

$$[\lambda_\iota(T),\rho_{\iota'}(T')]=\delta_{\iota\iota'}(P_o+P_{(\iota)})\lambda_\iota([T,T'])=\delta_{\iota\iota'}(P_o+P_{(\iota)})\rho_\iota([T,T']),$$

and note that $P_o+P_{(\iota)}$ is a reducing projection for $\lambda_\iota(L(H_\iota))$ and $\rho_\iota(L(H_\iota))$.

1.4. Consider now for each $\iota\in I$ a pair (A_ι,φ_ι) where A_ι is a C*-algebra and φ_ι a state of A_ι. Let $\pi_\iota:A_\iota\to L(H_\iota)$ be the representation associated with φ_ι and $\xi_\iota\in H_\iota$ the corresponding vector. On $(H,\xi)=\underset{\iota\in I}{\ast}(H_\iota,\xi_\iota)$ there are representations $\sigma_\iota=\lambda_\iota\circ\pi_\iota$ and we have

(i) $\omega_\xi\circ\sigma_\iota=\varphi_\iota$;

(ii) if $(\iota_1,\ldots,\iota_n)\in D_n(I)$ and $a_j\in A_{\iota_j}$, $\varphi_\iota(a_j)=0$ then $\omega_\xi(\sigma_{\iota_1}(a_1)\ldots$ $\ldots\sigma_{\iota_n}(a_n))=0$;

(iii) ξ is cyclic for $\bigcup_{\iota\in I}\sigma_\iota(A_\iota)$.

Consider A the C*-algebra generated by $\bigcup_{\iota\in I}\sigma_\iota(A_\iota)$ and $\varphi=\omega_\xi|A$. Then (A,φ) *will be called the reduced free product of the* (A_ι,φ_ι) and we shall write $(A,\varphi)=\underset{\iota\in I}{\ast}(A_\iota,\varphi_\iota)$.

1.5. It is easily seen that if the A_ι are unital, then (A,φ) can be characterized in the following way:

a) A is unital and there are unital *-homomorphisms $\sigma_\iota:A\to A$ such that A is generated by $\bigcup_{\iota\in I}\sigma_\iota(A_\iota)$;

b) $\varphi\circ\sigma_\iota=\varphi_\iota$;

c) for $(\iota_1,\ldots,\iota_n)\in D_n(I)$ and $a_j\in \operatorname{Ker}\varphi_{\iota_j}$ we have $\varphi(\sigma_{\iota_1}(a_1)\ldots\sigma_{\iota_n}(a_n))=0$;

d) the GNS construction applied to (A,φ) yields a faithful representation of A.

To see that these properties characterize (A,φ) one notes first that a)-c) determine φ on a dense subalgebra of A, so that in view of d) one gets A by completing the image of this subalgebra via the representation corresponding to φ.

1.6. If the A_ι are not unital one adjoins a unit and considers $(\tilde{A}_\iota,\tilde{\varphi}_\iota)$ and $\ast A_\iota$ is the subalgebra of $\ast\tilde{A}_\iota$ generated by the $\sigma_\iota(A_\iota)$.

1.7. Note also that we could have constructed $\underset{\iota\in I}{\ast}(A_\iota,\varphi_\iota)$ by using ρ_ι instead of λ_ι, so that in general A has a left and a right representiations denoted λ and respectively ρ on $\underset{\iota\in I}{\ast}H_\iota$ which are intertwined by the unitary $\Theta \in L(\underset{\iota\in I}{\ast} H_\iota)$ defined by $\Theta\xi=\xi$, $\Theta(h_1\otimes\ldots\otimes h_n)=h_k\otimes\ldots$ $\ldots\otimes h_1$ if $h_j\in H^o_{\iota_j}$ and $(\iota_1,\ldots,\iota_n)\in D_n(I)$.

1.8. LEMMA. *Let* $A_\iota \subset L(H_\iota)$ $(\iota\in I)$ *be von Neumann algebras and assume* ξ_ι *is a cyclic and separating vector for* A_ι. *Then we have*

$$(\bigcup_{\iota\in I}\lambda_\iota(A_\iota))'=(\bigcup_{\iota\in I}\rho_\iota(A'_\iota))''$$

PROOF. The proof is an application of Tomita's theory of Hilbert algebras along the same lines as the Cuculescu-Sakai proof for the commutation theorem for tensor products presented in 10.7 of [41]. Remark first that ξ is cyclic and separating for $(\bigcup_{\iota\in I}\lambda_\iota(A_\iota))''=A$. Consider the operators

S_ι=the closure of $[A_\iota\xi_\iota \ni a_\iota\xi_\iota \to a^*_\iota\xi_\iota \in H_\iota]$

S=the closure of $[A\xi\ni a\xi\to a^*\xi\in H]$.

Then $S_\iota\xi_\iota=\xi_\iota$ and $S_\iota(\mathcal{D}_{S_\iota}\cap\{\xi_\iota\}^\perp)\subset\{\xi_\iota\}^\perp$. In order to describe the relation between S_ι and S we must somewhat digress. Let T_ι be linear or antilinear operators densely defined on $\mathcal{D}_{T_\iota}\subset H_\iota$ and taking values in H'_ι. Assume $\xi_\iota\in\mathcal{D}_{T_\iota}$, $T_\iota\xi_\iota=\xi'_\iota$ and $T_\iota(\mathcal{D}_{T_\iota}\cap\{\xi_\iota\}^\perp)\subset\{\xi'_\iota\}^\perp$. Then we define $\underset{\iota\in I}{\ast} T_\iota$ on $\mathcal{D}_{\ast T_\iota}$ (= the vector subspace of $\ast H_\iota$ generated by ξ and $h_1\otimes\ldots$ $\ldots\otimes h_n$ where $(\iota_1,\ldots\iota_n)\in D_n(I)$ and $h_j\in\mathcal{D}_{T_{\iota_j}}\cap\{\xi_{\iota_j}\}^\perp$) by $(\underset{\iota\in I}{\ast} T_\iota)\xi=\xi'$

$$(\underset{\iota\in I}{\ast} T_\iota)(h_1\otimes\ldots\otimes h_n)=T_{\iota_1}h_1\otimes\ldots\otimes T_{\iota_n}h_n.$$

If the T_ι are closed then $\ast T_\iota$ is preclosed and its closure will be denoted by $\overline{\underset{\iota\in I}{\ast}} T_\iota$. We have

$$(\overline{\underset{\iota\in I}{\ast}} T_\iota)^*=\overline{\underset{\iota\in I}{\ast}} T^*_\iota.$$

This follows from the corresponding result for tensor products (see 9.33 of [41] applied to $(T_{\iota_1}|(\mathcal{D}_{T_{\iota_1}}\cap\{\xi_{\iota_1}\}^\perp))\otimes\ldots\otimes(T_{\iota_n}|(\mathcal{D}_{T_{\iota_n}}\cap\{\xi_{\iota_n}\}^\perp))$ and taking the direct sum of these operators.

It is easily checked that we have

$$S=\Theta(\overline{\underset{\iota\in I}{\ast}} S_\iota).$$

This implies

$$S^*=(\overline{\underset{\iota\in I}{\ast}} S^*_\iota)\Theta=\Theta(\overline{\underset{\iota\in I}{\ast}} S^*_\iota).$$

But from Tomita theory (see 10.6 in [41]) we have:

S^*=the closure of $[A'\xi\ni a'\xi\to a'^*\xi\in H]$

S^*_ι=the closure of $[A'_\iota\xi\ni a'_\iota\xi_\iota\to a'^*_\iota\xi_\iota\in H_\iota]$

$\Theta(\overline{\bigstar_{\iota\in I} S_\iota^*})$=the closure of $[B\xi \ni b\xi \to b^*\xi \in H]$

where $B=(\bigcup_{\iota\in I} \rho_\iota(A_\iota'))''$.

Therefore, with the notations of [41] for Tomita theory, we have:

$$S^{A'}=S^*=\Theta(\overline{\bigstar_{\iota\in I} S_\iota^*})=S^B.$$

Then, since $A'\xi \supset B\xi$ we can apply Lemma 3, from 10.5 of [41] to obtain

$$A'\xi=B\xi$$

so that $A'=B$. Q.E.D.

1.9. PROPOSITION. *Let* $A_\iota, B_\iota \subset L(H_\iota)$ $(\iota\in I)$ *be von Neumann algebras so that* $A_\iota'=B_\iota$ *and assume* ξ_ι *is a cyclic vector for* A_ι. *Then for* $A=\bigcup_{\iota\in I} \lambda_\iota(A_\iota))''$ *and* $B=(\bigcup_{\iota\in I} \rho_\iota(B_\iota))''$ *we have* $A'=B$.

PROOF. Let Q_ι denote the orthogonal projection of H_ι onto $\tilde{H}_\iota=\overline{B_\iota\xi_\iota}$ and consider Q the orthogonal projection of $H=\bigstar_{\iota\in I} H_\iota$ onto $\tilde{H}=\bigstar_{\iota\in I}\tilde{H}_\iota$. Clearly $Q_\iota \in A_\iota$ and we shall show that $Q\in A$. Let $\Omega_\iota=\bigcup_{n\geq 1}\{\lambda_{\iota_1}(w_1)\ldots\lambda_{\iota_n}(w_n) \mid (\iota_1\ldots\ldots\iota_n)\in D_n(I),\ \iota_n\neq\iota,\ w_k\in A_{\iota_k}, <w_k\xi_{\iota_k}, \xi_{\iota_k}>=0, 1\leq k\leq n\}\cup\{I\}$.

Then we have

$$(I-Q)H=\bigvee_{\iota\in I}\ \bigvee_{\omega\in\Omega_\iota} \omega(I-\lambda_\iota(Q_\iota))H$$

which shows that $Q\in A$.

Consider further $\tilde{A}_\iota=Q_\iota A_\iota Q_\iota|\tilde{H}_\iota$, $\tilde{B}_\iota=B_\iota|\tilde{H}_\iota$ so that $\tilde{A}_\iota'=\tilde{B}_\iota$ and ξ_ι is cyclic and separating for $\tilde{A}_\iota, \tilde{B}_\iota$. For $\tilde{A}=QAQ|\tilde{H}$, $\tilde{B}=B|\tilde{H}$ we have $\tilde{B}=(\bigcup_{\iota\in I} \rho_\iota(\tilde{B}_\iota))''$ (of course ρ_ι is relative to the $(\tilde{H}_\iota,\xi_\iota)$) and $\tilde{A}\subset\tilde{B}'$. Since $\tilde{A}\supset\lambda_\iota(\tilde{A}_\iota)$ we infer from Lemma 1.8 that $\tilde{A}=\tilde{B}'$. But $\tilde{H}$ is cyclic for A since $\xi\in\tilde{H}$. Thus from $\tilde{A}=\tilde{B}'$ we infer $A=B'$. Q.E.D.

1.10. REMARK. The preceding proposition cannot be improved by giving up the condition that the ξ_ι be cyclic for the A_ι, as the following example shows. Let $H_1=H_2=\mathbb{C}^n$ with $n>1$ (finite) and $A_1=L(H_1)$, $B_1=\mathbb{C}I_{H_1}$, $A_2=\mathbb{C}I_{H_2}$, $B_2=L(H_2)$ so that A and B are then isomorphic to A_1 and respectively B_2 and hence finite-dimensional. Since $\bigstar_{i=1,2} H_\iota$ is infinite-dimensional we cannot have $A'=B$.

1.11. The commutation relations 1.3 can also be used in case the commutators are compact, which is of interest in connection with the Brown-Douglas-Fillmore Ext-groups [6] or the related $KK^1(\bullet,\mathbb{C})$ of Kasparov [28]. The group with which we shall work will be denoted

$\mathrm{Ext}_s^{inv}(B)$, where B is separable and unital, and is the group of invertible elements in the semigroup Ext_s of strong equivalence classes of extensions. Elements of $\mathrm{Ext}_s^{inv}(B)$ are defined by pairs (μ,Q) where $\mu:B\to L(X)$ is a unital $*$-representation and Q a self-adjoint idempotent such that $[Q,\mu(B)]\subset K$. Such a pair is called trivial if $[Q,\mu(B)]=0$. Two pairs (μ,Q) and (μ',Q') are equivalent if for some trivial (μ'',Q''), (μ''',Q''') there is a unitary u such that $u(Q\oplus Q'')=(Q'\oplus Q''')u$ and $u(\mu\oplus\mu'')(b)-(\mu'\oplus\mu''')(b)u\in K$ for $b\in B$. Note that we may assume (μ,Q), (μ',Q') are such that

$$b\to p(\mu(b)Q),\qquad b\to p(\mu(b)(I-Q))$$
$$b\to p(\mu(b)Q')\qquad b\to p(\mu(b)(I-Q')).$$

are monomorphisms, where p is the map to the Calkin algebra L/K and then using [44], (μ,Q) and (μ',Q') will be equivalent iff there is a unitary u such that $uQ=Q'u$ and $u\mu(b)-\mu'(b)u\in K$. The addition is defined by taking direct sums. If strong and weak equivalence coincide for B, then Ext_s^{inv} coincides with $E^1(B,\mathbb{C})\approx KK^1(B,\mathbb{C})$ (see Definition 3, Lemma 2 and Remark 1 of [28]).

1.12. Let $(A,\varphi)=\underset{\iota\in I}{\ast}(A_\iota,\varphi_\iota)$ where the A_ι are separable and unital, and let $(\pi_\iota,H_\iota,\xi_\iota)$ be the cyclic representations corresponding to the states φ_ι and consider $J_\iota=\pi_\iota^{-1}(\pi_\iota(A)\cap K)$ and $q_\iota:A_\iota\to A_\iota/J_\iota$ the quotient map.

PROPOSITION. *There is a homomorphism*

$$r:\prod_{\iota\in I}\mathrm{Ext}_s^{inv}(A_\iota/J_\iota)\to\mathrm{Ext}_s^{inv}(A)$$

such that for $\sigma:\mathrm{Ext}_s^{inv}(A)\to\prod_{\iota\in I}\mathrm{Ext}_s^{inv}(A_\iota)$ *defined by* $\sigma(\alpha)=(\sigma_\iota^*(\alpha))_{\iota\in I}$ *we have*

$$\prod_{\iota\in I}q_\iota^*=\sigma\circ r.$$

PROOF. Let $\beta_\iota\in\mathrm{Ext}_s^{inv}(A_\iota/J_\iota)$ be defined by (μ_ι,Q_ι), $\mu_\iota:A_\iota/J_\iota\to L(X_\iota)$ such that $b\to p(\mu_\iota(b)Q_\iota)$ and $b\to p(\mu_\iota(b)(I-Q_\iota))$ are monomorphisms. By [44], we can find unitaries $v_\iota:H_\iota\oplus X_\iota\to H_\iota$ such that $v_\iota(\xi_\iota\oplus 0)=\xi_\iota$ and $\pi_\iota(a)v_\iota-v_\iota(\pi_\iota(a)\oplus\mu_\iota(q_\iota(a)))$ is compact for $a\in A_\iota$. We define $r((\beta_\iota)_{\iota\in I})$ as the equivalence class of the pair consisting of the left representation of A on $\ast H_\iota$ and the projection $Q=\sum_{\iota\in I}\rho_\iota(v_\iota(O\oplus Q_\iota)v_\iota^*)$. Using 1.3 one easily gets that $[\lambda(A),Q]\subset K$. It is also easily seen that $(\sigma\circ r)((\beta_\iota)_{\iota\in I})$ is then represented by the classes of the pairs $(\pi_\iota,v_\iota(O\oplus Q)v_\iota^*)$ so that what is left to be proved, is that r is a well-defined homomorphism. If $(\mu_\iota,Q_\iota,X_\iota)$ are replaced by $(\mu_\iota',Q_\iota',X_\iota')$ representing also β_ι and having the same additional properties, then consider unitaries $u_\iota : X_\iota \to X'_\iota$ such that $u_\iota Q_\iota u_\iota^* = Q'_\iota$ and $u_\iota \mu_\iota(a) - \mu'_\iota(a) u$ are compact for $a \in A_\iota$. Then for $U_\iota = v'_\iota (I \oplus u_\iota) v_\iota^*$ we have that $U_\iota \xi_\iota = \xi_\iota$ and so the unitaries $\rho_\iota(U_\iota)$ $(\iota \in I)$ commute and $U = \prod_{\iota \in I} \rho_\iota(U_\iota)$ makes sense since for $\iota \neq \iota'$, $(\rho_\iota(U_\iota) - I)(\rho_{\iota'}(U_{\iota'}) - I) = 0$. It is easily checked that $UQU^* = Q'$ and $[\lambda(A), U] \subset K$. To see that r is a homomorphism, note that we may represent β_ι, β'_ι, $\beta''_\iota = \beta_\iota + \beta'_\iota$ by (μ_ι, Q_ι), (μ_ι, Q'_ι), (μ_ι, Q''_ι) where $Q'' = Q + Q'$ and the additional properties required hold. Then we have $\rho_\iota(v_\iota(0 \oplus Q_\iota)v_\iota^*) + \rho_\iota(v_\iota(0 \oplus Q'_\iota)v_\iota^*) = \rho_\iota(v_\iota(0 \oplus Q''_\iota)v_\iota^*)$ from which it is easy to infer that r is a homomorphism. Q.E.D.

1.13. We want to record here a simple lemma, which extends to the situation we are considering, a construction, which has turned out to play an important role in all the succesive papers where the K-theory of the reduced C*-algebra of the free group has been considered ([35], [29], [15], [12]).

Let $(A, \varphi) = \mathop{\ast}_{\iota \in I} (A_\iota, \varphi_\iota)$ where the A_ι are unital, and let $(\pi_\iota, H_\iota, \xi_\iota)$ be the cyclic representations associated with the φ_ι and $(H, \xi) = \mathop{\ast}_{\iota \in I} (H_\iota, \xi_\iota)$.

LEMMA. *Let* $S_\iota \in L(H_\iota)$ $(\iota \in I)$ *be isometries such that* Ker $S_\iota^* = \mathbb{C}\xi_\iota$ *and assume* $[\pi_\iota(A_\iota), S_\iota] \subset K$ *for all* $\iota \in I$. *Let further* $S : H \otimes \ell^2(I) \to H$ *be defined by* $S(h \otimes e_\iota) = \rho_\iota(S_\iota)h$, *where* $(e_\iota)_{\iota \in I}$ *is the canonical basis of* $\ell^2(I)$. *Then* S *is an isometry with* Ker $S^* = \mathbb{C}\xi$ *and* $S(\lambda(a) \otimes I) - \lambda(a)S$ *is compact for* $a \in A$.

The proof is an easy application of 1.3 which we leave to the reader.

Under the assumptions of the Lemma the isometry S *can be used to define an element of* KK(A, $\mathbb{C}$). The element will be defined by $X = H \oplus (H \otimes \ell^2(I))$ with the grading $X_+ = H \oplus 0$, $X_- = 0 \oplus (H \otimes \ell^2(I))$ and $F = \begin{bmatrix} 0 & S \\ S^* & 0 \end{bmatrix}$.

On the other hand denoting as in 1.12 by J_ι the ideal $\pi_\iota^{-1}(\pi_\iota(A_\iota) \cap K)$ it is easily seen that *if "weak" and "strong"* Ext *coincide for* A_ι / J_ι *then there exist isometries* S_ι *as in the assumption of the Lemma and the conclusion of the Lemma is that "weak" and "strong"* Ext *coincide also for* A (compare with [32]).

§2

In this section we shall consider reduced free products of copies of the shift-algebra. This yields Cuntz-algebras acting on the full Fock-space as in ([17] see also [16]). We construct a unitary representation of a group U(1,n) on this Fock-space, which induces automorphisms of the reduced free product algebra.

2.1. Let H be a complex Hilbert space and let

$$\mathcal{T}(H)=\mathbb{C}1 \oplus \bigoplus_{n\geq 1} H^{\otimes n}$$

be the full Fock-space over H. Also, given a contraction $T:H_1 \to H_2$ there is a corresponding morphism $\mathcal{T}(T):(\mathcal{T}(H_1),1)\to(\mathcal{T}(H_2),1)$ such that

$$\mathcal{T}(T)(h_1\otimes\ldots\otimes h_m)=Th_1\otimes\ldots\otimes Th_m$$

2.2. It is easily seen that there is a canonical identification

$$\mathop{\ast}_{\iota\in I}(\mathcal{T}(H_\iota),1)=(\mathcal{T}(\bigoplus_{\iota\in I} H_\iota),1)$$

2.3. On $\mathcal{T}(H)$ there are left and right creation operators defined by

$$\ell(h)\eta=h\otimes\eta \text{ and } r(h)\eta=\eta\otimes h$$

where $h\in H$, $\eta\in\mathcal{T}(H)$. For $h_\iota\in H_\iota$, in the context of 2.2, we have:

$$\lambda_\iota(\ell(h_\iota))=\ell(h_\iota), \qquad \rho_\iota(r(h_\iota))=r(h_\iota)$$

where in the right-hand sides of these equalities h_ι is viewed as an element of $\bigoplus_{\iota\in I} H_\iota$.

2.4. We have:

$$[\ell(h),r(k)]=0$$

$$[\ell(h),r(k)^*]=-\langle h,k\rangle P_0$$

where P_0 is the projection of $\mathcal{T}(H)$ onto $\mathbb{C}1$.

2.5. When dim H=1 and $h\in H$, $||h||=1$, then $\ell(h)=r(h)$ is the familiar unilateral shift S of multiplicity 1. Thus for H of arbitrary dimension, we infer from 2.2 and 2.3 that the C*-algebra $C^*(\ell(H))$ generated by the $\ell(h)$, $h\in H$, can be identified with the reduced free product of dim H-copies of $(C^*(S),\varepsilon)$ where ε is the unique state of $C^*(S)$ (see [10]) such that $\varepsilon([S^*,S])=1$. As remarked in [17] (see also [16]) $C^*(\ell(H))$ is a Cuntz-algebra if H is infinite-dimensional and an extension of the compacts by a Cuntz-algebra ([34], [31]) if dim $H<\infty$. Also in [17] (see also [16]) it is remarked that the natural representation Γ of the unitary group of H on $\mathcal{T}(H)$ induces automorphisms of $C^*(\ell(H))$ since

$$\Gamma(U)\ell(h)\Gamma(U^{-1})=\ell(Uh).$$

We shall exhibit a larger group acting on $\mathcal{T}(H)$ which induces automorphisms of $C^*(\ell(H))$ and of course if dim $H<\infty$, also of $C^*(\ell(H))/\mathcal{K}$.

2.6. On $\mathbb{C}\oplus H$ we consider the operator

$$J=\begin{bmatrix} 1 & 0 \\ 0 & -I_H \end{bmatrix}$$

and the group $U(\mathbb{C},H)$ of invertible operators $A\in L(\mathbb{C}\oplus H)$ such that

$$A^*JA=I.$$

2.7. Instead of describing $C^*(\ell(H))$ by means of its generators $\ell(h)$ it will be more convenient (though obvious equivalent) to consider the isometry

$$L: H\otimes T(H) \to T(H)$$

given by

$$L(h\otimes\eta) = \ell(h)\eta.$$

We shall need the following generalization of the decomposition of an isometry into unitary and completely non-unitary part, which is quite probably folklore.

LEMMA. *Let*

$$L': H\otimes X \to X$$

be an isometry. Consider $X_n \subset X$ *defined by* $X_o = X, X_{n+1} = L'(H\otimes X_n)$. *Then defining*

$$M = X_o \ominus X_1,\quad X'' = \bigcap_{n>0} X_n,\quad X' = X\ominus X''$$

we have:

(i) $L'(H\otimes X'') = X''$, $L'(H\otimes X') \subset X'$;

(ii) *there is a unitary operator* $U: T(H)\otimes M \to X'$ *such that the diagramm*

$$\begin{array}{ccc} H\otimes T(H)\otimes M & \xrightarrow{\;L\otimes I_M\;} & T(H)\otimes M \\ \Big\downarrow{\scriptstyle I_H\otimes U} & & \Big\downarrow{\scriptstyle U} \\ H\otimes X' & \xrightarrow{\;L'\mid(H\otimes X')\;} & X' \end{array}$$

is commutative and $U(1\otimes\mu) = \mu$ *for* $\mu\in M$.

PROOF. The proof being quite straightforward we only sketch the main facts from which the lemma follows.

Defining $\ell'(h)x = L'(h\otimes x)$ the isometricity of L' implies that $\ell'(h_1)^*\ell'(h_2) = \langle h_2, h_1\rangle\, I$. Note also that $M = \operatorname{Ker} L'^* = \bigcap_{h\in H} \operatorname{Ker} \ell'^*(h)$ and $X_n = \bigvee_{h_1\dots h_n\in H} \ell'(h_1)\dots\ell'(h_n)X$. Also, if $\mu\in M$ for $h_1,\dots,h_m$, $h_1',\dots,h_n'\in X$ we have if $m\geq n$:

$$\langle \ell'(h_1)\dots\ell'(h_m)\mu,\ \ell'(h_1'),\dots,\ell'(h_n')\mu\rangle =$$
$$= \langle h_1,h_1'\rangle\dots\langle h_n,h_n'\rangle\langle\mu,\ \ell'^*(h_m)\dots\ell'^*(h_{m+1})\mu\rangle =$$
$$= \delta_{m,n}\langle h_1,h_1'\rangle\dots\langle h_n,h_n'\rangle$$

(clearly by symmetry this holds also for $m\leq n$). Since $L'(H\otimes(X_j\ominus X_{j+1})) = X_{j+1}\ominus X_{j+2}$ we easily infer that $\bigvee_{\substack{n\geq 0\\ h_1,\dots,h_n\in H}} \ell'(h_1)\dots\ell'(h_n)M = X'$. Thus we may define U by $U((\ell(h_1)\dots\ell(h_n)1)\otimes\mu) = \ell'(h_1)\dots\ell'(h_n)\mu$ for $\mu\in M$. It is now easy to check the assertions of the lemma. Q.E.D.

An application of the preceding lemma is the following fact which is not new (see for instance [14]).

By the simplicity of the Cuntz-algebras there is a representation of $C^*(\ell(H))$ on X" such that $\ell(h)$ is mapped into $\ell'(h)|K"$ and using the above lemma we infer the existence of a representation of $C^*(\ell(H))$ on X such that $\ell(h)$ is mapped into $\ell'(h)$.

2.8. Consider $\tilde{L}:(\mathbb{C}\oplus H)\otimes T(H)\simeq T(H)\oplus(H\otimes T(H))\to T(H)$ given by $\tilde{L}=(I,L)$. Then we have $\tilde{L}(J\otimes I)\tilde{L}^*=I-LL^*=P_o$. Consider also

$$A=\begin{bmatrix} a_o & <.,h_1> \\ h_2 & A_1 \end{bmatrix} \varepsilon U(\mathbb{C},H)$$

where $a_o\varepsilon\mathbb{C}$, $h_1,h_2\varepsilon H, A_1\varepsilon L(H)$. Then

$$\tilde{L}(A^{-1}\otimes I)^*(J\otimes I)(A^{-1}\otimes I)\tilde{L}=P_o$$

and putting $\tilde{L}(A^{-1}\otimes I)^*=(X_o,X_1)$ we have

$$X_oX_o^*-X_1X_1^*=P_o.$$

Computing, we find

$$X_o=a_oI-\ell(h_2)$$

$$X_1=-\ell(h_1)^*L+L(A_1\otimes I).$$

Now from $A\varepsilon U(\mathbb{C},H)$ it follows that $|a_o|^2-||h_2||^2=1$, is invertible and we may define $L'=X_o^{-1}X_1$. We have

$$I-L'L'^*=X_o^{-1}P_oX_o^{*-1}=<\bullet,X_o^{-1}1>X_o^{-1}1.$$

But $X_o^{-1}1=a_o^{-1}\sum_{j=0}^{\infty}(a_o^{-1}\ell(h_2))^j1$ so that $||X_o^{-1}1||^2=|a_o|^2.1/(1-|a_o|^{-2}||h_2||^2)=1$. Thus L' is a partial isometry with dim Ker $L'^*=1$. Since L' is a norm-continuous function of A and $U(\mathbb{C},H)$ is connected in the norm-topology, by the continuity of the index for semi-Fredholm operators we infer that Ker $L'=0$.

From 2.7 we have that L' defines a homomorphism $C^*(\ell(H))\to L(T(H))$. Since

$$\ell'(h)=(a_oI-\ell(h_2))^{-1}(-<h,h_1>\ I+\ell(A_1h))$$

we infer that this homomorphism is actually an endomorphism $\alpha_A:C^*(\ell(H))\to$ $\to C^*(\ell(H))$. By straightforward computations we have $\alpha_A\ \alpha_B=\alpha_{AB}$ and since $\alpha_I=\mathrm{id}$ it follows that the α_A are automorphisms.

2.9. To construct the representations of $U(\mathbb{C},H)$ on $T(H)$ we consider $\mathrm{Alg}(\ell(H))$ the algebra (un-starred) generated by I and the $\ell(h)$'s and consider the map

$$\mathrm{Alg}(\ell(H))\ni x\to x1\varepsilon T(H)$$

which is injective and has dense range. We define U_A on the dense subset $\mathrm{Alg}(\ell(H))1$ by

$$U_Ax1=\alpha_A(x)(a_oI-\ell(h_2))^{-1}1.$$

To prove that U_A is isometric recall that $(a_oI-\ell(h_2))^{-1}1=X_o^{-1}1$,

$\|X_o^{-1}1\|=1$, $\mathbb{C}X_o^{-1}1=\mathrm{Ker}\, L'^*$ so that by the proof of Lemma 2.7 we have:

$$\langle U_A\ell(h_1)\ldots\ell(h_m))1, U_A(\ell(h_1')\ldots\ell(h_n')1\rangle=$$
$$=\langle \ell'(h_1)\ldots\ell'(h_m)X_o^{-1}1,\ \ell'(h_1')\ldots\ell'(h_n')X_o^{-1}\rangle=$$
$$=\delta_{m,n}\langle h_1,h_1'\rangle\ldots\langle h_n,h_n'\rangle$$

which clearly implies that U_A is isometric. Having proved that U_A is isometric, note that this implies the validity of the formula defining U_A for x in the norm-closure of $\mathrm{Alg}(\ell(H))$. Consider now also

$$B=\begin{bmatrix} b_o & \langle\bullet,k_1\rangle \\ k_2 & B_1 \end{bmatrix}.$$

Then,

$$U_AU_Bx1=U_A\alpha_B(x)(b_oI-\ell(k_2))^{-1}1=$$
$$=\alpha_A\alpha_B(x)(b_oI-\alpha_A(\ell(k_2)))^{-1}(a_oI-\ell(h_2))^{-1}1=$$
$$=\alpha_{AB}(x)((a_ob_o+\langle k_2,h_1\rangle)I-\ell(b_oh_2+A_1k_2))^{-1}1=U_{AB}x1.$$

Since $U_I=I$ it follows that U_A is unitary.

Because for $x_1,x_2\in\mathrm{Alg}(\ell(H))$ we have $U_Ax_1x_21=\alpha_A(x_1)U_Ax_21$ we infer that $U_AtU_A^{-1}=\alpha_A(t)$ for all $t\in C^*(\ell(H))$.

2.10. Summing up we have the following theorem

THEOREM. *There exist homomorphisms* $U(\mathbb{C},H)\ni A\to\alpha_A\in\mathrm{Aut}\, C^*(\ell(H))$ *and* $U(\mathbb{C},H)\ni A\to U_A\in U(T(H))$ *such that for*

$$A=\begin{bmatrix} a_o & \langle\bullet,h_1\rangle \\ h_2 & A_1 \end{bmatrix}\in U(\mathbb{C},H)$$

we have

$$\alpha_A(\ell(h))=(a_oI-\ell(h_2))^{-1}(-\langle h,h_1\rangle I+\ell(A_1h))$$

and

$$U_Ax1=\alpha_A(x)(a_oI-\ell(h_2))^{-1}1$$

for x *in the unstarred algebra* $\mathrm{Alg}(\ell(H))$ *generated by* $\ell(H)$. *Endowing* $U(\mathbb{C},H)$ *with the topology of* $*$*-strong convergence,* $\mathrm{Aut}\, C^*(\ell(H))$ *with point-norm convergence and* $U(T(H))$ *with* $*$*-strong convergence, these homomorphisms are continuous.*

§3

Associating with a complex Hilbert space the Canonical Anticommutation Relations C*-algebra one gets a functor from complex Hilbert spaces and contractions to C*-algebras and completely positive maps (see for instance [26]). What we shall do in this section can be viewed as an analogue, we construct a functor from real Hilbert spaces and contractions to C*-algebras and completely positive maps the C*-algebras associated having unique trace states if the Hilbert spaces have dimension >1, and the corresponding von Neumann algebras are isomorphic to the type II_1 factors of free groups. In particular this yields a new type of automorphic action of the orthogonal group O(n) on the factor of the free group on n generators.

3.1. Consider H a real Hilbert space and $H^{\mathbb{C}}=H\otimes\mathbb{C}=H+iH$ its complexification. Let $s:H\to L(T(H^{\mathbb{C}}))$, $d:H\to L(T(H^{\mathbb{C}}))$ be the $\mathbb{R}$-linear maps defined by

$$s(h)=1/2(\ell(h)+\ell(h)^*)$$
$$d(h)=1/2(r(h)+r(h)^*).$$

We have

$$[s(h),\ d(h)]=0.$$

The C*-algebra $C^*(s(H))$ generated by $s(H)$ can be also viewed as generated by $(s(e_\iota))_{\iota\in I}$ where $(e_\iota)_{\iota\in I}$ is an orthonormal basis of H. Identifying as in §2 the C*-algebra $C^*(\ell(H^{\mathbb{C}}))$ with the reduced free product of copies of $(C^*(S),\varepsilon)$ indexed by I, then $C^*(s(H))\subset C^*(\ell(H^{\mathbb{C}}))$ identifies with the reduced free product of copies of $(C^*(\mathrm{Re}\ S),\varepsilon)$ indexed by I, where $\mathrm{Re}\ S=1/2(S+S^*)$ and $\varepsilon_o=\varepsilon|C^*(\mathrm{Re}\ S)$. The only thing that must be remarked to this end, is that for the representation of $C^*(\mathrm{Re}\ S)\simeq C^*(s(\mathbb{R}))$ on $T(\mathbb{C})$ the vector 1 is cyclic. Clearly, $(C^*(s(\mathbb{R})))''=(C^*(d(\mathbb{R})))''$ is maximal abelian and 1 is also separating. The measure on the spectrum of Re S, corresponding to ε_o, is absolutely continuous with respect to Lebesgue measure (a computation of this measure has been included in §4). Thus $(C^*(s(\mathbb{R})))''$ has no atoms and $(C^*(s(H)))''$ when dim H>1, is isomorphic to the type II_1-factor of the free group on dim H generators. Also, in view of Lemma 1.8 we have $(C^*(s(H)))'=(C^*(d(H)))''$ and $(C^*(s(H)))''$ is in standard form on $T(H^{\mathbb{C}})$ with trace vector 1.

3.2. REMARK. Since $(C^*(s(H)))''$ has no minimal projections, we infer that $C^*(s(H))\cap K=0$ and hence $C^*(s(H))$ is isomorphic with the C*-subalgebra of the Cuntz-algebra of $H^{\mathbb{C}}$ generated by the real parts of the dim H isometries which generate the Cuntz-algebra.

3.3. Consider a contraction $T:H_1\to H_2$ between real Hilbert space and let $H_1^{\mathbb{C}},H_2^{\mathbb{C}},T^{\mathbb{C}}$ be the complexifications. Then there is a morphism

$$T(T^{\mathbb{C}}):(T(H_1^{\mathbb{C}}),1)\to(T(H_2^{\mathbb{C}}),1).$$

LEMMA. *We have*

$$T(T^{\mathbb{C}})(C^*(s(H_1)))''1\subset(C^*(s(H_2)))''1$$
$$T(T^{\mathbb{C}})(C^*(s(H_1)))1\subset(C^*(s(H_2)))1$$

and the corresponding maps

$\tilde{\Phi}(T):(C^*(s(H_1)))''\to(C^*(s(H_2)))''$

$\Phi(T):C^*(s(H_1))\to C^*(s(H_2))$

are completely positive and unital.

PROOF. Note that if the above assertions are proved for $T_k:H_k\to H_{k+1}$ (k=1,2) then they also hold for $T_2\circ T_1$. Thus it will be sufficient to prove the Lemma in two cases: a) T is isometric and b) $H_1\supset H_2$ and T is the orthogonal projection of H_1 onto H_2.

a) Assume T is isometric. Then $\mathcal{T}(T^{\mathbb{C}})$ implements an isomorphism between $(s(H_1))''$ (respectively $C^*(s(H_1))$) and $(s_{H_2}(TH_1))''|X$ (respectively $C^*(s(H_1))|X$) where $X=\mathcal{T}(T^{\mathbb{C}})\mathcal{T}(H_1^{\mathbb{C}})=\overline{(s_{H_2}(TH_1))''1}$. Indeed, we have

$(s_{H_2}(Th_1)|X)\mathcal{T}(T^{\mathbb{C}})=\mathcal{T}(T^{\mathbb{C}})s_{H_1}(h_1)$.

But $1\in X$ is separating for $(s_{H_2}(H_2))''$ so that

$(s_{H_2}(TH_1))''\ni x\to x|X$

is faithful. In particular this yields the inclusions to be proved and $\tilde{\Phi}$, Φ are the $*$-homomorphisms ($\tilde{\Phi}$ is moreover normal)) which take $s_{H_1}(h_1)$ to $s_{H_2}(Th_1)$ and hence clearly completely positive.

b) When T is the orthogonal projection of H_1 onto H_2 then $\mathcal{T}(T^{\mathbb{C}})$ is the orthogonal projection of $\mathcal{T}(H_1^{\mathbb{C}})$ onto $\mathcal{T}(H_2^{\mathbb{C}})$. Also, $(s_{H_1}(H_2))''$ and $(s_{H_2}(H_2))''$ are isomorphic and what must be proved about the existence and complete positivity of $\tilde{\Phi}$ are well-known facts about the existence of a conditional expectation of $(s(H_1))''$ onto $(s_{H_1}(H_2))''$ constructed using the given faithful trace on $(s(H_1))''$. It is easily seen that Φ takes the $*$-algebra generated by $s(H_1)$ onto the $*$-algebra generated by $s(H_2)$ and hence by the norm-continuity of Φ (which is a projection of norm one) we get the existence of Φ.

3.4. Thus we have constructed two functors $\tilde{\Phi}$ and Φ from real Hilbert spaces and contractions to C*-algebras and completely positive maps and we shall use from now on the notations $\tilde{\Phi}(H)=(s(H))''$, $\Phi(H)==C^*(s(H))$. Note also that $\tilde{\Phi}(T)$ is normal.

Remark also that the map

$(\mathcal{L}(H_1,H_2))_1\ni T\to\Phi(T)\in CP_1(\Phi(H_1),\Phi(H_2))$

where CP_1 denotes completely positive unital maps, is continuous when $(\mathcal{L}(H_1,H_2))_1$ is endowed with the strong topology and CP_1 with the topology of pointwise norm-convergence. Correspondingly for Φ, we have that the map

$(\mathcal{L}(H_1,H_2))_1\ni T\to\tilde{\Phi}(T)\in CP_1(\tilde{\Phi}(H_1),\tilde{\Phi}(H_2))$

is continuous when $(\mathcal{L}(H_1,H_2))_1$ is given the strong topology and

$CP_1(\mathfrak{F}(H_1),\mathfrak{F}(H_2))$ the topology of pointwise *-strong convergences. In particular we have continuity of the actions of the orthogonal group $O(H)$ on $\Phi(H)$ and $\mathfrak{F}(H)$ in the appropriate topologies.

3.5. For concrete computations with the completely positive maps $\Phi(T)$ it is necessary to be able to compute the map

$\Phi(H)\ni x\to x1\in\mathcal{T}(H^{\mathbb{C}})$.

To this end we shall use Gegenbauer polynomials $P_n(t)$ which correspond to $C_n^1(t)$ in Ch.IX, §3 of [43] or to $P_n^{(1)}(t)$ in Ch.IV, §4.7, of [42]. These polynomials are defined by the generating function identity

$$(1-2rt+r^2)^{-1}=\sum_{n\geq 0} P_n(t)r^n$$

for $|r|<1$ and $|t|\leq 1$. On the other hand the $P_n(t)$ are obtained by orthonormalization of the sequence $1,t,t^2,\ldots$ in $L^2([-1,1], \frac{2}{\pi}(1-t^2)^{1/2}dt)$.

LEMMA. *Let* $\{e_\iota\}_{\iota\in I}$ *be an orthonormal basis of* H. *Then for* $(\iota_1,\ldots,\iota_n)\in D_n(I)$ *and* $m_j>0$ $(1\leq j\leq n)$ *we have*

$$P_{m_1}(s(e_{\iota_1}))\ldots P_{m_n}(s(e_{\iota_n}))1=$$

$$=\underbrace{e_{\iota_1}\otimes\cdots\otimes e_{\iota_1}}_{m_1\text{-times}}\otimes\cdots\otimes\underbrace{e_{\iota_n}\otimes\cdots\otimes e_{\iota_n}}_{m_n\text{-times}}$$

PROOF. The measure on the spectrum of $s(e_\iota)$ corresponding to the vector-state ω_1 $(1\in\mathcal{T}(H^{\mathbb{C}}))$ is $\frac{2}{\pi}(1-t^2)^{1/2}dt$ on $[-1,1]$ (see Lemma 4.5). Hence $\{P_n(s(e_\iota))1\}_{n\geq 0}$ is the result of the orthonormalization of $\{(s(e_\iota))^n 1\}_{n\geq 0}$. On the other hand $(s(e_\iota))^n 1=\sum_{k=0}^{n}\alpha_k e_\iota^{\otimes k}$ with $\alpha_n>0$ and hence orthonormalization of $\{(s(e_\iota))^n 1\}_{n\geq 0}$ yields $\{e_\iota^{\otimes n}\}_{n\geq 0}$. The assertion of the Lemma follows now from $\Phi(H)=\bigstar_{\iota\in I}\Phi(\mathbb{R}e_\iota)$ as is easily seen.

Q.E.D.

3.6. We turn now to proving that if $\dim H>1$ then $\Phi(H)$ has a unique trace-state. Let τ_0 be the trace on $\Phi(H)$ corresponding to the trace-vector $1\in\mathcal{T}(H^{\mathbb{C}})$ and let τ be some other trace-state on $\Phi(H)$. Consider further $(e_\iota)_{\iota\in I}$ an orthonormal basis of H. The spectrum of $s(e_\iota)$ is $[-1,1]$ and the measure on $[-1,1]$ induced by τ_0 is $\varphi d\lambda$ where $d\lambda$ is Lebesgue-measure and $\varphi(t)=\frac{2}{\pi}(1-t^2)^{1/2}$ (see Lemma 4.5). Let then $\psi:[-1,1]\to[-\pi,\pi]$ be given by $\psi(t)=2(t(1-t^2)^{1/2}+\arcsin t)$ so that $u_\iota=\exp(i\psi(s(e_\iota)))$ the measure induced by τ_0 on the 1-torus $\mathbb{T}$ which is the spectrum of u_ι, coincides with Haar-measure. This together with the fact that $\Phi(H)$ is isomorphic to the reduced free product of the $C^*(s(e_\iota))$ implies that the C^*-subalgebra $B\subset\Phi(H)$ generated by the u_ι is isomorphic to the reduced C*-algebra of a free group on dim H generators, the unitaries corresponding to the generators being precisely the u_ι. It is known [36] that B has a unique trace-state and hence $\tau|B=\tau_o|B$.

Since $\exp(i\psi(\bullet))$ is injective on $[-1,1)$ the measure induced by τ or equivalently τ_o on the spectrum of u_ι has no atoms, we infer that for π the representation of $\Phi(H)$ associated with $1/2(\tau+\tau_o)$ we have $(\pi(B))''\ni\pi(s(e_\iota))$ and hence $(\pi(B))''=(\pi(\Phi(H)))''$. Since τ and τ_o yield ultraweakly-continuous functionals on $(\pi(\Phi(H)))''$ which coincide on $\pi(B)$, we infer that they coincide on $(\pi(B))''$ and hence $\tau=\tau_o$.

§4

Independent random-variables arise from tensor-products of commutative C*-algebras with specified states. We shall consider here the analogue of independent random-variables with the tensor product replaced by the free product. We will be mainly concerned with the addition problem for random-variables and we shall prove what may be viewed as the central limit theorem in this situation (Thm.4.8). The limit law is not the usual gaussian distribution but a certain modified arcsine-law.

4.1. For (A,φ) a unital C*-algebra with specified state, we shall view $a\in A$ as a random-variable. This will mean that we shall be interested in the distribution of a which will be interpreted as the functional $\xi_a:\mathbb{C}[X]\to\mathbb{C}$ given by $\xi_a(1)=1$, $\xi_a(X^n)=\varphi(a^n)$. The assumption that our "random variable" a is given by a self-adjoint element of A, will be imposed only towards the end of this section.

4.2. DEFINITION. *Let (M,φ) be a unital C*-algebra with specified state φ and let $1\in A_\iota\subset M(\iota\in I)$ be subalgebras. The family $(A_\iota)_{\iota\in I}$ will be called free if $\varphi(a_1,a_2\ldots a_n)=0$ whenever $(\iota_1,\ldots,\iota_n)\in D_n(I)$, $a_j\in A_{\iota_j}$ and $\varphi(a_j)=0$. A family of subsets $X_\iota\subset M$ (elements $a_\iota\in M$) will be called free if the family of subalgebras A_ι generated by $\{1\}\cup X_\iota$ (respectively $\{1,a_\iota\}$) is free.*

4.3. PROPOSITION. *If $\{a,b\}$ is a free pair of elements of (M,φ) then ξ_{a+b} and ξ_{ab} depend only on ξ_a and ξ_b. There are universal polynomials with integer coefficients $P_n(x_1,\ldots,x_n,y_1,\ldots,y_n)$ and $Q_n(x_1,\ldots,x_n,y_1,\ldots,y_n)$ such that considering x_j,y_j as having degree j we have:*

(i) *P_n is homogeneous of degree n in the x-and y-variables taken together,*

(ii) *Q_n is homogeneous of degree n both in the x-and in the y-variables,*

(iii) $\xi_{a+b}(X^n)=P_n(\xi_a(X),\ldots,\xi_a(X^n),\ \xi_b(X),\ldots,\xi_b(X^n))$,

(iv) $\xi_{ab}(X^n)=Q_n(\xi_a(X),\ldots,\xi_a(X^n),\ \xi_b(X),\ldots,\xi_b(X^n))$,

(v) $P_n(x_1,\ldots,x_n,y_1,\ldots,y_n)=P_n(y_1,\ldots,y_n,x_1,\ldots,x_n)$,

(vi) $Q_n(x_1,\ldots,x_n,\ y_1,\ldots,y_n)=Q_n(y_1,\ldots,y_n,x_1,\ldots,x_n)$

(vii) $\Sigma=\{\xi:\mathbb{C}[X]\to\mathbb{C}\mid \xi(1)=1,\ \xi \text{ linear}\}$ *is an abelian group for the operation*

$(\xi\boxplus\eta)(X^n)=P_n(\xi(X),\ldots,\xi(X^n),\eta(X),\ldots,\eta(X^n))$,

(viii) Σ *defined in* (vii) *is a commutative semigroup for the operation,*

$(\xi \boxtimes \eta)(X^n)=Q_n(\xi(X),\ldots,\xi(X^n),\eta(X),\ldots,\eta(X^n))$.

PROOF. Let A,B be the algebras (unstarred) generated by {1,a} and respectively {1,b}. If $a_1,\ldots,a_k\varepsilon A$ and $b_1,\ldots,b_k\varepsilon P$ then we have

$\varphi(a_1b_1\ \ldots\ a_kb_k)=$

$=\varphi((a_1'+\varphi(a_1)1)(b_1'+\varphi(b_1)1)\ldots(a_k'+\varphi(a_k)1)(b_k'+\varphi(b_k)1)$

where $a_j'=a_j-\varphi(a_j)1, b_j'=b_j-\varphi(b_j)1$.

Since

$\varphi(a_1'b_1'\ \ldots\ a_n'b_n')=0$

$\varphi(b_1'a_2'b_2'\ \ldots\ a_n'b_n')=0$

$\varphi(a_1'b_1'\ \ldots\ a_n')=0$

we get that $\varphi(a_1b_1\ \ldots\ a_kb_k)$ may be written as a sum of terms of the form

$\varphi(a_{i_1})\ldots\varphi(a_{i_r})\ \varphi(b_{j_1})\ldots\varphi(b_{j_t})\ \varphi(\tilde{a}_1\tilde{b}_1\ \ldots\ \tilde{a}_\ell\tilde{b}_\ell)$ with $\tilde{a}_j\varepsilon A,\tilde{b}_j\varepsilon B$ and and $\ell<k$. Iterating this procedure one gets $\varphi(a_1b_1\ \ldots\ a_kb_k)$ computed from $\varphi|A$ and $\varphi|B$. Applying this to $\varphi((a+b)^n)$ and $\varphi((ab)^n)$ on easily gets (i)-(v) after remarking that $\varphi((\lambda a+\lambda b)^n)=\lambda^n\varphi((a+b)^n)$ and $\varphi(((\lambda a)(\mu b))^n)=\lambda^n\mu^n\varphi((ab)^n)$. To prove (vi) we must show that $\varphi((ab)^n)=$ $=\varphi((ba)^n)$ which is a particular case of

$\varphi(a_1b_1\ \ldots\ a_nb_n)=\varphi(b_na_n\ \ldots\ b_1a_1)$

which in turn is easily proved using the inductive process described above and taking into account that A and B are commutative. Now (viii) also follows easily. To get (vii) note that $P_n(x_1\ \ldots\ x_n,y_1\ \ldots\ y_n)=$ $=x_n+y_n+\tilde{P}_n(x_1,\ldots,x_{n-1},y_1,\ldots,y_{n-1})$. Thus the inverse η of ξ is obtained by computing $\eta(X^n)$ recurrently, the neutral element being $\xi_o(1)=1$, $\xi_o(X^n)=0$ $(n\geq 1)$. Q.E.D.

PROPOSITION. *There are universal polynomials* $R_n(x_1,\ldots,x_n)$ *such that considering the* x_j *as having degree* j *we have*

(i) R_n *is homogeneous of degree* n,

(ii) $R_n(x_1,\ldots,x_n)=x_n+\tilde{R}_n(x_1,\ldots,x_{n-1})$,

(iii) $R_n((\xi\boxplus\eta)(X),\ldots,(\xi\boxplus\eta)(X^n))=$

$=R_n(\xi(X),\ldots,\xi(X^n))+R_n(\eta(X),\ldots,\eta(X^n))$.

PROOF. The operation

$(x_1,\dots,x_n) ⊞_n (y_1,\dots,y_n)=$

$=(P_1(x_1,y_1),P_2(x_1,x_2,y_1,y_2),\dots,P_n(x_1,\dots,x_n,y_1,\dots,y_n))$

where $P_1,\dots,P_n$ are the polynomials defined in Proposition 4.3 turns $\mathbb{C}^n$ into a commutative algebraic group. As usual, considering the exponential map $\exp:\mathbb{C}^n\to\mathbb{C}^n$ where the first $\mathbb{C}^n$ is the Lie algebra of the second, one easily gets that $\exp(b_1,\dots,b_n)=(E_1(b_1),\dots,E_n(b_1,\dots,b_n))$ where

$$\partial E_j/\partial b_k=\partial P_j/\partial x_k\ (\underbrace{0,\dots,0}_{j\text{-times}},\ E_1,\dots,E_j)\ (1\le k\le j)\quad E_j(0,\dots,0)=0.$$

Since $\partial P_j/\partial x_j=1$, one easily gets recurrently that E_j is a polynomial in $b_1,\dots,b_j$

$$E_j(b_1,\dots,b_j)=b_j+\tilde{E}_j(b_1,\dots,b_{j-1})$$

and E_j is homogeneous of degree j when b_k is assigned degree k. Since $E_j=b_j+\tilde{E}_j$ it follows that the inverse of the exponential map is a map of the form $(R_1(x_1),\dots,R_n(x_1,\dots,x_n))$ with R_j satisfying (i) and (ii). The exponential map being an isomorphism between the groups $(\mathbb{C}_1^n,+)$ $(\mathbb{C}^n,⊞)$ we infer that (iii) also holds. Q.E.D.

4.5. The following lemma is clearly not new, it has been included for the sake of completeness and because the proof may be new.

LEMMA. *Let* $(C^*(S),\varepsilon)$ *be the* C^**-algebra generated by the non-unitary isometry* S *with the state* ε , *such that* $\varepsilon([S^*,S])=1$. *Let further* μ *denote the measure on* $\mathbb{R}$ *such that*

$$\varepsilon(f(\mathrm{Re}S))=\int f(t)\,d\mu(t)$$

for f *continuous functions on* $\mathbb{R}$. *Then we have*

$$d\mu/d\lambda(t)=\chi(t)$$

where $d\lambda$ *denotes Lebesgue measure and*

$$\chi(t)=\begin{cases}0 & \text{if } |t|\ge 1\\ \frac{1}{\pi}\sqrt{1-t^2} & \text{if } |t|<1\ .\end{cases}$$

PROOF. The Lemma follows from a more general fact about operators T with trace-class self-commutator $[T^*,T]$. Let ν be the Helton-Howe-measure of T on $\mathbb{R}^2$([23],[9]) and let ω be the measure on $\mathbb{R}$ which is the image of ν via the projection of $\mathbb{R}^2$ onto the first $\mathbb{R}$-factor. The Helton-Howe formula, gives for a polynomial F:

$$\mathrm{Tr}[F(\mathrm{Re}\ T),T]=\int_{\mathbb{R}^2} F'(x)\,d\nu(x,y)=\int_{\mathbb{R}} F'(x)\,d\omega(x)$$

and since

$$\mathrm{Tr}[(\mathrm{Re}\ T)^n,T]=\frac{1}{2}\sum_{k=0}^{n}\mathrm{Tr}((\mathrm{Re}T)^{n-k}[T^*,T](\mathrm{Re}\ T)^{n-k-1})=\frac{n}{2}\mathrm{Tr}((\mathrm{Re}\ T)^{n-1}[T^*,T])$$

we infer

$$\mathrm{Tr}[F(\mathrm{Re}\ T),T]=\frac{1}{2}\ \mathrm{Tr}(F'(\mathrm{Re}\ T)[T^*,T]$$

and hence

$$\mathrm{Tr}(F'(\mathrm{Re}\ T)[T^*,T])=2\int_{\mathbb{R}} F'(x)d\omega(x)dx.$$

The lemma follows from this equality applied to T=S after remarking that $\varepsilon(X)=\mathrm{Tr}(X[S^*,S])$ and taking into account that the Helton-Howe-measure of S is Lebesgue measure multiplied by $-(2\pi)^{-1}$ index $(S-(x+iy))$. Q.E.D.

4.6. Returning to the context of Proposition 4.4 for a functional ξ on $\mathbb{C}X$ such that $\xi(1)=1$, we shall call $R(\xi)=(R_1(\xi(X)),R_2(\xi(X),\xi(X^2))\ldots)$ its R-transform and we shall write $R_n(\xi)$ for $R_n(\xi(X),\ldots\ldots,\xi(X^n))$.

For $r\in\mathbb{C}$ let d_r denote the automorphism of $\mathbb{C}[X]$ such that $d_r(X)=rX$. Because of (i) in Proposition 4.4 we have $R_n(\xi\circ d_r)=r^n R_n(\xi)$. This implies also a corresponding relation for the action of $\mathbb{R}$ by dilations on the probability measures on $\mathbb{R}$.

4.7. PROPOSITION. (i) *For δ_t the probability-measure concentrated at $\{t\}\subset\mathbb{R}$ we have*

$$R_n(\delta_t)=\begin{cases} t & \text{if } n=1 \\ 0 & \text{if } n\geq 2; \end{cases}$$

(ii) *For the measure μ considered in Lemma 4.3 we have*

$$R_n(\mu)=\begin{cases} 0 & \text{if } n=1 \text{ or } n\geq 3 \\ 1/4 & \text{if } n=2. \end{cases}$$

PROOF. (i) Since $R_1(x_1)=x_1$ it is easy to see that $R_1(\delta_t)=t$. To see that $R_n(\delta_t)=0$ for $n\geq 2$, consider the free family $\{tI,tI\}$ which gives

$$2^n R_n(\delta_t)=R_n(\delta_{2t})=R_n(\delta_t)+R_n(\delta_t)=2R_n(\delta_t)$$

so that $(2^n-2)R_n(\delta_t)=0$.

(ii) It is easy to see that $R_2(x_1,x_2)=x_2-x_1^2$ and $\varepsilon(\mathrm{Re}\ S)=0$, $\varepsilon((\mathrm{Re}\ S)^2)=1/4$ so that $R_2(\mu)=1/4$.

To prove that $R_n(\mu)=0$ if $n\neq 2$, we shall consider as in §3 the real Hilbert space $\mathbb{R}^2$ with orthonormal basis $\{e_1,e_2\}$ and the algebra $\Phi(\mathbb{R}^2)=C^*(s(\mathbb{R}^2))$ with unique trace-state τ. Then $\{s(e_1),s(e_2)\}$ is a free family and by the results in §3 the automorphism $\Phi(\begin{bmatrix} 1/\sqrt{2} & 1/\sqrt{2} \\ -1/\sqrt{2} & 1/\sqrt{2}\end{bmatrix})$ transforms $s(e_1)$ into $s(1/\sqrt{2}\ (e_1+e_2))$ so that $\xi_{s(e_1)}=\xi_{s(2^{-1/2}(e_1+e_2))}$. Thus we shall have $R_k(\xi_{s(e_1)})=2R_k(\xi_{2^{-1/2}s(e_1)})=2^{1-k/2}R_k(\xi_{s(e_1)})$

and hence $R_k(\xi_{s(e_1)})=0$ for $k\neq 2$. In view of Lemma 4.5 this concludes the proof of (ii). Q.E.D.

4.8. THEOREM. *Let* $(A,\varphi)=\underset{n\in\mathbb{N}}{\bigstar}(A_n,\varphi_n)$ *and let* $a_n\in A_n$ *be such that* $a_n=a_n^*$, $\varphi_n(a_n)=0$, $\|a_n\|<C$ *and assume that* $\lim_{n\to\infty} n^{-1}(\varphi(a_1^2)+\ldots+\varphi_n(a_n^2))=\alpha^2/4$. *Consider* $S_n=n^{-1/2}(\sigma_1(a_1)+\ldots\sigma_n(a_n))$ *and* μ_n *the measure on* $\mathbb{R}$ *defined by*

$$\varphi(g(\alpha^{-1}S_n))=\int g\,d\mu$$

for g *continuous functions on* $\mathbb{R}$. *Then we have* $\|S_n\|\leq 2C$ *and the measures* μ_n *converge weakly to the measure* μ *considered in Lemma 4.5.*

PROOF. As in §1 consider $\underset{n\in\mathbb{N}}{\bigstar}(H_n,\xi_n)=(H,\xi)$ the Hilbert space used to construct (A,φ) and P_o the projection of H onto $\mathbb{C}\xi$. Let $H_n^o=H_n\ominus\mathbb{C}\xi_n$ and consider the matrix representation of a_n with respect to the decomposition $H_n=\mathbb{C}\xi_n\oplus H_n^o$

$$a_n=\begin{bmatrix}0 & X_n\\ X_n^* & Y_n\end{bmatrix}$$

and let

$$b_n=\begin{bmatrix}0 & 0\\ X_n^* & Y_n\end{bmatrix}\qquad c_n=\begin{bmatrix}0 & X_n\\ 0 & 0\end{bmatrix}.$$

Then we have $m\neq n\Longrightarrow(\lambda_m(b_m))^*\lambda_n(b_n)$ and $\lambda_m(c_m)(\lambda_n(c_n))^*=0$. This implies

$$\Big\|\sum_{k=1}^n\sigma_k(a_k)\Big\|\leq\Big\|\sum_{k=1}^n\lambda_k(b_k)\Big\|+\Big\|\sum_{k=1}^n\lambda_k(c_k^*)\Big\|\leq$$

$$\leq\Big(\sum_{k=1}^n\|\lambda_k(b_k)\|^2\Big)^{1/2}+\Big(\sum_{k=1}^n\|\lambda_k(c_k^*)\|^2\Big)^{1/2}\leq 2n^{1/2}C,\text{ and hence } S_n\leq 2C.$$

To prove that the measures μ_n converge weakly to μ it will be sufficient to prove that

$$\lim_{n\to\infty}\varphi((\alpha^{-1}S_n)^k)=\int t^k d\mu \text{ for all } k\geq 1.$$

In view of (ii) in Proposition 4.4 and of (ii) in Proposition 4.7 this is equivalent to proving that

$$\lim_{n\to\infty}R_k(\xi_{\alpha^{-1}S_n})=\begin{cases}0 & \text{if } k\neq 2\\ 1/4 & \text{if } k=2.\end{cases}$$

We have

$$R_1(\xi_{\alpha^{-1}S_n})=\alpha^{-1}n^{-1/2}(\varphi(\sigma_1(a_1))+\ldots+\varphi(\sigma_n(a_n)))=$$
$$=\alpha^{-1}n^{-1/2}(\varphi_1(a_1)+\ldots+\varphi_n(a_n))=0$$

$$\lim_{n\to\infty} R_2(\xi_{\alpha^{-1}s_n})=\lim_{n\to\infty} \alpha^{-2}n^{-1}(R_2(\xi_{a_1})+\ldots+R_2(\xi_{a_n}))=$$

$$=\lim_{n\to\infty} \alpha^{-2}n^{-1}(\varphi_1(a_1^2)+\ldots+\varphi_n(a_n^2))=1/4$$

and for $k\geq 3$

$$\overline{\lim_{n\to\infty}} |R_k(\xi_{\alpha^{-1}s_n})|=$$

$$=\overline{\lim_{n\to\infty}} n^{-k/2}\alpha^{-k}|R_k(\xi_{a_1})+\ldots+R_k(\xi_{a_n})|\leq$$

$$\leq\lim_{n\to\infty} n^{1-k/2}\alpha^{-k}C^kM_k=0$$

where M_k is a constant such that

$|R_k(\xi_a)|\leq M_k||a||^k$. Q.E.D.

§5

This section is a brief discussion of the definition of reduced free products with amalgamation and the corresponding generalization of the Hilbert space free products to the case of Hilbert modules over C*- algebras ([37], [30]). For Hilbert modules we shall use the terminology of ([27], see especially 1.10-1.16 and 2.8).

5.1. We consider first the generalization of the Hilbert space free product. Let B be a unital C*-algebra and let $H_\iota=B\oplus H_\iota^o$ be a Hilbert (right) B-module, where B is endowed with the inner product $\langle b_1,b_2\rangle=b_1^*b_2$. By $\xi_\iota\in H_\iota$ we shall denote $1_B\oplus 0\in B\oplus H_\iota^o$. We assume moreover that unital *-homomorphisms $\chi_\iota:B\to L(H_\iota)$, $\chi_\iota^o:B\to L(H_\iota^o)$ are given, so that $\chi_\iota(b')(b\oplus H)=b'b\oplus\chi_\iota^o(b')b$.

We define $(H,\xi)=\mathop{\ast}\limits_{\iota\in I}(H_\iota,\xi_\iota)$ by

$$H=B\oplus \bigoplus_{n=1}^{\infty}\bigoplus_{(\iota_1,\ldots,\iota_n)\in D_n(I)}(H_{\iota_1}^o\otimes_B\ldots\otimes_B H_{\iota_n}^o)=B\oplus H^o$$

$$\xi=1_B\oplus 0\in B\oplus H^o=H.$$

5.2. Let X,Y,Z be Hilbert B-modules and assume there are unital *-homomorphisms $\chi:B\to L(Y)$, $\chi':B\to L(Z)$ and let $T\in L(Y,Z)$ be such that $T\chi(b)=\chi'(b)T$. Then $I\otimes T\in L(X\otimes_B Y,X\otimes_B Z)$. Indeed, for ξ,η in the algebraic tensor product of X with Y and respectively Z over B, it is easy to check that

$$\langle(I\otimes T)\xi,\eta\rangle=\langle\xi,(I\otimes T^*)\eta\rangle$$

and

$$\langle(I\otimes T)\xi,(I\otimes T)\xi\rangle\leq||T||^2\langle\xi,\xi\rangle$$

by reducing the proof of the inequality (like in 5.3 of [37]) to that of

$$\langle(I\otimes S)\xi,(I\otimes S)\xi\rangle\geq 0$$

where $S=(\|T\|^2-T^*T)^{1/2}$ is in the commutant of $\chi(b)$.

5.3. Consider $H_\iota=B\oplus H_\iota^o$, $\xi_\iota=1_B\oplus 0\in H_\iota$, χ_ι,χ_ι^o as in 5.1 and consider also H_ι', $H_\iota'^o,\xi_\iota',\chi_\iota',\chi_\iota'^o$ satisfying the same assumptions. Let further $T_\iota\in L(H_\iota,H_\iota')$, $\|T_\iota\|\le 1$, $T_\iota\xi_\iota=\xi_\iota'$. It is easily seen that $T_\iota H_\iota^o\subset H_\iota'^o$ (use for instance the general Lemma 2.4 of [37]) and put $T_\iota^o=T_\iota|H_\iota^o$. Then combining 5.2 with (2.8 of [27]) we easily get that

$$T_{\iota_1}^o\otimes\ldots\otimes T_{\iota_n}^o:H_{\iota_1}^o\otimes_B\ldots\otimes_B H_{\iota_n}^o\to H_{\iota_1}'\otimes_B\ldots\otimes_B H_{\iota_n}'$$

for $(\iota_1,\ldots,\iota_n)\in D_n(I)$ is well-defined and in $L(H_{\iota_1}^o\otimes_B\cdots\otimes_B H_{\iota_n}^o,$ $H_{\iota_1}'^o\otimes_B\cdots\otimes_B H_{\iota_n}'^o)$. Thus $T=\underset{\iota\in I}{\bigstar}T_\iota$ is well-defined and in $L(H,H')$, where $H=\underset{\iota\in I}{\bigstar}H_\iota$, $H'=\underset{\iota\in I}{\bigstar}H_\iota'$. Note also that $T\chi(b)=\chi'(b)T$.

5.4. Consider

$$H(\ell,\iota)=B\oplus\bigoplus_{n=1}^{\infty}\bigoplus_{(\iota_1,\ldots,\iota_n)\in D_n(I)}(H_{\iota_1}^o\otimes_B\cdots\otimes_B H_{\iota_n}^o)\subset H.$$

Then using ([27], 2.8) we see there are isomorphisms

$V_\iota:H\to H_\iota\otimes_B H(\ell,\iota)$

such that

$$V_\iota(h_1\otimes\ldots\otimes h_n)=\begin{cases}h_1\otimes(h_2\otimes\ldots\otimes h_n) & \text{if } \iota_1=\iota,\ n\ge 2\\ h_1\otimes\xi & \text{if } \iota_1=\iota,\ n\ge 1\\ \xi_\iota\otimes(h_1\otimes\ldots\otimes h_n) & \text{if } \iota_1\ne\iota\end{cases}$$

$V_\iota\xi=\xi_\iota\otimes\xi$

where $h_{\iota_j}\in H_{\iota_j}^o$ and $(\iota_1,\ldots,\iota_n)\in D_n(I)$. Thus we may define

$\lambda_\iota:L(H_\iota)\to L(H)$

by

$\lambda_\iota(R)=V_\iota^{-1}(R\otimes I)V_\iota$.

5.5. Consider (A_ι,Φ_ι) where $A_\iota\supset B\ni 1_{A_\iota}$ and $\Phi_\iota:A_\iota\to B$ is a projection of norm one of the C*-algebra A_ι onto the C*-subalgebra B. We shall perform the analogue of the Gelfand-Segal construction for (A_ι,Φ_ι) (see 3.2 - 3.4 of [37]). By H_ι we denote the separation and completion of A_ι with respect to $\|\Phi_\iota(a^*a)\|^{1/2}$ and consider H_ι as a right B-module, the module structure being obtained from the right B-module structure of A_ι. Moreover the B-valued inner product $\langle a_1,a_2\rangle=\Phi_\iota(a_1^*a_2)$ on A_ι yields an inner product on H_ι, so that H_ι is a Hilbert B-module. Note that $A_\iota=B+\mathrm{Ker}\Phi_\iota$ the sum being a direct sum of right B-modules and it is easily seen that after separation and completion we get a corresponding orthogonal direct sum of right Hilbert B-modules $H_\iota=B\oplus H_\iota^o$. On the other hand, left multiplication on A_ι yields a unital *-homomorphism $\pi_\iota:A_\iota\to L(H_\iota)$ and for $\chi_\iota=\pi_\iota|B$ we have that $\chi_\iota(b')(b\oplus h)=b'b\oplus\chi_\iota^o(b')h$. Thus H_ι satisfies the condition specified in 5.1 and the $*$-homomorphism π_ι is such that $<\xi,\pi_\iota(a)\xi_\iota>=\Phi_\iota(a)$.

Passing to $(H,\xi)=\underset{\iota\in I}{\ast}(H_\iota,\xi_\iota)$ *we define* $\underset{\iota\in I}{\ast}(A_\iota,\Phi_\iota)$ *the reduced free product with amalgamation of the* (A_ι,Φ_ι) *as the C*-algebra A generated by* $\bigcup_{\iota\in I}(\lambda_\iota\circ\pi_\iota)(A_\iota)$ *in* $L(H)$ *and* $\Phi(a)=<\xi,a\xi>\in B$.

To see that this definition makes sense we have to remark that $(\lambda_\iota\circ\pi_\iota)(b)=(\lambda_{\iota'}\circ\pi_{\iota'})(b)$ for $\iota,\iota'\in I$ so, there is a natural identification of B with a subalgebra of A and it is easily seen that Φ is a conditional expectation.

5.6. With the notation of 5.5 define the $*$-homomorphism $\sigma_\iota:A_\iota\to A$ by $\sigma_\iota(a)=(\lambda_\iota\circ\pi_\iota)(a)$. Then, like in 1.5 we have that

a) B identifies with a subalgebra of A, so that $A\supset B\ni 1_A$ and $\sigma_\iota(b)=b$ for all $b\in B$ and $\iota\in I$. Moreover A is generated by $\bigcup_{\iota\in I}\sigma_\iota(A_\iota)$.

b) $\Phi\circ\sigma_\iota=\Phi_\iota$.

c) for $(\iota_1,\ldots,\iota_n)\in D_n(I)$ and $a_j\in\mathrm{Ker}\,\Phi_{\iota_j}$ we have $\Phi(\sigma_{\iota_1}(a_1)\ldots\sigma_{\iota_n}(a_n))=0$.

d) if $c\in A$ is such that $\Phi(a^*c^*ca)=0$ for all $a\in A$ then $c=0$.

Also, like in 1.5, being given (A,Φ) where Φ is a conditional expectation of A onto B and being given unital $*$-homomorphisms $\sigma_\iota:A_\iota\to A$ satisfying a)-d), then (A,Φ) is the reduced free product with amalgamation of the (A_ι,Φ_ι). The proof of this, like in the case $B=\mathbb{C}1_A$, consists in remarking that A, with the B-valued scalar product derived from Φ, yields after separation and completion a Hilbert B-module naturally isomorphic to H and the representation of A on this Hilbert module is faithful in view of d). Thus the norm on A can be reconstructed from this representation.

§6

In this section we study the representation of $U(\mathbb{C},H)$ on $T(H)$ constructed in §2. When dim $H<\infty$, the representation decomposes into a direct sum of irreducible representations with extreme weights, so that we are led to study the primitive vectors in $T(H)$. This also shows that when H is infinite-dimensional the irreducible representations which appear are parametrized by their primitive vectors and the study of these reduces to the case when dim $H<\infty$. Thus our main concern will be the case dim $H<\infty$.

6.1. Let

$$A=\begin{bmatrix} a_o & <\bullet,h_1> \\ h_2 & A_1 \end{bmatrix}\in U(\mathbb{C},H).$$

For $k_j\in H, 1\le j\le m$ we have

$$U_A k_1 \otimes \ldots \otimes k_m = U_A \ell(k_1)\ldots\ell(k_m)1=$$
$$=\alpha_A(\ell(k_1)\ldots\ell(k_m))(a_o I-\ell(h_2))^{-1}1=$$
$$=(a_o I-\ell(h_2))^{-1}(-<k_1,h_1>I+\ell(A_1k_1))\cdot$$
$$\cdot(a_o I-\ell(h_2))^{-1}(-<k_2,h_1>I+\ell(A_1k_2))\cdot$$
$$\cdots\cdots\cdots\cdots\cdots\cdots$$
$$\cdot(a_o I-\ell(h_2))^{-1}(-<k_m,h_1>I+\ell(A_1k_m))(a_o I-\ell(h_2))^{-1}1.$$

In particular when A is of one of the following forms:

$$A'=\begin{bmatrix}\lambda & 0\\ 0 & \lambda I\end{bmatrix}\qquad A''=\begin{bmatrix}\lambda & 0\\ 0 & A_1\end{bmatrix}\qquad A'''=\begin{bmatrix}(\det A_1)^{-1} & 0\\ 0 & A_1\end{bmatrix}$$

we have correspondingly:

$$U_{A'}k_1\otimes\ldots\otimes k_m=\lambda^{-1}k_1\otimes\ldots\otimes k_m$$
$$U_{A''}k_1\otimes\ldots\otimes k_m=\lambda^{-m-1}A_1k_1\otimes\ldots\otimes A_1k_m$$
$$U_{A'''}k_1\otimes\ldots\otimes k_m=(\det A_1)^{m+1}A_1k_1\otimes\ldots\otimes A_1k_m.$$

6.2. The formula for $U_A k_1\otimes\ldots\otimes k_m$ given in the preceding section shows that $U_A k_1\otimes\ldots\otimes k_m$ as a function of A extends as a complex analytic function to the set of those A for which $|a_o|>\|h_2\|$. Thus we may use the formula in 6.1 for the computation of dU(X) where X is in $gl(\mathbb{C}\oplus H)$ the complexification of the Lie algebra $u(\mathbb{C},H)$. This gives

$$dU(\begin{bmatrix}0 & 0\\ h_2 & 0\end{bmatrix})k_1\otimes\ldots\otimes k_m=$$
$$=h_2\otimes k_1\otimes\ldots\otimes k_m+k_1\otimes h_2\otimes\ldots\otimes k_m+$$
$$+\ldots+k_1\otimes\ldots\otimes h_2\otimes k_m+k_1\otimes\ldots\otimes k_m\otimes h_2$$

$$dU(\begin{bmatrix}0 & <\cdot,h_1>\\ 0 & 0\end{bmatrix})k_1\otimes\ldots\otimes k_m=$$
$$=-\sum_{j=1}^{m}<k_j,h_1>\,k_1\otimes\ldots\otimes k_{j-1}\otimes k_{j+1}\otimes\ldots\otimes k_m$$

6.3. Consider the unitary

$$\Theta(k_1\otimes\ldots\otimes k_m)=k_m\otimes\ldots\otimes k_1$$

so that $\Theta\ell(h)\Theta=r(h)$. It is easily seen from 6.2 that Θ commutes with dU(X) for $X\in gl(\mathbb{C}\oplus H)$ and hence $[U_A,\Theta]=0$ for $A\in U(\mathbb{C},H)$. This implies that U_A implements an automorphic representation of $U(\mathbb{C},H)$ not only on $C^*(\ell(H))$, but also on $C^*(r(H))$ and $U_A r(h)U_A^{-1}=\Theta\alpha_A(\ell(h))\Theta$.

6.4. Assume $\dim H=n<\infty$ and let $f_1,\ldots,f_n$ be an orthonormal basis of H. Consider the subgroup $K \subset U(1,n)$ consisting of elements of the form

$$A=\begin{bmatrix} a_o & 0 \\ 0 & A_1 \end{bmatrix}.$$

From the formulae in 6.1 we easily infer that restricting the representation of $U(1,n)$ on $\mathcal{T}(H)$ to K, the representation so obtained decomposes into the direct sum of irreducible representations, each irreducible representation having finite multiplicity. This implies that *the representation of* $U(1,n)$ *on* $\mathcal{T}(H)$ *is a discrete direct sum of irreducible representations each irreducible representation having finite multiplicity.*

6.5. Let $\mathcal{DT}(H) \subset \mathcal{T}(H)$ be the linear span of $\{1\} \cup \bigcup_{n=1}^{\infty} H^{\otimes n}$. Then, $\mathcal{DT}(H)$ coincides with the space of K-finite vectors for the representation of $U(n,1)$ on $\mathcal{T}(H)$. In view of 6.4 we have a decomposition

$$\mathcal{DT}(H)= \bigoplus_{\alpha \varepsilon J} X_\alpha$$

where the direct sum is orthogonal but does not involve completion and the representation of the universal enveloping algebra of $gl(n+1)$ is algebraically completely irreducible on each X_α (see 4.5.5.4 in [45]). Note also that each X_α is the direct sum of the $X_{\alpha,m}=X_\alpha \cap H^{\otimes m}$.

6.6. With the notation $f_o=1$ we have a basis $f_o, f_1, \ldots, f_n$ for $\mathbb{C} \oplus H$ and writing the elements of $gl(n+1)$ as matrices with respect to this basis we consider the Cartan subalgebra

$$\underline{j}=\{(a_{ij})_{0\le i,j\le n} \varepsilon gl(n+1) \mid a_{ij} \neq 0 \Rightarrow i=j\}$$

the Borel subalgebra

$$\underline{b}=\{(b_{ij})_{0\le i,j\le n} \varepsilon gl(n+1) \mid b_{ij} \neq 0 \Rightarrow i \le j\}$$

and the nilpotent radical

$$\underline{n}=[\underline{b},\underline{b}]=\{(b_{ij})_{0\le i,j\le n} \varepsilon gl(n+1) \mid b_{ij} \neq 0 \Rightarrow i<j\}.$$

The corresponding subgroups of $GL(n+1,\mathbb{C})$ will be denoted by J,B,N.

Let $m(\alpha)=\min\{m\ge 0 \mid X_{\alpha,m} \neq 0\}$. The formulae in 6.2 and 6.1 imply that $dU(\underline{b})X_{\alpha,m(\alpha)} \subset X_{\alpha,m(\alpha)}$. Hence by Lie's theorem we infer the existence of a highest weig $\Lambda_\alpha : \underline{b} \to \mathbb{C}$, $\Lambda_\alpha[\underline{b},\underline{b}])=0$ and of a $\underline{b}$-primitive element of weight Λ_α in X_α. Then the representations $U|\bar{X}_\alpha$ and $U|\bar{X}_\beta$ are unitarily equivalent if and only if $\Lambda_\alpha=\Lambda_\beta$ (see 2.4.1.4 and 4.5.5.3 in [45]) or equivalently $\Lambda_\alpha|\underline{j}= =\Lambda_\beta|\underline{j}$. Thus the problem of finding the irreducible representations into which U decomposes reduces to finding the $\underline{b}$-primitive element in $\mathcal{DT}(H)$ and the corresponding highest weights.

6.7. Consider $gl(n) \subset gl(n+1)$ where

$$gl(n)=\{(a_{ij})_{0\le i,j\le n}\in(gl(n+1) \mid i=0 \text{ or } j=0 \Longrightarrow a_{ij}=0\}$$

and let

$$\underline{b}_1=\underline{b} \cap gl(n),\ \underline{n}_1=\underline{n} \cap gl(n).$$

Let further T_h for $h\in H$ denote the map

$$dU\left(\begin{bmatrix} 0 & <\bullet,h> \\ & \\ 0 & 0 \end{bmatrix}\right): \mathcal{DT}(H)\to\mathcal{DT}(H),\text{ i.e.}$$

$$T_h k_1\otimes\ldots\otimes k_m=-\sum_{j=1}^{m} <k_j,h>\ k_1\otimes\ldots\otimes k_{j-1}\otimes k_{j+1}\otimes\ldots\otimes k_m.$$

We shall also use the notation $T_j=T_{f_j}$ where $1\le j\le n$.

A vector $\xi\in H^{\otimes m}$ is $\underline{b}$-primitive if and only if it is $\underline{b}_1$-primitive, and $T_1\xi=\ldots=T_n\xi=0$.

For $V\in U(n)$ let $\Gamma(V)=U_A$ where $A=\begin{bmatrix} 1 & 0 \\ 0 & V \end{bmatrix}$ i.e.

$$\Gamma(V)k_1\otimes\ldots\otimes k_m=Vk_1\otimes\ldots\otimes Vk_m.$$

We have $\Gamma(V)T_h\Gamma(V^{-1})=T_{Vh}$.

The determination of $\underline{b}_1$-primitive vectors in $\bigcap_{h\in H}\mathrm{Ker}(T_h|H^{\otimes m})$ is equivalent to decomposing the restriction of Γ to $\bigcap_{h\in H}\mathrm{Ker}(T_h|H^{\otimes m})$ into irreducible representations of $U(n)$.

6.8. It is useful to note, that T_h is a derivation of $\mathcal{DT}(H)$ viewed as the tensor-algebra over H, so that in particular if ξ_1,ξ_2 are $\underline{b}$-primitive corresponding to highest weights Λ_1,Λ_2 then $\xi_1\otimes\xi_2$ in also $\underline{b}$-primitive with highest weight $\Lambda_1+\Lambda_2$. Similarly $\bigcap_{h\in H}\mathrm{Ker}T_n$ is a subalgebra of $\mathcal{DT}(H)$.

6.9. It will be useful to consider also further reformulations of the problem of finding the $\underline{b}$-primitive vectors and their highest weights.

Thus, let

$$\Delta_m: H^{\otimes m}\to H^{\otimes m}$$

be defined by

$$\Delta_m\xi=\sum_{j=1}^{n} f_j\otimes T_j\xi.$$

Then we have $\mathrm{Ker}\Delta_m=\bigcap_{j=1}^{n}\mathrm{Ker}(T_j|H^{\otimes m})=\bigcap_{h\in H}\mathrm{Ker}(T_j)|H^{\otimes m})$.

Note also that

$$\Delta_m k_1\otimes\ldots\otimes k_m=-\sum_{j=1}^{m} k_j\otimes k_1\otimes\ldots\otimes k_{j-1}\otimes k_{j+1}\otimes\ldots\otimes k_m$$

so that Δ_m is in the commutant of $\Gamma(U(n))|H^{\otimes m}$. Denoting by τ_n the representation of the symmetric group $\mathfrak{S}(m)$ on $H^{\otimes m}$, we have

$$-\Delta_m = \sum_{j=1} \tau_n((1,2,\ldots,j)).$$

Denote also by τ_n the representation of $C^*(\mathfrak{S}(m))$ on $H^{\otimes m}$. The kernel of τ_n is known, so that our problem actually amounts to determine the largest projection $P\in C^*(\mathfrak{S}(m))$ such that $\alpha_m P=0$ where $\alpha_m = \sum_{j=1}^{m} (1,2,\ldots,j)\in C^*(\mathfrak{S}(m))$. Of course we would then have to decompose P into minimal projections and to determine their central support. Recall that $\tau_n(Q)\neq 0$ for a minimal projection Q, if and only if the signature $m_1\geq m_2\geq\ldots\geq m_m$ corresponding to its central support is such that $m_{n+1}=\ldots=m_m=0$, so that taking into account n=dim H once the problem in $C^*(\mathfrak{S}(m))$ is solved, is an easy operation.

6.10. The C^*-algebra $C^*(\mathfrak{S}(m))$ has an automorphism Θ such that $\Theta\sigma=\varepsilon(\sigma)\sigma$ where σ is a permutation and $\varepsilon(\sigma)$ its sign. Thus the problem we are studying for α_m is equivalent with the same problem for $\Theta\alpha_m = \sum_{j=1}^{m} (-1)^{j+1}(1,2,\ldots,j)$.

This leads us to consider D_h defined by

$$D_h k_1\otimes\ldots\otimes k_m = -\sum_{j=1}^{m} (-1)^{j+1}\langle k_j,h\rangle k_1\otimes\ldots\otimes k_{j-1}\otimes k_{j+1}\otimes\cdots\otimes k_m$$

so that for $\tilde{\Delta}_m$ defined by

$$\tilde{\Delta}_m k_1\otimes\ldots\otimes k_m = \sum_{j=1}^{m} f_j\otimes D_j\xi$$

we have $\tilde{\Delta}_m=\tau_n(\Theta\alpha_m)$.

Denoting for a signature $\mu=(m_1\geq m_2\geq\ldots\geq m_m)$ by $\check{\mu}=(\check{m}_1\geq\check{m}_2\geq\ldots\geq\check{m}_n)$ the conjugate signature where $m_1+\ldots+m_m=\check{m}_1+\ldots+\check{m}_n=m$ and $\check{m}_i=\mathrm{card}\{j:m_j\geq i\}$ we have from the preceding discussion:

LEMMA. *The multiplicity of the representation of* U(n) *with signature* $(m_1\geq\ldots\geq m_n)$ *in* $\bigcap_{h\in H}\mathrm{Ker}(T_h|H^{\otimes m})$ $(m=m_1+\ldots+m_n)$ *equals the multiplicity of the representation of* U(n') *with signature* $(m_1'\geq\ldots\geq m_{n'}')$ *in* $\bigcap_{h\in H'}\mathrm{Ker}(D_h|H^{\otimes m})$ *where* dim H'=n' *when* $\check{\mu}=\mu'$ *where* $\mu=(m_1\geq\ldots\geq m_{\min(m,n)}\geq 0\geq\ldots\geq 0_m)$, $\mu'=(m_1'\geq\ldots\geq m'_{\min(m,n')}\geq 0\geq\ldots\geq 0_m)$ *and* $m_1+\ldots+m_n=m_1'+\ldots+m_n'=m$.

6.11. Consider $\ell(n,1)=\{(a_{ij})_{0\leq i,j\leq n}|a_{ij}\in\mathbb{C}\}$ the vector space $L(\mathbb{C}\oplus H)$ with the $\mathbb{Z}/2\mathbb{Z}$ grading $(\ell(n,1))_{\bar{0}}=\{(a_{ij})_{0\leq i,j\leq n}|j\neq 0,i\neq 0\Longrightarrow a_{0j}=a_{i0}=0\}$ $(\ell(n,1))_{\bar{1}}=\{(a_{ij})_{0\leq i,j\leq n}|a_{ij}\neq 0\Longrightarrow(ij=0$ and $i+j>0)\}$ and the bracket $[x,y]=xy-(-1)^{\alpha\beta}yx$, which turn $\ell(n,1)$ into a Lie superalgebra ([25]). Consider the mapping

$\Phi: \ell(n,1) \to \mathrm{End}(\mathcal{DT}(H))$

where $\mathcal{DT}(H)$ is given the $\mathbb{Z}/2\mathbb{Z}$-grading derived from its natural $\mathbb{Z}$-grading, and Φ is defined by: $\Phi(x)=dU(x)$ if deg $x=0$

$$\Phi\begin{bmatrix} 0 & \langle\bullet,h\rangle \\ & \\ 0 & 0 \end{bmatrix}=D_h$$

$$\Phi\begin{bmatrix} 0 & 0 \\ & \\ h & 0 \end{bmatrix}=-D_h^* \quad .$$

Note that

$$D_h^* k_1\otimes\ldots\otimes k_m=\sum_{j=0}^{m}(-1)^j k_1\otimes\ldots\otimes k_{j-1}\otimes h\otimes k_{j+1}\otimes\ldots\otimes k_m.$$

Note also that denoting by $e_{k\ell}$ the matrix $(\delta_{ik}\delta_{j\ell})_{0\le i,j\le n}\in\ell(n,1)$ we have $\Phi(e_{0j})=D_j, \Phi(e_{j0})=-D_j^*$ for $1\le j\le n$.

We have

$$-(D_k^*D_j+D_jD_k^*)k_1\otimes\ldots\otimes k_m=$$

$$=D_k^*\sum_{s=1}^{m}(-1)^{s+1}\langle k_s,f_j\rangle k_1\otimes\ldots\otimes\hat{k}_s\otimes\ldots\otimes k_m+$$

$$+D_j\sum_{s=0}^{m}(-1)^s k_1\otimes\ldots\otimes k_s\otimes f_k\otimes k_{s+1}\otimes\ldots\otimes k_m=$$

$$=\sum_{s=1}^{m}(-1)^{s+1}\langle k_s,f_j\rangle(-1)^s k_1\otimes\ldots\otimes k_{s-1}\otimes f_k\otimes k_{s+1}\otimes\ldots\otimes k_m+$$

$$+\sum_{1\le t\le s}(-1)^t k_1\otimes\ldots\otimes k_{t-1}\otimes f_k\otimes k_t\otimes\ldots\otimes k_s\otimes\ldots\otimes k_m+$$

$$+\sum_{s+1\le t\le m}(-1)^t k_1\otimes\ldots\otimes\hat{k}_s\otimes\ldots\otimes k_t\otimes f_k\otimes k_{t+1}\otimes\ldots\otimes k_m)+$$

$$+\sum_{s=0}^{m}(-1)^s((-1)^{s+1}\delta_{k,j}\; k_1\otimes\ldots\otimes k_m+$$

$$+\sum_{1\le t\le s}(-1)^t\langle k_t,f_j\rangle k_1\otimes\ldots\otimes\hat{k}_t\otimes\ldots\otimes k_s\otimes f_k\otimes k_{s+1}\otimes\ldots\otimes k_m+$$

$$+\sum_{s+1\le t\le m}(-1)^{t+1}\langle k_t,f_j\rangle k_1\otimes\ldots\otimes k_s\otimes f_k\otimes k_{s+1}\otimes\ldots\otimes k_t\otimes\ldots\otimes k_m)=$$

$$=(\Phi(e_{k,j})-(m+1)\delta_{k,j}I)k_1\otimes\ldots\otimes k_m\; +$$

$$\sum_{1\le t<s\le m}((-1)^{s+t+1}+(-1)^{s+t})k_1\otimes\ldots\otimes f_k\otimes k_t\otimes\ldots\otimes\hat{k}_s\otimes\ldots\otimes k_m+$$

$$+\sum_{1\le s<t\le m}((-1)^{s+t+1}+(-1)^{t+s})k_1\otimes\ldots\otimes\hat{k}_s\otimes\ldots\otimes f_k\otimes k_{t+1}\otimes\ldots\otimes k_m=$$

$=\Phi(e_{k,j}+\delta_{k,j}\cdot e_{00})k_1\otimes\ldots\otimes k_m$ so that

$$\Phi(e_{0j})\Phi(e_{k0})+\Phi(e_{k0})\Phi(e_{j0})=\Phi([e_{0j},e_{k0}]).$$

Computing, we also have

$$D_h^2 k_1\otimes\ldots\otimes k_m = -D_h \sum_{j=1}^{m} (-1)^j <k_j,h> k_1\otimes\ldots\otimes\hat{k}_j\otimes\ldots\otimes k_m =$$

$$= \sum_{j=1}^{m} (-1)^j \Big(\sum_{1\le k<j} (-1)^{k+1} k_1\otimes\ldots\otimes\hat{k}_k\otimes\ldots\otimes\hat{k}_j\otimes\ldots\otimes k_m +$$

$$+ \sum_{j<k\le m} (-1)^k k_1\otimes\ldots\otimes\hat{k}_j\otimes\ldots\otimes\hat{k}_k\otimes\ldots\otimes k_m\Big) <k_j,h><k_k,h>=0.$$

This implies $D_jD_k+D_kD_j=0$ and $D_h^{*2}=0$ and $D_j^*D_k^*+D_k^*D_j^*=0$, so that

$$\Phi(e_{0j})\Phi(e_{0k})+\Phi(e_{0k})\Phi(e_{0j})=\Phi(e_{k0})\Phi(e_{j0})+\Phi(e_{j0})\Phi(e_{k0})=0.$$

Remark also that

$$\Gamma(V)D_h\Gamma(V^{-1})=D_{Vh}$$

which is easily seen to imply

$[\Phi(e_{ik}),\Phi(e_{0j})]=-\delta_{i,j}\Phi(e_{0k})$ for $1\le i,j,k\le n$ and passing to adjoints $[\Phi(e_{ik}),\Phi(e_{j0})]=\delta_{j,k}\Phi(e_{i0})$ where $1\le i,\ j,k\le n$. Since $\Phi(e_{00})k_1\otimes\ldots\otimes k_m = -(m+1)k_1\otimes\ldots\otimes k_m$ it is easily seen that

$$[\Phi(e_{00}),\Phi(e_{0k})]=\Phi(e_{0k})$$

$$[\Phi(e_{00}),\Phi(e_{k0})]=-\Phi(e_{k0})$$

where $1\le k\le n$.

Summarizing we have proved the following

LEMMA. *The map* Φ *is a representation of the Lie superalgebra* $\ell(n,1)$ *on* $\mathcal{DT}(H)$, *i.e.*

$$\Phi([x,y])=xy-(-1)^{\deg x.\deg y}yx$$

for $x,y\in\ell(n,1)$.

6.12. As remarked in 6.8 we have

$$T_h(\omega_1\otimes\omega_2)=(T_h\omega_1)\otimes\omega_2+\omega_1\otimes(T_h\omega_2)$$

for $\omega_1,\omega_2\in\mathcal{DT}(H)$. For the operators D_h it is easily seen that

$$D_h(\omega_1\otimes\omega_2)=(D_h\omega_1)\otimes\omega_2+(-1)^{\deg\omega_1}\omega_1\otimes(D_h\omega_2)$$

for $\omega_1,\omega_2\in\mathcal{DT}(H)$. In particular, $\bigcap_{h\in H}\mathrm{Ker}D_h$ is a subalgebra of $\mathcal{DT}(H)$.

6.13. We indicate now a way for constructing elements in $\bigcap_{h\in H}\mathrm{Ker}D_h$. Consider $D=D_n\ldots D_1$. Since $H^{\otimes n}\ni h_1\otimes\ldots\otimes h_n\to D_{h_n}\ldots D_{h_1}\in\mathrm{End}\mathcal{DT}(H)$ is an anti-symmetric linear map we infer that

$$D_{h_n}\ldots D_{h_1}=\lambda D_n\ldots D_1$$

where $\lambda\in\mathbb{C}$ is such, that

$$\lambda f_n\wedge\ldots\wedge f_1=h_n\wedge\ldots\wedge h_1.$$

We have

$$\mathrm{Im}D\subset\bigcap_{h\in H}\mathrm{Ker}D_h.$$

This gives the possibility of constructing elements in $\bigcap_{h\varepsilon H} \mathrm{KerD}_h$. If $\eta\varepsilon\mathcal{DT}(H)$ is a $\underline{b}_1$-primitive element the highest weight of which corresponds to $(m_1\geq\ldots\geq m_n)$ then $D\eta$ in case it is $\neq 0$, is also a $\underline{b}_1$-primitive element and the corresponding highest weight corresponds to $(m_1-1, m_2-1,\ldots,m_n-1)$.

6.14. Assume $1\leq k\leq n$ and consider

$\eta=(f_1\wedge\ldots\wedge f_k)\otimes(f_1\wedge\ldots\wedge f_n)\otimes(f_1\wedge\ldots\wedge f_n)$.

(Here $f_1\wedge\ldots\wedge f_k=\frac{1}{n!}\sum_\sigma \varepsilon(\sigma)f_{\sigma(1)}\otimes\ldots\otimes f_{\sigma(n)}$). It is easily seen that

$$D_j(f_1\wedge\ldots\wedge f_k)=\begin{cases}0 & \text{if } j>k\\ (-1)^{j+1}f_1\wedge\ldots\wedge \hat{f}_j\otimes\ldots\otimes f_k & \text{if } 1\leq j\leq k.\end{cases}$$

We will show that $D\eta\neq 0$. Writing $D\eta$ as a sum

$\sum c_{i_1\ldots i_{n+k}} f_{i_1}\ldots f_{i_{n+k}}$

we will prove that $c_{1\;1\;2\;3\ldots k\;2\;\ldots\;n}\neq 0$.

We have that $c_{1\;1\;2\;3\ldots k2\ldots n}$ coincides if $k\geq 2$ with the coefficient of $f_1\otimes f_1\otimes f_2\otimes\ldots\otimes f_k\otimes f_2\otimes\ldots\otimes f_n$ in $(D_n\ldots D_{k+1})(-1)^{n+k}(-1)^{k-1}f_1\otimes(f_1\wedge\ldots\wedge f_n)\otimes(f_2\wedge\ldots\wedge f_n)$ which is

$(1/(k!(n-1)!))(-1)^{n+k}.(-1)^{k-1}.(-1)^{(n-k)(k+1)}=\pm 1/(k!(n-1)!)\neq 0.$

In case $k=1$ it is easy to see that $c_{1\;1\;2\;3\ldots n}$ arises as the coefficient of $f_1\otimes f_1\otimes f_2\otimes\ldots\otimes f_n$ in $\sum_{j=1}^{n}(-1)^{(j-1)(n+1)+(n-j+1)}f_1\otimes((D_n\ldots\ldots D_j)(f_1\wedge\ldots\wedge f_n))\otimes((D_{j-1}\ldots D_1)(f_1\otimes\ldots\otimes f_n))=\sum_{j=1}^{n}(-1)^{jn}f_1\otimes((-1)^{(n-j+1)(j-1)}f_1\wedge\ldots\wedge f_{j-1})\otimes(f_j\wedge\ldots\wedge f_n)=\sum_{j=1}^{n}(-1)^{n+j+1}f_1\otimes(f_1\wedge\ldots\wedge f_{j-1})\otimes\otimes(f_j\wedge\ldots\wedge f_n)$

which is

$$\sum_{j=1}^{n}(-1)^{n-j-1}\,1/((n-j+1)!(j-1)!)=((1+(-1))^n-1)/n!\neq 0.$$

Thus we have proved that for $1\leq k\leq n$

$\xi=D((f_1\wedge\ldots\wedge f_k)\otimes(f_1\wedge\ldots\wedge f_n)\otimes(f_1\wedge\ldots\wedge f_n))\neq 0$

is a $\underline{b}_1$-primitive vector with highest weight

$(\underbrace{2\ldots2}_{k\text{-times}}\;\underbrace{1\ldots1}_{(n-k)\text{-times}})$ such that $\xi\varepsilon\bigcap_{h\varepsilon H}\mathrm{KerD}_h$.

6.15. Using 6.14 and 6.12 it is not hard to see that $\bigcap_{h\varepsilon H}\mathrm{KerD}_h$ contains $\underline{b}_1$-primitive vectors of highest weight $m_1\geq\ldots\geq m_n$ if $m_1\equiv 0\ (\mathrm{mod}\ 2)$ and $m_n\geq 0$. Note, that there are also other examples. For instance

$$\xi=\tau\begin{pmatrix}1&2&3&4&5&6\\1&3&2&5&4&6\end{pmatrix}((f_1\wedge f_2)\otimes(f_1\wedge f_2)\otimes(f_1\wedge f_2))$$

is a $\underline{b}_1$-primitive element for the highest weight (3,3), such that $\xi\varepsilon\bigcap_{h\varepsilon H}\operatorname{Ker}D_h$.

REFERENCES

1. Akemann, C.A.; Ostrand, P.A.: Computing norms in group C*-algebras, *Amer.J.Math.*, 98(1976), 1015-1047.

2. Arveson, W.B.: Notes on extensions of C*-algebras, *Duke Math.J.*, 44(1977), 329-355.

3. Behncke, H.: Automorphisms of $A(\Phi_1)$, unpublished notes.

4. Bratteli, O.; Robinson, D.W.: *Operator Algebras and Quantum Statistical Mechanics*. II, Springer-Verlag, 1981.

5. Brown, L.G.: Ext of certain free product C*-algebras, *J.Operator Theory*, 6(1981), 135-141.

6. Brown, L.G.; Douglas, R.G.; Fillmore, P.A.: Extensions of C*-algebras and K-homology, *Ann. of Math.*, 105(1977), 265-324.

7. Ching, W.M.: Free products of von Neumann algebras, *Trans.Amer. Math.Soc.*, 178(1973), 147-163.

8. Choi, M.D.: A simple C*-algebra generated by two finite-order unitaries, *Canad.J.Math.*, 31(1979), 867-880.

9. Clancey, K.: *Seminormal operators*, Springer Lecture Notes in Math., 742(1979).

10. Coburn,L.A.: The C*-algebra generated by an isometry, *Bull.Amer. Math.Soc.*, 73(1967), 722-726.

11. Cohen, J.: Operator norms on free groups, preprint.

12. Connes, A.: *Non-commutative differential geometry*, Chapter I, preprint.

13. Cuntz, J.: Simple C*-algebras generated by isometries, *Comm. Math. Phys.*,57(1977), 173-185.

14. Cuntz, J.: K-theory for certain C*-algebra. II, *J.Operator Theory*, 5(1981), 101-108.

15. Cuntz, J.: K-theoretic amenability for discrete groups, preprint.

16. Enomoto, M.; Takehana, H.; Watatani, Y.: Automorphisms on Cuntz-algebra, *Math.Japon*, 24(1979), 463-468.

17. Evans, D.: On O_n, *Publ.Res.Inst.Math.Sci.*, 16(1980), 915-927.

18. Fannes, M.; Quaegebeur, J.: Central limits of product mappings between CAR algebras, preprint.

19. Figa -Talamanca, A.; Picardello, M.A.: Spherical functions and harmonic analysis on free groups, preprint.

20. Graev, M.J.: Unitary representations of real simple Lie groups (in russian), *Trudi Mosk.Mat.Obsch.*, 7(1958), 335-389.

21. Haagerup, U.: An example of a non-nuclear C*-algebra which has the metric approximation property, *Invent.Math.*, 30(1979),

279-293.

22. de la Harpe, P.; Jhabvala, K.: Quelques propriétés des algèbres d'une groupe discontinu d'isometries hyperboliques, preprint.

23. Helton, J.W.; Howe, R.: Integral operators: commutators, traces, index and homology, *Proc.Conf. on Operator Theory*, Springer Lecture Notes in Math., 345(1973), 141-209.

24. Hudson, R.L.; Wilkinson, M.D.; Peck, S.N.: Translation-invariant integrals, and Fourier Analysis on Clifford and Grassmann algebras, *J.Functional Analysis*, 27(1980), 68-87.

25. Kac, V.G.; Lie superalgebras, *Advances in Math.*, 26(1977), 8-96.

26. Kadison, R.V.: Notes on the Fermi gas, *Symposia Math. vol.XX*, 425-431, Academic Press (1976).

27. Kasparov, G.G.: Hilbert C*-modules: theorems of Stinespring and Voiculescu, *J.Operator Theory*, 4(1980), 133-150.

28. Kasparov, G.G.: The operator K-functor and extensions of C*-algebras, *Izv.Akad.Nauk, Ser.Mat.*,44(1980), 536-571.

29. Lance, E.C.: K-theory for certain group C*-algebras, preprint.

30. Paschke, W.: Inner product modules over B*-algebras, *Trans.Amer. Math.Soc.*, 182(1973), 443-468.

31. Paschke, W.; Salinas, N.; Matrix algebras over O_n, *Michigan Math. J.*, 26(1979), 3-12.

32. Paschke, W.; Salinas, N.: C*-algebras associated with free products of groups, *Pacific J.Math.*, 82(1979), 211-221.

33. Phillips, J.:Automorphisms of full II_1 factors with applications to factors of type III, *Duke Math.J.*, 43(1976), 375-385.

34. Pimsner, M.; Popa, S.: The Ext-groups of some C*-algebras considered by J.Cuntz, *Rev.Roumaine Math.Pures Appl.*, 23(1978), 1069-1076.

35. Pimsner, M.; Voiculescu, D.: K-groups of reduced crossed products by free groups, *J.Operator Theory*, 8(1982), 131-156.

36. Powers, R.T.: Simplicity of the C*-algebra associated with the free group on two generators, *Duke Math.J.*, 42(1975), 151-156.

37. Rieffel, M.A.: Induced representations of C*-algebras, *Adv.in Math.*,13(1974), 176-257.

38. Rosenberg, J.: Amenability of crossed products of C*-algebras, *Comm.Math.Phys.*, 57(1977), 187-191.

39. Shale, D.: Linear symmetries of the free boson field, *Trans. Amer.Math.Soc.*, 103(1962), 149-167.

40. Skandalis, G.: Exact sequences for the Kasparov groups of graded algebras, preprint.

41. Strătilă, S.; Zsidó, L.: *Lectures on von Neumann algebras*, Editura Academiei, Abacus Press, 1979.

42. Szegö, G.: *Orthogonal polynomials*, AMS Colloquium Publications vol.XXIII, 1959.

43. Vilenkin, N.Ia.: *Special functions and the theory of group representations* (in russian), Moscow, 1965.

44. Voiculescu, D.: A non-commutative Weyl-von Neumann theorem, *Rev.Roumaine Math.Pures Appl.*, 21(1976), 97-113.

45. Warner, G.: *Harmonic Analysis of Semi-Simple Lie Groups*. I, Springer-Verlag, 1972.

46. Watatani, Y.: Clifford C*-algebras, *Math.Japon*, 24(1980), 533-536.

47. Zhelobenko, D.P.: *Compact Lie groups and their representations* (in russian), Nauka, Moscow, 1970.

48. Cuntz,J.: Automorphisms of certain simple C*-algebras.

Dan Voiculescu
Department of Mathematics, INCREST
Bd.Păcii 220, 79622 Bucharest
Romania.

VECTOR GLEASON MEASURES AND THEIR FOURIER TRANSFORMS

Kari Ylinen

1. Linear extension and dilation of vector Gleason measures

Let A be a W*-algebra and $P_A = \{e \in A \mid e = e^* = e^2\}$. For any dual system $\langle E,F \rangle$, we call a function $m: P_A \to E$ an $\langle E,F \rangle$-*Gleason measure*, if m is completely additive with respect to $\sigma(E,F)$ (i.e., for any orthogonal set $\Lambda \subset P_A$ $m(\sum_{e \in \Lambda} e)$ equals the limit of $\sum_{e \in L} m(e)$ with respect to $\sigma(E,F)$, L ranging over the finite subsets of Λ). A basic question is whether m extends to a $\sigma(A,A_*)$-$\sigma(E,F)$-continuous linear map on A; if this is the case, we call m $\langle E,F \rangle$-*extendible*. We say that A has the property (G), if every completely additive $m: P_A \to [0,\infty)$ extends to a (normal, see [12, p. 136]) positive linear form. It has recently been shown, generalizing the classical theorem of Gleason [3], that a very extensive class of W*-algebras have this property, see [2], [7], [9], [13], [14] and the survey [6]; one only needs to exclude A if it has a type I_2 direct summand. The following result follows from Theorem 2.2 in [17] or from a simple direct argument.

1.1. Lemma. If A has the property (G), B is a W-algebra and $m: P_A \to B_+$ a $\langle B,B_* \rangle$-Gleason measure, then m is $\langle B,B_* \rangle$-extendible.*

It is easy to show that in the situation of Lemma 1.1 m is a noncommutative projection valued measure (i.e., $m(P_A) \subset P_B$) if, and only if, its σ-weakly continuous linear extension $T: A \to B$ is a Jordan morphism. This observation yields a very short proof of a generalization (to the not necessarily separable case) of the main theorem of Paszkiewicz [8]. In particular, our proof is independent of [5]. All vector spaces we consider are over the complex field.

<u>1.2. Theorem</u>. (Essentially [8]) <u>Let</u> H_1 <u>and</u> H_2 <u>be Hilbert spaces</u>, $\dim(H_1) \geq 3$. <u>If for the full operator algebras</u> $L(H_i)$, $i = 1, 2$, $m: P_{L(H_1)} \to P_{L(H_2)}$ <u>is σ-weakly completely additive, then</u> H_2 <u>can be expressed as a Hilbert sum</u> $H_2 = K_o \oplus (\sum^{\oplus}_{i\in I_1} K_i') \oplus (\sum^{\oplus}_{j\in I_2} K_j'')$ <u>in such a way that for some linear isometric isomorphisms</u> $U_i: H_1 \to K_i'$ <u>and anti-linear isometries</u> $V_j: H_1 \to K_j''$ $m(e) = (\sum^{\oplus}_{i\in I_1} U_i e U_i^{-1}) \oplus (\sum^{\oplus}_{j\in I_2} V_j e V_j^{-1})$, $e \in P_A$ <u>(and so</u> $m(e)$ <u>is zero on</u> K_o<u>; of course</u> I_1 <u>or</u> J_2 <u>or both may be void</u>).

<u>Proof</u>. Since $L(H_1)$ has the property (G) (e.g., by [13]), as observed above m extends to a σ-weakly continuous Jordan morphism π from $L(H_1)$ into $L(H_2)$. Express π as a direct sum of a σ-weakly continuous *-homomorphism $\pi_1: L(H_1) \to L(H_2')$ and a σ-weakly continuous *-anti-homomorphism $\pi_2: L(H_1) \to L(H_2'')$ (see [11, p. 444]). We replace H_2' and H_2'' by the essential spaces of π_1 and π_2 and (and changing our notation to let H_2' and H_2'' denote these essential spaces) write $H_2 = K_o \oplus H_2' \oplus H_2''$. Fix a conjugation $J: H_1 \to H_1$ (i.e., an antilinear isometry with $J = J^* = J^{-1}$). Then $\pi_3: L(H_1) \to L(H_2'')$ defined by the formula $\pi_3(x) = \pi_2(Jx^*J)$ is a nondegenerate σ-weakly continuous *-representation. Now π_1 is known to be equivalent to a multiple of the identity representation: $\pi_1(x) = \sum^{\oplus}_{i\in I_1} U_i x U^{-1}$ (see e.g. [1, p. 20] and use the fact that $L(H_1)$ is the bidual of the space of compact operators on H_1). Similarly, $\pi_3(x) = \sum^{\oplus}_{j\in I_2} W_j x W_j^{-1}$ for some unitary W_j. The choice $V_j = W_j J$ completes the proof.

In what follows, H is a Hilbert space. We close this section with a dilation theorem and prepare it with a lemma.

1.3. Lemma. *Let $T: A \to H$ be a bounded linear map. There is a Hilbert space K with an isometric linear map $V: H \to K$ such that for some Jordan morphism $\pi: A \to L(K)$ and vector $\xi \in K$ $Tx = V^*\pi(x)\xi$ for all $x \in A$. If T is $\sigma(A,A_*)$-$\sigma(H,H^*)$-continuous, π may be taken to be σ-weakly continuous.*

Proof. It is possible to find a Hilbert space H_o, an isometric linear map $W_1: H \to H_o$, a bounded linear map $S: A \to H_o$ and a positive linear form $\omega: A \to \mathbf{C}$ such that $(Sx|Sy) = \omega(y^*x + x^*y)$ and $Tx = W_1^*Sx$ for all $x, y \in A$. This follows from Pisier's [10] Grothendieck type inequality (see [4, p. 256] or [16]). Now Theorem 3.1 in [17] shows that there is a Hilbert space K with an isometric linear map $W_2: H_o \to K$ such that for some Jordan morphism $\pi: A \to L(K)$ and $\xi_o \in H_o$ $Sx = W_2^*\pi(x)W_2\xi_o$, $x \in A$. We may now choose $\xi = W_2\xi_o$ and $V = W_2W_1$. If T is $\sigma(A,A_*)$-$\sigma(H,H^*)$-continuous, ω may be taken to be normal (see [4, p. 257]), and then the π constructed above is σ-weakly continuous (see [17, Theorem 3.1]).

1.4. Remark. Retaining only the first two sentences of the above lemma one gets a result which remains true if A is replaced by an arbitrary C*-algebra. The same proof applies, when Corollary 3.2 from [17] instead of Theorem 3.1 is used.

1.5. Theorem. *Let $m: P_A \to H$ be an $\langle H,H^*\rangle$-Gleason measure. Consider the following two statements:*

(i) *m is $\langle H,H^*\rangle$-extendible;*

(ii) *there is a Hilbert space K with an isometric linear map $V: H \to K$ such that for some $\langle L(K),L(K)_*\rangle$-Gleason measure $\pi: P_A \to P_{L(K)}$ and vector $\xi \in K$ $m(e) = V^*\pi(e)\xi$ for all $e \in P_A$.*

Statement (i) implies (ii), and if A has the property (G), then (ii) implies (i).

Proof. Lemma 1.3 shows that (i) implies (ii). If A has the property

(G), then the π in (ii) is $\langle L(K), L(K)_*\rangle$-extendible (Lemma 1.1) to $\Phi: A \to L(K)$, and $x \to V^*\Phi(x)\xi$ is a $\sigma(A,A_*)$-$\sigma(H,H^*)$-continuous linear extension of m.

2. Noncommutative Fourier transforms

Let G be a locally compact group, $C^*(G)$ its group C^*-algebra and $W^*(G)$ the bidual of $C^*(G)$. If $\langle E,F\rangle$ is a dual system such that E is sequentially complete with respect to the strong topology $\beta(E,F)$ and $m: P_{W^*(G)} \to E$ is finitely additive and $\sigma(E,F)$-bounded on commutative sets, then for any normal element $a \in W^*(G)$ with the spectral representation $a = \int_{\sigma(a)} \lambda de(\lambda)$ the integral $T(a) = \int_{\sigma(a)} \lambda d(m\circ e)(\lambda)$ exists in the weak sense (see [17]), so we may define the Fourier transform $\hat{m}: G \to E$ via $\hat{m}(x) = T(\omega(x))$ where $\omega: G \to W^*(G)$ is the continuous unitary representation of G corresponding to the universal representation of $C^*(G)$. (For a closely related notion, see [15].) The proof of our final theorem is based on results and techniques from [15] and [16]. We omit the details.

2.1. Theorem. For a function $\phi: G \to H$ the following conditions are equivalent:

(i) ϕ is the Fourier transform of some $\langle H,H^*\rangle$-extendible $\langle H,H^*\rangle$-Gleason measure $m: W^*(G) \to H$;

(ii) ϕ is weakly continuous, and the set

$\{\|\sum_{i=1}^{n} c_i\phi(s_i)\| \mid \|\sum_{i=1}^{n} c_i\omega(s_i)\|' \leq 1\}$ is bounded where $\|\cdot\|'$ denotes the norm in $W^*(G)$;

(iii) there is a Hilbert space K with a function $\psi: G \to K$ such that for some isometric linear map $V: H \to K$ and for some continuous positive definite function $\rho: G \to \mathbf{C}$ $\phi = V^*\circ\psi$ and $(\psi(s)|\psi(t)) = \rho(t^{-1}s) + \rho(st^{-1})$ whenever $s, t \in G$.

References

1. W. Arveson, An invitation to C*-algebras, Graduate Texts in Mathematics No. 39, Springer-Verlag, Berlin and New York, 1973.

2. E. Christensen, Measures on projections and physical states, Comm. Math. Phys. 86 (1982), 529-538.

3. A. M. Gleason, Measures on the closed subspaces of a Hilbert space, J. Math. Mech. 6 (1957), 885-893.

4. S. Goldstein and R. Jajte, Second order fields over W*-algebras, Bull. Acad. Polon. Sci. Sér. Sci. Math. 30 (1982), 255-259.

5. R. Jajte and A. Paszkiewicz, Vector measures on the closed subspaces of a Hilbert space, Studia Math. 58 (1978), 229-251.

6. P. Kruszyński, A review of extensions of Gleason's theorem (preprint).

7. M. S. Matveĭčuk, Description of the finite measures in semi-finite algebras, Funct. Anal. Appl. 15, 187-197 (1981). (Translated from Funkts. Anal. Prilozh. 15 No. 3 (1981), 41-53).

8. A. Paszkiewicz, On homomorphisms of projective lattices in complex Hilbert spaces, Colloq. Math. 43 (1980), 271-280.

9. A. Paszkiewicz, Measures on projections in W*-factors (to appear in J. Functional Anal.).

10. G. Pisier, Grothendieck's theorem for noncommutative C*-algebras, J. Functional Anal. 29 (1978), 397-415.

11. E. Størmer, On the Jordan structure of C*-algebras, Trans. Amer. Math. Soc. 120 (1965), 438-447.

12. M. Takesaki, Theory of operator algebras I, Springer-Verlag, New York, Heidelberg, Berlin, 1979.

13. J. Tischer, Gleason's theorem for type I von Neumann algebras, Pacific J. Math. 100 (1982), 473-488.

14. F. J. Yeadon, Measures on projections in W*-algebras of type II_1, Bull. London Math. Soc. 15 (1983), 139-145.

15. K. Ylinen, Fourier transforms of noncommutative analogues of vector measures and bimeasures with applications to stochastic processes, Ann. Acad. Sci. Fenn. Ser. A I 1 (1975), 355-385.

16. K. Ylinen, Dilations of V-bounded stochastic processes indexed by a locally compact group (to appear in Proc. Amer. Math. Soc.).

17. K. Ylinen, Vector measures on the projections of a W*-algebra (to appear in Ann. Univ. Turku).

Department of Mathematics
University of Turku
SF-20500 Turku 50
Finland